THE A-STAR PUZZLE

IAU SYMPOSIUM No. 224

COVER ILLUSTRATION (From Royer et al. p. 109):

The observed $v \sin i$ parameter is the projection of the equatorial velocity v of the star on the line of sight, i being the inclination between the stellar rotation axis and the line of sight. The assumption that stellar rotation axes are randomly oriented is adopted. This hypothesis has been tested many times (Gray 1992, Gaigé 1993) and is still considered to be valid.

The Probability Density Function (hereafter PDF) of the $v \sin i$ values is thus the result of the convolution between the distribution of "true" equatorial velocities v, the distribution of inclination angles i, and the observational error law. It is estimated from the observed histograms, using a Kernel method. The smoothed $v \sin i$ distributions are displayed in this Figure (dashed lines). The PDF of $v \sin i$ values is first deconvolved by the distribution of errors (chosen as log-normal) and then by the distribution of inclinations (axes randomly oriented). Both processes are carried out using the Lucy (1974) iterative method.

INTERNATIONAL ASTRONOMICAL UNION

UNION ASTRONOMIQUE INTERNATIONALE

THE A-STAR PUZZLE

PROCEEDINGS OF THE 224th SYMPOSIUM OF THE INTERNATIONAL ASTRONOMICAL UNION HELD IN POPRAD, SLOVAK REPUBLIC JULY 8–13, 2004

Edited by

JURAJ ZVERKO
Astronomical Institute, Slovak Academy of Sciences, Tatranská Lomnica, Slovak Republic

JOZEF ŽIŽŇOVSKÝ
Astronomical Institute, Slovak Academy of Sciences, Tatranská Lomnica, Slovak Republic

SAUL J. ADELMAN
Department of Physics, The Citadel, Charleston, SC, USA

and

WERNER W. WEISS
Institute for Astronomy, University of Vienna, Austria

CAMBRIDGE UNIVERSITY PRESS
The Edinburgh Building, Cambridge CB2 2RU, UK
40 West 20th Street, New York, NY 10011 4211, USA
477 Williamstown Road, Port Melbourne, VIC 3207, Australia
Ruiz de Alarcón 13, 28014 Madrid, Spain
Dock House, The Waterfront, Cape Town 8001, South Africa

First published 2005

Printed in the United Kingdom at the University Press, Cambridge

Typeset in System LaTeX 2_ε

A catalogue record for this book is available from the British Library

Library of Congress Cataloguing in Publication data

ISBN 0 521 85018 5 hardback
ISSN 1743-9213

Table of Contents

Section A. NORMAL A-STARS
Chair: Werner W. Weiss

Section B. STELLAR EVOLUTION, ROTATION, BINARITY
Chair: Nikolay Piskunov

Section C. CONVECTION IN STARS
Chair: J.D. Landstreet

Section D. DIFFUSION AS THE MECHANISM OF ELEMENT SEGREGATION
Chair: C.R. Cowley

Section H. EVOLVED A-TYPE STARS
Chair: S.J. Adelman

Section I. THE A-STAR LABORATORY
Chair: F. Kupka

Section J. OBSERVATIONAL CHALLENGES OF A-TYPE STARS
Chair: G.A. Wade

Preface

Nuclear processes, radiation transfer, opacities and many other important physical processes have been included in the modelling of stars during the last decades. As we learned more about all these processes, a lack of knowledge of certain other processes became more and more obvious. It was the goal of this conference to focus on some of these still insufficiently understood physical processes which are:

Convection - Diffusion - Rotation - Magnetic Fields

Of these four "disaster areas" for stellar modelling, probably diffusion is in better shape than convection. However, there is hope that the modelling of convection soon will be improved considerably so that sufficiently reliable simulations will be available for more than just a very few privileged stars. Stellar models certainly have to be improved further by a consistent implementation of rotation whose role has been clearly demonstrated in asteroseismology, in particular in the presence of a magnetic field.

An excellent region of the HR-diagram to study these processes is centered on spectral type A as it contains well identifiable stellar groups in the same temperature, mass and luminosity region for which at least one of our insufficiently understood physical mechanisms seems to be dominant or absent. For example, three different driving mechanisms have been suggested for the three families of pulsating stars (roAp-, δ Scuti- and γ Doradus stars) which cohabitate the same parameter space. Furthermore, different and largely independent techniques are available, including asteroseismology, to test the various concepts.

In conclusion, probably no other region in the HR-diagram can be compared in the diversity of observationally dominating physical effects in stars with otherwise similar temperature and luminosity! As the various groups of stars have similar values for the fundamental astrophysical parameters, temperature, luminosity and mass, effects due to rotation, diffusion, convection, magnetic fields, etc., can be isolated and studied.

This IAU Symposium is the hitherto latest highlight of activities linked to the IAU Working Group on Ap and Related Stars. It follows IAU Symposium 210 in Uppsala (Modelling of Stellar Atmospheres) which nicely prepared the grounds for our "Puzzle" conference, since understanding the atmospheres of Ap stars and comparing them with "normal" stars obviously is a key element to understanding the physical processes responsible for their characteristics.

Organizing such an international conference and making it to a success requires the collaboration and the efforts of many people. With the limited space available we cannot list all of them. The members of the Scientific Organizing Committee were S. Adelman, S. Bagnulo, L. Balona, C. Charbonnel, M. Cunha, F. D'Antona, I. Hubeny, F. Kupka, G. Mathys, G. Michaud, A. Noels, M. Parthasarathy, T.A. Ryabchikova, H. Shibahashi, C. Sterken, G.A. Wade, W.W. Weiss (co-chair), and J. Zverko (co-chair). The Local Organizing Committee chaired by J. Žižňovský was very successful in running the conference smoothly and in mastering all the various problems which always tend to emerge in the least appropriate moment. Unforgettable was Sunday's trip into one of the beautiful corners of nature, the Pieniny National Park.

The scientific programme of this Symposium consisted of 10 Sections each organized under the resposibility of a Key-note speaker (KNS). Invited speakers presented talks on topics carefully selected by the SOC and the KNS while a few contributed talks were accepted from the contributions submitted by participants. The remaining submitted contributions were accepted as posters for which generous space was made available. Panel discussions were held at the end of each section. The merit of these ideas culminated in a synergic "Section I" called "The A-star laboratory" organized by J.D. Landstreet. He and C.R. Cowley significantly helped the SOC when preparing the scientific programme of the Symposium.

As we realize that the posters contain much useful information we decided to make them accessible in their full extent via

http://www.ta3.sk/IAUS224/Proceedings/Posters

The support of the International Astronomical Union for this conference helped bring many young astronomers to Poprad. The financial and administrative support of the Ministry of Education of the Slovak Republic and the Astronomical Institute of the Slovak Academy of Sciences were indispensable for a smooth operation before, during and after the Symposium.

We thank the inestimable help of Carol Adelman, who contributed a great deal of language editing of all papers and posters, and Richard Komžík, who created and maintained the effective system of uploading of manuscripts. Without these two the editing would take much more time.

Last, but not least, we want to thank all the speakers, participants in the discussions and presenters of posters who made this Symposium to a very successful and remarkable meeting.

J. Zverko, J. Žižňovský, S.J. Adelman, and W.W. Weiss
The Editors

THE ORGANIZING COMMITTEE

Scientific

S.J. Adelman (USA)

S. Bagnulo (Chile)

L.A. Balona (South Africa)

C. Charbonnel (France)

M. S. Cunha (portugal)

F. D'Antona (Italy)

I. Hubeny (USA)

F. Kupka (Germany)

G. Mathys (Chile)

G. Michaud (Canada)

A. Noels (Belgium)

M. Parthasarathy (India)

T.A. Ryabchikova (Russia)

H. Shibahashi (Japan)

C. Sterken (Belgium)

G.A. Wade (Canada)

W.W. Weiss (co-chair, Austria)

J. Zverko (co-chair, Slovakia)

Local

L. Hric

Ľ. Klocok

R. Komžík

L. Neslušan

D. Novocký

V. Rušin

M. Vaňko

J. Žižňovský (chair)

J. Zverko

Acknowledgements

The symposium is sponsored and supported by the IAU Divisions IV (Stars), V (Variable Stars) and IX (Optical and IR Techniques); and by the IAU Commissions No. 9 (Instrumentation and Techniques), No. 27 (variable Stars), No. 29 (Stellar Spectra), No. 30 (Radial Velocities), No. 35 (Stellar Costitution), No. 36 (Theory of Stellar Atmospheres) and No. 45 (Stellar Classification)

The Local Organizing Committee operated under the auspices of the
Astronomical Institute, Slovak Academy of Sciences

Funding by the
International Astronomical Union,
Ministry of Education of the Slovak Republic,
SEPS, a.s. - Slovak Electric and Transmission System,
Matej bel University, Banská Bystrica, Poprad Branch,
Slovak Central Observatory, Hurbanovo,
HTC Computers, Poprad,
MPC - Mill & Bakery Company, Spišská Nová Ves,
Tatrakon - Cannery, Poprad,
and
Baliarne - Packhouse, Poprad,
is gratefully acknowledged.

1. J.D. Landstreet
2. L. Fraga
3. P. Reegen
4. H. Saio
5. S. Bagnulo
6. P. Padovani
7. P.J. Amado
8. N. Przybilla
9. S. Moehler
10. B. Freytag
11. B. Smalley
12. M.M. Dworetsky
13. J. Kubát
14. G. Wahlgren
15. P. Škoda
16. C.R. Cowley
17. I.Kh. Iliev
18. J. Krtička
19. H. Shibahashi
20. M. Aurière
21. G. Wade
22. F. LeBlanc
23. R. Monier
24. N. Grevesse
25. P. Lenz
26. M. Lemke
27. M. Vaňko
28. V. Elkin
29. M. Skulsky
30. J. Budaj
31. D. Moss
32. E. Michel
33. V. Khalack
34. C. Stütz
35. C. Maceroni
36. M. Browning
37. O. Richard
38. V. Antoci
39. M. Netopil
40. E. Guggenberger
41. J. Braithwaite
42. E. Paunzen
43. K.M. Bischof
44. M. Yıldız
45. T. Kallinger
46. J. Michaud
47. F. Kupka
48. M. Sachkov
49. N. Piskunov
50. E.A. Semenko
51. T. Ryabchikova
52. R. Arlt
53. A. Noels
54. O. Kochukov
55. S. Théado
56. G. Alecian
57. K. Kolenberg
58. S. Bernabei
59. T. Böhm
60. V.D. Bychkov
61. T. Tanriverdi
62. J. Zverko
63. I.I. Romanyuk
64. C. Corbally
65. S. Talon
66. F. Royer
67. M. Navrátil
68. Z. Mikulášek
69. S. Kocer
70. M. Netolický
71. D. Kocer
72. N. Nesvacil
73. W. Zima
74. I. Stateva
75. M. Rode-Paunzen
76. K. Zwintz
77. M. Marconi
78. J.D. Riley
79. L.V. Bychkova
80. M. Castro
81. Z. Kaňuchová
82. M. Hronovská
83. E. Gerth
84. Mrs. E. Gerth
85. Y. Frémat
86. A. Kawka
87. S. Vennes
88. A. Budovičová
89. O. Pintado
90. N. Drake
91. K. Yüce
92. S. Yushchenko
93. S.J. Adelman
94. V. Gopka
95. L. Balona
96. A.V. Shavrina
97. N. Polosukhina
98. T. Lüftinger
99. M. Breger
100. M. Cunha
101. M. Bazot
102. D.W. Kurtz
103. D. Kudryavtsev
104. D. Korčáková
105. J. Žižňovský
106. E. Kundra
107. H. Caliscan
108. M. Husárik
109. M. Feňovčík
110. J. Janík
111. G. Szász
112. M. Mocák
113. M. Tirpák
114. S. Vauclair
115. M.T. Çay
116. G.W. Preston
117. A. Teker
118. İ.H. Çay
119. J.V. Bychkova
120. D. Mkrtichian
121. R. Trampedach
122. W.W. Weiss
123. G. Mathys

Participants

Saul J. **Adelman**, Department of physics, The Citadel, USA — adelmans@citadel.edu
Georges **Alecian**, LUTH, Observatoire de Paris-Meudon, France — georges.alecian@obspm.fr
Pedro J. **Amado**, Instituto de Astrofisica de Andaluca - CSIC, Spain — pja@iaa.es
Victoria **Antoci**, Institute for Astronomy, Vienna University, Austria — antoci@astro.univie.ac.at
Rainer **Arlt**, Astrophysikalisches Institut Potsdam, Germany — rarlt@aip.de
Michel **Aurière**, Observatoire Midi Pyrénnées, France — michel.auriere@obs-mip.fr
Stefano **Bagnulo**, European Southern Observatory, Chile — sbagnulo@eso.org
Yury Yu. **Balega**, Special Astrophysical Observatory, Nizhnij Arkhyz, Russia — balega@sao.ru
Luis A. **Balona**, South African Astronomical Observatory, South Africa — lab@saao.ac.za
Michael **Bazot**, Observatoire Midi Pyrénnées, Toulouse, France — bazot@ast.obs-mip.fr
Stefano **Bernabei**, INAF-Osservatorio Astronomico di Bologna, Italy — bernabei@bo.astro.it
Katharina M. **Bischof**, Institute for Astronomy, Vienna University, Austria — bischof@astro.univie.ac.at
Torsten C. **Böhm**, Laboratoire Astrophysique de Toulouse, France — boehm@obs-mip.fr
Jonathan **Braithwaite**, University of Amsterdam, The Netherlands — jon@mpa-garching.mpg.de
Michel **Breger**, Institute for Astronomy, Vienna University, Austria — michel.breger@univie.ac.at
Matthew K. **Browning**, JILA, University of Colorado, USA — matthew.browning@colorado.edu
Ján **Budaj**, Astronomical Institute, Slovak Academy of Sciences, Slovakia — budaj@ta3.sk
Andrea **Budovičová**, Astronomical Institute, Czech Academy of Sciences, Czech Rep. — andrea@asu.cas.cz
Victor D. **Bychkov**, Special Astrophysical Observatory, Nizhnij Arkhyz, Russia — vbych@sao.ru
Larisa V. **Bychkova**, Special Astrophysical Observatory, Nizhnij Arkhyz, Russia — lbych@sao.ru
Hulya **Caliskan**, Istanbul University, Astronomy and Space Science, Turkey — caliskan@istanbul.edu.tr
Matthieu **Castro**, 5, bvd d'Arcole 31000 Toulouse — mcastro@ast.obs-mip.fr
İpek H. **Çay**, Istanbul University, Science Faculty, Turkey — ipek@istanbul.edu.tr
Taskin M. **Çay**, Istanbul University, Science Faculty, Turkey — taskin@istanbul.edu.tr
Christopher J. **Corbally**, Vatican Observatory Research Group, University of Arizona, USA — corbally@as.arizona.edu
Charles R. **Cowley**, Dept. of Astronomy, University of Michigan, USA — cowley@umich.edu
Margarida S. **Cunha**, Centro de Astrofísica da Universidade do Porto, Portugal — mcunha@astro.up.pt
Jadwiga **Daszynska-Daszkiewicz**, Copernicus Astronomical Center, Poland — daszynsk@camk.edu.pl
Natalia **Drake**, Observatório National/MTC, Rio de Janeiro, Brazil — drake@on.br
Michael M. **Dworetsky**, Dept. of Physics & Astronomy, University College London, UK — mmd@star.ucl.ac.uk
Vladimir **Elkin**, Special Astrophysical Observatory, Nizhnij Arkhyz, Russia — vgelk@sao.ru
Marian **Feňovčík**, Pavol Jozef Šafárik University, Košice, Slovakia — fenovcik@ta3.sk
Lucian **Fraga**, Departamento de Fisica – Universidade Federal de Santa Catarina, Brazil — luciano@astro.ufsc.br
Yves **Frémat**, Royal Observatory of Belgium, Brussel, Belgium — yves.fremat@oma.be
Bernd **Freytag**, GRAAL, University of Montpeliler, France — Bernd.Freytag@graal.univ-montp2.fr
Ewald **Gerth**, Potsdam, Gontardstr. 130, Germany — ewald-gerth@t-online.de
Vera **Gopka**, Odessa Astronomical Observatory, Ukraine — Gopka@arctur.tenet.odessa.ua
Nicolas **Grevesse**, Institut d'Astrophysique, Universit de Liège, Belgium — grevesse@astro.ulg.ac.be
R. Elizabeth M. **Griffin R.**, 5071 West Saanich Road, Victoria, Canada — Elizabeth.Griffin@nrc.gc.ca
Ladislav **Hric**, Astronomical Institute, Slovak Academy of Sciences, Slovakia — hric@ta3.sk
Mariana **Hronovská**, Pavol Jozef Šafárik University, Košice, Slovakia — zosma83@pobox.sk
Marek **Husárik**, Astronomical Institute, Slovak Academy of Sciences, Slovakia — mhusarik@ta3.sk
Ilian Kh. **Iliev**, National Astronomical Observatory Rozhen, Smolyan, Bulgaria — rozhen@mbox.digsys.bg
Jan **Janík**, Inst. of Theoret. Physics and Astrophysics, Masaryk Univ., Czech Rep. — honza@physics.muni.cz
Thomas **Kallinger**, Institute for Astronomy, Vienna University, Austria — kallinger@astro.univie.ac.at
Zuzana **Kaňuchová**, Astronomical Institute, Slovak Academy of Sciences, Slovakia — zkanuch@ta3.sk
Adela **Kawka**, Astronomical Institute, Czech Academy of Sciences, Czech Rep. — kawka@sunstel.asu.cas.cz
Victor **Khalack**, Main Astronomical Observatory, Kiev, Ukraine — khalack@mao.kiev.ua
Dursun **Kocer**, Dept. of Math. and Computer Sciences, Istanbul Kultur University, Turkey — d.kocer@iku.edu.tr
Oleg **Kochukhov**, Institute for Astronomy, Vienna University, Austria — kochukhov@astro.univie.ac.at
Katrien I. **Kolenberg**, Institute for Astronomy, Vienna University, Austria — kolenberg@astro.univie.ac.at
Daniela **Korčáková**, Astronomical Institute, Czech Academy of Sciences, Czech Rep. — kor@sunstel.asu.cas.cz
Jiří **Krtička**, Inst. of Theoret. Physics and Astrophysics, Masaryk Univ., Czech Rep. — krticka@physics.muni.cz
Jiří **Kubát**, Astronomical Institute, Czech Academy of Sciences, Czech Rep. — kubat@sunstel.asu.cas.cz
Dmitry **Kudryavtsev**, Special Astrophysical Observatory, Nizhnij Arkhyz, Russia — dkudr@sao.ru
Igor **Kudzej**, Vihorlat Observatory, Humenné, Slovakia — vihorlatobs1@stonline.sk
Emil **Kundra**, Pavol Jozef Šafárik University, Košice, Slovakia — emilkundra@hotmail.com
Friedrich **Kupka**, Max-Planck-Institute for Astrophysics, Garching, germany — fk@mpa-garching.mpg.de
Don **Kurtz**, Centre for Astrophysics, University of Central Lancashire, UK — dwkurtz@uclan.ac.uk
John D. **Landstreet**, University of Western Ontario, Canada — landstr@astro.uwo.ca
Francis **LeBlanc**, Départment de physique et d'astronomie, Université de Moncton, Canada — leblanfn@umoncton.ca
Michael **Lemke**, Sternwarte Bamberg, Germany — ai26@sternwarte.uni-erlangen.de
Patrick **Lenz**, Institute for Astronomy, Vienna University, Austria — lenz@astro.univie.ac.at
Theresa **Lüftinger**, Institute for Astronomy, Vienna University, Austria — lueftinger@astro.univie.ac.at
Carla **Maceroni**, INAF, Osservatorio Astronomico di Roma, Italy — maceroni@mporzio.astro.it
Marcella **Marconi**, INAF, Osservatorio Astronomico di Capodimonte, Italy — marcella@na.astro.it
Gautier **Mathys**, European Southern Observatory, Chile — gmathys@eso.org
Zdeněk **Mikulášek**, Inst. of Theoret. Physics and Astrophysics, Masaryk Univ., Czech Rep. — mikulas@ics.muni.cz
David **Mkrtichian**, Sejong University, Seoul, Korea — david@arcsec.sejong.ac.kr
Miroslav **Mocák**, Pavol Jozef Šafárik University, Košice, Slovakia — mmocak@ta3.sk
Sabine **Moehler**, Institut für Theoretische Physik und Astrophysik, Kiel, Germany — moehler@astrophysik.uni-kiel.de
Richard A. M. **Monier**, Universite Montpellier II, France — monier@graal.univ-montp2.fr
Josefina **Montalban**, Institut d'Astrophysique et Geophys., Universite de Liège, Belgium — J.Montalban@ulg.ac.be
David L. **Moss**, Department of Mathematics, University of Manchester, UK — moss@ma.man.ac.uk
Parthasarathy **Mudumba**, Indian Institute of Astrophysics, Bangalore, India — partha@iiap.res.in
Martin **Navrátil**, Observatory and Planetarium, Hradec Králové, Czech Rep. — navratil@astrohk.cz
Nicole **Nesvacil**, European Southern Observatory, Chile — nnesvaci@eso.org
Martin **Netolický**, Inst. of Theoret. Physics and Astrophysics, Masaryk Univ., Czech Rep. — netol@physics.muni.cz
Martin **Netopil**, Institute for Astronomy, Vienna University, Austria — netopil@astro.univie.ac.at
Arlette **Noels**, Institut d'Astrophysique et Geophysique, Universite de Liège, Belgium — Arlette.Noels@ulg.ac.be
Paolo **Padovani**, ST-ECF, European Southern Observatory, Germany — ppadovan@eso.org
Ernst **Paunzen**, Institute for Astronomy, Vienna University, Austria — ernst.paunzen@univie.ac.at
Olga I. **Pintado**, Departamento de Física - FACET -Universidad Nacional de Tucumán — opintado@tucbbs.com.ar
Nikolai **Piskunov**, Dept. of Astronomy and Space Physics, Uppsala University, Sweden — piskunov@astro.uu.se
Nina S. **Polosukhina**, Crimean Astrophysical Observatory, Nauchny, Ukraine — polo@crao.crimea.ua
George W. **Preston**, Carnegie Observatories, Pasadena, USA — gwp@ociw.edu
Norbert **Przybilla**, Dr. Remeis-Sternwarte Bamberg, Germany — przybilla@sternwarte.uni-erlangen.de
Peter **Reegen**, Institute for Astronomy, Vienna University, Austria — reegen@astro.univie.ac.at
Olivier **Richard**, Universite Montpellier II, France — richard@graal.univ-montp2.fr

Jonathan D. **Riley**, Centre for Astrophysics, University of Central Lancashire, UK — jdriley@uclan.ac.uk

Vincenzo **Ripepi**, INAF-Osservatorio Astronomico di Capodimonte, Italy — ripepi@na.astro.it

Monika **Rode-Paunzen**, Institute for Astronomy, Vienna University, Austria — rode@astro.univie.ac.at

Iosif I. **Romanyuk**, Special Astrophysical Observatory, Nizhnij Arkhyz, Russia — roman@sao.ru

Frédéric **Royer**, Observatoire de Genéve, Sauverny, Switzerland — frederic.royer@obs.unige.ch

Tatiana A. **Ryabchikova**, Institute of Astronomy, Russian Academy of Science, Russia — ryabchik@inasan.rssi.ru

Mikhail E. **Sachkov**, Institute of Astronomy, Russian Academy of Science, Russia — msachkov@inasan.ru

Hideyuki **Saio**, Astronomical Institute, Tohoku University, Sendai, Japan — saio@astr.tohoku.ac.jp

Eugene A. **Semenko**, Special Astrophysical Observatory, Nizhnij Arkhyz, Russia — sea@tiger.sao.ru

Angelina V. **Shavrina**, Main Astronomical Observatory, Kiev, Ukraine — shavrina@mao.kiev.ua

Hiromoto **Shibahashi**, Dept. of Astronomy, Univ. of Tokyo, Japan — shibahashi@astron.s.u-tokyo.ac.jp

Petr **Škoda**, Astronomical Institute, Czech Academy of Sciences, Czech Rep. — skoda@sunstel.asu.cas.cz

M. Yu. Mykhailo **Skulsky**, Lviv Polytechnical University, Ukraine — mysky@polynet.lviv.ua

Barry **Smalley**, Astrophysics Group, School of Chemistry & Physics, Keele University, UK — bs@astro.keele.ac.uk

Ivanka **Stateva**, National Astronomical Observatory Rozhen, Smolyan, Bulgaria — stateva@astro.bas.bg

Christian **Stütz**, Institute for Astronomy, Vienna University, Austria — stuetz@jan.astro.univie.ac.at

Gabriel **Szász**, Comenius University, Bratislava, Slovakia — g.szasz@nextra.sk

Suzanne **Talon**, Département de Physique, Université de Montréal, Canada — talon@astro.umontreal.ca

Taner **Tanriverdi**, Ankara University, Faculty of Science, Turkey — taner@astro1.science.ankara.edu.tr

Aysegul **Teker**, Dept. of Math. and Computer Sciences, Istanbul Kultur University, Turkey — a.teker@iku.edu.tr

Sylvie **Théado**, Institut d'Astronomie de Liège, Belgique — theado@astro.ulg.ac.be

Mikuláš **Tirpák**, Pavol Jozef Šafárik University, Košice, Slovakia — mikit1@post.sk

Regner **Trampedach**, Mt Stromlo Observatory, Australia — art@mso.anu.edu.au

Martin **Vaňko**, Astronomical Institute, Slovak Academy of Sciences, Slovakia — vanko@ta3.sk

Sylvie **Vauclair**, Laboratoire d'astrophysique de Toulouse, France — sylvie;vauclair@obs-mip.fr

Stéphane **Vennes**, Dept. of Physics and Astronomy, Johns Hopkins University, USA — vennes@pha.jhu.edu

Gregg A. **Wade**, Royal Military College of Canada — wade-g@rmc.ca

Werner W. **Weiss**, Institute for Astronomy, Vienna University, Austria — weiss@astro.univie.ac.at

Mutlu **Yıldız**, Ege University, Izmir, Turkey — yildiz@astronomy.sci.ege.edu.tr

Kutluay **Yüce**, Ankara University, Faculty of Science, Turkey — kyuce@astro1.science.ankara.edu.tr

Alexander V. **Yushchenko**, Odessa Astronomical Observatory, Ukraine — yua@odessa.net

Wolfgang **Zima**, Institute for Astronomy, Vienna University, Austria — zima@astro.univie.ac.at

Jozef **Žižňovský**, Astronomical Institute, Slovak Academy of Sciences, Slovakia — ziga@ta3.sk

Juraj **Zverko**, Astronomical Institute, Slovak Academy of Sciences, Slovakia — zve@ta3.sk

Konstanze **Zwintz**, Institute for Astronomy, Vienna University, Austria — zwintz@astro.univie.ac.at

Vera Lvovna Khokhlova

*1927 − †2003

Vera Lvovna Khokhlova a well known and admired astrophysicist unexpectedly died September 24, 2003. She was 77 year old.

Vera Khokhlova was born in the Moscow region on July 25, 1927. Graduating from the Moscow State University in 1950 she was one of the lucky students who received a chance to start a professional life in a new, still under construction, observatory in the Crimea. These were very joyful and productive years in Vera Khokhlova's life. She met new people and made new friends many of whom remained in her life for many decades. Her first teacher and mentor was Academician S. B. Severny and very soon she finished her PhD thesis on the analysis of the active solar regions.

Later her family situation changed. Vera Khokhlova had to make the very difficult decision to move back to Moscow in 1961. Her new job was at the Astronomical Council (now the Institute of Astronomy RAS) under the leadership of Academician Mustel. Her new professional interest became the Magnetic Ap stars. Perhaps this choice was connected with her previous scientific activities. Ap stars and active solar regions have one common feature - magnetic fields. The idea that the surface structure of Ap stars was spotty had just appeared, and Vera Lvovna began to work actively in this direction. First, it was an application of Deutsch's harmonic analysis of the average spectral line characteristics: equivalent widths and radial velocity variations. Very quickly she developed her own method of stellar surface modelling based on these averaged characteristics. It was when she made her first contacts with French (C. Megessier) and Canadian (W. Wehlau) astrophysicists working in the same field.

Significant improvements of the accuracy and resolution of spectroscopic observations gave her the idea to apply solar methods to stellar spectra. She saw the possibility of extracting more precise information from the spectral lines using the line profile variations to study the surface structure of Ap stars. She proposed a formalism of reconstruction of stellar surface structures from line profile variations. But Vera Lvovna Khokhlova understood clearly that formulation of an Inverse problem is only a part of the job., New mathematical approaches were needed to solve it. She was always very persistent and successful in organizing scientific cooperations and she approached mathematicians at the Moscow State University. This very fruitful collaboration resulted in a new method of stellar surface structure investigation, known now as the Doppler Imaging technique, and is widely used with various modifications today. This made Vera Lvovna a leader in Ap star spectroscopy.

We mentioned above the organizational capabilities of Vera Lvovna. It was another of her very important and very productive activities. In the 1970s she organized a scientific cooperation program with the group of Werner Schoneich from the former DDR. It resulted in Subcommission No4 'Magnetic Stars', a multi-sided cooperation between the Academies of the former socialist countries. For many years Vera Lvovna chaired this Subcommission 'Magnetic Stars'. It is the only subcommission which survived after the destruction of the socialistic system and prospers at present under the leadership of the astronomers from the Special Astrophysical Observatory in Zelentchuk. As before we have International conferences every 3 years.

Vera Lvovna gave a lot of her time and energy working with the scientists from the different republics of the former Soviet Union. She was a teacher and guru for many young astronomers in Shemakha Observatory. She loved this place and spent a lot of time there. Her first Ap results were also obtained at the 2-m telescope in Shemakha.

Vera Lvovna was a very devoted person and she taught us to work without reservations and hesitations. Astrophysics was her love and life, and everything else was dependent on it. She practically lived in her office at work. At the beginning, we, her students and colleagues, did not understand this practice, then we started following her model. She was always on top of all major events and news in her area. She never missed the issues of current Astronomy journals. However, she felt almost ashamed that she did not have time for reading novels or going to the movies. We still cannot imagine how she could find time for her children and her mother.

Vera Lvovna was working until the very end. She planned to take part in last September's 'Magnetic field' meeting at the SAO, and only her death prevented it. Being in a hospital, during the two last weeks of her life she tried to read the PhD thesis of her student.

We would like to finish with the words of Margherita Hack:

'I am very sorry to know about the death of Vera Khokhlova. She was an example of very active and enthusiastic scientist. I was very much impressed with her method to detect the position of the spots on the surface of magnetic stars, a very informative method that I think, she was the first person to propose. It is a great loss for astronomy, and a great loss for her friends.'

N. Polosukhina, T. Ryabchikova, A. Krivosheina

Pasadena, California 1917

Horace

Horace Welcome Babcock

*1912 − †2003

Horace Welcome Babcock was introduced to astronomy at an early age by his father, Harold, who was a first-generation astronomer of the Mount Wilson Observatory. He earned a bachelor degree in engineering at Caltech, and then undertook graduate studies at UC Berkeley. His PhD dissertation about the rotation of M31 provided the first evidence for the existence of dark matter.

Following temporary positions at Lick and McDonald Observatories and the MIT Radiation laboratories Horace joined the staff of the Mount Wilson and Palomar Observatories in January, 1946. In that same year he proposed to measure stellar magnetic fields, constructed a polarizing analyzer for that purpose, made the first successful detection of a stellar magnetic field in the CP2 star 78 Virginis, and published his results in the Astrophysical Journal. In a survey that followed this detection Babcock measured magnetic fields in some 70 CP stars, discovered a number of variable periodic magnetic CP stars, and searched for stellar magnetic fields elsewhere in the HR diagram.

While engaged in his studies of stellar magnetism, Babcock found time to develop a successful solar magnetograph that could map the magnetic field of the sun on a daily basis. With this instrument he detected the reversing polar magnetic fields of the sun, and he devised a model of the solar cycle that survives to this day. An improved version of his magnetograph is still in operation at the 150-foot solar telescope of the Mount Wilson Observatory. In 1953 Babcock introduced astronomers to the concepts of adaptive optics, a scheme whereby the deleterious effects of atmospheric turbulence can be removed during the observation of astronomical sources. In this period he also supervised the Mount Wilson Grating Laboratory where he improved the performance of a second-generation ruling engine, and used it to provide diffraction gratings for more than sixty astronomical institutions around the world.

In 1964 Babcock accepted the directorship of the Mount Wilson and Palomar Observatories, set aside his personal research, and devoted the remainder of his scientific life to development of the Las Campanas Observatory in Chile. Babcock chose that location after testing a number of sites in New Zealand, Australia, and North-Central Chile with a seeing monitor of his own design. At Las Campanas he brought two telescopes, the Swope 1-meter and du Pont 2.5-meter, to completion. Today LCO is also the home for the Polish OGLE telescope, and two 6.5-meter telescopes that are operated by a five-institution Magellan Consortium.

Altogether, Horace Welcome Babcock made a big contribution to astronomical science.

George W. Preston

The A-Star Puzzle
Proceedings IAU Symposium No. 224, 2004
J. Zverko, J. Žižňovský, S.J. Adelman, & W.W. Weiss, eds.

© 2004 International Astronomical Union
DOI: 10.1017/S1743921304004314

The physical properties of normal A stars

Saul J. Adelman

Department of Physics, The Citadel, 171 Moultrie Street, Charleston, SC 29409, USA email:
adelmans@citadel.edu

Abstract. Designating a star as of A-type is a result of spectral classification. After separating the peculiar stars from those deemed to be normal using the results of a century of stellar astrophysical wisdom, I define the physical properties of the "normal" stars. The hotter A stars have atmospheres almost in radiative equilibrium. In the A stars convective motions can be found which increase in strength as the temperature decreases.

Keywords. Stars: abundances, stars: fundamental parameters, stars: rotation, convection, diffusion, magnetic fields

1. What are normal A stars?

Each star examined in sufficient detail is unique. But astronomical progress requires that we try to group stars so that patterns emerge as functions of fundamental parameters. A century ago the Harvard College Observatory spectral classifiers defined A-type stars using order 100 Å mm^{-1} spectra. They have strong Balmer lines and many faint to moderately strong lines. The number of lines and degree of line blending in the photographic region grows rapidly towards the F stars. A stars lack the He I lines seen at classification dispersions in B stars. F stars have even more numerous and stronger metal lines. MK classifications use pairs of spectral lines in the photographic region. Morgan and some other spectral classifiers opened the slits of their spectrographs to minimize the differences due to rotation. So their resolutions are less than one would infer from their inverse dispersions. Stellar rotation makes weak lines disappear, causes blending, and affects the continuum location. Photometric criteria are used to define some A-star classes, but these violate the spirit of the pure classification process.

Normal single A stars can be completely classified or are normal white dwarfs. Understanding the marginal and peculiar stars may be an exercise in the physical processes involved with convection, diffusion, rotation, and magnetic fields. Many assert that the varieties of A stars are distinct from one another. However, many types probably merge into one another. Processes associated with some kinds of photometric variability just disturb the outer layers while leaving the other properties essentially alone. We exclude SB2 and those binary stars that have participated in a substantial mass exchange.

Charles Cowley defined the superficially normal stars as those whose classification spectra look normal, but are not with those used by abundance analyzers. Many Mercury-Manganese and marginal metallic-lined stars belong to this category. Finding such stars also depends on the signal-to-noise ratio of the spectra and on the effective temperature as the mean metallic line strength in the optical region increases with decreasing temperature in the A stars. If one can photometrically calibrate classification dispersion spectra as the ASTRA spectrophotometer (Adelman *et al.* 2004) will do, then one may be able to more easily separate these stars from the normal stars.

Are the classification standards well chosen? Abt & Morrell (1995) note that some standards, in particular Vega and α Dra, are abnormal. Using them may result in similar stars being classified as normal. Further many normal stars near A2 IV have characteristics similar to those of peculiar stars and the mean rotational velocities of normal stars are depressed for those types where Am and Ap stars are most frequent. Sharp-lined and broad-lined standards for a given spectral type may not be equivalent.

The hotter main sequence band A stars have almost purely radiative atmospheres. In the mid-A stars convection begins to play a role in energy transport. The degree to which this happens increases with decreasing temperature and decreasing surface gravity.

Old metal-poor evolved A-type Population II stars often are horizontal-branch stars, e.g., RR Lyrae and the Field Horizontal Branch A Stars, which are found in globular clusters and the Galactic halo. Some Population I A stars are metal poor relative to the Sun, e.g., Vega and α Dra (Adelman & Gulliver 1990; Adelman et al. 2001), but many more should be known. Abt & Morrell (1995) suggest as candidates superficially normal stars with low rotational velocities and weak Mg II $\lambda4481$ and Ca II K lines. The most extreme metal-poor stars, the λ Boo stars (Morgan et al. 1943, Hauck & Slettebak 1983, Gray 1988), are extremely obvious at classification dispersions. Attempts to define them as a tightly defined group of rapidly rotating metal-poor stars (see, e.g., Gerbaldi et al. 2003) have problems. It might be more useful to look at the all metal poor A stars and recognize ways they are related to the prototype. Fast rotators seen nearly pole-on are not necessarily metal-poor (Gulliver, Hill & Adelman 1994), e.g, ν Cap.

Some main sequence band chauvinists exclude from the normal stars those outside of the main sequence band and not deriving all of their radiated energy from the conversion of hydrogen into helium in their cores. Pre-main sequence stars are evolving towards the Zero Age Main Sequence. Population I A supergiants evolved from more massive progenitors than main sequence band stars. Dredge-up events may have affected their envelope abundances especially of C, N, and O. Photospheric abundances can be altered by mass exchange events during the evolution of binary stars.

Some A stars have nearby circumstellar or interstellar material. For example, β Pic and Vega have outlying dust disks surrounding them, perhaps as remnants of star formation. Hot inner disks around rapidly rotating B and A dwarf or shell stars are due to material accreted from the interstellar medium in dense interstellar regions (Abt 2004). Also the Be star sequence extends into the early A stars, the Herbig Ae stars.

The magnetic CP (mCP) (Preston's (1974) CP2) stars include those peculiar B and A stars with detected magnetic fields. Many show spectrum, photometric, and magnetic variability. Stars with similar spectra showing abnormally strong lines of Si, Cr, Sr, and the rare earths are considered class members especially if they exhibit variability.

Roman et al. (1948) defined the Am or metallic-lined (Preston's (1974) CP1) stars as having Sp type (Ca II K line) < Sp type (Balmer lines) < Sp type (metallic lines). Conti (1970) extended the class to include stars with weak Sc II lines in place of or in addition to a weak K line. The Mercury-Manganese (HgMn) (Preston's (1974) CP3) stars are late peculiar B stars for which Hg II $\lambda3984$ and/or Mn II lines are the most prominent enhanced features. These types form a sequence of non-magnetic CP stars, but some have speculated that they have far more complex fields than the usual mCP stars. Identifying those with some rotation can be difficult. Recently Adelman et al. (2003) showed that the coolest HgMn stars evolve into the hottest Am stars. This unifies these classes and provides a test for theoreticians concerning the abundance of mercury. There is a full range of abundance patterns from the marginal Am stars whose abundances are merely suggestive of peculiarity through those who show the classical Am abundance pattern. Similar effects should occur for the HgMn stars. But due to the paucity of lines in optical region spectra, this may well be a project for high dispersion ultraviolet spectroscopy.

Table 1. Parameters of normal main sequence stars

Sp. Type	$B-V$	$b-y$	T_{eff} (K)	M_{V}	Mass $(M_\odot)$	Radius $(R_\odot)$	$\log g$	$<v\sin i>$ (km s^{-1})
A0	0.000	0.00	9727	0.7	2.40	1.87	4.27	149
A2	0.055	0.04	8820	1.3	2.19	1.78	4.28	145
A5	0.143	0.09	7880	2.0	1.86	1.69	4.25	137
A6	0.170	0.10	7672	2.1	1.80	1.66	4.25	134
A7	0.198	0.12	7483	2.3	1.74	1.63	4.25	132
A8	0.228	0.14	7305	2.4	1.66	1.60	4.25	128
A9	0.265	0.16	7112	2.5	1.62	1.55	4.26	124
F0	0.300	0.17	6949	2.6	1.55	1.51	4.27	117

Abt & Morrell (1995) noted that virtually all Am and mCP stars have equatorial rotational velocities $< 120\ \text{km s}^{-1}$ while normal A0-F0 main sequence stars have larger values. The latter include those with weak Ca II K lines or belong to short-period binaries. Abt (2000) concluded that rotation alone can explain the appearance of an A star at classification dispersions as either normal or abnormal. The few sharp-lined normal A stars are just fast rotators seen nearly pole-on. Gulliver *et al.* (1994) gave a few examples that exhibited weak line profiles with flat cores.

Preston's (1974) CP4 or Helium-weak stars are peculiar B stars whose He I line strengths are weaker than one would expect based on their *UBV* colors. They are a mixture of hotter analogues to the mCP (CP2) and to the HgMn (CP3) stars.

The classical variability strip crosses the main sequence between A2 V and F0 V (Wolff 1983). There is a paucity of sharp-lined normal stars here. Observations from the Earth show that 1/3 of these main sequence band stars are δ Scutis. Recent space observations suggest (Breger 2004) that the rest are variable with smaller amplitudes. Besides the δ Scuti stars, the roAp stars and the γ Dor stars also pulsate with each class having its own driving mechanism. The peculiar roAp stars have abnormal abundance distributions. They are cooler than most of the other mCP stars (Ryabchikova, private communication). Members of the two classes with solar type abundances should be considered normal stars. Waelkens *et al.* (1998) proposed an analogous strip to connect the β Cep stars, the slowly pulsating B stars, and the α Cyg variables.

Thus normal A stars have surficial abundances close to "solar" (supergiants and white dwarfs excepted), lack detectable magnetic fields and emission lines, have equatorial rotational velocities $>120\ \text{km s}^{-1}$, and have not participated in a substantial mass exchange event with a companion.

2. Properties of main sequence band normal A stars

Table 1 extracted from Gray (1992) provides a framework of useful values for main sequence A stars. Some sources cited later use slightly different values. In part this reflects differences between how the selections are made. Averages of the type given in Table 1 are absent for luminosity class III and IV A stars. Only the Sun is sufficiently close that we can determine its fundamental properties with good accuracies and precisions. Recently substantial progress has been and is continuing to be made for other nearby stars including main sequence band A stars. The Hipparcos satellite improved our knowledge of parallaxes. Most recent improvements in parallaxes made from the ground relate to stars not observed by Hipparcos.

Hipparcos photometry is a consistent database with order 100 observations of a large number of relatively bright stars with classified spectra. Adelman (2001) investigated the mean amplitude as a function of MK spectral type of apparently single normal stars in the Bright Star Catalogue (Hoffleit & Warren 1991). The most stable stars (ampl. < 0.026 mag.) were spectral types B9-A8 III and F0 III, B9-F0 IV, and B9-A3 V, A5 V, and A6 V. Then come A4 V, A8-F0 V, and A9 III stars, and the B9 II and A6-F0 II supergiants (ampl. 0.027-0.032 mag.), A0-A5 II and A5-F0 I supergiants (ampl. 0.035-0.036 mag.), and finally B9-A4 I supergiants (ampl.> 0.050 mag.). The increase in the mean amplitudes of late A dwarfs reflects that they are low amplitude δ Scuti stars. Other major photometric databases exist especially for faint stars. Small-automated telescope observers usually perform differential photometry of variable stars. Problems with comparison stars are often obliquely noted in publication. I use those near the variable on the sky with similar colors and small Hipparcos photometric data amplitudes (0.01-0.02 mag.). At best their ground-based standard deviations of the mean values in Strömgren $uvby$ photometry are <0.005 mag. However, some stars that were constant for the three years of Hipparcos observations may be variable now.

Photometrically constant normal A stars which are not in binary systems have constant radial velocities. Such stars also lack detectable magnetic fields (Shorlin *et al.* 2002). Substantial mass loss does not occur in the main sequence band.

Eclipsing binaries have directly measurable rotational velocities. But for most stars, the inclination angles are unknown. Only for a few single normal A stars with peculiar flat-bottomed profiles of weak lines such as Vega (Gulliver, Hill & Adelman 1994) can we find the equatorial velocity exactly. Modelling shows $v_{eq} = 160 \pm 10\,\mathrm{km\,s^{-1}}$ and $i = 7.9 \pm 0.5°$ (Hill, Gulliver & Adelman 2004). Several other stars can be similarly analyzed.

I concur with Abt (2000) that an equatorial velocity less than some particular value guarantees that a star is peculiar. But some fast rotating A stars are peculiar, e.g. CU Vir and 56 Ari. He I $\lambda4471$ and Mg II $\lambda4481$ are not the best lines for measuring $v \sin i$ as they are blends. Gray's (1992) use of Fourier techniques is very elegant, but requires high signal-to-noise and resolution. I fit clean single metal lines on the linear part of the curve of growth with Gaussian or rotational profiles and correct for the instrumental profile (using, e.g., VLINE (Hill & Fisher 1986)) or synthesize the spectrum, rotate, and convole it with the instrumental profile (with a spectrum synthesis program such as SYNTHE (Kurucz & Avrett 1981). Still the photospheric and outer envelope rotation rates do not necessarily tell us how fast the core rotates as the degree of coupling is uncertain.

Whether a star rotate as a solid body or differentially relates to the onset of atmospheric convection. When Abt & Morrell (1995) deconvolved their rotational velocities, normal A0-A1 V, A2-A4 V, and A5-F0 V stars had peak rotational velocity distributions at about 225, 190, and 170 $\mathrm{km\,s^{-1}}$, respectively. For other luminosity classes these distributions have smaller mean values since as the stars become more luminous, their radii increase and angular momentum is conserved. Between spectral types A4-F0 V there may not be any sharp-lined normal stars (see, e.g., Abt & Moyd 1973). Abt (2003) studied B and A dwarfs and giants. Combining normal and peculiar types, he found a statistically constant $v \sin i$ at 127±15 $\mathrm{km\,s^{-1}}$ for B0-A5 V stars. Then using interiors models, he predicted the giants' rotational velocities and obtained good results when he assumed rigid-body rotation. Reiners & Royer (2004) investigated spectral type A0-F1 stars with $v \sin i$ values between 60 and 160 $\mathrm{km\,s^{-1}}$. For 74 stars using a Fourier technique they found results consistent with rigid-body rotation. One star was probably seen near pole-on. Three stars appear to be rotating differentially with the spectral type of the earliest of these being A6.

The middle to late A-type stars show nonexistent to moderately strong chromospheric activity (see, e.g. Simon & Landsman's (1991) investigation of the C II $\lambda1335$ emission line). Using He I $\lambda5876$ as an activity indicator in the Hyades, Praesepe, and Coma

Clusters, the earliest stars detected range from $B - V = 0.26$ (Hyades) to $B - V = 0.31$ (Praesepe). But not all stars close to these in temperature are active (Rachford 1998). Schmitt (1997) studied X-ray data and found coronal emission is universal for spectral types A7 to G9 suggesting an onset of activity occurs near spectral type A7. Turner-Bey *et al.* (2003) found using Chandra satellite data that main sequence stars become X-ray emitters at $V - R = 0.15$, corresponding to spectral types A6 to A9. Simon & Landsman (1997) using HST/Goddard High Resolution Spectrograph) found N v $\lambda1239$ transition region emission for the A7 V stars α Aql and α Cep. They and τ^3 Eri (A4 IV) (T_{eff} = 8200 K) have chromospheric emission in Si III $\lambda1206$. The later is the hottest main sequence band star with a chromosphere and hence an outer convection zone.

Am stars have a high while the mCP stars a low binary frequency. Abt (1965) who studied 55 normal A4-F2 IV and V stars did not find any binary with a period of less than 100 days. Still the frequency of binaries with long periods may be normal. Abt & Bidelman (1969) found some normal A-type stars in binaries with P < 2.5 days.

Monnier (2003) reviews optical stellar interferometers for measuring angular diameters (θ). Lunar occultations have also been used for this purpose. With a parallax one can derive the radius. Bright stars with large θ's are currently studied. van Belle *et al.* (2001) measured the oblateness of Altair (A7 IV-V) ($v\sin i = 210 \pm 13\,\text{km}\,\text{s}^{-1}$; axial ratio = 1.140). Both Altair and Vega (Gulliver *et al.* 1994) indicate that rotating single stars are ellipsoids of revolution with smaller polar than equatorial radii. These effects may not be small for the fastest rotating normal stars, in which case one should give both polar and equatorial values. For Vega whose T_{eff} range is 350 K, if one uses mean values of T_{eff} and $\log g$ for a model atmosphere to derive elemental abundances the results will usually be close to those derived using spectral synthesis, but cannot reproduce the weak line profiles.

Astronomers can derive stellar parameters with and without the use of model atmospheres. The later leads to fundamentally determined values. Fundamental values of T_{eff} (the effective temperature) can be obtained from measurements of the angular diameter and of the total integrated flux (f_{earth}) of a star. We use these to obtain the total emergent flux at the stellar surface $F_{\text{star}} = 4f_{\text{earth}}/\theta^2$ and thus $T_{\text{eff}} = (F_{\text{star}}/\sigma)^{1/4}$ where σ is the Stefan-Boltzmann constant. Alternatively rearranging in terms of directly observable quantities $T_{\text{eff}} = 2341(F_{\text{bol}}/\theta_R^2)^{1/4}$ where F_{bol} is the total bolometric flux (in 10^{-8} ergs cm^{-2} s^{-1}) and θ_R is the angular diameter in milli-arcseconds.

Code *et al.* (1976) combined angular diameter measurements with ultraviolet, optical, and infrared fluxes to get such temperatures (see Table 2). Sirius (α CMa) might be anomalous. The 2MASS and DENIS surveys contain new consistent infrared fluxes. Bohlin & Gilliand (2004) give values on the HST White Dwarf flux scale from the far-uv to the ir and show discrepancies relative to the fundamentally calibrated values in the optical region. This pragmatic approach dependent on a theory may well be correct. Although they have achieved repeatability, systematic errors may be more important than random errors. Pagano *et al.* (2004) provides solar flux errors for a variety of instruments that are much greater than the White Dwarf flux method results indicate. I prefer direct calibrations with errors derived from the measurements for all spectral regions. The $\log g$ values for Table 2 are from Smalley & Dworetsky (1995). The value for ϵ Sgr is larger than expected and with respect to values in Table 1 three of five single luminosity class V stars have evolved beyond the ZAMS. To supplement these values I use binary star values from Smalley *et al.* (2002).

The optical region fluxes at selected wavelengths use the Hayes & Latham (1975) or Tüg, White & Lockwood (1977) calibrations of Vega which are at best good to 1%. Other grating scanner measurement errors are due to limited signal-to-noise and the use of

Table 2. Some representative effective temperatures and surface gravities

Name	Sp. Type	$T_{\mathrm{eff}}(K)$	$\log g$
α Car	F0II	7520±460	1.50
ϵ Sgr	B9.5III	9420±240	4.50
α Oph	A5III	7960±330	3.80
γ Gem	A0 IV	9220±330	3.50
β Car	A2IV	9150±240	3.50
β Lyr	A0Va	9600±180	4.00
α CMa	A1Vm	9940±210	4.33
β Leo	A3 V	8870±350	4.00
α PsA	A3 V	8760±310	4.20
α Aql	A7 V	7990±210	4.00
β Aur A	A1V	9131±257	3.93
β Aur B	A1V	9015±182	3.96
V624 Her A	A3m	8288±497	3.83
V624 Her B	A7V	8092±474	4.02
RS Cha A	A8V	7525±307	4.05
RS Cha B	A8V	7178±225	3.96
γ Vir A	F0V	7143±451	4.21
γ Vir B	F0V	7143±451	4.21

Table 3. Masses, radii, and bolometric magnitudes of normal main sequence stars

Sp. Type	Mass ($M_\odot$)	Radius ($R_\odot$)	M_{bol}
B9.5	2.38	2.17	0.68
A0	2.24	2.09	0.97
A1	2.14	2.03	1.18
A2	2.04	1.97	1.41
A3	1.98	1.92	1.57
A4	1.92	1.88	1.72
A5	1.86	1.84	1.88
A6	1.79	1.79	2.04
A7	1.74	1.75	2.19
A8	1.66	1.69	2.42
A9	1.58	1.62	2.67
F0	1.50	1.56	2.90

mean extinction coefficients. John Lester's Fourier Transform Spectrophotometer should provide absolute calibrations without the systematic errors of previous work. I hope he will calibrate a grid of stars. The ASTRA spectrophotometer (Adelman *et al.* 2004) should perform fast accurate absolute spectrophotometry for $\lambda\lambda$3300-9000 using a grid of well-calibrated standards, only one of which initially has to be absolutely calibrated.

Binary star studies can lead to masses and also for eclipsing systems radii. Popper (1980) and Andersen (1991) provide lists of eclipsing binary stars with both stars near the main sequence with well-determined masses and radii. Detached main-eclipsing systems whose spectral lines are at times well separated relative to their widths and with light curves having two well-defined minima, provide the greatest number of reliable masses and radii. There are a also a few visual binaries with well-determined masses and radii. Table 3 is from Harmanec's (1988) summary based on modern binary star data.

As the radii of components can be determined in binary systems, one can find the apparent stellar diameters if we know the distance. Fundamental values of T_{eff} can be obtained by determining the relative contributions of both components to the observed

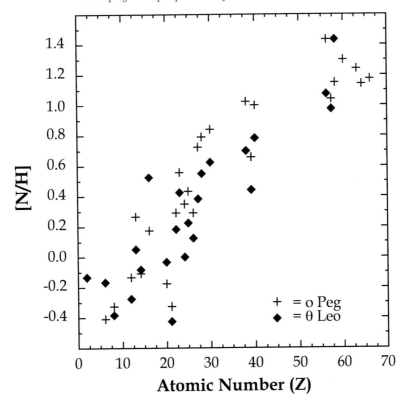

Figure 1. The abundance anomalies of the hot Am stars o Peg and θ Leo relative to solar values [N/H] as a function of atomic number Z.

flux. Smalley & Dworetsky (1995) studied 4 systems while Smalley *et al.* (2002) an additional 11. With better spectrophotometric fluxes many more systems can be studied. If one knows the mass and the radius for a star then the surface gravity $g = GM/R^2$.

If both fundamental and non-fundamental values of the effective temperatures and surface gravities for stars agree or the offsets are well determined, then many workers will use non-fundamental techniques as they are easier to perform. One matches the observed flux distribution and Balmer line profiles with those predicted by model atmospheres. For stars earlier than about B9, the energy distribution gives the temperature and the Balmer line profiles the surface gravity. For spectral types A4 and cooler the roles are reversed. For B9-A3 stars some ambiguity arises, but via iteration using both fluxes and Balmer line profiles satisfactory fits can be found.

The IRFM (infrared flux method) (Blackwell & Lynas-Gray 1994) can determine the effective temperature, T_{eff}, and angular diameter, θ of a star, based on absolute measurements of stellar monochromatic fluxes in the infrared region, F_λ, and the integrated flux F. If R is the ratio of the integrated to monochromatic flux, then when there is no interstellar absorption, one can find the stellar temperature and angular diameter from

$$R = F/F_\lambda = \sigma T_{\text{eff}}^4/\Phi(T_{\text{eff}}, g, \lambda, A) \text{ and } F = \theta^2 \sigma T_{\text{eff}}^4/4$$

where $\Phi(T_{\text{eff}}, g, \lambda, A)$ is the monochromatic emergent flux from a photosphere which is a function of the effective temperature (T_{eff}), surface gravity (g), wavelength of the monochromatic flux (λ), and the atomic abundances (A) determined by model atmosphere theory. The IRFM is designed for maximum accuracy in finding T_{eff} and θ, as θ values with parallaxes can lead to absolute stellar radii for studies of stellar evolution. One

Table 4. Stellar parameters

Star	Spectral Type	T_{eff}	$\log g$	mean Z	$\log \text{Si}/\log \text{Sr}$	$\log \text{Fe/H}$	$v \sin i$ [km s^{-1}]	ξ
29 Cyg	A7Vp	7900	3.75	-1.50±0.46	6.31	-6.34	80	3.2
Vega	A0Va	9400	4.03	-0.60±0.14	...	-5.08	21	0.6
γ Lyr	B9III	9550	2.75	-0.39±0.40	4.87	-5.41	70	0.8
σ Boo	F2V	6744	3.88	-0.31±0.38	4.85	-5.86	8	1.6
α Dra	A0III	10000	3.60	-0.23±-0.26	4.72	-4.56	25	0.1
η Lep	F1V	6850	4.05	-0.22±-0.36	4.83	-4.98	13	3.8
ι And	B9V	11600	3.35	-0.19±-0.15	...	-4.68	60	1.3
28 And	A7III	7358	4.65	-0.16±0.24	4.68	-4.24	18	3.5
14 Cyg	B9III	10750	3.55	-0.12±0.23	4.54	-4.57	31	0.0
134 Tau	B9IV	10825	3.88	-0.09±0.16	4.63	-4.72	27	0.0
29 Vul	A0IV	10200	4.10	-0.09±0.33	4.49	-3.81	49	1.2
ϵ Aqr	A1.5V	9050	3.75	-0.08±0.35	4.60	-5.32:	110	2.8
2 Lyn	A1Va	9295	4.10	-0.04±0.20	4.65	-4.51	44	2.1
α Sex	B9.5III	9875	3.55	-0.03±0.18	4.50	-4.46	21	0.3
ν Cap	B9.5V	10250	3.90	-0.02±0.23	4.53	-4.52	23	0.0
21 Lyn	A0mA1V	9500	3.75	-0.02±0.37	4.66	-3.77	18	1.6
κ Cep	B9III	10325	3.70	-0.01±0.20	4.56	-4.23	23	0.3
HR 6559	A7IV	7900	3.20	0.07±0.40	4.70	-3.71	18	3.5
γ Gem	A0IV	9260	3.60	0.09±0.23	4.36	-4.78	10	2.0
HD 60825	A0	8750	3.59	0.12±0.16	...	-4.41	14	2.5
5 Aqr	B9.5III	11150	3.35	0.13±0.20	...	-4.27	25	0.7
β UMa	A0mA1IV-V	9600	3.83	0.14±0.40	4.31	-3.52	45	2.5
60 Leo	Am	9250	4.25	0.15±0.35	4.04	-3.32	17	3.4
7 Sex	A0Vs	10135	3.69	0.17±0.38	4.27	-4.72	22	1.8
15 Vul	Am	7700	3.50	0.19±0.47	4.74	-3.99	10	4.0
σ Aqr	A0IV	10125	4.00	0.24±0.43	4.32	-3.59	21	1.0
ν Cnc	A0III	10375	3.50	0.25±0.88	4.51	-3.42	13	0.1
θ Leo	A2V	9325	3.65	0.25±0.51	4.38	-3.79	22	2.2
θ And	A2IV	9000	4.00	0.26±0.40	4.22	-4.16	93	3.6
ω UMa	A0IV-V	10026	3.88	0.27±0.63	4.27	-3.40	46	1.1
68 Tau	A2IV-Vs	9000	4.00	0.40±0.52	4.16	-3.43	9	2.3
ϵ Ser	Am	8422	4.30	0.44±0.54	4.26	-3.93	33	5.2
32 Aqr	Am	7700	3.65	0.44±0.63	4.42	-3.72	6	4.5
π Dra	A2A3IV	9125	3.80	0.46±0.66	4.18	-3.42	26	3.5
o Peg	A1IV	9535	3.74	0.49±0.60	4.21	-3.47	6	1.3
ϕ Aql	A1IV	9534	4.05	0.54±0.74	4.09	-3.99	28	3.1
6 Lyr	Am	8155	3.90	0.60±0.70	4.10	-3.59	31	5.6
λ UMa	A1IV	9000	3.75	0.66±0.66	4.26	-3.31	50	2.8
59 Her	A3III	9325	3.65	0.79±0.61	4.08	-3.62	27	2.8

begins with the apparent visual magnitude and the effective temperature. The apparent visual magnitude is converted to the absolute visual magnitude using the distance (a Lutz-Kelker (1973) correction may need to be applied). The luminosity ($L = 4\pi R^2 \sigma T_{\text{eff}}^4$) is found by applying a Bolometric Correction (see, e.g., Bessell, Castelli & Plex 1998) to account for the energy not included in the visual magnitude. Some widely used evolutionary tracks for solar composition stars ($Z = 0.019, Y = 0.273$) are those of Girardi et al. (2000) and ($Z = 0.020$) Schaller et al. (1992).

Landstreet (1998) obtained the spectra of several ultra-sharp-lined stars at high resolution and signal-to-noise ratios. He found that the A1 V star HD 72660 with $\xi = 2.2$ km s^{-1} and two Am stars with much larger values (4 to 5 km s^{-1}) had strong spectral lines with significant asymmetries while hotter stars did not. Such effects must be accounted for in the analyses of stars where they are found.

Table 4 summarizes A star abundances and gives the stellar name, spectral type, effective temperature, surface gravity, mean metallicity with respect to the solar, the log Fe/H value, the log Si/log Sr ratio, $v \sin i$ and the microturbulence ξ. Values are from Adelman (1986, 1987, 1989, 1991, 1994, 1996, 1999), Adelman & Albayrak (1998), Adelman & Gulliver (1990), Adelman & Philip (1996). Adelman et al. (1997, 1999, 2000, 2001, 2004), Caliskan & Adelman (1997), Kocer *et al.* (2003), and Pintado & Adelman (2003). My conceptual model is that stars rotating sufficiently fast have normal abundances over all of their photospheres. At some point when the local rotation rate decreases below a critical value, polar regions begin to show peculiar abundances. As rotation further decreases the portion with normal abundances decreases until the abundances over the entire surface are peculiar, but not necessarily uniform.

The Si/Sr is a measure of the change of the anomalies with respect to atomic number (Z). The solar value of log Si/log Sr is 4.58. A sign of the Am phenomena is the increase in the anomalies with Z as illustrated in Figure 1 from Adelman *et al.* (2004) for the hot Am stars θ Leo and o Peg. Lemke (1989) estimates errors in T_{eff} of ± 150 K and in $\log g$ of ± 0.2 dex. This suggests abundance errors of order 0.25 dex. The mean metallicities are not completely uniform as although they usually are based on most of the common elements there are wide differences in the detection and measurement of values for the less common elements and the spectral coverage differs from star to star. Although most of the HgMn stars are classified as B9 stars or close to this class, I include only ν Cnc, the coolest known HgMn star as an example since at abundance analysis spectral resolution this class is separable from the normal stars.

The stars are in order of increasing metallicity which correlates with log Fe/H and log Si/log Sr, expectations of the conceptual model. 29 Cyg is a λ Boo star. Next are metal-poor stars, Vega, γ Lyr, and σ Boo. I break the distribution here and again at β UMa, a known Am star. Stars with higher metallicities are Am stars except for 7 Sex, which behaves dynamically as a Population II star, and ν Cnc. The normal band is slightly asymmetric with respect to solar values. A few stars show non-solar Si/Sr ratios. Thus 29 Vul, 21 Lyn, and HR 6559 are incipient Am stars. The ϵ Aqr value may not be quite as well known due to its $v \sin i$ value. The errors in the analyses are sufficiently small for us to know that α Dra, at the low metallicity boundary is slightly metal-poor relative to the Sun and 5 Aqr at the high metallicity boundary is marginally metal-rich relative to the Sun. Table 4 samples stars with $v \sin i$ values between 6 and 110 $\mathrm{km\,s^{-1}}$. Within the normal star band the mean microturbulence is 1.6 ± 1.3 $\mathrm{km\,s^{-1}}$ with 6 out 17 stars having values close to 0.0.

The iron abundances of most of the stars agree with the previous classification. But η Lep now is metal-poor while 15 Vul has a mixture of underabundant and overabundant elements. This is also true of the 3 cited incipient Am stars. Thus among the sharp-lined A stars most do not have "solar" abundances.

Acknowledgements

I thank my fellow astronomers and collaborators who have guided me on the right path for the study of A and related type stars through discussions and by example. My adventures in A stars have been supported by grants from NSF, NASA, and The Citadel Foundation. I appreciate the observing time in the last 15 years particularly at the Dominion Astrophysical Observatory and with the Four College Automated Photoelectric Telescope, Fairborn Observatory, and at CASELO Observatory.

References

Abt H. A. 1965, *ApJS*, 11, 37.
Abt H. A. 2000, *ApJ*, 544, 933.
Abt H. A. 2003, *ApJ*, 582, 420.
Abt H. A. 2004, *ApJ*, 603, L109.
Abt H. A. 1969, *ApJ*, 158, 1091.
Abt H. A., Morrell N. I. 1995, *ApJS*, 99, 135.
Abt H. A., Moyd K. I. 1973, *ApJ*, 182, 809.
Adelman S. J. 1986, *A&AS*, 64, 173.
Adelman S. J. 1987, *A&AS*, 67, 353.
Adelman S. J. 1989, *MNRAS*, 239, 487.
Adelman S. J. 1991, *MNRAS*, 252, 1161.
Adelman S. J. 1994, *MNRAS*, 271, 355.
Adelman S. J. 1996, *MNRAS*, 280, 130.
Adelman S. J. 1999, *MNRAS*, 310, 146.
Adelman S. J. 2001, *IBVS*, 5050.
Adelman S. J., Albayrak B. 1998, *MNRAS*, 300, 359.
Adelman S. J., Gulliver A. F. 1990, *ApJ*, 348, 712.
Adelman S. J., Philip A. G. D. 1996, *MNRAS*, 252, 1181.
Adelman S. J., Gulliver A. F., Heaton R. J. 2004, *A&A*, in press.
Adelman S. J., Adelman A. S, Pintado O. I. 2003, *A&A*, 397, 267.
Adelman S. J., Caliskan H., Kocer D., Bocal, C. 1997, *MNRAS*, 288, 470.
Adelman S. J., Caliskan H., Kocer D., Cay I. H., Tektunali H. G. 2000, *MNRAS*, 316, 514.
Adelman S. J., Caliskan H., Cay T., *et al.* 1999, *MNRAS*, 305, 601.
Adelman S. J., Caliskan H., Kocer D. *et al.* 2001, *A&A*, 371, 1078.
Adelman S. J., Gulliver A. F., Smalley B., *et al.* 2004, *These Proceedings*, JP2
Andersen J. 1991, *A&AR*, 3, 91.
Bessell M. S., Castelli F., Plex B. 1998, *A&A,*, 333, 231.
Blackwell, D. E., Lynas-Gray, A. E. 1994, *A&A,*, 282, 899.
Bohlin R. C., Gilliland R. L. 2004, *AJ*, 127, 3508.
Breger, M. 2005, *These Proceedings*, 335.
Caliskan H., Adelman, S. J. 1997, *MNRAS*, 288, 501.
Code A. D., Davis J., Bless R. C, Hanbury Brown R. 1976, *ApJ*, 203, 417.
Conti P. S. 1970, *PASP*, 82, 781.
Gerbaldi M., Faraggiana R., Lai O. 2003, *A&A*, 412, 447.
Girardi L., Bressan A., Bertelli, G., Chiosi C. 2000, *A&AS*, 141, 371.
Gray D. F. 1992, The Observation and Analysis of Stellar Photospheres, Cambridge University
    Press, Cambridge.
Gray R. O. 1988, *AJ,*, 95, 220.
Gulliver A. F., Hill G., Adelman S. J. 1994, *ApJ*, 429, L81.
Harmanec P. 1988, *Bull. Astron. Inst. Czechosl.*, 39, 329.
Hauck B., Slettebak A. 1983, *A&A*, 127, 231.
Hayes D. S., Latham D. W. 1975, *ApJ*, 197, 593.
Hill G., Fisher W. A. 1986, Pub. Dom. Astrophys. Obs.*16*, No. 13.
Hill G., Gulliver A. F., Adelman S. J. 2005, *These Proceedings*, 35
Hoffleit D., Warren W. H., Jr. 1991, The Bright Star Catalogue, 5th Rev. Ed., ADC Selected
    Astronomical Catalogs, Vol. *1*, NASA Goddard Space Flight Center.
Kocer D., Adelman S. J., Caliskan H., Gulliver A. F., Tektunali H. G. 2003, *A&A*, 406, 975.
Kurucz R. L., Avrett E. H. 1981, SAO Special Report No. *191*.
Landstreet J. D. 1998, *A&A*, 338, 1041.
Lemke M. 1989, *A&A*, 225, 125.
Lutz T. F., Kelker D. H. 1973, *PASP*, 85, 573.
Monnier J. D. 2003, *Rep. Prog. Physics*, 66, 789.

Morgan W. W., Keenan P. C., Kellman E. 1943, Atlas of Stellar Spectra, University of Chicago Press, Chicago.
Pagano I., Linsky J. L., Valenti J., Duncan D. K. 2004, *A&A*, 415, 331.
Pintado O. I., Adelman S. J. 2003, *A&A*, 406, 987.
Popper D. M. 1980, *ARA&A*, 18, 115.
Preston G. W. 1974, *ARA&A*, 12, 257.
Rachford B. L. 1998, *ApJ*, 505, 255.
Reiners A., Royer F. 2004, *A&A*, 415, 325.
Roman N. G., Morgan W. W., Eggen O. J. 1948, *ApJ*, 110, 205.
Schmitt J. H. M. M. 1997, *A&A*, 318, 215.
Schaller G., Schaerer D., Meynet G., Maeder A. 1992, *A&AS*, 96, 269.
Shorlin S.L.S., Wade G. A., Donat J.-F., Landstreet J. D., *et al.* 2002, *A&A*, 392, 637.
Simon T., Landsman W. B. 1991, *ApJ*, 380, 200.
Simon T., Landsman W. B. 1997, *ApJ*, 483, 435.
Smalley B., Dworetsky M. M. 1995, *A&A*, 293, 446.
Smalley B., Gardiner R. B., Kupka, F., Bessell M. S. 2002, *A&A*, 395, 601.
Tüg H., White N. M., Lockwood G. W. 1977, *A&A*, 61, 679.
Turner-Bey D. E., Kashyap V., Evans N. *et al.* 2003, *AAS Meeting 203*, #85.03.
van Belle G. T., Ciardi D. R., Thompson R. R. *et al.* 2001, *ApJ*, 559, 1155.
Waelkens C., Aerts C., Kestens E. *et al.* 1998, *A&A*, 330, 215.
Wolff S. C. 1983, The A-Stars: Problems and Perspectives, NASA SP-463, Washington, DC.

Discussion

L. BALONA: You mentioned that one of the main characteristics separating normal and peculiar stars are the $v \sin i$ values. Yet a brief scan of $v \sin i$ in your table does not suggest any difference in $v \sin i$ between them.

S. ADELMAN: My stars were chosen to be relatively sharp-lined to make the elemental abundance analyses easily done. I agree that is very important to obtain abundances of stars with rotation. In this regard, I note the pioneering work of Grant Hill.

T. RYABCHIKOVA: If we find a low $v \sin i$ normal abundance star, does it mean that star should necessarily be pole-on fast rotator like Vega?

S. ADELMAN: Most likely. It would be useful to examine the model of Vega to see for what values of i there are flat-bottomed weak lines.

J. LANDSTREET: You have chracterized convection as being first clearly apparent around A5. However, as you decrease effective temperature along the main sequence, non-zero values of microturbulence parameter are first seen about A0. This seems to me to be pretty clear indication of convection at least in the deep atmosphere.

S. ADELMAN: I agree that the non-zero microturbulence found in the early A stars is an indicator of incipient convection.

C. CORBALLY: What kinds of differences are found for published $v \sin i$ values derived by the Fourier technique and your line profile determinations.

S. ADELMAN: For most cases the agreement is good to excellent. But it is the outlyers which worry Austin Gulliver and me.

The A-Star Puzzle
Proceedings IAU Symposium No. 224, 2004
J. Zverko, J. Žižňovský, S.J. Adelman, & W.W. Weiss, eds.

© 2004 International Astronomical Union
DOI: 10.1017/S1743921304004326

Standard model atmospheres for A-type stars and non-LTE effects

Jiří Kubát and Daniela Korčáková

Astronomický ústav, Akademie věd České republiky, CZ-251 65 Ondřejov, Czech Republic
email: kubat@sunstel.asu.cas.cz, kor@sunstel.asu.cas.cz

Abstract. The current status of NLTE model atmosphere calculations of A type stars is reviewed. During the last decade research has concentrated on solving the restricted NLTE line formation problem for trace elements assuming LTE model atmospheres. There is a general lack of calculated NLTE line blanketed model atmospheres for A type stars, despite the availability of powerful methods and computer codes that are able to solve this task. Some directions for future model atmosphere research are suggested.

Keywords. Stars: atmospheres, methods: numerical, radiative transfer, stars: early-type

1. Introduction

Compared to the atmospheres of hotter O and B stars that have stellar winds and to the atmospheres of cooler F and G stars that have convective atmospheres, the atmospheres of A-type stars are relatively quiet, which enables the presence of various interesting phenomena. This atmospheric calmness leads to the development of a number of different chemical peculiarities and magnetic field structures on long time scales.

There exist widely used grids of line blanketed LTE ATLAS9 model atmospheres (Kurucz 1993, Castelli & Kurucz 2003). These models are extensively used for A stars. In addition several attempts to remove the inconsistent assumption of LTE have been performed. The first NLTE calculation of a cool B star, which can be somehow understood as a first step toward A star modelling, was performed by Auer & Mihalas (1970). The domain of A star temperatures was first investigated by Kudritzki (1973) and Frandsen (1974). A grid of NLTE model atmospheres of stars with $10000\,\mathrm{K} \leqslant T_{\mathrm{eff}} \leqslant 15000\,\mathrm{K}$ (thus also including a part of the A star domain) was calculated by Borsenberger & Gros (1978). A great improvement was made by Hubený (1980, 1981), who studied in detail the effect of the Lα line on the NLTE model atmospheres of A stars and applied his code to Vega. A detailed analysis of the physics needed to replace the assumption of LTE was presented by Hubený (1986). The first NLTE line blanketed model atmospheres of A stars were calculated using the method of superlevels and superlines by Hubeny & Lanz (1993). Unfortunately, although there has been great progress in calculating NLTE model atmospheres for hot stars and for white dwarfs, calculations of NLTE model atmospheres of A stars are very rare. The only recent NLTE model atmosphere of an A-type star was the model calculated by Hauschildt *et al.* (1999) for parameters corresponding to Vega. Their announced grid of NLTE models has not appeared.

The inherent reason which stands tacitly behind the lack of calculated NLTE A star model atmospheres is the different scaling properties of the atmospheres of this stellar type. The atmospheres of A stars have very extended line formation regions so that different lines form at very different depths (cf. Fig. 1 in Lanz & Hubeny 1993). Therefore they are very difficult to describe by a simple standard depth discretization as compared to O stars and white dwarfs. Hence a much more elaborate depth scale has to be used. Our own experience shows that there are severe convergence problems caused by improper depth scaling and only a tremendous number of depth points helped.

2. Basic equations for NLTE model atmospheres

A solution of a model atmosphere problem means taking the basic input values of stellar luminosity L, stellar radius R, and stellar mass M (or, equivalently for the usual plane-parallel case effective temperature T_{eff} and gravitational acceleration at the stellar surface g), adding physical laws that are important in the stellar atmosphere (a principal ingredient), and calculating the spatial distribution of various physical quantities (temperature $T(\vec{r})$, electron density $n_e(\vec{r})$, population numbers $n_i(\vec{r})$, density $\rho(\vec{r})$, velocity field $\vec{v}(\vec{r})$, etc. This procedure is often being performed in substeps. The first step is the determination of T, n_e, ρ, and n_i for the most important opacity sources. The second (optional) step is the determination of n_i for the minor abundant species, often referred to as trace elements. The final step in this process is the calculation of the theoretical emergent radiation which may be then compared to the observed one.

A standard model atmosphere here is a static one-dimensional (plane-parallel or spherically symmetric) atmosphere in hydrostatic, radiative, and statistical equilibrium. To calculate such a model atmosphere we need to solve the radiative transfer equation (to determine the radiation field I), the hydrostatic equilibrium equation (to determine ρ), the energy equilibrium equation (to determine T), and the statistical equilibrium equations (to determine n_i).

2.1. *Radiative transfer equation*

The radiative transfer equation is the principal equation for model atmosphere calculations. Regardless of the geometrical approximation which is adopted, the basic solution of the radiative transfer equation is performed along a ray:

$$\frac{\mathrm{d}I_\nu(r)}{\mathrm{d}s} = -\chi_\nu(r)I_\nu(r) + \eta_\nu(r). \tag{2.1}$$

Usually the simplest possibility, i.e., the plane-parallel atmosphere, is being used for the modelling of stellar atmospheres, basically due to its relative simplicity.

2.1.1. *Formal solution*

By formal solution of Equation (2.1) we mean the solution for a *given* opacity and emissivity. Such a solution may be performed using either differential or integral methods. The latter are especially useful for the multidimensional case. The inconsistency of the formal solution is that opacity χ_ν and emissivity η_ν are not given, but in the process of solving the model atmosphere problem they depend on temperature, density, and radiation field, which are to be determined when solving the radiative transfer equation. However, the formal solution remains at the heart of each model atmosphere code (see Auer 2003).

2.1.2. *Approximate solution using ALO*

The process of the formal solution may be expressed using the Λ-operator

$$J_\nu = \Lambda_\nu S_\nu. \tag{2.2}$$

This expression may be used in an iterative process, where the source function S_ν is determined by solving the other constraint equations of hydrostatic, energy, and statistical equilibria. As has been shown, such a process, however, effective for optically thin molecular clouds (cf. Dickel & Auer 1994), fails to produce a convergent solution for optically thick stellar atmospheres (see Auer 1984). This problem has been overcome using the Newton-Raphson method for model stellar atmospheres by Auer & Mihalas (1969). As an efficient alternative, an approximate Λ-operator Λ_ν^* may be used,

$$J_\nu = \Lambda_\nu^* S_\nu + (\Lambda_\nu - \Lambda_\nu^*) S_\nu \qquad (2.3)$$

The operator Λ_ν^* is constructed so that it contains the basic physics of the problem and is simply calculable. The most efficient method is to calculate this operator consistently with the numerical method used for the formal solution, as has been described by Rybicki & Hummer (1991) and by Puls (1991). Then the radiation field is determined by an iterative process,

$$J_\nu^{(n)} = \Lambda_\nu^* \left[S_\nu \left(J_\nu^{(n)} \right) \right] + (\Lambda_\nu - \Lambda_\nu^*) \left[S_\nu \left(J_\nu^{(n-1)} \right) \right] \qquad (2.4)$$

where (n) means the current iteration step and $(n-1)$ the previous one.

2.2. *Energy equilibrium*

The energy equilibrium equations determine the temperature structure. They describe the basic energy balance in the atmosphere. For static atmospheres, the equation of radiative equilibrium,

$$\int_0^\infty (\kappa_\nu J_\nu - \eta_\nu) \, d\nu = \int_0^\infty \kappa_\nu (J_\nu - S_\nu) \, d\nu = 0, \qquad (2.5)$$

which simply describes the radiative energy balance, is usually used. However, this equation does not guarantee radiative flux conservation, which results in an incorrect temperature structure at large optical depths. Therefore, another form, which comes from $\nabla \cdot \vec{\mathcal{F}} = 0$, is used at large optical depths. Better numerical properties of the scheme may be achieved if a linear combination of these equations is considered, as was first done by Hubeny & Lanz (1995). Another improvement of convergence properties, especially useful at small continuum optical depths where strong lines are still present, was introduced by Kubát *et al.* (1999) where a thermal balance of electrons instead of the radiative equilibrium equation is used.

Thus three different equations may be used in the stellar atmosphere to take into account energy equilibrium. In the deepest layers, the differential form of radiative equilibrium is used. In the middle parts, where continuum radiation is formed, the integral form of radiative equilibrium works well. In the outer parts where optically thick lines coexist with an optically thin continuum (which is the case of A star atmospheres), the thermal balance of electrons works best. A more detailed discussion with the numerical properties of these methods may be found in Kubát (2003a,b).

2.3. *Equations of statistical equilibrium*

The equations of statistical equilibrium are the key equations both for the NLTE model atmosphere and line formation problems. The set of equations for statistical equilibrium for the static case may be written as

$$n_i \sum_l (R_{il} + C_{il}) - \sum_l n_l (R_{li} + C_{li}) = 0 \qquad (2.6)$$

Radiative rates R_{il} are those responsible for the NLTE effects, since they introduce nonlocal interactions and alter the level population irrespective of local equilibrium conditions.

They cause the gas not to be in thermodynamic equilibrium. On the other hand, for an equilibrium (Maxwellian) electron velocity distribution (which is the common case), collisional rates reintroduce thermodynamic equilibrium. The balance between collisional and radiative rates determines the applicability of the LTE approximation. If collisions dominate, then the LTE approximation is feasible. If radiative transitions dominate, which is the case in A star atmospheres, then the equations of statistical equilibrium need to be solved to obtain the correct population numbers.

   Since the system of rate equations is linearly dependent, we have to close it with some other condition. For a reference (usually, but not necessarily hydrogen) atom either the particle conservation equation $\sum_k N_k = N - n_e$ or the charge conservation equation $\sum_k \sum_j q_j N_{jk} = n_e$ is used. For other atoms, the abundance equation $N_k = Y_k N_r$ is usually used. The detailed forms of the equations may be found, e.g., in Kubát (1997).

2.4. Solution of the system of equations

The whole system of equations is usually solved using a Newton-Raphson iteration scheme, which in modeling stellar atmospheres used to be referred to as linearization (Auer & Mihalas 1969). The radiation field is linearized as well, or it can be included using approximate lambda operators (see Eq. 2.4), as was first done by Werner (1986). Excellent reviews of ALI methods of model atmosphere solutions are provided by Hubeny (2003) and Werner et al. (2003). In addition, an implicit linearization of b-factors saves additional computer time and memory (Anderson 1987). The whole process may be accelerated using either Kantorovich (see Hubeny & Lanz 1992) or Ng acceleration (see Auer 1991).

2.5. Line blanketing

Line blanketing is caused by the enormous number of spectral lines in the UV region ($\sim 10^7$), mostly of iron and nickel, which causes radiation to be absorbed in the UV and reemitted in the visible region, thus changing the emergent flux and atmospheric structure significantly. There are two main approaches to line blanketing under the simplifying assumption of LTE, the ODF (Opacity Distribution Function) and OS (Opacity Sampling).

   If we do not use the assumption of LTE, we have to solve the equations of statistical equilibrium to obtain the correct population numbers for all levels. This task is relatively simple for simple atoms (like He), but becomes difficult for atoms with a complicated level structure, like iron. To cope with the complexity of these atoms, Anderson (1989) introduced the concept of superlevels and superlines. A superlevel is a level which is created by grouping several individual levels. To achieve a real simplification, it is useful to group the levels in such a manner that the relative population distribution inside the superlevel obeys the LTE distribution. Thus, this grouping is done according to the energy, E_i, of individual levels (Anderson 1989, Dreizler & Werner 1993). A more sophisticated grouping according to E_i and parity was done by Hubeny & Lanz (1995). Dreizler & Werner (1993) used opacity sampling in their calculations, whereas Hubeny & Lanz (1995) used the NLTE opacity distribution function.

2.6. LTE versus NLTE

Since the beginning of NLTE calculations there is a battle between 'LTE people' and 'NLTE people' concerning which approach is better. If NLTE model atmospheres were calculated as easily as the LTE variety, there would be no discussion and everybody would use NLTE. However, this is not the case and it is much more difficult to calculate a NLTE model than the LTE model. Two conflicting aims enter the scene. The first is to analyse as many stars as possible, which can be hardly done using NLTE models. The second is to study stars as accurately as possible, which can hardly be done using LTE models.

For the case of LTE it is relatively easy to handle line blanketing, since one neglects the effect of radiation on atomic population numbers, which saves enormous computing time. On the other hand the more general case of NLTE, including line blanketing, is computationally expensive, albeit now it is possible to handle NLTE calculations using sophisticated numerical methods and contemporary computers. However, for extended atmospheres the reliability of LTE decreases and one is forced to use NLTE.

To avoid repeated calculations of complicated model atmospheres one may take the advantage of precomputed grids. However, one has to keep in mind that using such grids, which were calculated under certain physical assumptions, always limit the results by their assumptions.

3. NLTE line formation calculations

Calculations of full NLTE model atmospheres of A type stars are a difficult task. Due to a large span of line formation regions down to optical depths in the continuum of about 10^{-9}, the numerical procedure becomes extremely unstable and the calculations require special care. That is why people started to solve the easier task of NLTE line formation for a given model atmosphere. A necessary condition for such calculation to be reasonable is negligible influence of the ion on the ionization balance and, consequently, on the global structure of the atmosphere. Such elements are being referred to as trace elements.

Contemporary NLTE analysis of A stars are mainly concerned with solving the statistical equilibrium equations for trace elements for a given (mostly LTE) model atmosphere. These calculations have already been reviewed by Hubený (1986) and Hubeny & Lanz (1993), so only new calculations that appeared after the last review are listed in Table 1. We did our best to mention all published calculations. In the case we unintentionally omitted some, we apologize in advance.

There are several codes available for this purpose. They use slightly different numerical techniques. The basic differences are in the treatment of the radiation field. Both the accelerated lambda iteration and complete linearization methods are used. A comparison of results from different codes for Vega (and Sun) was presented by Kamp *et al.* (2003), who found large differences in the results of different codes. Therefore, we indicate which code was used.

4. Beyond standard model atmospheres

4.1. *Nonthermal collisional rates*

In standard NLTE calculations only the radiation field and level populations are allowed to deviate from their equilibrium values. The velocity distributions of individual species are still assumed to be in equilibrium. Collisional rates are then calculated as an average over the Maxwellian electron velocity distribution,

$$C_{ij} = n_e \int_{v_0}^{\infty} \sigma_{ij}(v) f(v) v dv. \tag{4.1}$$

Thus in equilibrium there is a relation between collisional rates up and down (excitation/deexcitation or ionization/recombination),

$$n_i^* C_{ij} = n_j^* C_{ji}. \tag{4.2}$$

Table 1. List of NLTE calculations for a given LTE model atmosphere for A stars

Ion	Reference	Code	Comment on included levels and transitions
He I	Takeda (1994)	3	calculations for Deneb; 88 levels
Li I	Mashonkina et al. (2002)	4	
Li I/II	Shavrina et al. (2001)		20 levels
C I	Takeda (1992b)	3	129 levels, 2351 radiative transitions
	Venn (1995)	2	83 levels
	Rentzsch-Holm (1996b)	2	83 levels, 63 transitions
	Paunzen et al. (1999)	2	83 levels, 63 transitions
C I/II	Przybilla et al. (2001b)	1	C I levels with $n \leqslant 9$, C II levels with $n \leqslant 4$
N I	Takeda (1992b)	3	119 levels, 2119 radiative transitions
	Sadakane et al. l (1993)	3	used model atom of Takeda (1992b)
	Takeda & Takada-Hidai (1995)	3	used model atom of Takeda (1992b)
	Takada-Hidai & Takeda (1996)	3	used model atom of Takeda (1992b)
	Venn (1995)	2	93 levels, 189 transitions
	Lemke & Venn (1996)	2	93 levels, 189 radiative transitions
	Rentzsch-Holm (1996a)	2	96 levels, 82 transitions
N I/II	Przybilla & Butler (2001)	1	N I levels with $n \leqslant 7$, N II levels with $n \leqslant 6$
O I	Takeda (1992a, 1997)	3	86 levels, 294 radiative transitions
	Takeda & Takada-Hidai (1998)	3	86 levels, 294 radiative transitions
	Takeda et al. (1999)	3	calculations from Takeda (1997) for CP stars
	Paunzen et al. (1999)	2	15 levels
	Przybilla et al. (2000)	1	all levels below excitation energy
Na I	Takeda & Takada-Hidai (1994)	3	calculations for A supergiants, 92 levels, 178 radiative transitions
	Mashonkina et al. (2000)	4	solution for $T_{\text{eff}} = 4000 - 12\,000$ K, 21 levels
Mg I	Shimanskaya et al. (2000)	4	solution for $T_{\text{eff}} = 4500 - 12000$ K, 50 levels
	Idiart & Thévenin (2000)	5	104 levels, 980 radiative transitions
Mg I/II	Przybilla et al. (2001a)	1	all levels with $n \leqslant 10$
S I	Takeda & Takada-Hidai (1995)	3	56 levels, 173 radiative transitions
	Takada-Hidai & Takeda (1996)	3	56 levels, 173 radiative transitions
K I	Ivanova & Shimanskii (2000)	4	solution for $T_{\text{eff}} = 4000 - 10\,000$ K, 36 levels
Ca I	Idiart & Thévenin (2000)	5	84 levels, 483 transitions
Ti II	Becker (1998)	1	complete model atom, using superlevels
Fe II	Becker (1998)	1	complete model atom, using superlevels
Fe I/II	Rentzsch-Holm (1996b)	2	79+20 levels, 52+23 lines
	Thévenin & Idiart (1999)	5	256+190 levels, 2117+3443 lines
Sr II	Belyakova et al. (1999)	4	solution for $T_{\text{eff}} = 4000 - 12\,000$ K, 41 levels
Nd II/III	Mashonkina et al. (2005)	1	247+69 levels

Computer codes used : 1 – DETAIL (Giddings 1981); 2 – Kiel (Steenbock & Holweger 1984); 3 – Takeda (1991); 4 – NONLTE3 (Sakhibullin 1983); 5 – MULTI (Carlsson 1986).

If the velocity distribution is non-Maxwellian (e.g., in electron beams formed in the Sun during flares – cf. Kašparová & Heinzel 2002), then equation (4.2), which expresses the equilibrium condition, is invalid. In such a case the collisions may cause level population numbers to differ from the LTE values, which has an observable effect on the line profiles. The nonthermal collisional term may be considered as another source of NLTE effects. Such effects are present in the Sun, where electron beams are formed after magnetic reconnection. There is also a possibility that they may be present in magnetic A stars, where releasing magnetic energy in eruptive events like flares may be expected as well.

4.2. Magnetic fields

Inclusion of magnetic fields into model atmosphere calculations have been done only occasionally. An important attempt in calculating a model atmosphere with a magnetic field was done by Carpenter (1985), who found changes in the net gravity and pressure distribution due to the magnetic field. The most consistent model so far was recently developed by Valyavin *et al.* (2004), who assumed LTE and included the Lorentz force into the equation of hydrostatic equilibrium. A historical overview of magnetic model atmosphere calculations is also presented there.

4.3. *Diffusion*

An important aspect of A star atmospheres is that atmospheres of a good fraction of stars of this stellar spectral type are extremely quiet. Such quiet atmospheres without significant global motions such as global convection and stellar winds enable long time scale processes of diffusion and gravitational settling to take place. Diffusion occurs not only in the presence of magnetic field (note the effect of enhancing radiative acceleration in a polarized radiation field and the effect of ambipolar diffusion, Babel & Michaud (1991a, 1991b), but also due to different sensitivities of various atoms and ions to incoming radiation. Radiative diffusion is able to explain various chemical peculiarities in CP stars and is also responsible for the isotopic shift effect (Aret & Sapar 2002). However, consistent NLTE model atmospheres that take into account radiative diffusion processes are still missing for A type stars. NLTE model atmospheres with diffusion were calculated by Dreizler & Shuh (2003) for white dwarfs, where diffusion processes are also important. The physical process of diffusion is reviewed in detail by Michaud (2005).

4.4. *Extended atmospheres*

Plane-parallel model atmospheres are also being used for NLTE modeling of atmospheres of A-type supergiants (e.g., Kudritzki 1973). However, they do not describe the limb darkening correctly. A-type supergiants have also stellar winds and show P Cyg line profiles. Therefore it is impossible to describe them using plane parallel atmospheres. Also, wind models of A stars are not calculated very often. Multicomponent hydrodynamic radiatively driven wind models for A stars were calculated by Babel (1995). Another radiation driven wind model was calculated by Achmad *et al.* (1997), but they did not take into account for NLTE effects. Generally, for extended atmospheres NLTE effects are stronger. An attempt to calculate spherically symmetric NLTE model atmospheres with a stellar wind for Deneb was done by Aufdenberg *et al.* (2002). However, as the authors note, they were not able to fit the observed Hα line profile. Work on developing consistent NLTE wind models of A supergiants is currently in progress (Krtička & Kubát 2005).

4.5. *Limb darkening*

An important property of the emergent radiation from stellar atmospheres is the angular dependence of the specific intensity $I(\theta)$, usually called limb darkening. Approximate limb darkening laws are often used (see, e.g., Allen 1963 or Gray 1976). These approximate laws do not describe the angular dependence of the specific intensity very well even in the case of thin stellar atmospheres. In addition, limb darkening is strongly frequency dependent (Hadrava & Kubát 2003), and in the center of a line, even limb brightening instead of darkening may appear, as can be seen in Figure 1. This effect becomes very strong especially for extended stellar atmospheres.

For a correct description of limb darkening it is necessary to use model atmospheres with a better geometry than plane-parallel. This has been done by Claret & Hauschildt (2003), who, using a spherically symmetric code, calculated limb darkening for a grid of model atmospheres from A to G spectral types.

Knowledge of an accurate limb darkening law is necessary for the detection of stellar spots (see Kjurkchieva 1989). It is also very important for the analysis of interferometric measurements, as well as for the correct treatment of stellar rotation. Fortunately,

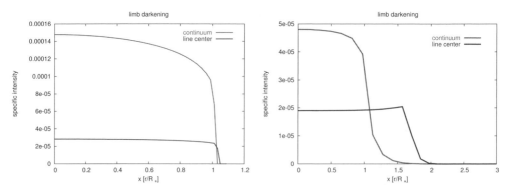

Figure 1. The comparison of limb darkening for the thin atmosphere (left panel) and the extended atmosphere (right panel). These results were obtained from the Korčáková & Kubát (2003) code.

eclipses from binaries (e.g., Twigg 1979) and gravitational microlensing (Heyrovský 2003) measurements of limb darkening are possible for at least some distant stars.

4.6. *Rotation*

Stellar rotation is usually being accounted by convolving the rotation profile with the static line profile (see Gray 1976). This technique is inaccurate not only due to using only approximate limb darkening laws, but also because the dependence of limb darkening on frequency as well as gravity darkening are neglected. A more accurate method is to integrate the static intensity over the rotating disk of the star, where gravity darkening can also be included. A "simple" way of including gravity darkening in this integration is described by Collins (1963). This method was used by Gulliver *et al.* (1994) to determine the rotation of Vega.

Stellar rotation is not like rigid body rotation, as it is usually assumed, but it is probably differential. Until now, there are only few measurements of differential rotation. Unfortunately, in many investigations, the authors use the simplest form of the limb darkening law and their results can be used only with reservations.

For an exact description of stellar rotation, plane-parallel model atmospheres are insufficient. It is necessary to use a hydrodynamic multidimensional model atmosphere, at least an axially symmetric one together with the correct description of the radiation field. Recently, we started to work on such a model (Korčáková *et al.* 2005).

5. Conclusions

Full NLTE model atmospheres of A stars are difficult to calculate, since the lines of different ions can form at very different depths, and, therefore, there is a need for a huge number of depth points to resolve all line formation regions sufficiently. Only LTE model atmospheres together with NLTE calculations for individual ions have been calculated in significant numbers. This underlines the need for more intensive calculations of NLTE model atmospheres of A stars. Adding consistently new different physical processes to NLTE model atmospheres calculations, such as diffusion (see Michaud 2005), magnetic fields (see Moss 2005), and more accurate treatments of convection (see Kupka 2005), are necessary. This is a real challenging task for the near future.

Acknowledgements

The authors would like to thank Dr. Adéla Kawka for her comments on the manuscript. This research has made an extensive use of the ADS. This work was supported by grants GA ČR 205/02/0445, 205/04/P224, and 205/04/1267. The Astronomical Institute Ondřejov is supported by projects K2043105 and Z1003909.

References

Achmad, L., Lamers, H. J. G. L. M., Pasquini, L. 1997, *A&A* 320, 196

Allen, C. W. 1963, *Astrophysical Quantities*, University of London, The Athlone Press

Anderson, L. S. 1987, in W. Kalkofen (ed.), *Numerical Radiative Transfer*, Cambridge Univ. Press, p. 163

Anderson, L. S. 1989, *ApJ* 339, 558

Aret, A. & Sapar, A. 2002, *AN* 323, 21

Auer, L. H. 1984, in W. Kalkofen (ed.), *Methods in Radiative Transfer*, Cambridge Univ. Press., p. 237

Auer, L. H. 1991, in L. Crivellari, I. Hubený, & D. G. Hummer (eds.), *Stellar Atmospheres: Beyond Classical Models*, NATO ASI Series C 341, Kluwer Academic Publishers, p. 9

Auer, L. H. 2003, *ASPC* 288, 3

Auer, L. H. & Mihalas, D. 1969, *ApJ* 158, 641

Auer, L. H. & Mihalas, D. 1970, *ApJ* 160, 233

Aufdenberg, J. P., Hauschildt, P. H., Baron E., *et al.* 2002, *ApJ* 570, 344

Babel, J. 1995, *A&A* 301, 823

Babel, J. & Michaud, G. 1991a, *A&A* 241, 493

Babel, J. & Michaud, G. 1991b, *A&A* 248,155

Becker, S. R. 1998, in I. Howarth (ed.), *Boulder-Munich II: Properties of Hot, Luminous Stars*, ASP Conf. Ser. Vol. 120, Astron. Soc. Pacific, San Francisco, p. 137

Belyakova, E. V., Mashonkina, L. I., Sakhibullin, N. A. 1999, *Astron. Zh.* 76, 929

Borsenberger, J. & Gros, M. 1978, *A&AS* 31, 291

Carlsson, M. 1986, Uppsala Astronom. Obs. Rep. 33

Carpenter, K. G. 1985, *ApJ* 289, 660

Castelli, F. & Kurucz, R. L. 2003, *IAUS* 210, A20

Claret, A. & Hauschildt, P. H. 2003, *A&A* 412, 241

Collins, G., W. 1963, *ApJ* 138, 1134

Dickel, H. R. & Auer, L. H. 1994, *ApJ* 437, 222

Dreizler, S. & Schuh, S. 2003, *IAUS* 210, p. 33

Dreizler, S. & Werner, K. 1993, *A&A* 278, 199

Frandsen, S. 1974, *A&A* 37, 139

Giddings, J. 1981, PhD thesis

Gray, D. F. 1976, *Observation and Analysis of Stellar Photospheres*, John Wiley & Sons, New York

Gulliver, A. F., Hill, G., Adelman, S. J., 1994, *ApJ* 429, 81

Hadrava, P. & Kubát, J. 2003, *ASPC* 288, 149

Heyrovský, D. 2003, *ApJ* 594, 464

Hauschildt, P. H., Allard, F., Baron, E. 1999, *ApJ* 512, 377

Hubený, I. 1980, *A&A* 86, 225

Hubený, I. 1981, *A&A* 98, 96

Hubený, I. 1986, in C. R. Cowley, M. M. Dworetsky & C. Mégessier (eds.), *Upper Main Sequence Stars with Anomalous Abundances*, IAU Coll. 90, D. Reidel Publ. Comp., Dordrecht, p. 57

Hubeny, I. 2003, *ASPC* 288, 17

Hubeny, I. & Lanz, T. 1992, *A&A* 262, 501

Hubeny, I. & Lanz, T. 1993, *ASPC* 44, 98

Hubeny, I. & Lanz, T. 1995, *ApJ* 439, 875

Hubeny, I., Hummer, D. G., Lanz, T. 1994, *A&A* 282, 151

Idiart, T. & Thévenin, F. 2000, *ApJ* 541, 207

Ivanova, D. V., Shimanskii, V. V. 2000, *Astron. Zh.* 77, 447

Kamp. I., Korotin, S., Mashonkina, L., Przybilla, N., Shimansky, S. 2003, *IAUS* 210, p. 323

Kašparová, J. & Heinzel, P. 2002, *A&A* 382, 688

Kjurkchieva, D. P. 1989, *Ap&SS* 159, 333

Korčáková, D., Kubát J. & D.,Krtička, J. 2005, *These Proceedings*, AP4

Korčáková, D., & Kubát J. 2003, *IAUS* 210, B8

Krtička, J. & Kubát J. 2005, *These Proceedings*, 23

Kubát, J. 1997, *A&A* 326, 277

Kubát, J. 2003a, *ASPC* 288, 91

Kubát, J. 2003b, *IAUS* 210, A8

Kubát, J., Puls, J., Pauldrach, A. 1999, *A&A* 341, 587

Kudritzki, R.-P. 1973, *A&A* 28, 103

Kupka, F. 2005, *These Proceedings*, 119

Kurucz, R. L. 1993, *Solar Abundance Model Atmospheres*, Kurucz CD-ROM No.19

Lanz, T., & Hubeny, I. 1993, *ASPC* 44, 517

Lemke, M. & Venn, K. A. 1996, *A&A* 309, 558

Mashonkina, L. I., Shimanskii, V. V., Sakhibullin, N. A. 2000, *Astron. Zh.* 77, 893

Mashonkina, L. I., Shavrina, A., Khalack, V., et al. 2002, *Astron. Zh.* 79, 31

Mashonkina, L. I., Ryabchikova, T. A., Ryabtsev, A. N. 2005, *These Proceedings*, 315

Moss, D. 2005, *These Proceedings*, 245

Michaud, G. 2005, *These Proceedings*, 173

Paunzen, E., Kamp, I., Iliev, I. K., et al. 1999, *A&A* 345, 597

Przybilla, N. & Butler, K. 2001, *A&A* 379, 955

Przybilla, N., Butler, K., Becker, S. R., Kudritzki, R. P., Venn, K. A. 2000, *A&A* 359, 1085

Przybilla, N., Butler, K., Becker, S. R., Kudritzki, R. P. 2001a, *A&A* 369, 1009

Przybilla, N., Butler, K., Kudritzki, R. P. 2001b, *A&A* 379, 936

Puls, J. 1991, *A&A* 248, 581

Rentzsch-Holm, I. 1996a, *A&A* 305, 275

Rentzsch-Holm, I. 1996b, *A&A* 312, 966

Rybicki, G. B. & Hummer, D. G. 1991, *A&A* 245, 171

Sadakane, K., Takeda, Y., Okyudo, M. 1993, *PASJ* 45, 471

Sakhibullin, N. A. 1983, *Trudy Kaz. Obs.* 48, 9

Shavrina, A. V., Polosukhina, N. S., Zverko, J., et al. 2001, *A&A* 372, 571

Shimanskaya, N. N., Mashonkina, L. I., Sakhibullin, N. A. 2000, *Astron. Zh.* 77, 599

Steenbock, W. & Holweger, H. 1984, *A&A* 130, 319

Takada-Hidai, M. & Takeda, Y. 1996, *PASJ* 48, 739

Takeda, Y. 1991, *A&A* 242, 455

Takeda, Y. 1992a, *PASJ* 44, 309

Takeda, Y. 1992b, *PASJ* 44, 649

Takeda, Y. 1994, *PASJ* 46, 181

Takeda, Y. 1997, *PASJ* 49, 471

Takeda, Y. & Takada-Hidai, M. 1994, *PASJ* 46, 395

Takeda, Y. & Takada-Hidai, M. 1995, *PASJ* 47, 169

Takeda, Y. & Takada-Hidai, M. 1998, *PASJ* 50, 629

Takeda, Y., Takada-Hidai, M., Jugaku, J., Sakaue, A., & Sadakane, K. 1999, *PASJ* 51, 961

Thévenin, F. & Idiart, T. 1999, *ApJ* 521, 753

Twigg, L. W. 1979, *MNRAS* 189, 907

Venn, K. A., 1995, *ApJ* 449, 839

Valyavin, G., Kochukhov, O., Piskunov, N. 2004, *A&A* 420, 993

Werner, K. 1986, *A&A* 161, 177

Werner, K., Deetjen, J. L., Dreizler, S., et al. 2003, *ASPC* 288, 31

The A-Star Puzzle
Proceedings IAU Symposium No. 224, 2004
J. Zverko, J. Žižňovský, S.J. Adelman, & W.W. Weiss, eds.
© 2004 International Astronomical Union
DOI: 10.1017/S1743921304004338

NLTE wind models of A supergiants

Jiří Krtička[1] and Jiří Kubát[2]

[1]Institute of Theoretical Physics and Astrophysics, Masaryk University, CZ-611 37 Brno, Czech Republic, email: krticka@physics.muni.cz

[2]Astronomical Institute, Academy of Sciences of the Czech Republic, CZ-251 65 Ondřejov, Czech Republic, email: kubat@sunstel.asu.cas.cz

Abstract. We present new numerical models of line-driven stellar winds of A supergiants. Statistical equilibrium (NLTE) equations of the most abundant elements are solved, and properly obtained occupation numbers are used to calculate consistent radiative force and radiative heating terms. Wind density, velocity, and temperature are calculated as solutions of the model's hydrodynamical equations. Our models allow for the calculation of the wind mass-loss rate and terminal velocity.

Keywords. Stars: atmospheres, stars: early-type, stars: winds, outflows, radiative transfer

1. Introduction

The stellar winds of A supergiants have properties similar to those of their hotter counterparts, i.e., the stellar winds of OB supergiants. These winds are accelerated by the absorption of radiation mainly in resonance lines of elements such as carbon, nitrogen, oxygen, and iron. However, the domain of A supergiant winds seems to be overlooked both by the wind theorists and by the observers. Reliable data on wind mass-loss rates and terminal velocities are scarce and there is a lack of detailed wind models. To fill this gap we present our NLTE wind models which are capable of predicting wind parameters (i.e., mass-loss rates and terminal velocities) of A supergiants.

2. Model assumptions

The preceding version of our wind code is described in detail by Krtička & Kubát (2004). Here we only summarise basic model properties and describe improvements with respect to the published version.

The basic model assumptions are the following:

• We assume a stationary spherically-symmetric flow.

• We solve the continuity equation, the momentum equation and the energy equation for each component of the flow, namely for absorbing ions, nonabsorbing ions (hydrogen and helium), and electrons. Nonabsorbing ions are ions for which the effect of a radiative force may be neglected (e.g., hydrogen and helium if Population I/II stars are considered). The possibility of helium decoupling is not accounted for (cf. Krtička & Kubát 2005).

• Level occupation numbers of model ions are obtained from NLTE rate equations with the inclusion of a superlevel concept (Anderson 1989).

• Radiation transfer is split into two parts. Line radiative transfer is solved in the Sobolev approximation (Rybicki & Hummer 1978) neglecting continuum opacity sources, line overlaps, and multiple scattering. Continuum radiative transfer is formally solved using the Feautrier method by including all free-free and bound-free transitions of the model ions, however neglecting line transitions. Thus, we neglect line-blocking.

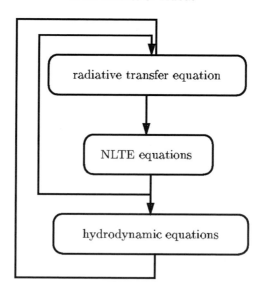

Figure 1. The scheme of the model calculation procedure. Note that the outer iteration cycle which determines the correct value of the mass-loss rate to fulfil the critical condition is not displayed here.

• Derived occupation numbers are used to calculate the radiative force in the hydro-dynamic equations.

• The radiative cooling/heating term is obtained by adding the contributions from all free-free, bound-free, and collisional bound-bound transitions of the model ions. For the calculation of this term we use the thermal balance of electrons method (Kubát *et al.* 1999).

• The description of the transition region between the quasi-static stellar atmosphere and the supersonic stellar wind is simplified. This region is important for continuum formation (Pauldrach 1987). To obtain the correct ionizing flux we included this subsonic region in our models, i.e., the boundary velocity at the stellar surface is set to some low value well below the sonic speed.

• The calculation of models below and above the critical point is split into two parts. Thus we neglect the influence of the radiation emitted in the region above the critical point on the region below the critical point.

• We neglect Gayley-Owocki heating (Gayley & Owocki 1994) which is not important for stars considered in this paper (see Krtička & Kubát 2001).

• We also neglect X-ray radiation which may influence the ionization balance.

3. Model description

The basic procedure for the wind model calculation is displayed in Figure 1. In the inner iteration cycle statistical equilibrium equations together with the radiative transfer equation are solved. Calculated level occupation numbers and mean radiative intensities are used to solve the hydrodynamic equations. Since the derived hydrodynamic structure influences the occupation numbers, another iteration cycle is necessary to account for this coupling. An additional (outer) iteration cycle, which is not displayed in the Figure 1, is used to fulfil the critical condition. During this outer iteration cycle the correct value of the wind mass-loss rate is determined.

3.1. *Continuum radiative transfer*

In the radiative transfer equation in continuum we neglect wind motion and bound-bound opacity sources. Thus, the radiative transfer equation in the spherical coordinates has the form

$$\mu\frac{\partial I(r,\nu,\mu)}{\partial r} + \frac{1-\mu^2}{r}\frac{\partial I(r,\nu,\mu)}{\partial\mu} = \eta(r,\nu,\mu) - \chi(r,\nu,\mu)I(r,\nu,\mu), \qquad (3.1)$$

where $I(r,\nu,\mu)$ is the specific intensity of radiation, $\mu = \cos\theta$ is the direction cosine, ν is the frequency and $\chi(r,\nu,\mu)$ and $\eta(r,\nu,\mu)$ are the emission and absorption coefficients, respectively. The solution of this equation is obtained using the Feautrier (1964) method (for a description, see Mihalas & Hummer (1974) or Kubát (1993)). The resulting tridiagonal system of equations is solved using the numerical package LAPACK (Anderson *et al.* 1999) (`http://www.cs.colorado.edu/~lapack`) instead of the classical Gaussian scheme. All bound-free transitions of explicit ions, free-free transitions of hydrogen and helium and light scattering due to free electrons contribute to the source function. The radiative transfer equation is solved only for frequencies above the hydrogen ionization edge. For frequencies lower than the hydrogen ionization limit we neglect the radiation-matter interaction and use an approximation $J_\nu(r) = 4W(r)H_c$, where $W(r) = \frac{1}{2}\left(1-\mu_*\right)$ is the dilution factor.

The lower boundary flux is taken either from model atmospheres of Kubát (2003) or of Hubeny & Lanz (1995).

3.2. *Line radiative transfer*

The line radiative transfer is solved in the Sobolev approximation with the neglect of continuum radiation as (Castor 1974, Rybicki & Hummer 1978)

$$\bar{J}_{ij} = (1-\beta)S_{ij} + \beta_c I_c, \qquad (3.2)$$

where $\bar{J}_{ij} = \int_0^\infty \mathrm{d}\nu \int_{-1}^1 \mathrm{d}\mu\phi_{ij}(\nu)I(r,\nu,\mu)$ is the mean intensity, $\phi_{ij}(\nu)$ is the line profile, I_c is the specific intensity emerging from the star, $\beta = \frac{1}{2}\int_{-1}^1 \mathrm{d}\mu\frac{1-e^{-\tau_\mu}}{\tau_\mu}$, $\beta_c = \frac{1}{2}\int_{\mu_*}^1 \mathrm{d}\mu\frac{1-e^{-\tau_\mu}}{\tau_\mu}$, $\mu_* = \left(1-R_*^2/r^2\right)^{1/2}$, and the source function $S_{ij} = \eta_{ij}/\chi_{ij}$.

3.3. *Statistical equilibrium equations*

Occupation numbers N_i of atoms in the state i are given by the solution of static statistical equilibrium (NLTE) equations

$$\sum_{j\neq i} N_j P_{ji} - N_i \sum_{j\neq i} P_{ij} = 0, \qquad (3.3)$$

where the P_{ij} are rates of all processes that transfer an atom from a given state i to a state j. For the calculation of these terms we account for radiative excitation and deexcitation, radiative ionization and recombination and corresponding collisional processes. We solve statistical equilibrium equations (3.3) for the elements listed in Table 1. Model atoms are taken mostly from TLUSTY (Hubeny 1988, Hubeny & Lanz 1992, 1995). The set of model atoms is slightly larger for hot star wind modelling. For this purpose we used data from the Opacity Project (Seaton 1987) and from the Iron Project (Hummer *et al.* 1992). For an acceleration of the convergence of the system of statistical equilibrium equations together with the radiative transfer equations we apply the Newton-Raphson method for the Sobolev radiative transfer in lines and accelerated lambda iterations in continuum (or, in other words, approximate Newton-Raphson iterations based on Rybicki & Hummer (1992)).

Table 1. Atoms and their ionization stages included into the NLTE equations

H I-II	He I-III	C I-IV	N I-IV
O I-IV	Ne I-IV	Na I-III	Mg II-IV
Al I-V	Si II-V	S II-V	Ar III-IV
Ca II-IV	Fe II-V	Ni II-V	

3.4. Hydrodynamic equations

The set of hydrodynamic equations has the following form:

$$\frac{\mathrm{d}}{\mathrm{d}r}\left(r^2\rho v_r\right) = 0 \qquad \text{(the continuity equation)}, \tag{3.4a}$$

$$v_r\frac{\mathrm{d}v_r}{\mathrm{d}r} = g^{\mathrm{rad}} - g - \frac{1}{\rho}\frac{\mathrm{d}}{\mathrm{d}r}\left(a^2\rho\right) \qquad \text{(the equation of motion)}, \tag{3.4b}$$

$$\frac{3}{2}v_r\rho\frac{\mathrm{d}a^2}{\mathrm{d}r} + \frac{a^2\rho}{r^2}\frac{\mathrm{d}}{\mathrm{d}r}\left(r^2 v_r\right) = Q^{\mathrm{rad}} \qquad \text{(the energy equation)}, \tag{3.4c}$$

where ρ is the wind density, v_r is the radial velocity, g is the acceleration of gravity, a is the isothermal sound speed, $g^{\mathrm{rad}} = g_{\mathrm{lines}}^{\mathrm{rad}} + g_{\mathrm{el}}^{\mathrm{rad}}$ is the radiative acceleration consisting of contributions from the lines $g_{\mathrm{lines}}^{\mathrm{rad}}$ and from electron scattering $g_{\mathrm{el}}^{\mathrm{rad}}$, and Q^{rad} is the radiative heating/cooling term calculated using the thermal balance of electrons method (Kubát *et al.* 1999). Note that although the model solves multicomponent equations, the hydrodynamic equations above are written in their one-component forms since multicomponent effects can be neglected for A supergiants as a result of our calculations (except for helium decoupling, see Krtička & Kubát 2005).

The radiative force in the Sobolev approximation is calculated as the sum of the contributions of individual lines after Castor (1974),

$$g_i^{\mathrm{rad}} = \frac{8\pi}{\rho_i c^2}\frac{v_r}{r}\sum_{\mathrm{lines}}\nu H_c\int_{\mu_c}^1\mu\left(1+\sigma\mu^2\right)\left(1-e^{-\tau_\mu}\right)\mathrm{d}\mu. \tag{3.5}$$

The line parameters used for the calculation of the radiative force are taken from the VALD database (Kupka *et al.* 1999). Occupation numbers determined in the course of the solution of the statistical equilibrium equations are then used for the radiative force calculation.

The hydrodynamic equations are solved using Newton-Raphson method. For more details we refer interested readers to Krtička (2003). To obtain faster model convergence, derivatives of occupation numbers with respect to fluid variables (i.e. v_{ri}, ρ_{p}, ρ_{e} and T_{e}) are calculated from the statistical equilibrium equations (3.3). These derivatives are accounted for in the Newton-Raphson iteration step. This is similar to the implicit linearization of b-factors discussed in Kubát (1997).

4. Stellar wind model of HD 12953

To show the ability of our code to predict correct wind parameters of A supergiants we present a wind model of HD 12953. The parameters of this star are (after Kudritzki *et al.* 1999) $T_{\mathrm{eff}} = 9\,100\,\mathrm{K}$, $R = 145\,R_\odot$, and $M = 9.7\,M_\odot$. The wind mass-loss rate derived from observation $\dot{M} = 4.3\times10^{-7}\,M_\odot\,\mathrm{year}^{-1}$ and the observed wind terminal velocity $v_\infty = 150\,\mathrm{km\,s}^{-1}$ (Kudritzki *et al.* 1999) are in a good agreement with the calculated wind parameters $\dot{M} = 1.3\times10^{-7}\,M_\odot\,\mathrm{year}^{-1}$ and $v_\infty = 140\,\mathrm{km\,s}^{-1}$. The model radial velocity and the temperature structure are displayed in Figure 2.

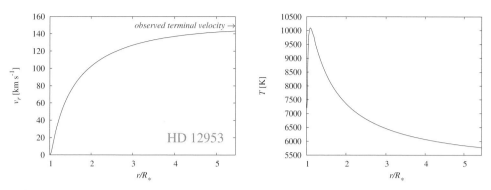

Figure 2. The calculated model of HD 12953. *Left panel:* Wind radial velocity. *Right panel:* Wind temperature structure.

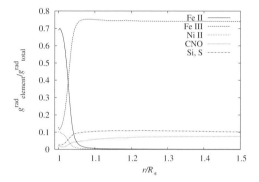

Figure 3. Relative contribution of individual atoms and their ionization states to the radiative acceleration

As can be seen from Figure 3, iron is the most important atom for the radiative acceleration. However, the contributions of the iron ionization states depend on the radius. While Fe II is important close to the star, Fe III is important mostly in the outer wind regions. The contributions of the light elements such as C, N, O, Si and S are much smaller.

5. Conclusions

We have presented our NLTE wind models of A supergiants which are able to predict both detailed wind structure and wind parameters (i.e. mass-loss rates and terminal velocities). Although our models are still preliminary, they are able to predict correct wind parameters of A supergiants, as was demonstrated in the case of HD 12953.

Acknowledgements

This work was supported by grants GA ČR 205/02/0445, 205/03/D020, 205/04/1267, MVTS SR-CR 128/04. JKr is grateful to IAU for travel grant. The Astronomical Institute Ondřejov is supported by projects K2043105 and Z1003909.

References

Anderson, L. S. 1989 *ApJ* **339**, 558–578

Anderson, E., Bai, Z., Bischof, C., *et al.* 1999 *LAPACK Users' Guide*, 3rd ed., SIAM, Philadelphia

Castor, J. I. 1974 *MNRAS* **169**, 279–306

Feautrier P. 1964 *C. R. Acad. Sci. Paris* **258**, 3189

Gayley, K. G. & Owocki, S. P. 1994 *ApJ* **434**, 684–694

Hubeny, I. 1988 *Comput. Phys. Commun.* **52**, 103

Hubeny, I. & Lanz, T. 1992 *A&A* **262**, 501–514

Hubeny, I. & Lanz, T. 1995 *ApJ* **439**, 875–904

Korčáková, D., Kubát J. & D., Krtička, J. 2005, *These Proceedings*, AP4

Krtička, J. 2003 In *Stellar Atmosphere Modelling* (eds. I. Hubeny, D. Mihalas & K. Werner) ASP Conf. Ser., Vol. **288**, 259–262

Krtička, J. & Kubát, J. 2001 *A&A* **377**, 175–191

Krtička, J. & Kubát, J. 2004 *A&A* **417**, 1003–1016

Krtička, J. & Kubát, J. 2005, *These Proceedings*, 201

Kubát, J. 1993 Ph.D. thesis, Astronomický ústav AV ČR, Ondřejov

Kubát, J. 1997, *A&A* **326**, 277–286

Kubát, J. 2003 In *Stellar Atmosphere Modelling* (eds. I. Hubeny, D. Mihalas & K. Werner) ASP Conf. Ser., Vol. **288**, 87–90

Kubát, J., Puls, J. & Pauldrach, A. W. A. 1999 *A&A* **341**, 587–594

Kudritzki, R. P., Puls, J., Lennon, D. J., Venn, K. A., Reetz, J., Najarro, F., McCarthy, J. K., Herrero, A. 1999 *ApJ* **350**, 970–984

Kupka F., Piskunov, N. E., Ryabchikova, T .A., Stempels, H. C., & Weiss, W. W. 1999 *A&AS* **138**, 119–133

Mihalas, D. & Hummer, D. G. 1974 *ApJS* **28**, 343–372

Pauldrach, A. W. A. 1987, *A&A* **183**, 295–313

Rybicki, G. B. & Hummer, D. G. 1978 *ApJ* **219**, 654–675

Rybicki, G. B. & Hummer, D. G. 1992 *A&A* **262**, 209–215

Seaton, M. J. 1987 *J. Phys. B* **20**, 6363–6378

Discussion

LANDSTREET: The radiative forces are exerted mainly on metal ions rather than on H and He. There is presumably some difference in the outflow speed with frictional coupling. Is frictional heating of the wind important?

KRTIČKA: The wind components are well coupled in these models. Thus frictional heating is not important in this case. However, for simplicity the helium decoupling was not allowed in these models (see Krtička & Kubát 2005) for a discussion of helium decoupling.

PRZYBILLA: Can you comment on the match between observed P-Cygni lines and your predictions?

KRTIČKA: There is a poster (Korčáková *et al.* 2005) which discusses this issue. However, the results are not yet satisfactory.

The A-Star Puzzle
Proceedings IAU Symposium No. 224, 2004
J. Zverko, J. Žižňovský, S.J. Adelman, & W.W. Weiss, eds.
© 2004 International Astronomical Union
DOI: 10.1017/S174392130400434X

Stellar model atmospheres with magnetic line blanketing

S. Khan[1,2], O. Kochukhov[2] and D. Shulyak[1,2]

[1]Tavrian National University, Yaltinskaya 4, 95007 Simferopol, Crimea, Ukraine
email: serg@starsp.org, dan@starsp.org

[2]Institut für Astronomie, Universität Wien, Türkenschanzstraße 17, 1180 Wien, Austria
email: oleg@astro.univie.ac.at

Abstract. Model atmospheres of A and B stars are computed taking into account magnetic line blanketing. These calculations are based on the new stellar model atmosphere code LLMODELS which implements a direct treatment of the line opacities and ensures an accurate and detailed description of the line absorption. The anomalous Zeeman effect was calculated for field strengths between 1 and 40 kG and a field vector perpendicular to the line of sight. The magnetically enhanced line blanketing changes the atmospheric structure and leads to a redistribution of energy in the stellar spectrum. The most noticeable feature in the optical region is the appearance of the λ5200 broad, continuum feature. However, this effect is prominent only in cool A stars and disappears for higher effective temperatures. The presence of a magnetic field produces an opposite variation of the flux distribution in the optical and the UV regions. A deficiency of the UV flux is found for the whole range of considered effective temperatures, whereas the "null wavelength region" where the flux remains unchanged shifts towards the bluer wavelengths for higher temperatures.

Keywords. Stars: chemically peculiar, stars: magnetic fields, stars: atmospheres

1. Introduction

Magnetic chemically peculiar (CP) stars are upper and middle main sequence stars characterized by anomalous chemical abundances and unusual distribution of energy in their spectra. A strong magnetic field is expected to have important effects on atmospheres and spectra of CP stars. Several characteristic features of the energy distribution of magnetic CP stars, such as the broad, continuum features in the visual (Kodaira 1969) and the UV flux deficiency (Leckrone 1973), are suspected to be a result of the enhanced line blanketing due to the magnetic intensification of spectral lines. This emphasizes the necessity to consider magnetic line blanketing in the atmospheres of CP stars.

In general, a magnetic field influences the energy transport, hydrostatic equilibrium, the diffusion processes, and the formation of spectral lines. Previous studies (Stępień 1978, Muthsam 1979, Carpenter 1985, Leblanc *et al.* 1994, Valyavin *et al.* 2004) have attempted to model some of these factors. However, due to a limitation of computer resources, it was impossible to fully account for the Zeeman effect on the line absorption. In early model atmospheres calculations magnetic splitting was treated very approximately by introducing a pseudo-microturbulent velocity (Muthsam 1979) or by adopting an identical Zeeman triplet pattern for all lines (Carpenter 1985). Recent investigation by Stift & Leone (2003) demonstrated that the magnetic intensification of spectral lines depends primarily on the parameters of anomalous Zeeman splitting pattern, in particular on the number of Zeeman components. In the light of these results it becomes clear that previous attempts to simulate magnetic line blanketing by an enhanced microturbulence or using a simple triplet pattern are insufficient.

2. Calculation of magnetic model atmospheres

2.1. *Stellar model atmosphere code LLModels*

The stellar model atmosphere code LLMODELS developed by Shulyak *et al.* (2004) uses a direct method, the so-called line-by-line or LL technique, for the line opacities calculation. This approach allowed us to account for the individual anomalous Zeeman splittings of spectral lines.

In all calculations presented here we employed two criteria to achieve the convergence of the models: the constancy of the total flux *and* the conservation of the radiative equilibrium. The LLMODELS uses either atomic line lists compiled by Kurucz (1993) or the VALD (Kupka *et al.* 1999) line list. The code relies on a preselection procedure to choose spectral lines contributing significantly to the total absorption coefficient at each of $(3–5)\times10^5$ frequency points considered in the flux calculation. The code selects spectral lines for which $\ell_\nu/\alpha_\nu \geqslant \varepsilon$, where ε is the adopted selection threshold and ℓ_ν and α_ν are line and continuous opacities.

2.2. *Line list*

Magnetic line blanketing was accounted for all spectral lines except those of hydrogen according to the individual anomalous Zeeman patterns. The initial line lists were extracted from VALD using the preselection threshold $\varepsilon = 1\%$.

Landé factors of the lower and upper atomic levels necessary for the calculation of the Zeeman splitting of lines are provided for majority of lines in VALD. For 4–10% of the preselected spectral lines which lack information on Landé g factors the LS coupling approximation was employed for light elements (He to Sc) and a classical Zeeman triplet with $g_{\rm eff} = 1.2$ was assumed for other lines.

2.3. *Zeeman effect in the line opacity*

In general, to calculate a stellar spectrum in the presence of a magnetic field one has to consider the polarized radiative transfer equation for the Stokes *IQUV* parameters. Solution of this problem requires special numerical techniques and is rather computationally expensive (e.g., Piskunov & Kochukhov 2002) and, hence, cannot be easily included in the routine model atmosphere calculation of magnetic stars. Nevertheless, the problem of the magnetic line blanketing can be simplified considerably by assuming that the magnetic field vector is oriented perpendicular to the line of sight. In this case we can neglect effects produced by the polarized radiative transfer and use the transfer equation for non-polarized radiation treating individual Zeeman components as independent lines. The method is fully justified for weak lines (Stenflo 1994) and produces reasonable results for moderate and strong spectral features. Thus, we modified the original line list by inserting additional spectral lines which correspond to individual Zeeman components of the anomalous splitting patterns. This procedure resulted in $(5–14)\times10^6$ (depending on the metallicity) individual transitions included in evaluation of opacity during the model atmosphere calculations.

3. Numerical results

We have calculated a set of model atmospheres with $T_{\rm eff} = 8000\,{\rm K}$, $11000\,{\rm K}$, and $15000\,{\rm K}$, and $\log g = 4.0$, and magnetic field moduli 0, 1, 5, 10, 20 and 40 kG. This model atmosphere grid covers the range of stellar parameters typical of Ap stars. In this first exploratory investigation we assumed scaled solar abundances with [M/H]= 0.0, +0.5 and +1.0, although accounting for individual stellar abundances is straightforward

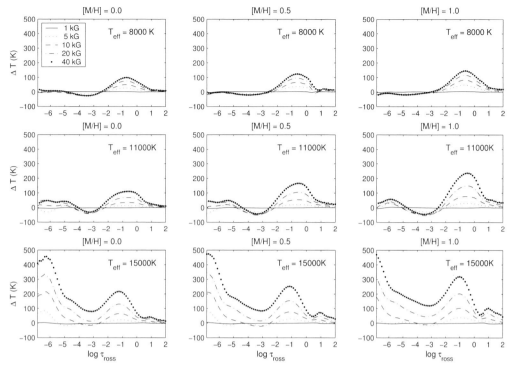

Figure 1. Temperature differences between magnetic and non-magnetic model atmospheres for effective temperatures $T_{\rm eff} = 8000\,{\rm K}$, $11000\,{\rm K}$, and $15000\,{\rm K}$ and [M/H]$= 0.0, +0.5, +1.0$.

with our code. Convection was neglected because the strong magnetic field is expected to prevent turbulent motions of plasma in the stellar atmosphere.

3.1. *Model structure*

The temperature difference between magnetic and nonmagnetic model atmospheres is presented in Fig. 1 as a function of Rosseland optical depth. The temperature anomaly due to the magnetically enhanced line opacity increases with magnetic field strength and metallicity. Pronounced temperature anomalies occur at the optical depth $\log \tau_{\rm ross} \approx -2$ and in the uppermost atmospheric layers. Additional magnetic line blanketing leads to a redistribution of the absorbed energy back to the underlying atmospheric layers and, thus, results in the heating of the main line forming region (near $\log \tau_{\rm ross} \approx -2$) by 100–300 K.

3.2. *Energy distribution*

The energy distributions for [M/H]$= +0.5$ and different $T_{\rm eff}$ are presented in Fig. 2. Flux distributions for other metallicity values show similar behaviour. We note three interesting features in theoretical energy distributions. The first one is a flux deficit in the UV region. Its magnitude depends on $T_{\rm eff}$ and increases with the magnetic field strength. Second, the magnitude of the well-known broad, continuum feature (flux depression) at $\lambda 5200$ increases with magnetic field intensity and metallicity. However, appreciable $\lambda 5200$ features are visible only for low $T_{\rm eff}$ but become negligibly small for higher temperatures. Finally, the presence of the magnetic field changes the flux distribution in the visual and the UV regions in opposite directions. Magnetic stars appears to be cooler in the UV and hotter in the visual than non-magnetic stars of similar $T_{\rm eff}$. The "null wavelength" where flux remains unchanged shifts to the shorter wavelengths with increasing $T_{\rm eff}$.

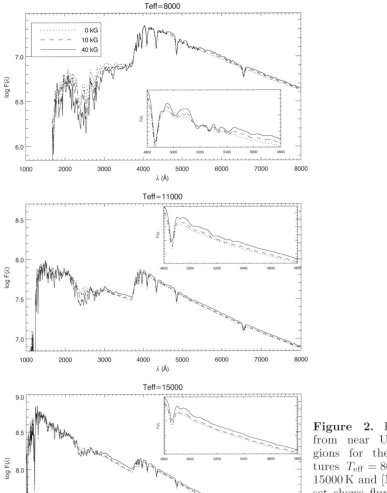

Figure 2. Energy distributions from near UV to near IR regions for the effective temperatures $T_{\rm eff} = 8000\,$K, $11000\,$K, and $15000\,$K and [M/H]= +0.5. The inset shows flux near the $\lambda5200$ region. For display purposes energy distributions have been smoothed with a Gaussian profile corresponding to the resolving power $R = 150$.

3.3. *Colors*

We studied the influence of magnetic line blanketing on the photometric colors in the $uvby\beta$, UBV, Geneva and Δa systems. Relations between the anomalies of all photometric indices except for c_1 and the intensity of magnetic field are strong functions of the effective temperature. For low $T_{\rm eff}$ a change relative to a non-magnetic case is clearly present, whereas for high $T_{\rm eff}$ this anomaly disappears. Enhanced metal abundances increase the anomalies of all photometric indices. In general, we find that saturation effects become important in all photometric indices for field strengths above $\geqslant 10$ kG and, hence, there is no linear dependence of any photometric index with the magnetic field intensity.

3.4. *Hydrogen line profiles*

We calculated the profiles of Hα, Hβ and Hγ hydrogen lines using the SYNTH code (Piskunov 1992) and recent Stark broadening computations by Stehlé (1994).The changes in the atmospheric structure of magnetic stars do not influence the Balmer line profiles. The maximum deviation is only about 3% of the continuum level for the greatest considered field strength.

4. Discussion

Our study shows that magnetically modified line blanketing heats the atmospheres of magnetic stars for certain range of optical depths, an effect which increases with effective temperature and metallicity. We found that magnetic model atmospheres have flux deficits in the UV. The presence of a magnetic field leads to the flux redistribution from the UV to the visual region. This property of our theoretical models is in agreement with observations of Ap stars (Leckrone 1974, Molnar 1973, Jamar 1977).

Theoretical energy distributions show that an increase of T_{eff} is accompanied by the blueward shift of the null wavelength. This agrees with the observational results and numerical experiments presented by Leckrone *et al.* (1974).

Unlike normal stars, magnetic CP stars show several broad, continuum features in the visual and ultraviolet regions. The most prominent feature in the visual is the one centered on λ5200. Our numerical results reproduce this feature for cool Ap stars and show that the magnitude of the depression increases with magnetic field strength and metallicity.

Historically, photometric data were widely used to distinguish peculiar star from normal counterparts and to determine atmospheric parameters. We found that the magnetically induced anomalies of photometric indices are noticeable for lower effective temperatures and negligible for higher T_{eff}. This is in compliance with the properties of flux distribution and the study of Cramer & Maeder (1980), who noted that energy distribution becomes less sensitive to magnetic field effects for hotter stars.

The peculiarity indices Δa, Z and $\Delta(V_1 - G)$ are used for identification and determination of the properties of magnetic CP stars. It is interesting to examine the influence of magnetic field on their values. We found a relatively small effect (the change in Δa is as large as +0.075 mag for a 40 kG field). In generally the relations between the indices and the magnetic field strength, which saturates for strong fields, is nonlinear. In any case, Δa appears to be more sensitive to the magnetic field modulus, in compared with Z. We note, that Cramer & Maeder (1980) emphasized that a saturation of the photometric effects with strong fields is an observational fact and needs to be explained in terms of model atmospheres. Our numerical results provide a theoretical explanation of this effect.

Finally, we investigated how magnetic opacity affects the determination of the atmospheric parameters of Ap stars from their photometric indices. We found that the deviation of T_{eff} and $\log g$ derived with standard photometric calibrations are within usual error bars (100–300 K and 0.1 dex respectively). In other words, magnetic opacity are not expected to produce significant errors of photometric estimates of stellar parameters.

Acknowledgements

We are grateful to Prof. V. Tsymbal for enlightening discussion.

This work was supported by the Lise Meitner fellowship to OK (FWF project M757-N02), INTAS grant 03-55-652 to DS and by the Austrian *Fonds zur Förderung der wissensc haftlichen Forschung* (P–14984), the *BM:BWK* and *ASA* (project COROT).

Financial support came from the *RFBR* (grant 03-02-16342) and from a *Leading Scientific School*, grant 162.2003.02.

References

Carpenter, K. 1985, *ApJ*, **289**, 660

Cramer, N., & Maeder, A. 1980, A&A, 88, 135

Jamar, C. 1977, A&A, 56, 413

Kodaira, K. 1969, ApJ, 157, 59

Kupka, F., Piskunov, N., Ryabchikova, T. A., Stempels, H. C., & Weiss, W. W. 1999, A&AS, 138, 119-133

Kurucz, R. L. 1993, Kurucz CD-ROM 13, Cambridge, SAO

Leckrone, D. 1973, ApJ, 185, 577

Leckrone, D., Fowler, W., & Adelman, S. 1974, A&A, 32, 237

Leckrone, D. 1974, ApJ, 190, 319

LeBlanc, F., Michaud, G., & Babel, J. 1994, ApJ, 431, 388

Molnar, M. R. 1973, ApJ, 190, 319

Muthsam, H. 1979, A&A, 73, 159

Piskunov, N., & Kochukhov, O. 2002, A&A, 381, 736

Piskunov, N. 1992, in Stellar Magnetism, ed. Yu. V. Glagolevskij, I. I. Romanyuk (St. Petersburg: Nauka), 92

Shulyak, D., Tsymbal, V., Ryabchikova, T., Stütz Ch., & Weiss, W. 2004, A&A, in press.

Stenflo, J. O. 1994, *Solar Magnetic Fields*, Dordrecht: Kluwer Academic Publishers

Stehlé, C. 1994, A&AS, 104, 509

Stępień, K. 1978, A&A, 70, 509

Stift, M., & Leone, F. 2003, A&A, 398, 411

Valyavin, G., Kochukhov, O., & Piskunov, N. 2004, A&A, 420, 993

Discussion

WADE: For the field strengths >10 kG, one would expect Zeeman components of most lines to be fully split and resolved, and so that the influence of very large fields on the flux should be not so much more than intermediate fields. Can you explain then why the effect of a 40 kG field is so much greater than for a 10 kG field?

KOCHUKHOV: The major contribution to the influence of the magnetic line blanketing on the model atmosphere structure and flux distribution comes from many strong UV spectral lines. In that spectral region magnetic splitting is factor 2-3 smaller than in the optical due to the λ^2 dependence of the Zeeman effect. Furthermore, real spectral lines often exhibit complex *anomalous* Zeeman patterns. Even in the visual region the magnetic intensification of such a line does not reach maximum until $\approx$ 50–100 kG (see Stift & Leone, 2003, A&A, 398, 411). Thus, a 40 kG field is not sufficient to resolve all metal lines (especially in the UV) and result in a saturation of the magnetic effect in our model atmosphere calculations.

DWORETSKY: It is good to see a theoretical explanation in detail of the $\lambda5200$ feature. I recall that H. Maitzen using photographic spectra to look for spectroscopic clues (absorption lines) to its origin many years ago. Your work clearly provides a neat explanation.

KOCHUKHOV: Indeed, our modelling suggests that the $\lambda5200$ depression in cool A stars appears due to the magnetic line blanketing. However, we do not find depressions in the theoretical flux distributions computed for hot magnetic stars. It seems that some other effects (vertical chemical stratification being the best candidate) are responsible for the $\lambda5200$ feature at these higher temperatures.

The A-Star Puzzle
Proceedings IAU Symposium No. 224, 2004   © 2004 International Astronomical Union
J. Zverko, J. Žižňovský, S.J. Adelman, & W.W. Weiss, eds.   DOI: 10.1017/S1743921304004351

A spectral study of Vega: A rapidly-rotating pole-on star

Graham Hill[1,4], Austin F. Gulliver[2,4] and Saul J. Adelman[3,4]

[1]Department of Physics, University of Auckland, Auckland, New Zealand 1001
email: gp-hill@xtra.co.nz

[2]Department of Physics and Astronomy, Brandon University, Brandon, MB R7A 6A9, Canada
email: gulliver@brandonu.ca

[3]Department of Physics, The Citadel, 171 Moultrie Street, Charleston, SC 29409, USA
email: adelmans@citadel.edu

[4]Guest Investigator, Dominion Astrophysical Observatory, Herzberg Institute of Astrophysics, National Research Council of Canada, 5071 W. Saanich Road, Victoria, BC V9E 2E7, Canada

Abstract. Ultra-high signal-to-noise, high dispersion spectroscopy over the wavelength range $\lambda 4487 - 4553$ shows Vega to be a rapidly rotating star ($V_{eq} \sim 160\,\mathrm{km\,s^{-1}}$) seen almost pole-on. These data, analyzed anew, are combined with analyses of the hydrogen lines (Hγ, Hβ and Hα) and the latest absolute continuum flux for Vega to yield the following results: $V \sin i = 21.9 \pm 0.1\,\mathrm{km\,s^{-1}}$, polar $T_{eff} = 9680 \pm 10\,\mathrm{K}$, polar $\log g = 4.00 \pm 0.02\,\mathrm{dex}$, $V_{eq} = 160 \pm 10\,\mathrm{km\,s^{-1}}$, $\xi_T = 1.08 \pm 0.02\,\mathrm{km\,s^{-1}}$ and $i = 7.9 \pm 0.5°$. The variations in T_{eff} and $\log g$ over the photosphere total 350 K and 0.06 dex, respectively. The mean $T_{eff} = 9510 \pm 10\,\mathrm{K}$ and mean $\log g = 3.97 \pm 0.02\,\mathrm{dex}$ agree with the spherical model values derived here and by others.

Keywords. Line: profiles, techniques: spectroscopic, stars: fundamental parameters, stars: individual (Vega), stars: rotation

---

1. Introduction

Since Johnson & Morgan (1953) included Vega (α Lyrae = HD 172167 = HR 7001) as one of the ten primary UBV standards the star has been used extensively as a primary and secondary spectrophotometric standard in the near infrared, optical, and ultraviolet regions (Hayes 1985, Bohlin *et al.* 1990). Studies of its elemental abundances show that it is metal-weak by [−0.6] dex (Adelman & Gulliver 1990) compared to the Sun.

Castelli & Kurucz (1994) compared blanketed LTE models for Vega calculated with the new ATLAS9 and ATLAS12 codes (Kurucz 1993) with the observed energy distribution and Balmer line profiles. Their preferred model had $T_{eff} = 9550\,\mathrm{K}$, $\log g = 3.95\,\mathrm{dex}$ and a microturbulent velocity $\xi = 2\,\mathrm{km\,s^{-1}}$ for a metallicity [M/H] $= -0.5$.

Aumann *et al.* (1984) found that Vega is surrounded by a dust shell or disk of diameter about 30 arcsec. Recent research has concentrated on imaging Vega in the IR and near IR. Van der Bliek *et al.* (1994) measured diameters of 22±2 and 35±5 arcsec. The latter value was confirmed by Heinrichsen, Walker & Klaas (1998) who found 36 ± 3 arcsec. Hanbury Brown *et al.* (1974) measured a limb-darkened angular diameter of 3.24 ± 0.07 arcsec, a result confirmed by Ciardi et al. (2001) who obtained 3.28 ± 0.01 arcsec. These results are not without uncertainty since they depend on a limb darkened model for interpretation, but they lead to a radius of $2.73\,R_{\odot}$.

Vega's weak line profiles as seen in high S/N spectra are clearly flat-bottomed resulting in a trapezoidal appearance while the strong lines exhibit normal rotationally-broadened profiles (Gulliver *et al.* 1994). Gray (1985, 1988) suggested that Vega might be a star

seen pole-on because of its excessive luminosity. The program STELLAR was developed
to demonstrate that a rotating model with a temperature gradient over the photosphere
could produce flat-bottomed profiles and found that Vega could be modelled as a pole-on
star seen at an angle $i = 5°$. Our present work, based in part on a criticism by Peterson
(2003, private communication), shows that i is closer to 8°.

Various techniques have been developed to determine the fundamental parameters of
stars from observational data. These have included photometric calibrations; Balmer
line matching and continuous flux matching to obtain T_{eff} and $\log g$; fine analysis to
obtain abundances and ξ_T, spectral line fitting to obtain $V \sin i$ and Doppler imaging to
obtain the inclination of the rotational axis. Given the proper conditions the program
STELLAR is capable of determining all of these parameters for a star by performing a
grand simultaneous fit to the continuous flux, the Balmer lines and the line spectrum.
In practice, however, combining all the data in such a way results in a serious loss of
weight for the parameters because no data set is synthesized accurately as judged by
the fit of the model to the various observational data. Generally we treat each data set
independently, fixing parameters for some data where the dependency is weak (such as
T_{eff} in a hydrogen profile) and solving for others.

2. Observations

Vega was observed with the Dominion Astrophysical Observatory 1.2-m telescope using the coudé spectrograph and a 1872 pixel bare Reticon with 15μ pixels and more
recently with a variety of CCDs (2K and 4K format). The portion of the spectrum of
Vega discussed here extends from $\lambda 4487 - 4553$ in wavelength steps of 0.035 Å and has
a mean $S/N = 3300$ for the continuum regions after the co-addition of two spectra.
Digitized instrumental profiles of the spectrograph plus detector were constructed by
co-adding intensity-weighted lines from the comparison spectra. The resultant FWHM
for the instrumental profile was 0.080 Å. The comparison star, o Peg, used to correct the
laboratory gf values was treated similarly and had a mean $S/N = 840$.

The continuous energy calibration of Vega is a combination of the ultraviolet fluxes
from IUE (see Bohlin *et al.* 2001) combined with the Hayes (1985) visual-near IR mean
energy distribution . The UV data are updated in the web site

ftp.stsci.edu/instruments/cdbs/cdbs2/calspec.html

under the name alpha_lyr_stis_001.fits and are sampled at steps of 1.2 and 1.8 Å respectively for the short wavelength ranges ($\lambda 1149 - 2000$ and $\lambda 2000 - 3300$) and with
a resolution of about 6 Å for the long wavelength region. We used a revised version of
these data (Bohlin *et al.* 2001).

Profiles of Hγ, Hβ and Hα were taken from Peterson (1969). Unfortunately these
profiles only extended 40 Å from each line center and in A stars the wings of these lines
have yet to reach continuum levels even 40 Å from line center. We had to deal with this
observational limitation by including the height and slope of the continuum as unknowns.

3. Modelling

We modelled Vega with the program STELLAR which is in turn based on the eclipsing
binary modelling code for LIGHT2 (Hill 1979). As the first step in the modelling process,
a grid of ATLAS9 models (Kurucz 1993) was calculated. This grid includes the range
of stellar parameters relevant for Vega, T_{eff} in steps of 500 K, $\log g$ in steps of 0.25,
a metallicity of $[M/H] = -0.5$, and $\xi_T = 0, 1$ and 2 $km s^{-1}$. These models used the

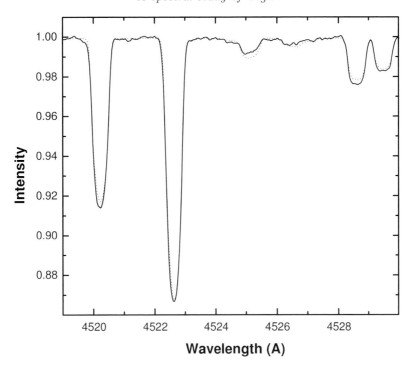

Figure 1. Vega spectrum showing the final synthetic spectrum (dashed line) and the observed line profiles (solid line) for the $\lambda 4519 - 4530$ region. The model parameters are $V \sin i = 21.9\,\mathrm{km\,s}^{-1}$, polar $T_{\mathrm{eff}} = 9680$ K, polar $\log g = 4.00\,\mathrm{dex}$, $V_{\mathrm{eq}} = 160\,\mathrm{km\,s}^{-1}$, $\xi_{\mathrm{T}} = 1.08\,\mathrm{km\,s}^{-1}$ and $i = 7.9°$.

improved continuum opacities, increased line opacities and finer sampling of the latest opacity distribution functions of Kurucz (1993).

As a given stellar application is addressed, input files of synthetic specific intensity line spectra are calculated using the SYNTHE program (Kurucz & Avrett 1981) for the 17 values of $\mu = 1.0$ to 0.01 normally used in ATLAS9 and for each T_{eff}, $\log g$, ξ_{T} and [M/H] point in the above grid. These are then sampled every $0.01\,\text{Å}$ for the relevant wavelength intervals. The wavelengths of interest for Vega are $\lambda 4480 - 4580$ for the metallic line spectra and $250\,\text{Å}$ intervals centered on Hγ, Hβ and Hα for the hydrogen line profiles. Continuum flux files were extracted using ATLAS9 for fitting to Vega's observed continuous flux distribution.

The line parameters used were those provided with SYNTHE supplemented by astrophysical gf values determined from o Peg for the $\lambda 4519 - 4540$ interval. A portion of the high resolution spectral region we are analyzing is shown in Figure 1. Anticipating our later results we have included the synthetic spectrum as well as the observed data. The discrepancies between the observed and synthetic spectra arise from individual abundance errors as well as errors in the $\log gf$ values. Although the model could have produced a better fit by allowing T_{eff} and $\log g$ to be free parameters this was not viewed as being appropriate.

Independent astrophysical gf values were determined for 31 lines in the $\lambda 4519 - 4540$ region using the narrow-lined ($V \sin i = 6.6\,\mathrm{km\,s}^{-1}$), A1 IV slightly metal-rich star, o Peg, assuming $T_{\mathrm{eff}} = 9600$ K, $\log g = 3.60$, $\xi_{\mathrm{T}} = 1.3$ km s^{-1} and the appropriate abundances (Adelman 1988). In light of new values for the stellar parameters of o Peg, $T_{\mathrm{eff}} = 9500$ K, $\log g = 3.75$ (Adelman *et al.* 2004), this analysis could be redone.

For the analysis of the continuous flux, separate input files of continuum specific intensity were generated by ATLAS9 at the standard values of μ and at 1141 wavelengths from $506\,\text{Å}$ - $160\,\mu$ for the grid of models are used. All input files, the continuous flux, the Balmer lines and the line spectra, are convolved with the appropriate instrumental profile. Because it has proven impossible to match the hydrogen profiles in the published continuous energy distribution, we removed the data in the vicinity of the Balmer and Paschen series confluences and for other individual hydrogen lines.

The physical model generates the run of temperature and gravity over the surface of a star with these values defined initially at the pole and changing as a function of rotational velocity and a gravity brightening exponent given as 0.25 for a radiative atmosphere. The rotating star is then viewed at some inclination angle. The integrations are performed using Gauss-Legendre quadrature with typically 32 by 32 integration points in the θ and ϕ axes, respectively. Thus we find intensity as a function of wavelength for each surface integration point T_{eff}, $\log g$ and μ are calculated by successive parabolic interpolations in the database within λ, T_{eff}, $\log g$, μ, ξ_T and [M/H]. By summing these weighted intensities we generated a theoretical spectrum that could be compared with observation.

Obviously, depending on our needs, some of these variables are held constant. For example in our analysis of Vega we fixed the abundance at -0.5 solar. In addition to the models we incorporated results of parallax measurements from Hipparcos (ESA 1997) and interferometry (Ciardi *et al.* 2001) to aid us in completely specifying the system in terms of mass, radius, temperature and luminosity.

4. Results

Our approach to solve for Vega's stellar parameters uses the continuous energy flux to give a temperature and the hydrogen lines to determine the gravity. By solving these data jointly we arrived at polar values of T_{eff} and $\log g$. These were used in the high resolution analysis to yield $V \sin i$, ξ_T and i. Then the whole process was iterated again.

4.1. *Effective temperature and surface gravity*

Naturally we tried to see if the continuous data would yield a value of i independent of the line profile data. In this we were disappointed though ironically the earlier calibration of continuous flux (Bohlin *et al.* 1990) was successful (Gulliver *et al.* 1994). The observed continuum data we used were converted to H_ν but left as vacuum wavelengths for comparison with the ATLAS9 output. We edited the hydrogen lines as noted earlier and tried to match the resolution of the observational data to the output from ALTLAS9. ATLAS9 used 'boxcar' smoothing and we matched this in STELLAR.

Test solutions of various combinations of the hydrogen profiles and continuous data revealed that the best approach was to combine these data sets. This produced excellent fits to both sets of data (see Figures 2 and 3) in which the fits are almost indistinguishable from the data (see also Table 1). A close examination of the UV calibration near $3200\,\text{Å}$ (see Figure 3) shows poor continuity between the observations of Hayes and the data at shorter λ but we have chosen to accept the revised calibration at face value.

4.2. *$V \sin i$, microturbulence and inclination*

The original solution for the two flat-bottomed lines given by Gulliver *et al.* (1994) yielded parameter values for the polar T_{eff}, polar $\log g$, $V \sin i$, and i, of 9695 K, 3.75, 21.8 km s^{-1} and 5.1°, respectively. The rms error of 5.0×10^{-4} approximates a $S/N \sim 2000$, close to that of the observations themselves. Clearly a reasonable solution had been reached but a subsequent analysis of the hydrogen line and continuum data showed that the

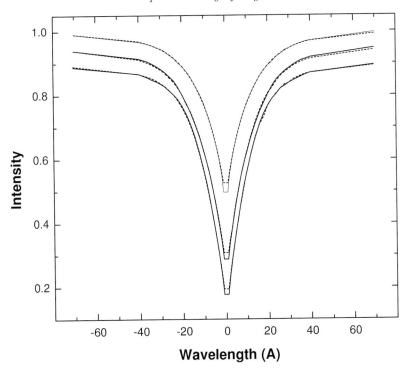

Figure 2. STELLAR model fit (dashed line) to the observed Hα, Hβ, and Hγ hydrogen line profiles (solid lines) from top to bottom, respectively. The values of i, $V \sin i$ and ξ were fixed at the values in Figure 1; only polar T_{eff} and polar $\log g$ were free to vary.

Table 1. A Summary of Results

	Mean Values			Polar Values						
Sln	$<T>$	$<\log g>$	T	$\log g$	ω/ω_c	i	$V \sin i$	ξ_{T}	rms^2
1	9424	3.91	9602 ± 35	$3.94 \pm .02$	$.48 \pm .06$	7.7	$21.9 \pm .2$	$1.08 \pm .02$	1.33e-6
2	9424	3.91	9602	3.94	$.47 \pm .01$	7.9	$21.9 \pm .1$	1.08	1.32e-6
3	9506	3.97	9680	4.00	$.56 \pm .01$	6.4	$21.5 \pm .1$	$1.11 \pm .01$	1.37e-5
4	9506	3.97	9680 ± 3	$4.00 \pm .01$		7.8	21.9	1.08	8.11e-4
5	9506	3.96	9680	$3.99 \pm .01$		7.8	22.0	1.08	3.60e-5
6	9506	3.97	9680 ± 4	4.00		7.8	21.8	1.08	7.63e-5

Analysis/Source: Sln 1 - Fe+Ti lines/R; Sln 2 - λ4528.29/R; Sln 3 - Fe+Ti lines/R; Sln 4 - H & continuum/P,H,B; Sln 5 - H lines/P; Sln 6 - continuum/H,B
B - Bohlin *et al.* (2001)
H - Hayes (1985)
P - Peterson (1969)
R - Reticon

value of polar $\log g$ was too low. Peterson (2003) observed that our rotational velocity of 250 km s^{-1} implied a rotational breakup fraction $\omega/\omega_{\mathrm{crit}}$ of ~ 0.75 as opposed to our breakup fraction ~ 0.5. This has a profound effect on the run of temperature and $\log g$ over the star since for the Roche model, $V_{\mathrm{crit}} = GM/R_{\mathrm{eq}}^2$ where R_{eq} is the equatorial radius at breakup. For a given polar radius R_{p}, $R_{\mathrm{eq}} = 1.5R_{\mathrm{p}}$ at breakup and using radii from Ciardi *et al.* (2001) we have $R_{\mathrm{p}} \sim 2.62R_{\odot}$ and a mass of $M \sim 2.3M_{\odot}$ calculated from $M = gR^2/G$ where g is the surface gravity and G the gravitational constant. These values imply a $V_{\mathrm{crit}} \sim 340$ km s^{-1} as opposed to the value of 500 km s^{-1} we had

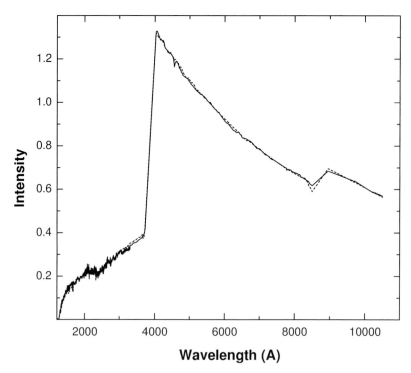

Figure 3. STELLAR model fit (dashed line) to the ultraviolet IUE fluxes of Bohlin *et al.* (1990) and the visual fluxes of Hayes (1985) (solid line). The Balmer and Paschen lines have been removed throughout because of incompatibilities. The values of i, $V \sin i$ and ξ were fixed at the values in Figure 1; only polar $T_{\rm eff}$ and polar $\log g$ were free to vary.

erroneously adopted. These differing values imply a value of i near $8°$ rather than the previously determined $5°$ quoted by Gulliver *et al.* (1994). In addition the earlier study yielded a polar $\log g$ near 3.75 dex considerably at variance with the canonical value of 3.95 dex. Correcting for $V_{\rm crit}$ produced consistent $\log g$'s for all our varied data. We are grateful to Peterson for pointing out this error.

The solution proceeded iteratively since the interferometry ($\Theta = 3.28$ mas) and parallax ($\pi = 0.1289$) yielded $R = 2.73 \pm 0.01 R_{\odot}$. Our original solution implied $\omega/\omega_{\rm crit} \sim 0.45$ and an equatorial radius of $\sim 2.62 R_{\odot}$. Given the polar radius and the value of $\log g$ ~ 3.94 dex our solutions predicted a mass $\sim 2.42 M_{\odot}$ and hence a breakup velocity of 340 km s^{-1}. This value was used throughout our analysis. Rather than solve for inclination we solved for the breakup fraction $\omega/\omega_{\rm crit}$ and derived i from that. The rms of the fit implies a $S/N \sim 1000$ but the results are better than that since the presence of unidentified lines are biasing this result. The results are shown in Table 1. Note that in Table 1, if an error is quoted the associated variable was allowed to be free in the fit and if no error is present the variable was fixed. In the last two columns of the table the nature of the analysis and the data fitted are shown. The final set of stellar parameters are an amalgamation of the six solutions given.

The best values of $T_{\rm eff}$ and $\log g$ from the continuum and hydrogen line results were adopted and for all subsequent solutions these polar values were fixed at to 9680 K and 4.00, respectively. Having fixed these values we concentrated on the two flat bottomed lines and solved for $V \sin i$ and i after determining the microturbulence from the fit to the $\lambda4519 - 4530$ region of Figure 1. Note that microturbulence is a minor component

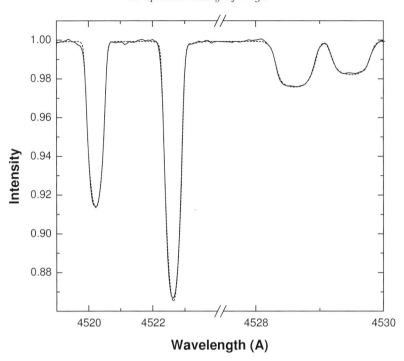

Figure 4. Vega spectrum showing the synthetic spectrum (dashed line) and the observed line profiles (solid line) for the $\lambda 4519 - 4530$ region. The model parameters are $V \sin i = 21.9 \, \text{km s}^{-1}$, polar $T_{\text{eff}} = 9600 \, \text{K}$, polar $\log g = 3.94 \, \text{dex}$, $V_{\text{eq}} = 160 \, \text{km s}^{-1}$, $\xi_{\text{T}} = 1.07 \, \text{km s}^{-1}$ and $i = 7.7°$.

of the weak lines. The fit to the $\lambda 4528 - 4530$ region is given in Table 1 and has an rms $\sim 1.3x10^{-6}$ about one half of the expected $S/N \sim 2000$. The results are illustrated in Figure 4. The reader should take note that the T_{eff} and $\log g$ of this model are not the final values but have been decreased to allow for the abundance and $\log gf$ errors surmised above.

As in our earlier study we tested the uniqueness of the solution by generating extensive grids to explore the multidimensional surface in detail and we are confident that our solution is unique. Thus the five parameters for Vega are polar $T_{\text{eff}} = 9680 \pm 3$ K, polar $\log g = 4.00 \pm 0.01$ dex, $V \sin i = 21.9 \pm 0.1 \, \text{km s}^{-1}$, $\xi = 1.09 \pm 0.01 \, \text{km s}^{-1}$ and $w/w_c = 0.47\pm0.01$. T_{eff} and $\log g$ come from the iterations between the continuous energy flux and the hydrogen-line profiles and the other parameters from the line analysis. The breakup fraction implies an inclination $i = 7.8 \pm 0.1°$. The value of $\xi_{\text{T}} = 1.1 \, \text{km s}^{-1}$ can be compared to that of 0.6 km s^{-1} from Adelman & Gulliver (1990) using classical techniques or 1.0 km s^{-1} of Hill & Landstreet (1993) using a spectrum fitting technique. The derived mean log g value of 3.97 is close to the value normally used to model Vega ($\log g = 3.95$, Castelli & Kurucz 1994) and the mean $T_{\text{eff}} = 9506 \pm 10$ K is within the errors of Ciardi *et al.* 's (2001) direct determination of $T_{\text{eff}} = 9555 \pm 111$ K.

5. Conclusions

It is unfortunate that the premiere photometric standard upon which our spectrophotometry is anchored should be the object it is - a rapidly-rotating star seen almost pole-on and surrounded by a ring of dust - but this circumstance need not be an impediment to its use as a standard. It simply means that the fitting model should be more complex than a spherical one and in any case is far more testing of the models than a simple spherical comparison.

Acknowledgements

SJA and AFG thank Dr. James E. Hesser, Director of the Dominion Astrophysical Observatory for the observing time. Financial support was provided to AFG by the Natural Sciences and Engineering Research Council of Canada. SJA's contribution to this paper was supported in part by grants from The Citadel Foundation.

References

Adelman, S.J. 1988, *MNRAS*, 230, 671
Adelman, S.J., & Gulliver, A.F. 1990, *ApJ*, 348, 712
Adelman, S. J., Gulliver, A. F., & Heaton, R. J. 2004, submitted to A&A
Aumann, H. H. *et al.* 1984, *ApJ*, 96, L23
Bohlin, R.C., Harris, A.W., Holm, A.V., & Gry, C. 1990, *ApJS*, 73, 413
Bohlin, R.C., Dickinson, M.E. & Calzetti, D. 2001, *AJ*, 122, 2118
Castelli, F., & Kurucz, R.L. 1994, *A&A*, 281, 817
Ciardi, D.R., van Belle, G.T., Akeson, R.L., Thompson, R.R., & Lada, E.A., 2001, *ApJ*, 559, 1147
ESA, 1997, The Hipparcos and Tycho Catalogs, SP-1200
Gray, R.O. 1985, *JRASC*, 79, 237
Gray, R.O. 1988, *JRASC*, 82, 336
Gulliver, A.F., Hill, G., & Adelman, S.J. 1994, *ApJ*, 429, L81
Hanbury Brown,R., Davis, J., & Allen, L.R. 1974, *MNRAS*, 167, 121
Hayes, D.S. 1985, in "Calibration of Fundamental Stellar Quantities", eds. D. S. Hayes, L.E. Pasinetti, & A.G.D. Philip, (Dordrecht: Reidel), 225
Heinrichsen, I., Walker, H.J., & Klaas, U. 1998, *MNRAS*, 293, L78
Hill, G. 1979, *Publ. Dominion Astrophysical Obs.*, 15, 279
Hill, G.M., & Landstreet, J.D. 1993, *A&A*, 276, 142
Johnson, H.L., & Morgan, W.W. 1953, *ApJ*, 117, 313
Kurucz, R.L. 1993, *ASP Conference Series*, 44, 87 (see also CD-ROM distribution)
Kurucz, R.L., & Avrett, E.H. 1981, *SAOSpecRep*, 391
Peterson, D.M. 1969, *SAOSpecRep*, 293
Peterson, D.M. 2003, private communication
van der Bliek, N.A., Prusti, T., & Waters, L.B.F.M. 1994, *A&A*, 285, 229

Discussion

E. GRIFFIN: Do you remove the IS components from your observed spectra?

S. ADELMAN: We will have to do this for modelling these lines with them. But for the lines we have analyzed so far this is not necessary. For our in process atlases we will leave the IS components in the data.

The A-Star Puzzle
Proceedings IAU Symposium No. 224, 2004
J. Zverko, J. Žižňovský, S.J. Adelman, & W.W. Weiss, eds.
© 2004 International Astronomical Union
DOI: 10.1017/S1743921304004363

Panel discussion section A

CHAIR: W.W. Weiss

SECTION ORGANIZER & KEY-NOTE SPEAKER: Saul J. Adelman
INVITED SPEAKERS: J. Kubát, D. Korčáková
CONTRIBUTION SPEAKERS: J. Krtička, O. Kochukhov, S.J. Adelman, F. Royer

Discussion

COWLEY: For O. Kochukov: This concerns the $\lambda5200$ feature which you could reproduce in the cooler Ap stars, but not in the hotter ones: 1) Do you know which atoms or ions cause the depression in the cooler stars?

2) In the hotter stars, can you suggest a cause for the observed depressions in these objects?

KOCHUKOV: 1) The model atmosphere calculations for magnetic stars described in my talk were carried out for solar and for scaled solar compositions. In this case neutral iron lines are the main contributor to the $\lambda5200$ feature.

2) As for the feature in hotter Ap stars, other effects related to the strong surface magnetic fields, for instance vertical abundance stratification, may be responsible. Indeed opacity sample model atmosphere calculations by Shulyak et al. (2004, A&A, in press) demonstrated that by introducing a fairly moderate increase at $\log \tau_{5000} = -1$) the vertical stratification of iron can reproduce the feature in the hot magnetic star CU Vir.

BREGER: I would like to comment on the division into normal and abnormal stars by rotation. It was mentioned that the low-$v \sin i$ normal stars a really rapid rotators seen pole-on. Nonradial pulsation permits one to determine uniquely the angle of inclination of the rotation axis to the line of sight. For the δ Scuti star FY Vir we have determined the angle i to be $20°$, making the true rotation $65\ \mathrm{km\,s^{-1}}$. But the star is normal! Maybe the rotation rule is not absolute, as is also shown by a few other known stars.

BALONA: Rapid rotation inhibits difusion and may explain why normal A stars are mostly fast rotators. Similarly a star born with a high magnetic field could end up as a slow rotator due to magnetic braking. Assuming that stars in a cluster have about the same angular momentum and magnetic field distribution one expects to see that A stars in open clusters would either be nearly all normal (fast rotation) or nearly all peculiar (slow rotation). Do observations confirm this?

MONIER: In answer to the Balona's question to the audience: I will present tomorrow abundance results for A and F stars in various open clusters (these Proceedings, 209) and discuss the abundance patterns.

MATHYS: If there was a signicant number of fast rotating stars with chemical peculiarities that have remained undetected so far in spectroscopic abundance studies, should we not expect a fraction of them at least to show photometric variations that could and should have been found, e.g., by the Hipparcos survey?

ADELMAN: Hipparcos photometry variability is dependent on its bandpass which is most of the optical region. What remains is the possibility of a) Stars which are low amplitude variables only in narrow wavelength regions, and b) Late B stars where the optical spectrum has few lines. But when the amplitude of variability drops below 0.02 mag., one cannot do much with this data.

LANDSTREET: I have two comments on using the Hipparcos photometry to find Ap photometric variables: 1) The Hipparcos photometry is not extremely accurate, with typical errors of more than 0.01 mag. Some known photometric variable Ap stars are not detected as such by Hipparcos. 2) I have search the Hipparcos variable star archive for all the A stars in several clusters (several tens of stars). No new photometric variables of Ap type were found.

WADE: Because most of these stars are probably magnetic, a relatively robust way to identify them, essentialy insensitive to rotation, is polarimetry within Balmer lines (using instruments like John Landstreet's photopolarimeter or FORS1 @ VLT).

RYABCHIKOVA: Does including the magnetic field in line-blanketing calculations explain an inconsistency of about 1 dex in $\log g$, obtained for highly magnetic stars from optical spectrophotometry and from Balmer line profiles?

KOCHUKOV: The model structure does not change dramatically when the magnetic field is included. Changes in the flux distribution and in the H-line profiles relatve to a nonmagnetic model are definitely too small to explain a 1 dex discrepancy of $\log g$. Including the indirect effects of magnetic field (i.e., chemical stratification) seem to be more promising for deriving a consistent $\log g$ from energy distributions and Balmer lines.

CUNHA: Given the amount of physics that goes into model atmospheres, how much can we trust the global parameters derived through spectroscopic techniques, particularly T_{eff} in Ap stars? What are the main improvements needed for a more precise determination of these parameters?

ADELMAN: The temperature calibrations for normal stars from say Strömgren photometry are not correct for the CP stars. The models have to fit the observed energy distributions including the broad, continuum features and the Balmer line profiles. It is the features which have proven problematic. In part they are due to the stars being metal-rich, but not in a scaled solar sense. I believe that opacities which include the effects of auto-ionization will help. As soon as the ASTRA spectrophotometer begins to produce data (hopefully within two years) much better fluxes will become available and this should help. An effort will be made to get T_{eff} values for as many fundamental stars as possible.

KUPKA: Two comments: 1) On O. Kochukov's answer to C. Cowley's question concerning the $\lambda 5200$ feature: I agree that Fe is the most important line opacity source in the cool mCP stars which have an important ifluence on the flux redistribution. Moreover, from analyses done with E. Paunzen & collaborators on the $\lambda 5200$ feature, we found Cr not to be evenly distributed throughout the feature in the wavelength range measured by the Δa-system. Cr is certainly a key contributor to the feature at least for the cool mCP stars and the feature and its variation over the HRD is more likely caused by several contributing species, even if Fe is much more important for the temperature structure and flux distribution of these stars as a whole.

2) About M. Cunha's question on T_{eff} determinations: Recently it has become popular to use a modified version of D. F. Gray's procedure based on line strength ratios to determine extremely accurate absolute T_{eff} values. However, these all have to rely on some calibration using model atmospheres and T_{eff} values based on fundamental methods. As a result these errors propagate into the determination: accurate T_{eff} values for fundamental stars are scarce and in the 100 K to 400 K range for the most cases. As trends over the H-R Diagram are not known well enough, it is in my opinion questionable to claim ABSOLUTE T_{eff} values to be more accurate than say 150 K.

LANDSTREET: Dave Gray's efforts to measure very accurate temperatures in cool stars (uncertainities of a few K) have been discussed. I simply want to remark that Dave's very high precision is claimed for detecting small CHANGES in temperature, not for absolute temperatures.

WADE: Theme introduced by L. Balona: Returning to the issue of open clusters, we find from our survey of magnetic stars in open clusters that there exists a tremendous HETEROGENEITY of magnetic and abundance characteristics amongst Ap stars in individual open clusters, even at very young ages (few $\times 10^6$ y). The situation is very complex indeed.

WEISS: I remember a conference in Bormio where C. Cowley mentioned in his talk that the influence of NLTE effects seem to be much overestimated and it is not really so important to do NLTE calculations in chemical peculiar stars. I also remember a comment of Sydney Wolff in the "The A-Type Stars: Problems and Perspective"(1983, NASA SP-463) and I asked her about Hydrogen line profiles and metallic line profiles and she answered she was also quite surprised. We have heard today quite a bit about NLTEs effects.

COWLEY: Just to clarify my comments on the relative importance of NLTE vs. LTE. This comes from a time when those who firmly believed that NLTE was the ONLY way to compute spectra. They did not believe in ANY anomalous abundances in CP stars. Today the question for us to ask is which lines are the most sensitive to NLTE. Then those of us who still use LTE codes can avoid these features.

The A-Star Puzzle
Proceedings IAU Symposium No. 224, 2004   © 2004 International Astronomical Union
J. Zverko, J. Žižňovský, S.J. Adelman, & W.W. Weiss, eds.   DOI: 10.1017/S1743921304004375

A–type stars: Evolution, rotation and binarity

Arlette Noels[1], Josefina Montalbán[1] and Carla Maceroni[2]

[1]Institut d'Astrophysique et Géophysique Université de Liège,
Allée du 6 Août, B-4000 Liège, Belgium
email: first.last@ulg.ac.be

[2]INAF - Osservatorio Astronomico di Roma,
via Frascati 33, I-00040 Monteporzio C. (RM) - Italy
email: maceroni@coma.mporzio.astro.it

Abstract. We discuss the internal structure of stars in the mass range 1.5 to $4\,M_\odot$ from the PMS to the subgiant phase with a particular emphasis on the convective core and the convective superficial layers. Different physical aspects are considered such as overshooting, treatment of convection, microscopic diffusion and rotation. Their influence on the internal structure and on the photospheric chemical abundances is briefly described.

The role of binarity in determining the observed properties and as a tool to constrain the internal structure is also introduced and the current limits of theories of orbital evolution and of available binary datasets are discussed.

Keywords. Stars: evolution, binaries: general, stars: rotation

1. Introduction

The theoretical evolution of A stars is extremely simple to compute if one ignores complex phenomena such as gravitational settling, radiative forces, rotation, turbulent mixing, magnetic fields, binarity... We shall briefly introduce the infuence of different treatments of convection, diffusion and rotation after a presentation of a "conservative" situation in which the only mixings come from convection and overshooting. Binarity as a tool to constrain stellar models is then discussed.

2. A–type stars: evolution

As A stars are located on or near the main sequence, we shall start this discussion with the core hydrogen burning phase.

2.1. *Core hydrogen burning*

Figure 1 (lp) shows the evolutionary track of a $3\,M_\odot$ with a solar chemical composition.

The evolution with time of the radiative gradient, given by

$$\nabla_{\rm rad} \sim \frac{L}{m}\,\kappa \qquad (2.1)$$

is shown in Figure 2 (lp) in the inner 30% of the mass for selected models on the evolutionary path.

The adiabatic gradient, $\nabla_{\rm ad}$, is also drawn on this Figure and its decrease towards the centre shows that the influence of the radiation pressure increases with time. Starting from the right of the Figure, the radiative gradient increases when reaching the CNO burning central layers, due to the increasing value of L/m, and hydrogen burning thus

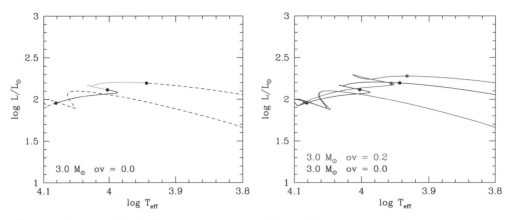

Figure 1. Left panel (lp): The evolutionary track for a 3 $M_\odot$ star computed without overshooting. Right panel (rp): Same Figure with an overshooting parameter of 0.2 showing a wider MS track than in the lp.

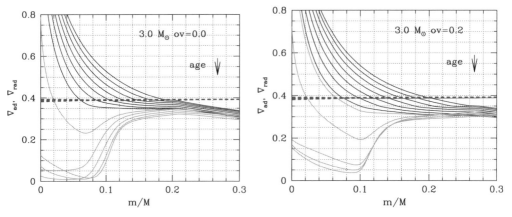

Figure 2. Left panel (lp): Radiative gradient for different models on the main sequence (black curves - first two dots starting from the left in Figure 1) and in the early H shell burning phase (gray curves - last two dots starting from the left in Figure 1). Right panel (rp): Same Figure with an overshooting parameter of 0.2.

takes place in a convective core. Inside this core, the hydrogen abundance, X, decreases. The opacity, κ, being proportional to $(1+X)$, decreases as time goes on. This means that the convective core has its maximum extension in mass on the ZAMS and then slowly recedes as hydrogen is transformed into helium. The difference in X between the core and the non-burning regions thus increases with time without any discontinuity at the edge since the convective core recedes. A gradient of hydrogen builds up and a plateau in $\nabla_{\rm rad}$ appears (see Figure 2 (lp)).

The decrease of X leads to an increase in the temperature (Figure 3 (lp)) which is limited to the central layers while the other layers become cooler and cooler.

The central part of the star is contracting while the other layers are expanding. Once X comes close to zero in the core, the increase in temperature becomes more pronounced (see Figure 3 (lp)). This is done by a global contraction (the so-called *second gravitational contraction*).

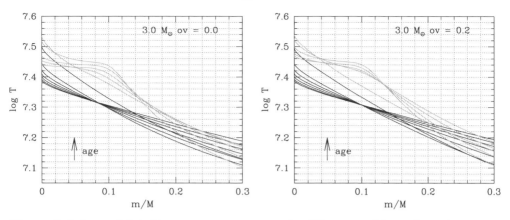

Figure 3. Left panel (lp): Temperature versus mass for different models on the main sequence (black curves - first two dots starting from the left in Figure 1) and in the early H shell burning phase (gray curves - last two dots starting from the left in Figure 1). Right panel (rp): Same Figure with an overshooting parameter of 0.2.

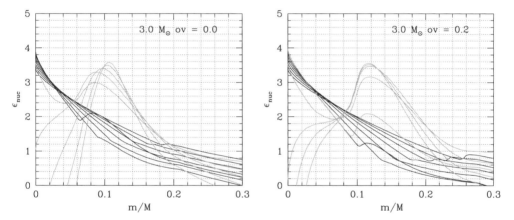

Figure 4. Left panel (lp): Nuclear energy rate versus mass for different models on the main sequence (black curves - first two dots starting from the left in Figure 1) and in the early H shell burning phase (gray curves - last two dots starting from the left in Figure 1). Right panel (rp): Same Figure with an overshooting parameter of 0.2.

2.2. *Hydrogen shell burning*

In the μ-gradient zone, a hydrogen burning shell starts to develop. A secondary maximum is seen in $\nabla_{\rm rad}$ (Figure 2 (lp)) as well as in the nuclear enregy rate, ϵ (Figure 4 (lp)). Soon after, $\nabla_{\rm rad}$ becomes smaller than $\nabla_{\rm ad}$ and the convective core vanishes. As the energy production stops, an isothermal core appears. The models are now located between the last two dots on the evolutionary track in Figure 1 (lp). As hydrogen is burned in the shell, the isothermal core increases in mass. However, its mass cannot exceed the limiting mass of Schönberg-Chandrasekhar, given by

$$\left(\frac{m}{M}\right)_{\rm SC} \simeq 0.37\,\frac{\mu_{\rm e}}{\mu_{\rm i}} \tag{2.2}$$

which is of the order of 0.08. When this value is reached, the core must contract to create a temperature gradient (see Figure 3 (lp)) and this is accompanied by an expansion of the envelope where the temperature decreases.

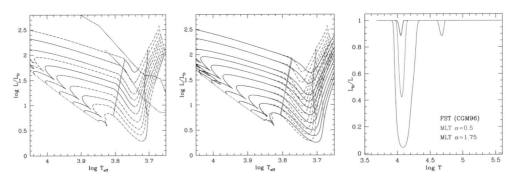

Figure 5. Left panel (lp): PMS tracks (see text) with two birthlines (bottom one: Palla & Stahler (1993), top one: Behrend & Maeder (2001). Center panel (cp): Evolutionary tracks computed with FST (dashed curves) and with MLT ($\alpha = 1.6$) (faded curves). Right panel (rp): The fraction of energy transported by radiation versus $\log T$ in the external layers. (1) *With FST* : Only one (the smallest) minimum near $\log T = 4.1$ (H I and He I) and no minimum near $\log T = 4.7$. (2) *With MLT ($\alpha = 0.5$)* : One minimum (the intermediate one) near $\log T = 4.1$ and no minimum near $\log T = 4.7$. (3) *With MLT ($\alpha = 1.75$)* : One (the largest) minimum near $\log T = 4.1$ and a second small one near $\log T = 4.7$ (He II).

2.3. *Effect of overshooting*

To fit CM diagrams of open clusters as well as eclipsing binaries, overshooting seems to be needed with an extent increasing with stellar mass. We have computed a similar evolution with an overshooting parameter of 0.2. A larger hydrogen reservoir due to the increase of the mixed region evidently translates into a longer core hydrogen burning phase as well as in a longer track in the HR diagram (Figure 1 (rp)). However, after the turn-off, for a similar variation in effective temperature, the evolution with overshooting is much more rapid. Figure 2 (rp) shows that $\nabla_{\rm rad}$ never reaches zero so there is no plateau in temperature (Figure 3 (rp)). The reason lies in the fact that with the increase of the mixed region, the exhaustion of hydrogen takes place in a core whose mass is already greater than $\left(\frac{m}{M}\right)_{\rm SC}$. This results in a much quicker contraction of the central regions. The star moves more rapidly and for a similar change in effective temperature, the shell, shown by a maximum in the distribution of ϵ, is still at the same mass fraction with overshooting while the shell has already moved significantly in mass without overshooting (compare lp with rp in Figure 4).

   Although the μ-gradient has a smaller slope with overshooting than without overshooting, the presence or absence of an isothermal core leads to a similar extent of the helium core at a similar effective temperature.

2.4. *Pre-main sequence evolution*

Let us now turn toward the pre-main sequence evolution (Figure 5). The two quasi-vertical lines on the Hayashi tracks and soon after show where the star becomes partly radiative and then completely radiative. The loops near the MS are the signatures of the CNO reactions evolving toward equilibrium. We have checked the effect of the new (smaller) cross-section of the slowest of the CNO reactions, i.e., ^{14}N$(p, \gamma)^{15}$O (Formicola & LUNA 2004). Although the approach toward equilibrium is slower, the difference in the main sequence lifetime is not significant.

   The points in the HR diagram where the stars become visible define the birthline. Two birthlines are drawn in Figure 5 (lp)).

   The bottom one comes from Palla & Stahler (1993) and is constructed assuming a constant accretion rate of $10^{-5}\,M_\odot$/yr. This value was obtained by fitting the Herbig Ae

Figure 6. Illustration of diffusion. See text

and Be objects in the HR diagram. The top one comes from an analysis by Behrend &
Maeder (2001) where the accretion rate

$$\dot{M} = \frac{1}{3}\,\dot{M}_{\mathrm{disk}} \qquad (2.3)$$

has been derived to reproduce the observations in the HR diagram, in particular for
the most massive stars. The time spent on the Hayashi track depends on the adopted
birthline. This results in a different transfer of angular momentum from the star to the
disk, which in turn can change the angular momentum at the beginning of the MS phase.

In these fully or partly convective phases, the treatment of convection affects the
location in the HR diagram, creating a significant difference in effective temperature.
Figure 5 (cp) shows the location of the Hayashi tracks for models computed with the
FST treatment of convection (dashed curves) and with the MLT treatment ($\alpha = 1.6$)
(faded curves). The difference in effective temperature is of the order of 200 K. As the
convective envelope recedes, the tracks become undistinguishable.

However, the extremely thin surface convection zone which remains on the main sequence is affected (Figure 5 (rp)). The more superadiabatic the temperature gradient the
less efficient is the convection. For a $1.8\,M_\odot$, it can make the convective He II ionization
zone appear or disappear. This is important since the thickness of the mixed superficial
layers is crucial in explaining the abundance anomalies.

2.5. *Gravitational settling*

With such thin convective envelopes, gravitational settling is very efficient. Figure 6
illustrates this point.

In the left panel, a duck-star has all its chemical elements (chicken) showing at the
surface. In the middle panel, the MS phase is symbolized by the crossing of a grid.
If radiative forces are not taken into account together with gravitational settling, the
duck-star ends its MS phase with only one element (chick) remaining at the surface.

However, radiative forces can in turn be too efficient and another mechanism must enter
the game. It is the turbulent mixing which will not only affect the surface abundances
but will induce changes in the internal structure as well, especially in the thickness of
the convective envelope (Richard *et al.* 2001). One striking example is the formation of
a convective iron zone at a temperature of 200000 K caused by an accumulation of iron-
peak elements as a result of radiative forces. According to the mass of the star, this zone
can merge with the H-He ionisation zone during the main sequence or remain detached,
which affects the surface abundances of the iron-peak elements.

3. A–type stars: rotation

From their location in the HR diagram, A-type stars are expected to rotate rapidly and not to be affected by magnetic braking. They can reach indeed rotational velocities up to 300 km s^{-1}. Abt & Morrel (1995, hereafter AM95) showed, however, that the distribution of rotational velocities has a bimodal shape, with virtually all the Am and Ap stars having equatorial rotational velocities (v_{rot}) less than 120 km s^{-1}, and most of the normal A0-F0 main-sequence stars having $v_{rot} > 120$ km s^{-1}. They concluded (see also Abt (2000)) that rotation alone can explain the occurrence of abnormal or normal main-sequence A stars, and that the apparent overlap between their v_{rot} distributions is only due to our inability to distinguish marginal Am stars from normal ones, or to disentangle rotational and evolutionary effects.

Considerable observational data indicates that during the MS evolution of A-type stars: 1) mass loss is limited to $\sim 2.10^{-10}$ $M_\odot$yr^{-1}. (Lanz & Catala 1992); 2) rotational velocity does not depend on age; and 3) no significant angular momentum loss by magnetic braking is observed (e.g. Wolff & Simon 1997, Hubrig et al. 2000). Consequently, the observed v_{rot} distribution must be determined by the angular momentum evolution during the pre-main sequence (PMS) phase. In a study of the distribution of angular momentum in Orion stars, Wolff *et al.* (2004) find this hypothesis consistent with a simplified model of PMS evolution in which angular momentum is lost by an interaction with the protostellar disk during the convective phase, and conserved in the radiative one. Furthermore a core-envelope decoupling occurs during the convective-radiative transition (see also Stępień (2000) and Stępień & Landstreet (2002) for Ap PMS angular momentum evolution). The understanding of angular momentum evolution during the PMS and the MS is fundamental to understand the Ap and the Am phenomena.

3.1. *Modelling the evolution of a rotating star*

Rotation has several different effects on stellar evolution: change of the internal hydrostatic equilibrium, changes in the apparent effective temperature and luminosity (Pérez Hernández *et al.* 1999 and references therein), transport of chemicals and of angular momentum by shears in differentially rotating stars, by meridional circulation and by horizontal turbulence (e.g., Zahn 1974, Knobloch & Spruit 1982).

The effect of rotation on stellar evolution has been treated with different approaches: Endal & Sofia (1981) assigned to each transport process (including meridional circulation) a diffusion coefficient, while Zahn (1992a) and Maeder & Zahn (1998) hypothesized that differential rotation in the radiative zone of a nonmagnetic star gives rise to anisotropic turbulence (much stronger in the horizontal direction than in the vertical one due to the stratification) and results in "shellular" rotation. In this model the effective diffusion coefficient for the chemicals is:

$$D_{\text{eff}} = \frac{|rU(r)|^2}{30\, D_{\text{h}}}, \text{with } D_{\text{h}} = \frac{r}{C_{\text{h}}} \left| \frac{1}{3\rho r} \frac{\mathrm{d}(\rho r^2 U)}{\mathrm{d}r} - \frac{U}{2} \frac{\mathrm{d}\ln r^2 \Omega}{\mathrm{d}\ln r} \right|. \tag{3.1}$$

Where U is the meridional circulation velocity, C_{h} is a free parameter related to the turbulent horizontal viscosity, and the other quantities have the habitual meaning. Palacios *et al.* (2003) have applied Zahn's modeling of rotation with the formulation by Maeder & Zahn (1998) (including also microscopic diffusion but not radiative accelerations) to masses from 1.35 to 2.2 M$_\odot$. We recall two results of this application: 1) the rotation profile inside the star shows differential rotation mainly close to the convective core. 2) The models (computed without overshooting), show a wider MS when rotation is included. However, while the lower mass models increase their MS lifetime by a 20%, the largest masses increase it by only 10% (see their Fig. 1). This trend is opposite to that observed in open clusters and in binary systems, where the fitting of observations requires to increase the overshooting parameter with the mass (i.e. Andersen *et al.* 1990, Ribas et al. 2000).

Recently, Maeder (2003) and independently Richard & Zahn (1999) have updated the value for the horizontal diffusion coefficient, C_h by a factor of 10^2 with respect to the value used in Maeder & Zahn (1998). Its implementation in a stellar evolution code (Mathis *et al.* 2004) leads to: 1) enhanced mixing, and 2) significant changes in the profile of chemical mixing with depth (see their Fig. 2). To what degree will the new D_{eff} (with a non-zero value until the boundary of the convective core) affect the MS width and the surface chemical abundances?

The scheme proposed by AM95 matches qualitatively well the predictions of the microscopic diffusion models (e.g., Richer *et al.* 2000), in the hypothesis that the extra-mixing required to decrease the microscopic diffusion is induced by rotation. Some problems are, however, left, for instance: 1) the "normal" late B and early A-type main-sequence stars show a true bimodal v_{rot} distribution (Royer *et al.* 2005) ; 2) the binary V392 Carinae has a $v \sin i$ of 27 km s^{-1} and no peculiar abundances (Debernardi & North 2001). 3) There is no correlation between the strength of chemical peculiarities and v_{rot} (Erspamer & North 2003). Models including microscopic diffusion and radiative accelerations show that the chemical abundances are very sensitive to the thickness of the mixed layer (e.g., Alecian 1996, Hui-Bon-Hoa, 2000, Richard *et al.* 2001), so that we should expect a signature of v_{rot} in the abundances, if rotation is the responsible of the extra mixing below the convective envelope. A first approach to explicitly include a model of rotationally induced mixing and a complete treatment of microscopic diffusion is in progress (Richard et al. 2005) .

Finally, there are also other effects that should be taken into account in the study of the evolution of a rotating A-type star:

1. *Interaction convection-rotation.* Rotation can reduce the efficiency of convection and is able to reduce the extension of the overshooting region (e.g., Julien *et al.* 1997). Recent numerical simulations of a rotating convective core (Browning *et al.* 2004) show that rotation leads to a variation of convective overshooting and penetration, and induces internal waves, meridional circulation, and differential rotation at the core boundary.

2. *Magnetic field.* Maeder & Meynet (2003) have shown that the Tayler-Spruit magnetic instability (Spruit 2002) could take place in the interior of stars with small magnetic fields and differential rotation, induce a process of angular momentum transport much more efficient than that due to the meridional circulation and horizontal turbulence, and lead to solid body rotation on a short timescale. Furthermore, in Ap stars, which have undergone strong braking during their PMS phases (Stępień 2000), differential rotation together with a magnetic field could lead to magnetorotational instabilities. These could transport angular momentum from the interior to the surface (Arlt *et al.* 2003, and Artl 2005), and result in solid body rotation of these stars, as expected from some observational data (Hubrig et al. 2000).

4. A–type stars: binarity

Binarity plays a fundamental role in the origin and definition of the chemical peculiarities of A-type stars. The pioneering work of Abt (1961, 1965) showed that most (he even suggested all) Am type stars are relatively short period binaries and that the period distribution of binaries with Am and with non-peculiar A-type components are complementary. Normal A-type components are found in systems with period shorter than ~ 2.5 or longer than ~ 100 days, while Am stars form binaries with period in the range 2.5–100 days.

Abt's statements have somewhat lost their strength with time and with the increasing size of studied binary samples. Some overlap between the period distribution of normal and Am binaries was found by the same author (AM95) and the binary frequency among Am stars has steadily decreased with time (see Abt & Levy (1985) and North *et al.* 1998, Debernardi (2000) for results based on CORAVEL surveys). The latter author finds a frequency as low as 57% (but CORAVEL samples are certainly biased against fast rotation and, because of the limited time span of the survey, against longer period–eccentric orbit binaries). The – at any rate – high frequency has a straightforward explanation: in close binaries the spin–orbit synchronization by tidal mechanisms can efficiently brake the stellar rotation to values compatible with the Am phenomenon.

The other relevant connection between binarity and A-type star peculiarity is in the low frequency of spectroscopic binaries among the magnetic Ap stars (Abt & Snowden 1973, Gerbaldi *et al.* 1985, North *et al.* 1998). The current explanation is that the strong magnetic fields prevent close binary formation (however, Budaj (1999) suggested instead that it is binarity which affects magnetism and not the other way around).

While the general outline of the binarity–peculiarity connection is well established, there are still shortcomings in the theory of tidal synchronization. Besides, possible drawbacks in the interpretation of the observations are caused by the origin and composition of the observed samples (biases, selection effects).

Two competing theories were developed to explain the observed levels of orbital circularization and spin–orbit synchronization in close binaries: the tidal theory of Zahn (1992b) and references therein and the hydrodynamical theory of Tassoul & Tassoul (1992) and references therein. The necessary ingredients of the first are tidal bulges and an efficient dissipation mechanisms; in absence of synchronism, dissipation causes a lag of the bulge and hence a torque, which tends to establish synchronization (and orbit circularization). The dissipation mechanisms at work are different for late and early type stars, in the first case it is turbulent dissipation in the convective envelope retarding the equilibrium tide, in the second it is radiative damping acting on the dynamical tide (forced gravity waves are emitted from the lagging convective core and are damped in the outer layers).

Tassoul's theory is based, instead, on the idea that while the torque due to dissipation processes is negligible, transient strong meridional currents are produced by the tidal action and transfer angular momentum between the stellar interior and an "Ekman layer" close to the surface. As a consequence if the rotation period is shorter than the orbital one, the star is braked.

Both theories yield time-scales for synchronization (t_{ms}) and circularization (t_{mc}). For early type stars, in Zahn's case:

$$\frac{1}{t_{ms}} \propto \left(\frac{GM}{R^3}\right)^{\frac{1}{2}} q^2 (1+q)^{\frac{5}{6}} \frac{MR^2}{I} E_2 \left(\frac{R}{a}\right)^{\frac{17}{2}} ; \quad \frac{1}{t_{mc}} \propto \left(\frac{GM}{R^3}\right)^{\frac{1}{2}} q(1+q)^{\frac{11}{6}} E_2 \left(\frac{R}{a}\right)^{\frac{21}{2}}$$

(4.1)

where q and a are the mass ratio and the semi-axis of the binary, M, R, I the mass, the radius and the inertial moment of the braked star and E_2 a constant related to the size of its convective core. In Tassoul's theory:

$$\frac{1}{t_{ms}} \propto 10^{\frac{N}{4}-\gamma} q(1+q)^{\frac{3}{8}} \left(\frac{ML^2}{R^9}\right)^{\frac{1}{8}} \left(\frac{R}{a}\right)^{\frac{33}{8}} ; \quad \frac{1}{t_{mc}} \propto 10^{\frac{N}{4}-\gamma}(1+q)^{\frac{11}{8}} \beta^2 \left(\frac{ML^2}{R^9}\right)^{\frac{1}{8}} \left(\frac{R}{a}\right)^{\frac{49}{8}}$$

(4.2)

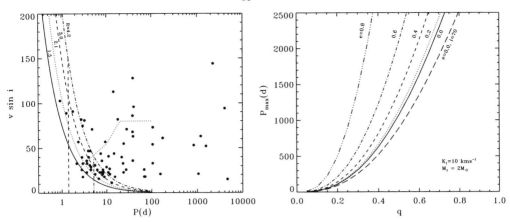

Figure 7. Left panel: The rotational velocities ($v \sin i$) of spectroscopic binaries with Am components versus orbital period. The hyperbolic curves give the theoretical relation ($i = 90°$) for synchronized systems of $2\,M_\odot$ and different radii. The discontinuous dotted line is the lower boundary of the "avoidance zone", according to Budaj (1996), the vertical lines are the expected upper boundary for circularization, according to North & Zahn (2003): dotted, $R = 2.1\,R_\odot$, $q = 1$; dash-dotted, $R = 3\,R_\odot$, $q = 0.2$. Right panel: The longest orbital period of a SB1 binary, corresponding to the assumed minimum observable radial velocity amplitude $K_1 = 10\,\mathrm{km\,s}^{-1}$, is plotted as function of mass ratio and eccentricity for primary mass $M = 2\,M_\odot$ and inclination $i = 90°$

with L being the star luminosity, β its fractional gyration radius, N the ratio between eddy and radiative viscosity ($N = 0$ for radiative envelopes). The other factor, γ, was introduced by Claret et al. 1995) to take somehow into account the fact that the orbital evolution is derived (as in Zahn's case) under the hypothesis of small deviations from synchronism and from circular orbit. The circularization time is in general two/three orders of magnitude longer than that of synchronization, due to the larger amount of angular momentum stored in the orbit. It is evident by comparing Eqs. 4.1 with 4.2 that Tassoul's mechanism has a longer range and a much higher efficiency, especially for early-type stars. The timescales are – at any rate – *only an indication* of the process speed, and cannot replace the full integration of the orbital evolution equations.

In principle the comparison between the expected and the observed degree of synchronization/circularization in binaries of known age and accurate dimensions could discriminate between the two theories, and early type stars are the best choice for such a test. However, the detailed treatment of Claret et al. 1995) and Claret & Cunha (1997) (with simultaneous computation of orbital and stellar evolution and a comparison sample of observed binaries with the best known parameters) showed that the results are still inconclusive, as both models can explain only a part of the observations. Zahn's mechanism is not efficient enough for early type stars (but the pre-MS phases were not considered in the abovementioned papers) while Tassoul's one is too efficient, unless a high value of γ is introduced (to shorten the time-scales by a factor ~ 40). The most promising scenario remains, therefore, that suggested in a study of circularization in binaries with A-type components by Matthews & Mathieu (1992): a process imposed by an important PMS phase of orbital evolution followed by a MS phase, which however still awaits full modeling.

The recent survey of A-type star rotational velocities by Royer et al. (2002), hereafter R02, provides excellent material for the study of synchronization in binaries with Am components. Figure 7 shows the updated $v \sin i$ values from R02 versus orbital period, for the Am binaries listed in Budaj (1996), and the relations expected in case of spin-orbit synchronization. The vertical lines are the periods corresponding to the fractional radius (R/a) that should assure circularization (North & Zahn (2003)) – and presumably circularization – the case of a $2\,M_\odot$ MS-primary. The data are generally in good agreement with the expectations, with a marginal indication of radii larger than in normal stars (as the synchronization relations should be upper envelopes, having assumed $i = 90°$).

The R02 velocities are generally higher than those measured by AM95 and used by Budaj; the reason is a systematic effect in the velocity of standard stars used by AM95. As a consequence the "avoidance zone", found by Budaj (1996), for periods $4 < P < 20$ days (i.e., the region above the dotted line in Fig. 7) is no longer empty and it is not necessary for an ad-hoc mechanisms to explain it, such as the "tidal mixing" introduced by the abovementioned author; the decreasing number of systems found in proximity of the synchronism lines can be due to a dependence on period of the braking efficiency.

The other feature appearing in Figure 7, the lack of systems between $180 < P < 800$ days, is quite probably a selection effect due to the low probability of discovery *as a spectroscopic binary*. An idea of the relevance of selection effects can be derived from the right panel of Fig. 7. This shows, for a $M = 2\,M_\odot$ primary, the largest orbital period corresponding to a given minimum observable radial velocity amplitude, K_1. This is a function of mass ratio, q, and eccentricity, e (in the hypothesis of inclination $i = 90°$). Even with the conservative hypothesis $K_1 = 10$ km s^{-1} (but Hogeveen (1992), on the basis of a detailed study of selection effects for spectroscopic binaries, suggests 25 km s^{-1} for A-type stars) Figure 7 indicates that the largest period of binaries detectable by spectroscopic surveys is of some hundred days (and many systems will escape detection, as the distribution of mass ratio of single lined spectroscopic binaries is peaked around $q = 0.2$). The gap in the period distribution could, therefore, be due to the transition between spectroscopic binaries, discovered as such, and visual binaries with known radial velocity.

In conclusion, a proper treatment of selection effects is a primary issue to solve if we want to extract correctly the rich information that A-type stars can provide on binary secular evolution and that, in their turn, binaries can yield on A-type star properties.

Acknowledgements

The authors thank the IAU for grants covering the registration fees. A.N and J.M. acknowledge ESA-PRODEX contract 15448/01/NL/SFe(IC)-C90135 and IAP P5/36, and C.M. a F-INAF program for funding.

References

Abt, H.A. 1961, *ApJS* 6, 37
Abt, H.A. 1965, *ApJS* 11, 429
Abt, H.A. 2000, *ApJ*, 544, 933
Abt, H.A. & Levy, S.G. 1985, *ApJS* 59, 229
Abt, H.A. & Morrell, N.I. 1995, *ApJS* 99, 135 (AM95)
Abt, H.A. & Snowden, M.S. 1973, *ApJS* 25, 137
Andersen, J., Nordström, B., & Clausen, J.V. 1990, *ApJ* 363, L33
Arlt. R. 2005, *These Proceedings, 103*
Arlt, R., Hollerbach, R., & Rüdiger, G. 2003, *A&A* 401, 1087
Behrend, R. & Maeder, A. 2001, *A&A* 373, 190
Browning, M.K., Brun, A.S., & Toomre, J. 2004, *ApJ* 601, 512
Budaj, J. 1996, *A&A* 313, 523
Budaj, J. 1999, *MNRAS* 310, 419

Chan, K.L. 1996, in *Cool Stars, Stellar Systems, and the Sun, 9th Cambridge Workshop*, ASP Conf. Ser. 109, p. 561

Claret, A. & Cunha, N.C.S. 1997, *A&A* 318, 187

Claret, A., Gimenez, A., & Cunha, N.C.S. 1995, *A&A* 299, 724

Debernardi, Y. 2000, *IAU Symposium 200*, p. 161

Debernardi, Y. & North, P. 2001, *A&A* 374, 204

Endal, A.S. & Sofia, S. 1981, *ApJ* 243, 625

Erspamer, D., & Norht, P. 2003, *A&A* 398, 1121

Formicola, A. & LUNA 2004, *Physical Letters B* 591, 61

Gerbaldi, M., Floquet, M., & Hauck, B. 1985, *A&A*, 146, 341

Hogeveen, S.J. 1992, *Ap&SS* 196, 299

Hubrig, S., North, P., & Medici, A. 2000, *A&A*, 359, 306

Hui-Bon-Hoa, A. 2000, *A&ASS* 144, 203

Julien, K., Werne, J., Legg, S., & McWilliams, J. 1997, in *SCORe'96 : Solar Convection and Oscillations and their Relationship*, ASSL Vol. 225, p. 227

Knobloch, E. & Spruit, H.C. 1982, *A&A* 113, 261

Lanz, T. & Catala, C. 1992, *A&A*, 257, 663

Mathis, S., & Zahn J.-P. 2004, *A&A* in press

Mathis, S., Palacios, A., & Zahn J.-P. 2004, *A&A* in press

Maeder, A. 2003, *A&A* 399, 263

Maeder, A., & Meunet, G. 2003, *A&A* 411, 543

Maeder, Z., & Zahn, J.-P. 1998, *A&A* 334, 1000

Matthews, L.D. & Mathieu, R.D. 1992, in *IAU Colloq. 135: Complementary Approaches to Double and Multiple Star Research*, ASP Conf. Ser. 32, p. 244

North, P., Ginestet, N., Carquillat, J.-M., Carrier, F., & Udry, S. 1998, *Contributions of the Astronomical Observatory Skalnate Pleso* 27, 179

North, P. & Zahn, J.-P. 2003, *A&A* 405, 677

Palacios, A., Talon, S., Charbonnel, C., & Forestini, M. *A&A* 399, 603

Palla, F. & Stahler, S. W. 1993, *ApJ* 418, 414

Pérez Hernández, F., Claret, A., Hernández, M.M., & Michel, E. 1999, *A&A* 346, 586

Ribas, I., Jordi, C., Tora, Torra, J., Giménez, Á. 2000, *MNRAS*, 313, 99

Richard, D. & Zahn, J.-P. 1999, *A&A* 347, 734

Richard, O., Michaud, G., & Richer, J. 2001, *ApJ* 558, 377

Richard, O., Talon, S., Michaud, G., 2005, *These Proceedings, 215*

Royer, F., Grenier, S., Baylac, M.-O., Gómez, A.E., & Zorec, J. 2002, *A&A* 393, 897 (R02)

Royer, F., Zorec, J., Goméz, A. E., 2005, *These Proceedings, 109*

Spruit, H.C. 2002, *A&A*, 381, 923

Stępień, K. 2000, *A&A* 353, 227

Stępień, K. & Landstreet, J.D. 2002, *A&A* 385, 554

Tassoul, J. & Tassoul, M. 1992, *ApJ* 395, 259

Wolff, S.C., & Simon, T. 1997, *PASP* 109, 759

Wolff, S.C., Strom, S.E., & Hillenbrand, L.A. 2004, *ApJ* 601, 979

Zahn, J.-P. 1974, in *IAU Symp. 59: Stellar Instability and Evolution*, p. 185

Zahn, J.-P. 1992a, *A&A* 265, 115

Zahn, J.-P. 1992b in A. Duquennoy & M. Mayor (eds), *Binaries as Tracers of Stellar Formation*. Cambridge: Cambridge University Press, p. 253

Discussion

Discussion postponed to the panel discussion B (see pp. 115 – 118, These Proceedings).

The A-Star Puzzle
Proceedings IAU Symposium No. 224, 2004
J. Zverko, J. Žižňovský, S.J. Adelman, & W.W. Weiss, eds.
© 2004 International Astronomical Union
DOI: 10.1017/S1743921304004387

Mass loss, meridional circulation and turbulence in contemporary stellar evolution models

Suzanne Talon

Département de Physique, Université de Montréal, Montréal, PQ, H3C 3J7, Canada
email: talon@astro.umontreal.ca

Abstract. In this review, I will briefly discuss the hypotheses made in the treatment of modern rotating stellar models and review the expected efficiency of mixing along the HR diagram. The role of mixing in the localization of abundance anomalies will also be discussed. Finally, I will show how mass loss and gravitational settling of helium may influence the evolution of rotating stars, and how A stars can play a unique role in constraining our models.

Keywords. Turbulence, hydrodynamics, stars: abundances, stars: evolution, stars: chemically peculiar, Hertzsprung-Russell diagram, stars: interiors, stars: mass loss, stars: rotation

1. Introduction

Rotation plays an important role in determining stellar structure. It modifies surface properties such as the temperature and the local gravity and might even influence mass loss. Furthermore, it can lead to a significant amount of mixing due to meridional circulation and various hydrodynamical instabilities. The initial angular momentum contained in a star will thus be a third parameter guiding a star's structure and evolution. It is now widely recognized that a complete description of the evolution of the angular momentum distribution within the star is required and that rotational mixing is linked to the internal rotation profile.

The first modelization of this type was undertaken by Endal & Sofia (1976, 1978, 1981) and pursued by Pinsonneault *et al.* (1989). However, in these early investigations meridional circulation was treated merely as a diffusive process.

The competition of meridional circulation and microscopic diffusion as competing processes in the formation of Am/Fm stars was quantified for the first time by Michaud (1982) and Charbonneau & Michaud (1991). But the postulated solid body rotation was not derived from a self consistent calculation.

Here I will describe how meridional circulation can be described as a truly advective process (§ 2) and present a modelization of anisotropic turbulent transport in stars (§ 3). I will further describe the expected efficiency of mixing along the HR diagram (§ 4), discuss the impact of various boundary conditions on this efficiency (§ 5) and comment on the effect of mass loss on other physical processes (§ 6). I will conclude with the examination of the constraints obtained from rotational mixing models of A and Am stars (§ 7).

2. Meridional circulation in 1D

The evolution of the rotation state within a star subject to meridional circulation should in principle be treated at least with a 2D model. However, one can take advantage of the fact that in stratified fluids turbulence is expected to be strongly anisotropic. Assuming this is the case and that the horizontal turbulent viscosity ν_h is much larger than the vertical turbulent viscosity ν_v (or more specifically $\nu_h/l_h^2 \gg \nu_v/l_v^2$) allows one to by-pass the difficulty. This approach was followed by Zahn (1992) to derive equations guiding the evolution of rotating stars. He assumes that the horizontal turbulence is efficient enough to enforce a shellular rotation state where $\Omega = \Omega(r)$.

A first order expansion leads to an advection-diffusion equation for the transport of angular momentum

$$\rho \frac{d}{dt}\left[r^2\Omega\right] = \frac{1}{5r^2}\frac{\partial}{\partial r}\left[\rho r^4 \Omega U\right] + \frac{1}{r^2}\frac{\partial}{\partial r}\left[\rho \nu_v r^4 \frac{\partial\Omega}{\partial r}\right] \tag{2.1}$$

ρ being the density and ν_v the vertical component of the turbulent viscosity. The amplitude of the vertical circulation velocity $u(r,\theta) = U(r)\left(P_2(\cos\theta) + 1/5\right)$ is given by

$$U(r) = \frac{L}{mg}\left(\frac{P}{C_P\rho T}\right)\frac{1}{(\nabla_{\rm ad} - \nabla + \nabla_\mu)}\left[E_\Omega + E_\mu\right], \tag{2.2}$$

where L is the luminosity, m the mass, P the pressure, C_P the specific heat at constant pressure, T the temperature, $\nabla_{\rm ad} = (\partial\ln T/\partial\ln P)_{\rm ad}$, $\nabla = d\ln T/d\ln P$, and $\nabla_\mu = d\ln\mu/d\ln P$: the adiabatic, radiative and mean molecular weight gradients, respectively (Maeder & Zahn 1998). E_Ω and E_μ, derived from first principles, depend on the rotation profile (neglecting any horizontal fluctuations) and on the horizontal variations of the mean molecular weight Λ, respectively (see Zahn 1992 and Maeder & Zahn 1998 for details). Let us only remark that to first order $E_\Omega \propto \frac{\Omega^2 R^3}{GM}$, which is a measure of the star's oblateness.

The presence of a strong horizontal turbulence D_h modifies the evolution of chemicals. The combination of an advective meridional circulation with this horizontal turbulence is equivalent to considering an effective vertical diffusion given by

$$D_{\rm eff} = \frac{|rU(r)|^2}{30\,D_h} \tag{2.3}$$

(Chaboyer & Zahn 1992). This equation expresses the fact that horizontal turbulence erodes the advective process and thus reduces the efficiency of meridional circulation in the transport of chemicals†. The concentration of a given element c_i then obeys

$$\rho\frac{dc_i}{dt} = \dot c_i + \frac{1}{r^2}\frac{\partial}{\partial r}\left[r^2\rho U_{ip}c_i\right]\frac{1}{r^2}\frac{\partial}{\partial r} + \left[r^2\rho\left(D_{\rm eff} + D_v\right)\frac{\partial c_i}{\partial r}\right], \tag{2.4}$$

where $\dot c_i$ is the nuclear production/destruction rate and U_{ip} is the microscopic diffusion velocity of the element of interest with respect to protons. D_v is the vertical turbulent diffusivity.

In this model, the rotation profile can reach a stationary state, in which the inward‡ advection of angular momentum by meridional circulation balances the outward turbulent or viscous diffusion of momentum. This stationary state obeys

$$U(r) = -\frac{5\nu_v}{\Omega}\frac{\partial\Omega}{\partial r}. \tag{2.5}$$

---

† The advection of momentum is not affected since the horizontal average is performed on $r^2\Omega$ while turbulence homogenizes Ω. See Chaboyer & Zahn (1992) for details.

‡ It can be shown that the advected flux will be directed inward as long as differential rotation is not too large. If differential rotation is large with the core rotating faster, the direction of the circulation reverses and both turbulent viscosity and meridional circulation carry angular momentum towards the surface.

Calculations show that this equilibrium leads to a core rotating ~ 1.2 to 1.4 times faster than the surface in a homogeneous ZAMS star (this ratio rises in an evolved star). In a static model, there is thus no net momentum flux. However, there always exists a flux of concentration, as both the effective and turbulent diffusivity lead to concentration fluxes that are directed opposite to the concentration gradient. Moreover, while horizontal diffusion acts to reduce the efficiency of diffusion for the chemicals, it does not have the same effect on angular momentum; in a turbulent region, $D_v \simeq \nu_v$ is thus much larger than D_{eff}.

As the model star evolves, the circulation must adjust itself to provide the momentum flux required to reconstruct the equilibrium profile. This readjustment is made more drastic by mass loss, as the upward migration of the surface layers will create an even larger deficit of momentum. Mass loss is thus a key ingredient in the determination of the amount of large scale mixing taking place in the outer layers of stars. This is especially true of slow rotators in which turbulence is not large at the surface.

3. Turbulence

To complete the model, one has to obtain an estimate of the turbulent viscosities ν_v and ν_h and turbulent diffusivities D_v and D_h. We will assume that $D_v = \nu_v$ and $D_h = \nu_h$. The source of turbulence considered here is the shear instability†.

Shear instabilities develop because energy is stored in differential rotation. A linear stability analysis shows that instability may grow as soon as there is an inflection point in the profile (Rayleigh 1880, Watson 1981). However, laboratory experiments indicate that destabilization occurs even if this is not the case, and in general one may assume the presence of turbulence as soon as the Reynolds number is large enough (see, e.g., Richard & Zahn 1999). This is the point of view we shall adopt.

In the case of the vertical viscosity ν_v, work must be done against the stable stratification for the instability to occur. In the case relevant to stellar models, this will be made possible only if the thermal stratification measured by the Brunt-Väisälä frequency $N_T^2 = \frac{g\delta}{H_P}(\nabla_{\text{ad}} - \nabla)$ is reduced by the thermal diffusivity (Townsend 1958, Dudis 1974, Lignières *et al.* 1999). Stratification in stars is also due to mean molecular weight gradients. The composition part of the Brunt-Väisälä frequency $N_\mu^2 = \frac{g\phi}{H_P}\nabla_\mu = -g\frac{d\ln\mu}{dr}$ is unaffected by thermal diffusivity. However, it can be weakened through the action of strong horizontal diffusion (Talon & Zahn 1997). Meynet & Maeder (1997) have shown that such an effect has to be taken into account if rotational mixing is to account for abundance anomalies in massive stars.

Small eddies are easier to render unstable by these diffusive processes, but they lead to a limited amount of mixing. Very large eddies remain stable. A turbulent diffusion coefficient can be deduced which corresponds to the largest unstable eddies. Taking into account both thermal diffusivity and horizontal diffusion yields

$$\nu_v = \frac{8Ri_c}{5} \frac{(r\,d\Omega/dr)^2}{N_T^2/(K+D_h) + N_\mu^2/D_h}. \tag{3.1}$$

for the vertical viscosity, where $Ri_c \simeq 1/4$ is the critical Richardson number. The leading factor $8Ri_c/5$ depends on the geometry (here assumed spherical) of the turbulent eddy and on the exact value of Ri_c. It is thus not well constrained only from theory and could

† For a discussion on the magnitude of the baroclinic instability which will be neglected here, see, e.g., Spruit & Knobloch (1984).

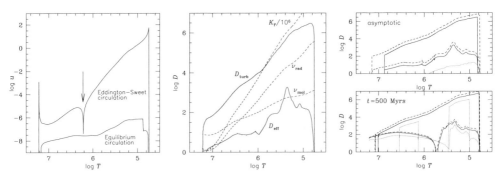

Figure 1. (*left*) Comparison of the Eddington-Sweet circulation velocity, including the reversed cell due to the Gratton-Öpik term (indicated by the arrow) and of the asymptotic solution described by (2.5), for a turbulent diffusion given by (3.1) and a surface velocity of 100 km s^{-1} in a 1.7 $M_\odot$ star. The magnitude of the circulation is reduced by several orders of magnitudes, while the reversed cell disappears totally, since the circulation must transport momentum outward to balance the turbulent diffusion. (*center*) Turbulent and effective diffusion corresponding to the asymptotic rotation state for a surface velocity of 100 km s^{-1}. (*right*) Turbulent and effective diffusion for various surface velocities (dashed - 150, full - 100, dotted - 50 km s^{-1}) in the asymptotic state, and in an evolving model. These models include a cutoff based on the critical Reynolds number (*cf.* Eq. 3.2).

vary by a factor of a few either way. Furthermore, this instability will grow only provided

$$\nu_v \gtrsim (\nu_{\mathrm{mol}} + \nu_{\mathrm{rad}}) Re_c \qquad (3.2)$$

where the critical Reynolds number is $Re_c \simeq 10$.

For horizontal shears there is no competing force preventing the growth of the instability. The magnitude of the horizontal turbulent viscosity is related to the magnitude of the horizontal differential rotation $\widehat{\Omega}$, which is constantly regenerated by meridional circulation. Zahn (1992) originally suggested that it should be proportional to the horizonal shear

$$\nu_h \simeq \frac{1}{c_h} r \, |2V - \alpha U| \qquad (3.3)$$

and that c_h could be determined from observations, but it did not rely on a specific model. This prescription has recently been reviewed by Maeder (2003) and Mathis *et al.* (2004). The first paper proposes a formulation in which the viscosity is based on the dissipation and feeding of turbulent energy while the second relies on experimental results. Both lead to the same order of magnitude for ν_h. However, the second formulation is smoother and will thus be preferred here:

$$\nu_h = \left(\frac{\beta}{10}\right)^{1/2} \left(r^2 \Omega\right)^{1/2} \left(r \, |2V - \alpha U|\right)^{1/2} \qquad (3.4)$$

with $\beta \simeq 1.5 \times 10^{-5}$ derived from laboratory measures (α is a measure of differential rotation). Here again, one may wonder whether this coefficient is correct for compressible fluids and a modification by a factor of a few cannot be ruled out *a priori*. Recent models by Maeder (private com.) suggest that this prescription is actually too strong and produces too little mixing in massive stars; a value between (3.3) and (3.4) should be preferred. This also suggest that horizontal differential rotation might have to be considered in future work (Mathis & Zahn 2004).

4. Rotational mixing along the HR diagram

To understand the varying importance of mixing along the HR diagram, one may discuss various time-scales involved with mixing. First, let us note that the time spent on the main sequence τ_{ms} varies as $\tau_{ms} \propto M^{-0.6}$ for $M \gtrsim 10\,M_\odot$ and $\tau_{ms} \propto M^{-3}$ for $M \lesssim 10\,M_\odot$ (Schaller *et al.* 1992). The time-scale for turbulent mixing is of order

$$\tau_{turb} \simeq \frac{R^2}{D}. \tag{4.1}$$

If turbulence is dominated by shears, as suggested here, then we have $D \propto K$. Classical scalings (see, e.g., Hansen & Kawaler 1994) yield $R \propto M^{0.75}$ and $K \propto M^{3.25}$ for electron scattering opacities and $K \propto M^{5.3}$ for Kramers opacities. Thus, we get

$$\tau_{turb} \propto M^{-1.75} \text{ to } M^{-3.8} \tag{4.2}$$

(Maeder 1998). This implies that the level of mixing (measured by τ_{ms}/τ_{turb}) rises with mass above $\sim 10M_\odot$ and stays about constant in lower mass stars for a given rotation profile. In the framework of rotational mixing, this explains why evidences of deep mixing are observed only in massive stars while "surface" abundance anomalies driven by microscopic diffusion occur in lower masses.

One must also consider the Eddington-Sweet time-scale, which gives an estimate of the time required to reach the asymptotic rotation profile defined by (2.5)

$$\tau_{E.S.} \simeq \frac{R}{U} \propto \frac{1}{\alpha} \frac{M^2}{LR}. \tag{4.3}$$

Using the relation $L \propto M^{2.25}$ for $M \gtrsim 10M_\odot$ and $L \propto M^4$ for $M \lesssim 10M_\odot$, we get

$$\tau_{E.S.} \propto M^{-2.5} \text{ for low mass} \quad \propto M^{-0.75} \text{ for high mass stars} \tag{4.4}$$

for a given rotation velocity $V = \Omega R$. These time-scales vary with mass similarly to the main sequence lifetime.

For a $M = 1.7\,M_\odot$ star rotating at $100\,\mathrm{km\,s^{-1}}$, $\tau_{E.S.} \simeq 10\%\,\tau_{ms}$. Thus, for fast rotators, the initial rotation profile has little impact on the main sequence evolution. However, if the rotation velocity drops to $30\,\mathrm{km\,s^{-1}}$, we get $\tau_{E.S.} \simeq \tau_{ms}$. This has received little attention from theoreticians of rotational mixing so far; it implies that, in very slow rotators, turbulent diffusion should not drop as fast as Ω^2.

Figure 2 illustrates the impact of the initial rotation state on the diffusion coefficient. Two cases are shown:

(*a*) asymptotic rotation state according to (2.5) on the main sequence;
(*b*) solid body rotation in a fully convective star at the top of the Hayashi track.

If meridional circulation and turbulence are the only processes contributing to the transport of momentum, condition (*b*) should apply. The initial rotation profile could be made somewhat smoother by including angular momentum transport by internal gravity waves, since the convective envelope is deep during the pre-main sequence phase (see Talon & Charbonnel 2003) or some large scale magnetic field (see, e.g., MacGregor & Brenner 1991). It would also be smoother in a star that completed its accretion phase after the formation of its radiative core (see, e.g., Stahler 1983). Condition (*a*) which would require strong coupling between the radiative and the convective zone during the pre-main sequence phase is not expected to be realistic from a theoretical point of view. Furthermore, recent observations of the evolution of the surface velocity from the birth line to the ZAMS favor core-envelope decoupling during the pre-main sequence as a general rule (Wolff, Strom & Hillenbrand 2004).

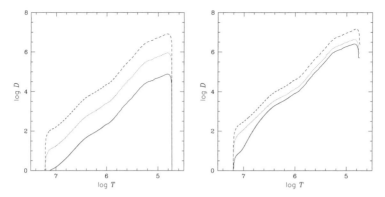

Figure 2. (*left*) The equilibrium turbulent diffusion profile on the ZAMS. (*right*) The turbulent diffusion profile at 500 Myrs with angular momentum evolution from the PMS. No cutoff based on the Reynolds number has been applied here (*cf.* Eq. 3.2) and the velocities shown here are denoted by different lines (dashed - 150, dotted - 50, and full - 15 km s^{-1}).

When differential rotation on the PMS is considered, mixing in slow rotators increases. This is related to the larger differential rotation, linked to the longer Eddington-Sweet time-scale. This could explain why abundance anomalies do not depend on the rotation velocity for slow rotators (see, e.g. Varenne & Monier 1999).

5. Boundary conditions

To complete the mathematical system of equations, one must also specify boundary conditions. Since the problem is described by a fourth order differential equation, four conditions are required. The first two impose momentum conservation at convective boundaries

$$\frac{\partial}{\partial t}\left[\Omega \int_{r_t}^{R} r^4 \rho \, dr\right] = -\frac{1}{5} r^4 \rho \Omega U + \mathcal{F}_\Omega \qquad \text{for } r = r_t$$

$$\frac{\partial}{\partial t}\left[\Omega \int_{0}^{r_b} r^4 \rho \, dr\right] = \frac{1}{5} r^4 \rho \Omega U \qquad \text{for } r = r_b$$

where r_t and r_b are respectively the top and the bottom of the convection zone, which are assumed to rotate as solid bodies†. Here, we also allow explicitly for a torque $\mathcal{F}_\Omega$ to be applied at the stellar surface, as would be the case in an asynchronous binary system, or in a star subject to braking due to a magnetic wind. If such braking occurs, it will lead to an increase of both meridionnal circulation and shear turbulence and enhance significantly mixing close to the surface. For A stars, we will assume $\mathcal{F}_\Omega = 0$.

For the other two conditions,

$$\frac{d\Omega}{dr} = 0 \quad \text{for } r = r_t, \ r_b \qquad\qquad\qquad (5.1)$$

has generally been used. This condition actually states that the differential rotation just below the convection zone should be the same as in the convection zone; in a 1D model, it is equivalent to the stress-free condition of hydrodynamics. However, it is unclear whether there is no radial differential rotation at the boundary of convective regions, and thus the impact of the choice of boundary condition should also be examined.

† This is motivated by the fact that the Sun presents little differential rotation in radius in its convection zone.

6. Mass loss

Another parameter to take into account in modeling is the presence of some amount of mass loss at the stellar surface. If the star looses some mass, the inner layers will constantly have to move outward to replace those that were lost to the stellar wind. This will have two impacts on the transport of chemicals:

• Firstly, the homogeneous drift of those surface layers will reduce the efficiency of element separation by microscopic diffusion;

• Secondly, rotational mixing will be enhanced by the required transport of momentum to the surface layers that are spun down by local angular momentum conservation as they drift upward.

As far as theory is concerned, the nature of mass loss in A stars is not well understood. It should not come from a hot corona as in the case of the Sun (Parker 1960) due to the absence of a deep surface convection zone†; radiative accelerations are also believed to be too small to produce significant mass loss (Abbott 1982) in main sequence A stars. Babel (1995) studied the case of A stars in more details. He concludes that, for an effective temperature between $8000 \lesssim T_{\mathrm{eff}} \lesssim 14000$ K, only metals can be lifted by radiative forces, giving rise to fully seperated winds, consisting only of metals. From complete calculations made for an A star at 10000 K, he finds a wind of $\dot{M} \lesssim 10^{-16} \ M_{\odot} \ \mathrm{yr}^{-1}$. Such winds would lead to very different anomalies, as the star could be drained only of some metals (Michaud & Charland 1986).

From an observational viewpoint, mass loss is very difficult to measure directly in A stars. Bertin *et al.* (1995) derived a mass loss rate in Sirius A between 2×10^{-13} and $1.5 \times 10^{-12} \ M_{\odot}\mathrm{yr}^{-1}$ based on Mg II lines. However, only the lower limit would permit surface abundance anomalies to form; for higher mass loss rates, matter would leave the surface of the star before microscopic diffusion had time to act (Babel & Michaud 1991). It has been suggested that surface abundance anomalies in A stars could represent a means to measure mass loss rates, if they are produced solely by the competition of mass loss and microscopic diffusion (see, e.g., Landstreet *et al.* 1998). Babel (1993) suggests that, in Ap stars, inhomogeneous mass loss modulated by surface magnetic fields can be responsible for "patches" in the surface abundance of various elements.

7. Observational constraints from Am stars

Rotation plays an important role in the appearance of the Am phenomenon. Indeed, only slow rotators bear the characteristic signature of radiative forces and settling (see, e.g., the review by Conti (1970)). Furthermore, it seems that slow rotation (below $100 \ \mathrm{km\,s^{-1}}$) is a necessary and sufficient condition for late Am stars, and a necessary (but not sufficient) condition for early Am stars.

This must be reflected in the rotating models and give strong constraints on the coefficients for D_v (*cf.* Eq. 3.1) and D_h (*cf.* Eq. 3.4) as well as on the initial rotation profile (*cf.* Fig. 2).

7.1. *Meridional circulation and mixing in A stars*

The efficiency of rotational mixing changes during the main sequence. It is initially rather large, since the pre-main sequence contraction leads to the formation of a rapidly rotating core. This phase lasts for an Eddington-Sweet time-scale (4.3). After the asymptotic state is reached, differential rotation in the stellar interior rises between the ZAMS and the TAMS. This is related to the increase of the ratio of stellar luminosity to surface gravity

---

† Recent observational evidence indicate there could be some activity in late A stars.

during the same period (see Eq. 2.2) which induces an enhancement of the meridional circulation. We thus expect the presence of a minimum in the efficiency of rotational mixing occurring at $t = \tau_{\mathrm{E.S.}}$.

Another delicate issue to consider is the role of mean molecular weight gradients. It is well known that these gradients act to reduce the growth of instabilities, leading to a reduction of the turbulent diffusion coefficients presented here. This could be an important ingredient in the formation of Am stars.

Another point that has been neglected here is the role of horizontal diffusion in the reduction of the efficiency of vertical turbulence in the transport of chemicals (*cf.* Vincent *et al.* 1996). The importance of this effect will be discussed in a future study (Talon, Richard & Michaud, in preparation).

Acknowledgements

S.T. is supported by NSERC of Canada.

References

Abbott, D. C., 1982 *ApJ* **259**, 282.
Babel, J., 1983 In *Peculiar versus Normal Phenomena in A-type and Related Stars*, IAU Coll. 138, PASP, ed. M.M. Dworetsky, F. Castelli, R. Faraggiana, 458.
Babel, J., 1995 *A&A* **301**, 823.
Babel, J. & Michaud, G., 1991 *ApJ* **241**, 493.
Bertin, P., Lamers, H.J.G.L.M., Vidal-Madjar, A., Ferlet, R. & Lallement, R., 1995 *A&A* **302**, 899.
Chaboyer, B. & Zahn, J.-P., 1992 *A&A* **253**, 173.
Charbonneau, P. & Michaud, G., 1991 *ApJ* **370**, 693.
Conti, P. S., 1970 *PASP* **82**, 488.
Dudis, J. J., 1974 *J. Fluid Mech.* **64**, 65.
Endal, A. S. & Sofia, S., 1976 *ApJ* **210**, 184.
Endal, A. S. & Sofia, S., 1978 *ApJ* **220**, 279.
Endal, A. S. & Sofia, S., 1981 *ApJ* **243**, 625.
Hansen, C. J. & Kawaler S. D. , 1994 *Stellar Interiors* Ed. Springer-Verlag, A&A Library.
Landstreet, J. D., Dolez, N. & Vauclair, S., 1998 *A&A* **333**, 977.
MacGregor, K. B. & Brenner, M., 1991 *ApJ* **376**, 204.
Maeder, A., 1998 In *Properties of Hot Luminous Stars*, ASP Conf. Series, **131**, ed. Ian D. Howarth, 85.
Maeder, A., 2003 *A&A* **399**, 263.
Maeder, A. & Zahn, J.-P., 1998 *A&A* **333**, 977.
Mathis, S., Palacios, A. & Zahn, J.-P., 2004 *A&A* submitted.
Meynet, G. & Maeder, A., 1997 *A&A* **321**, 465.
Michaud, G., 1982 *ApJ* **258**, 349.
Michaud, G. & Charland, Y., 1986 *ApJ* **311**, 326.
Pinsonneault, M. H., Kawaler, S. D., Sofia, S. & Demarque, P., 1989 *ApJ* **338**, 424.
Rayleigh, Lord, 1880 *Proc. London Math. Soc.* **11**, 57.
Richard, D. & Zahn, J.-P., 1999 *A&A* **347**, 734.
Schaller, G., Schaerer, D., Meynet, G. & Maeder, A., 1992, *A&AS* **96**, 269.
Stahler, S. W., 1983 *ApJ* **822**, 274.
Spruit, H. C. & Knobloch, E., 1984 *A&A* **132**, 89.
Talon, S. & Charbonnel, C., 2003 *A&A* **405**, 1025.
Talon, S. & Zahn, J.-P., 1997 *A&A* **317**, 749.
Towsend, A. A., 1958 *J. Fluid Mech.* **4**, 361.
Varenne, O. & Monier, R., 1999 *A&A* **351**, 247.
Vincent, A., Michaud, G. & Meneguzzi, M., 1996 *Phys. Fluids* **8** (**5**) 1312.

Watson, M., 1981 *Geophys. Astrophys. Fluid Dyn.* **161**, 285.
Wolff, S. C., Strom, S. E. & Hillenbrand, L. A., 2004 *ApJ* **601**, 979.
Zahn, J.-P., 1992 *A&A* **265**, 115.

Discussion

KHALACK: Have you investigated the influence of the magnetic field on the shellular structure of a modelled star? How will such a star evolve?

TALON: The set of equations presented here does not take into account internal magnetic forces. However, one can take into account the combined effect of mass loss and the surface magnetic field that will act to brake the star's surface. This leads to an increase of the internal (surface) differential rotation, and thus, produces extra mixing.

KUBÁT: How could the stellar oblateness of rapidly rotating stars affect the processes in stellar envelopes?

TALON: The oblateness becomes significant (of order 10%) for stars rotating faster than about 150 $\mathrm{km\,s}^{-1}$. In that case, the temperature of the atmosphere will vary accoring to latitude.

The A-Star Puzzle
Proceedings IAU Symposium No. 224, 2004
J. Zverko, J. Žižňovský, S.J. Adelman, & W.W. Weiss, eds.
© 2004 International Astronomical Union
DOI: 10.1017/S1743921304004399

Pre-Main-Sequence A-type stars

Marcella Marconi[1] and Francesco Palla[2]

[1]INAF-Osservatorio Astronomico di Capodimonte, Via Moiariello 16, 80131 Napoli, Italy
email: marcella@na.astro.it

[2]INAF-Osservatorio Astrofisico di Arcetri, Largo E. Fermi 5, 50125 Firenze, Italy
email: palla@arcetri.astro.it

Abstract. Young A-type stars in the pre–main-sequence (PMS) evolutionary phase are particularly interesting objects since they cover the mass range ($\sim$1.5-4 $M_\odot$) which is most sensitive to the internal conditions inherited from the protostellar phase. In particular, they undergo a process of thermal relaxation from which they emerge as fully radiative objects contracting towards the Main Sequence. A-type stars also show intense surface activity (including winds, accretion, pulsations) whose origin is still not completely understood, and infrared excesses related to the presence of circumstellar disks and envelopes. Disks display significant evolution in the dust properties, likely signalling the occurrence of protoplanetary growth. Finally, A-type stars are generally found in multiple systems and small aggregates with lower mass companions.

Keywords. Stars: evolution, stars: Pre–Main-Sequence, stars: oscillations (including pulsations), stars: variables: δ Scuti

1. Introduction

Young A-type stars do not resemble their mature siblings. As a class, they are known as Herbig Ae stars (Herbig 1960) because of the presence of optical emission lines in their spectra, their association with nebulosities, and conspicuous infrared excesses in the spectral energy distributions. Their early evolution is marked by the occurrence of a variety of phenomena that disappear in the course of time. When they emerge from the protostellar phase, their internal structure is highly unrelaxed and must undergo a global readjustment. Unlike low-mass protostars which are fully convective, intermediate mass protostars (in the mass range $\sim$1.5 to 4 $M_\odot$) have developed a radiatively stable core and an outer convective region where deuterium burns in a shell (Palla & Stahler 1990). As a result, stars in this mass range begin with a modest surface luminosity and undergo thermal relaxation in which the central regions contract while transferring heat to the expanding external regions. Then, in a short time, the star acquires its full luminosity and begins contracting towards the Main Sequence (Palla & Stahler 1993).

Another distinction of young A stars is their surface activity. Winds are relatively common at rates $\sim 10^{-8}-10^{-7}\ M_\odot\,\mathrm{yr}^{-1}$, but their origin is not understood (Corcoran & Ray 1997, Catala & Boehm 1994). Evidence for accretion at similar rates is also available from ultraviolet lines and redshifted Lyman α lines (Deleuil *et al.* 2004). Equally unexplained is the X-ray emission with luminosities intermediate between those in T Tauri and massive stars (Hamaguchi *et al.* 2001). Also, young A stars rotate more rapidly than lower mass objects, but still significantly below breakup and below the values observed in Main Sequence stars of the same spectral type (Boehm & Catala 1995). Thus, they should spin up considerably during (PMS) contraction, assuming conservation of angular momentum. Additionally, their outer layers are subject to a δ Scuti-like pulsational instability, albeit for a limited amount of time (Marconi & Palla 1998). Finally, and quite importantly,

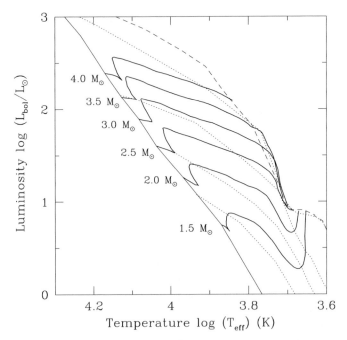

Figure 1. HR diagram of stars in the mass range 1.5–4 $M_\odot$. The evolutionary tracks start at the birthline (dashed line) and end on the ZAMS (thin solid line). Selected isochrones (dotted lines) are for 0.1, 1, 3, and 10 Myr (from top to bottom). (Adapted from Palla & Stahler 1999)

circumstellar disks are almost invariably found around them with characteristics that indicate significant evolution during the PMS phase.

Once they reach the Main Sequence, A-type stars continue to show interesting properties, but most of the excitement is gone. The major changes take place in a very short time, from ~17 Myr for a 1.5 $M_\odot$ star to ~1 Myr for a 3.5 $M_\odot$ object. Now we shall provide an overview of the main properties and the many puzzles that characterize the brief, but intense early life of A stars.

2. Evolutionary properties

As stated above, the initial phases of PMS evolution are marked by the persistence of the conditions inherited from protostellar accretion. Intermediate mass stars are the most affected by the prior history and their evolution departs significantly from the standard results established in the works of Hayashi, Iben and collaborators in the 1960s. A more modern description of PMS evolution in the Hertzprung-Russell (HR) diagram from the birthline to the ZAMS is shown in Figure 1 (Palla & Stahler 1999). The birthline is the locus where stars first appear in the HR diagram after the main accretion phase and the results displayed here have been obtained assuming a protostellar mass accretion rate of $\dot{M}_{\rm acc} = 10^{-5}\ M_\odot\,{\rm yr}^{-1}$, typical of the observed values in dense molecular cores undergoing dynamical collapse (e.g., Lee *et al.* 2004). Although quantitatively the initial conditions change with $\dot{M}_{\rm acc}$, the qualitative behavior of protostars is not changed substantially (see a discussion in Palla 2002).

The main feature of the new tracks of Figure 1 is the complete absence of a fully convective phase. Starting at about 1.5 $M_\odot$, PMS stars have almost fully depleted their initial deuterium reservoir and, due to their small radius (~4 $R_\odot$), do not have any

Table 1. Evolution of intermediate-mass PMS stars

Property	1.5 $M_\odot$	2.0 $M_\odot$	2.5 $M_\odot$	3.0 $M_\odot$	3.5 $M_\odot$	4.0 $M_\odot$
R_{init} ($R_\odot$)	5.0	4.4	4.1	4.0	4.2	7.8
R_{ZAMS} ($R_\odot$)	1.7	1.8	2.1	2.3	2.5	2.7
$\Delta M_{con}^{\mathrm{init}}$ ($M_\odot$)	1.5	2.0	1.8	1.1	0.7	...
$\Delta M_{con}^{\mathrm{ZAMS}}$ ($M_\odot$)	0.2	0.3	0.5	0.6	0.7	0.9
t_{rad} (Myr)	9.4	3.4	1.3	0.4	0.07	...
t_{ZAMS} (Myr)	17	8.4	3.9	2.0	1.3	0.8

energy sources (nuclear or cooling from the surface) to maintain convection. Thus, their evolution starts near the bottom of the Hayashi phase and quickly joins the radiative tracks of homologous contraction.

Stars in the range 2.5–3.5 $M_\odot$ begin their evolution at low luminosity and move up to join the radiative track. In fact, the tracks loop behind the birthline and then recross it before pursuing the horizontal paths. These stars undergo a nonhomologous contraction and a thermal relaxation during which the radius, luminosity, and effective temperature all increase, while the outer convection disappears. Note also the crowding of the tracks that all start from about the same position before moving up vertically. Thus, assigning a correct mass to stars in this part of the HR diagram is tricky.

Finally, stars of mass $\gtrsim 4$ $M_\odot$ appear immediately on the radiative track and begin to contract homologously under their own gravity. As the star contracts, the average interior luminosity rises as $R_*^{-1/2}$, as is evident from the slope of the tracks.

Some important properties of the evolution, both at the beginning and the end of the PMS phase, are given in Table 1. The first two rows list the stellar radii on the birthline and on the ZAMS. Note the similarity of the values of R_{init} that explains the crowding of the tracks in Figure 1. The table also lists ΔM_{con}, the mass in the convection zone. For the ZAMS models, this convection is located in the center and is due to the strongly temperature-sensitive CN burning. Finally, t_{rad} is the time (in units of 10^6 yr) when the star becomes fully radiative, measured relative to the star's appearance on the birthline. Note the dramatic reduction of t_{rad} for stars more massive than ~ 3 $M_\odot$. The last entry, t_{ZAMS}, gives the total duration of the PMS phase.

3. Circumstellar disks around A stars

Once the main phase of accretion is completed, the stellar core emerges as an optically visible star along the birthline. The circumstellar matter around the star, partly distributed in a disk and the rest in an extended envelope, still emits copiously at infrared and millimeter wavelengths. The spectral energy distributions of Herbig Ae stars resemble those observed in classical T Tauri stars (CTTSs), thus suggesting that the dust responsible for the thermal emission has similar properties and geometrical distribution (e.g., Hillenbrand *et al.* 1992). However, there are important differences. Unlike CTTSs, a significant fraction of the luminosity is emitted at short wavelengths with a prominent peak at 2–3 μm (Meeus *et al.* 2001). These properties cannot be explained in terms of standard disk models where the dust is heated by viscosity and stellar radiation, and is distributed in optically thick and geometrically thin or flared disks. Thus, some basic modifications of the disk structure are required. For example, Dullemond *et al.* (2001) have suggested that the innermost regions where the dust sublimates (few tenths of AU)

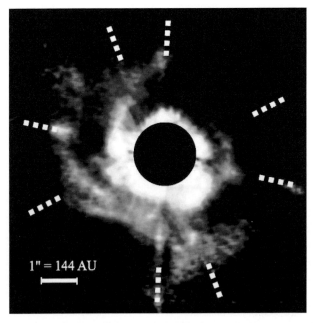

Figure 2. Disk around AB Aur. H-band image at resolution 0.1 arcsec obtained with a coronagraphic imager on Subaru (Fukagawa *et al.* 2004).

are distributed in an optically thick puffed-up rim that shadows the optically thin dust at larger distances (which remains cool), thus explaining the strong NIR emission.

The predictions of these models and the geometrical shape of the disks can now be directly probed by means of near- and mid-IR interferometry and adaptive optics on large telescopes that can resolve regions ∼0.1 to a few AU in size at the typical distances of 0.3–1 kpc. The initial results indicate that indeed the data are best reproduced by flared passive disks with puffed-up inner rims (Eisner *et al.* 2004, Leinert *et al.* 2004). A particularly striking example is the circumstellar disk around AB Aurigae shown in Figure 2. This A0 star is one of the closest (d=144 pc) and best studied Herbig Ae objects, with mass ∼2.5 $M_\odot$ and age ∼4 Myr. Millimeter observations of ^{13}CO gas have revealed the presence of a rotating disk ∼450 AU in radius and an estimated mass ∼0.02 $M_\odot$ (Mannings & Sargent 1998). The disk is immersed in an extended envelope (>1000 AU) visible in scattered light. The image in Figure 2 shows that the extended emission has a double spiral structure almost the same size of the CO disk and is possibly associated with the latter rather than the envelope (Fukagawa *et al.* 2004).

That disks are common around Ae stars is directly seen at millimeter wavelengths where spatially resolved images reveal disk features at scales of few hundred AU in CO emission (Dutrey 2004). Interestingly, while Keplerian disks are found between 75 and 100% of Ae stars, the percentage drops dramatically for the more massive Herbig Be stars. This is shown in the upper panel of Figure 3 that covers two orders of magnitude in stellar mass (from CTTSs to B0 type stars). A likely explanation of this trend is related to the rapid dispersal of the disks around more massive and luminous stars which are subject to strong UV radiation fields and more powerful outflows. Finally, the lower panel of Figure 3 reveals that although the disk mass increases with stellar mass, the ratio of the disk-to-star mass remains basically constant in the range 0.2–3 $M_\odot$, implying that all PMS stars have low-mass disks.

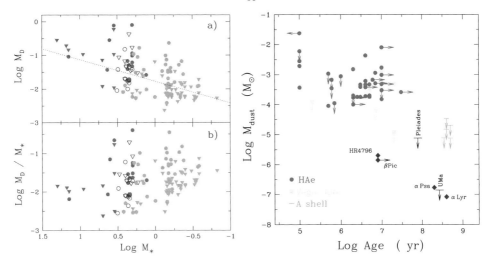

Figure 3. Left: Properties of disks around low- and intermediate-mass stars. *Upper panel*: disk mass as a function of the stellar mass as inferred from mm-interferometric observations. Filled and empty circles are detections, while triangles are upper limits. *Lower panel*: ratio of the disk-to-star mass *vs.* stellar mass (Natta 2004). Right: Evolution of the disk mass with time for A-type stars (Natta *et al.* 2000).

As a last property of disks, let us point out that A-type stars provide the strongest evidence for disk evolution during the PMS phase and on the MS. The time evolution of the disk mass is displayed in the right panel of Figure 3 covering more than 4 decades in stellar ages. For ages less than ~10 Myr, the disk mass is about constant. Then, in objects on the ZAMS, such as HR 4796 and β Pic, the disks are significantly reduced in mass and almost completely devoid of gas (so-called *debris disks*). At this stage, there is also evidence for grain growth in which dust gradually agglomerates into larger, more crystalline structures that should favor planetesimal condensation. The dust grains are subject to the Poyting-Robertson drag from stellar photons which causes them to spiral inward. Others must be resupplied, presumably through the mutual collisions of larger orbiting bodies. In fully evolved stars (α Lyr and α PsA, $\gtrsim 200$ Myr) the dust becomes undetectable in scattered light and visible only in the far-IR and its actual luminosity has declined to only $\sim 10^{-5} L_*$.

4. Interaction with the environment

Most Herbig Ae stars are found close to or embedded within their natal gas clouds. Studies of the distribution of the ^{13}CO emission reveal the pattern shown in Figure 4 where the youngest stars are still immersed in dense cores (top) while the older ones are found in cavities created during PMS evolution (Fuente *et al.* 2002). The physical process responsible for the removal of the dense gas is likely to be related to the activity of bipolar outflows during the protostellar phase. Bipolar outflows sweep out matter along the poles and creat a biconical cavity. Material in the envelope accretes onto a circumstellar disk that feeds the growing star. In this way, the cloud evolves toward a more centrally peaked morphology while a significant fraction of the core material is dispersed. Assuming a typical accretion rate of 10^{-5} $M_\odot$ yr^{-1}, an A-type star (2-4 $M_\odot$) will be formed in a few×0.1 Myr. By this time, about 90% of the dense core is dispersed. From a few×0.1 Myr to $\simeq 1$ Myr, the star begins the PMS contraction and only a small amount of circumstellar matter is removed because (a) the outflow activity fades rapidly

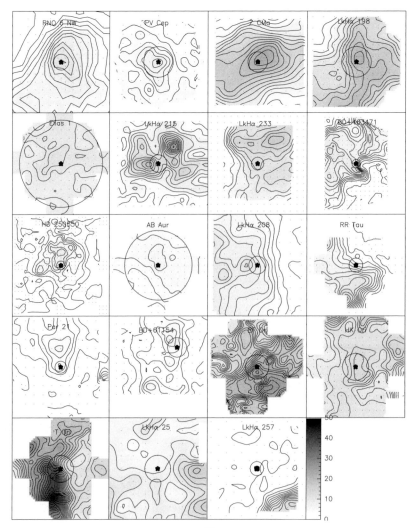

Figure 4. The birth sites of Ae stars seen in ^{13}CO emission. The size of the box varies from source to source. In each panel, a circle of 0.08 pc in radius is drawn around the central star (Fuente *et al.* 2002).

in time, and (b) the stellar surface is still too cold to generate UV photons that can photodissociate the surrounding gas. Finally, from $\geqslant 1$ Myr to ~ 10 Myr the star reaches the ZAMS and completes the creation of the cavity thanks to the UV radiation, albeit at a slow rate due to the low effective temperatures. This evolutionary sequence highlights once more the transitional character of A-type stars. The late-type A stars behave like T Tauri stars and their associated weak winds, whereas the early-type A stars begin to show the phenomenology (both radiative and mechanical) associated with the more luminous Be-type objects.

In addition to the properties of the gas, it is interesting to consider the issue of stellar companionship. It is well known that most stars are found in binary or multiple systems. Herbig Ae stars are no exception. The binary frequency of A0-A9 stars with a semi-major axis less than 2000 AU (corresponding to log P=5-7 days) is 42%, or 50-60% when corrected for incompleteness (Bouvier & Corporon 2001). For comparison, the binary

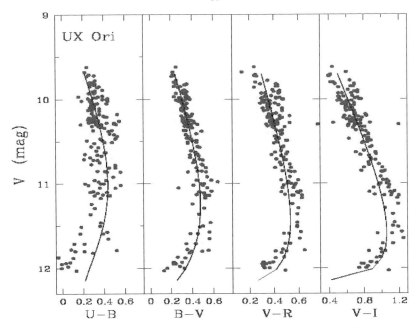

Figure 5. Color-magnitude diagrams of UX Ori showing the blueing effect at minimum
(Rostopchina *et al.* 1999).

frequency of field G dwarfs in the same period interval is only 20%. Thus, binary Herbig
Ae stars exceed binary G-type stars by a factor of 2 and show an higher percentage than
binary T Tauri stars. This difference certainly has some bearing on the formation process.
In particular, it is possible that Herbig Ae stars form in more crowded environments
than lower mass stars and that dynamical interactions lead to capture of neighboring
stars. Indeed, models of the evolution of small clusters predict an increase of the binary
frequency with the mass of the primary. Interestingly, Ae stars tend to be found in
aggregates with density $\lesssim 10^2$ stars pc^{-3} (Testi *et al.* 1999), higher than the isolated
regime of T Tauri stars ($\lesssim 10$ stars pc^{-3}), but lower than the clusters associated with
massive stars ($\gtrsim 10^3$ stars pc^{-3}).

5. Variability

Young A-type stars show evidence of a dusty environment, intense stellar activity
and strong stellar winds. As a result they are photometrically, spectroscopically and
polarimetrically variable on very different time scales and wavelength domains (e.g.,
Herbst & Shevchenko 1999). The typical long-term variability due to obscuration by
circumstellar dust is called *UX Ori* variability from the name of the prototype object
(see Figure 5). The periodicity has a time scale on the order of ~1 yr and can be as large
as 3 magnitudes in the V band. After a few days at minimum, the star resumes its normal
brightness in several weeks. During the approach to minimum, the color becomes bluer
and the polarization increases dramatically. Both effects suggest that the origin of the
dimming is due to scattering by circumstellar dust distributed in optically thick clouds
that partially occult the star (e.g., Grinin *et al.* 1991). The other type of variability takes
place on shorter time scales (typically, few hours) and involves much smaller variations
(few thousandths to few hundredths mag). Since the variability is associated with the
κ mechanism, its study allows to peer into the the intrinsic properties (and through
asteroseismology also the internal structure) of A-type stars, as we will describe next.

Table 2. Pulsation properties of known or suspected PMS δ Scuti stars.

Name	F1 ($\mathrm{c\,d^{-1}}$)	F2 ($\mathrm{c\,d^{-1}}$)	F3 ($\mathrm{c\,d^{-1}}$)	F4 ($\mathrm{c\,d^{-1}}$)	ΔV (mag)	V (mag)	Sp.type	Ref.
V588 Mon	7.1865 ±0.0006	?	?		0.04	9.7	A7	1
V589 Mon	7.4385 ±0.0006	?	?		0.04	10.3	F2	1
HR 5999	4.812±0.010				0.02	7.0	A7	2
HD 104237	33 ± 0.2				0.02	6.6	A7	3
HD 35929	5.10 ± 0.13				0.02	8.1	A5	4
V351 Ori	15.687 ±0.002	13.337 ±0.002	16.868 ±0.002	11.780 ±0.002	0.1	8.9	A7	5
BL 50	13.9175 ±0.0005	9.8878 ±0.0009			0.02	14.5	–	6
HP 57	12.72557 ±0.0002	15.52437 ±0.0003			0.03	14.6	–	6
HD 142666	21.43±3				0.01	8.8	A8	7
V346 Ori	35.3±2.3	22.6±2.7	45.5±2.5	18.3±2.5	0.015	10.1	A5	8
H254	7.406±0.008				0.02	10.6	F0	9
NGC 6383-4	14.376	19.436	13.766	8.295	0.014	12.61	A7	10
NGC 6383-T55	19.024				0.002	20.9	–	10
IP Per	22.887	34.599	30.449	48.227	0.004	10.35	A7	11
V375 Lac	5.20±0.14	2.40±0.14	9.90±0.14		0.004	13.56	A7	12
VV Ser	5.15±0.01	8.61±0.01	4.46±0.01		0.005	11.87	A2(?)	12
BN Ori	10-12.7				0.002	9.67	F2 (?)	12
BF Ori	5.7±0.3				0.006	10.41	A5 (?)	12
HD34282	79.5±0.06	71.3±0.06			0.011	9.873	A0-A3	13
IC 4996-37	31.87				0.005	15.302	A5	14
IC 4996-40	42.89				0.008	15.028	A4	14

Sources: (1) Breger (1972), Peña *et al.* (2002), (2) Kurtz & Marang (1995), Kurtz & Catala (2001), (3) Donati *et al.* (1997), Kurtz & Muller (1999), (4) Marconi *et al.* (2000), (5) Ripepi *et al.* (2003), (6) Pigulski *et al.* (2000), (7) Kurtz & Müller (2001), (8) Pinheiro *et al.* (2003), (9) Ripepi *et al.* (2002), (10) Zwintz *et al.* (2004), (11) Ripepi *et al.* (2004), (12) Bernabei *et al.* (2004), (13) Amado *et al.* (2004), (14) Zwintz (2004, private communication)

5.1. *The δ Scuti type pulsation*

During the contraction phase toward the Main Sequence, intermediate-mass stars cross the pulsation instability strip of more evolved variables, suggesting that, despite the relatively short time spent in the strip ($\sim 10^5$-10^6 years), at least part of the observed activity could be due to intrinsic variability.

The first observational evidence for such variability was due to Breger (1972) who detected δ Scuti-like pulsations in two Herbig stars of the young cluster NGC 2264, namely V588 Mon and V589 Mon. The issue was reconsidered more than 20 years later by Kurtz & Marang (1995) and Donati *et al.* (1997) who found δ Scuti-like variability in the Herbig stars HR 5999 and HD 104237. Since then there has been a renewed interest in the study of these young pulsators, both from the observational and the theoretical points of view. For the latter, using convective nonlinear models, Marconi & Palla (1998) computed the first theoretical instability strip for PMS δ Scuti stars (see Figure 6). They also identified a list of candidates with spectral types in the range of the predicted instability region. This theoretical investigation stimulated new observational programs

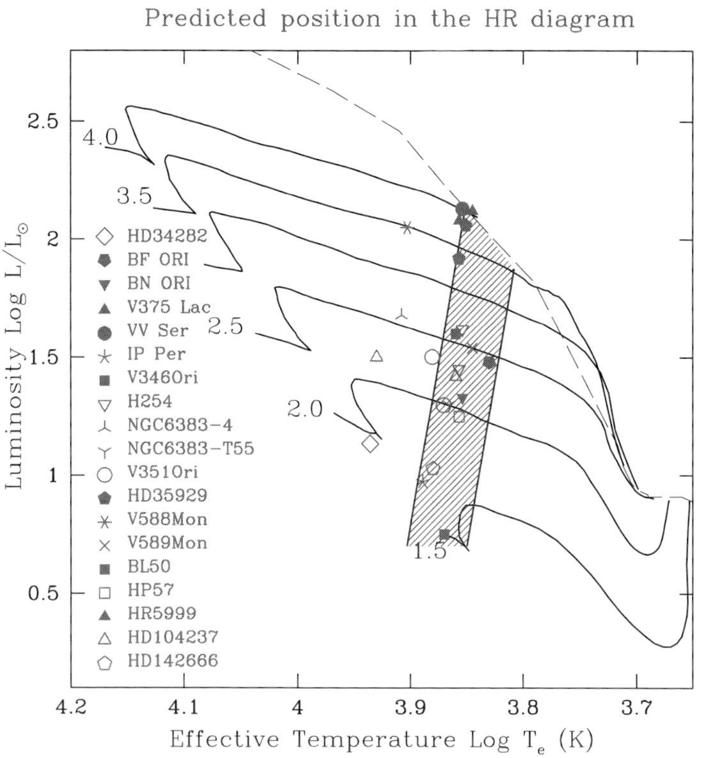

Predicted position in the HR diagram

Figure 6. The position of PMS δ Scuti stars in the HR diagram as predicted on the basis of the comparison between the observed periodicities and linear nonadiabatic radial pulsation models. The dashed region is the theoretical instability strip for the first three radial modes (Marconi & Palla 1998), that is the region between the second overtone blue edge and the fundamental red edge.

carried out by various groups with the result that the current number of known or suspected candidates amounts to about 20 stars.

A census of known or suspected PMS δ Scuti stars is reported in Table 2, whereas the position of these pulsators in the HR diagram, as resulting from the comparison between the observed and predicted pulsation frequencies, is shown in Figure 6. We notice the good agreement between the predicted instability strip and the location of observed pulsators. The only deviating objects are hotter than the theoretical second overtone blue edge and are indeed predicted to pulsate in higher overtones, whereas no pulsating object is found to the right of the theoretical red edge. The main limitations of this approach are: 1) the uncertainties still affecting many of the observed frequencies, due to poor data quality and/or the aliasing problem; 2) the difficulty of discriminating between PMS and post-MS evolutionary phases on the basis of radial models, in particular for pulsators that are predicted to be located close to the MS; and 3) that the likely presence of nonradial modes is not taken into account.

Concerning the first point, significant improvements can be obtained by means of multisite campaigns and will certainly be obtained with future satellite missions (e.g., EDDINGTON and COROT). As for the last two issues, it is clear that both radial and nonradial models should be computed to better understand the intrinsic properties of the rather unexplored class of young variable stars. To cope with this problem, we have started a project to apply the adiabatic nonradial code by Christensen-Dalsgaard (available on the web page http://astro.phys.au.dk/~jcd/adipack.n) to PMS evolutionary models. Preliminary results seem to suggest the coexistence of radial and nonradial frequencies at least in the young pulsators V351 Ori and IP Per.

6. Concluding remarks

Although PMS A-type stars cover a small mass interval ($\sim$1.5–4 $M_\odot$), they represent an interesting laboratory for the study of a variety of physical processes, many of which are still poorly understood. As a class, they share many of the characteristic properties of both the lower mass T Tauri stars and of the more massive Herbig Be objects, but depart from them in significant ways. Under the influence of protostellar evolution, A-type stars begin their PMS evolution as thermally unrelaxed structures that undergo a global readjustment from partially convective to fully radiative interiors in a short time scale. Circumstellar disks are much more frequent than in Herbig Be stars, and at least as common as in T Tauri stars. However, the more intense radiation field from the hotter stars makes standard viscous disk models insufficient to explain the observed infrared emission, requiring important modifications in the disk structure and geometry. Unlike Herbig Be stars, PMS A-type stars are generally found in relative isolation or in aggregates containing $\sim$10-100 lower mass members. Finally, their youth is marked by an intense surface activity, which includes winds, accretion, fast rotation, X-ray emission, variable obscuration from circumstellar dust and also intrinsic δ Scuti type pulsation. The latter, with its asteroseismological implications, can provide a unique tool to study their internal structure, to test evolutionary models, and to obtain independent estimates of the stellar mass, the fundamental parameter governing stellar evolution.

References

Breger, M. 1972, *ApJ*, 171, 539

Boehm, T. & Catala, C. 1995, *A&A*, 301, 155

Bouvier, J. & Corporon, P. 2001, *in The Formation of Binary Stars, IAU Symp. 200* 155

Catala, C. & Boehm, T. 1994, *A&A*, 290, 167

Corcoran, M. & Ray, T.P. 1997, *ApJ*, 324, 263

Deleuil, M., Lecavalier des Etangs, A., Bouret, J.-C. *et al.* 2004, *A&A*, 418, 577

Donati, J.-F., Semel, M., Carter, B.D., Rees, D.E., Collier Cameron, A. 1997, *MNRAS*, 291, 658

Dullemond, C.P., Dominik, C., Natta, A. 2001, *ApJ*, 560, 957

Dutrey, A. 2004, *in Star Formation at High Angular Resolution* IAU Symp. 221

Eisner, J.A., Lane, B.F., Hillenbrand, L.A., Akeson, R.L., Sargent, A.I., *astro-ph/0406356*

Fuente, A., Martin-Pintado, J., Bachiller *et al.* 2002, *A&A*, 387, 977

Fukagawa, M., Hayashi, M., Tamura, M., *et al.* 2004, *ApJ*, 605, L53

Grinin, V.P. *et al.* 1991, *Ap&SS*, 186, 283

Hamaguchi, K. *et al.* 2001, *in From Darkness to Light, A.S.P. Conf. Ser.* 243, 627

Herbig, G.H. 1960, *ApJS*, 4, 337

Herbst, W., Shevchenko, V.S. 1999, *AJ*, 118, 1043

Hillenbrand, L.A., Strom, S.E., Vrba, F.J., Keene, J. 1992, *ApJ*, 397, 613

Kurtz, D.W. & Marang, F. 1995, *MNRAS*, 276, 191

Lee, C.W., Myers, P.C., Plume, R. 2004, *ApJS*, 153, 523

Leinert, Ch., van Boekel, R., Waters, L.B.F.M., *et al.* 2004, *A&A*, 423, 537

Mannings, V. & Sargent, A.I. 1998, *ApJ*, 490, 792

Marconi, M. & Palla, F. 1998, *ApJ*, 507, L141

Meeus, G., Waters, L.B.F.M., Bouwman, J. *et al.* 2001, *A&A*, 365, 476

Natta, A. 2003, *in Debris Disks and the Formation of Planets, ASP Conf. Ser.*, 324, in press

Natta, A., Grinin, V.P., Mannings, V. 2000, *in Protostars and Planets IV* 559

Palla, F. 2002, in *Physics of Star Formation in Galaxies, Saas-Fee Adv. Cour. 39*, Springer 9

Palla, F. & Stahler, S.W. 1990, *ApJ*, 360, 47

Palla, F. & Stahler, S.W. 1993, *ApJ*, 418, 414

Palla, F. & Stahler, S.W. 1999, *ApJ*, 525, 772

Rostopchina, A.N., Grinin, V.P., Shakhovskoi, D.N. 1999, *AstL*, 25, 243

Testi, L., Palla, F. & Natta A. 1999, *A&A*, 342, 515

Discussion

BUDOVIČOVÁ: Why do Be stars lack gaseous disks? It is one of their basic properties. Why is it different from Ae stars (having a gas disk)?

MARCONI: We are talking of PMS A-type and B-type stars. The lower frequency of disks around the latter may be due to a rapid distruction due to photoevaporation and/or winds. However, there are recent detections of circumstellar disks around Herbig Be stars (e.g., Fuente *et al.* 2003 ApJ 598, 39).

PISKUNOV: T Tauri stars come in two flavours: classical and weak-line. Do we see a similar phenomenon in pre-MS A stars?

MARCONI: No, the same distinction does not hold for Herbig Ae (or Be) stars. Although many young clusters contain a large population of intermediate mass PMS stars, only a small minority of these objects qualify as being genuine Herbig-type stars. As George Herbig suggested, this may indicate that either the Ae/Be phenomenon is a temporary one, or that some stars in that mass range do not show it at all. In this sense there might be two classes of A-type PMS stars.

The A-Star Puzzle
Proceedings IAU Symposium No. 224, 2004
J. Zverko, J. Žižňovský, S.J. Adelman, & W.W. Weiss, eds.

© 2004 International Astronomical Union
DOI: 10.1017/S1743921304004405

Binarity as a tool for determining the physical properties and evolutionary aspects of A-stars

Mutlu Yıldız

Ege University, Dept. of Astronomy and Space Sciences, Bornova, 35100 Izmir, Turkey
email: yildiz@astronomy.sci.ege.edu.tr

Abstract. Double-lined-eclipsing binaries are the essential systems for the *measurement* of stellar masses and radii. About 50-60 systems have components (mostly A-stars) for which these values are known with an uncertainty less than 1-2%. Therefore, these systems are very suitable to improve our understanding of stellar structure and evolution. In this paper, special attention is given to the assessment of the role of internal rotation of the early-type stars in selected double-lined binaries (i.e., EK Cep, PV Cas, and θ^2 Tau): it is shown that adoption of rapidly rotating cores for such stars permits the models to be in very good agreement with the observational results including the apsidal advance rates.

Keywords. Stars: rotation, stars: interiors, stars: evolution, stars: early-type, binaries: eclipsing, stars: individual (EK Cep, PV Cas, θ^2 Tau)

---

1. Introduction

Life, death, and the evolution of a star depend mostly on its mass. This dependence occurs since the physical conditions in the central regions of stars are primarily determined by the total mass of the overlying layers and its distribution. While outgoing radiation (luminosity) is determined by cumulative effect of the physical conditions in the central regions, the radius is entirely a function of matter-radiation and matter-matter interactions and their consequences. In addition to the mass, the comprised stellar material and physical mechanisms, which affect the stellar structure equations or macroscopic quantities, also influence the stellar structure and consequently the evolution. In this regard, the observables of a model are also functions of chemical composition, rotation and magnetic field, which have secondary roles in comparison with the stellar mass. The latter two properties are difficult to handle in model computations and are, therefore, either entirely neglected or incorporated using extremely simple forms, such as solid-body rotation, despite their importance for the early-type stars. Therefore, this study concentrates on the internal rotational properties of these stars.

The most suitable systems for the assessment of the role of the secondary effects are the stellar systems for which the masses of the components are accurately determined. These systems are double-lined eclipsing binaries (DLEBs). We have very accurate observational data of more than 100 stars in these systems. Many of these binaries show apsidal motion, which puts further constraints on the models. Uncertainties in the mass and the radius are believed to be less than 1-2% (Andersen 1991, Harmanec 1988, Lacy *et al.* 2003, Popper 1980). There are many theoretical papers on the structure and the evolution of these stars (e.g., Claret & Gimenez 1993, Lastennet & Valls-Gabaud 2002, Pols *et al.* 1997, Yıldız 2003, 2004, Young *et al.* 2001), which mostly discussing the efficiency of the overshooting process (see Sect. 2).

For some double-lined visual binaries, we have good quality observational data. The masses of their components determined from observations, of course, are not as accurate as those of the double-lined eclipsing binaries. In addition, for modelling, these systems have two disadvantages: 1) no direct information about the radii of the components, and 2) the problem of dividing the total brightness between the components. But, if we have several such binaries as members of an open cluster, these disadvantages can be removed. In this respect, the Hyades is a very fruitful cluster (Lebreton *et al.* 2001): the individual masses of the components of its five double-lined binaries are observationally known (one eclipsing and four visual binaries). Two of the visual binaries, namely, θ^2 Tau and 51 Tau, have early-type components (Torres *et al.* 1997a). The primary (massive) component of θ^2 Tau is a δ Scuti star at the top of the cluster main-sequence. Hence, it is crucial for the determination of the cluster's age. However, its model properties should be determined together with a binary which has late-type components, to limit the chemical composition and to determine its physical properties and evolutionary phase (see section 5). ϕ 342 whose components are F-type stars is the most suitable binary system for this task (Torres *et al.* 1997b). F-type stars are most suitable stars for the standard modelling: Their convective zones are 1) fairly shallow which does not cause as many complications in the theoretical and the observational results, and 2) simultaneously are sufficiently deep for the effective loss of angular momentum.

A typical problem encountered in the determination of physical properties and evolutionary aspects of stars in binary systems by constructing models for the stellar interior, is due to the degeneracy in the HR diagram. That is to say, putting rotation and magnetic fields aside, at each point of the HRD, there are many different models with different chemical compositions and ages. This nonuniqueness is valid for many of the binary systems since the sensitivities of the models with different masses have very similar sensitivities to both the hydrogen and the heavy element abundances. Only in very special cases we can get (more or less) precise information concerning the structure and the evolutionary phases of the component stars: For example, binary systems with very different components (EK Cep, TY CrA, TZ Men, etc.) and systems with very small apsidal motion periods (PV Cas, GG Lup, ζ Phe, V760 Sco, etc.).

In this paper, we present and discuss results concerning the internal structure and evolutionary aspects of A- and B-type stars by constructing a rich variety of rotating models for them. The rest of the paper is organized as follows: The overshooting paradigm is discussed in Sect. 2. In Sect. 3, differential rotation is introduced for later use and the basic steps of our method for the construction of such models are presented. While model properties of the component stars in the two DLEBs (EK Cep and PV Cas) are discussed in Sect. 4, Sect. 5 is devoted to the visual binary θ^2 Tau of the Hyades. Finally, conclusions are drawn in Sect. 6.

2. The overshooting paradigm

As emphasized above, for the most part, the overshooting process is discussed in the studies on the early-type stars, especially in those of eclipsing binaries. The reason for this may be that it is easy in principle to apply it in model computations, although the phenomenon itself is surely very complex (see Roxburgh 1992) if not chaotic. The main effect of this process on the models looks essentially the same as that of a mixing process operating efficiently above the nuclear core. It is supposed that the convective cells penetrate into the radiative region and their length in this region is a *small* fraction of the pressure scale height (H_P) at the top of the convective core (r_{cc}). The adopted maximum value of this *small* fraction is 0.6 (Guinan *et al.* 2000), that is, a sphere with radius $r_{cc} + 0.6H_P$ is chemically homogeneous. The misleading point is that 0.6 may be considered as a small number, but, H_P is not. If things were so easy, in other words, if there was a single overshooting distance for a given stellar mass (or, in general, for all masses), then, after many applications to stellar systems, we should already found the proper value(s). Therefore, the problems we are facing are much more complicated and change star by star.

The seismic investigations of Aerts *et al.* (2003) and Dupret *et al.* (2004) show that the role of overshooting is at least exaggerated: they find the overshooting distance as $0.1\,H_P$. Furthermore, these seismic studies rule out solid-body rotation: the inner regions rotate faster than the outer regions.

3. Differential rotation

We should remember that differential rotation is accompanied by contraction due to gravity, provided that the angular momentum transfer in the stellar interior conditions is not a sudden process. In this respect, the most suitable stars, to determine the effect of differential rotation on the stellar structure, are the stars very close to either the zero-age-main-sequence (ZAMS) or the terminal-age-main-sequence (TAMS). While the primary component of EK Cep is a good example of a ZAMS star, θ^2 Tau A is a post-MS star, burning its hydrogen in the shell surrounding the He core.

Here we discuss the calibration method of a differentially rotating model for a given chemical composition. We assume that the model at the threshold of stability point, where it is completely convective (except for the photosphere), rotates like a solid-body, say, with an angular velocity Ω_0. As a radiative region forms and as a result of contraction, differential rotation begins, the most rapidly rotating region is that which has contracted the most, namely the central region. If angular momentum is conserved (or equivalently not transfered instantaneously in radiative regions), the luminosity and the radius of a model with such rotational properties are functions of Ω_0, for a given time.

As the synchronization for an early type star in a binary system starts from the surface regions (Zahn 1977, Goldreich & Nicholson 1989), the rotation profile determined by contraction is changed in the outermost regions whose mass fraction is M_s. If M_s is very small, then the luminosity is independent of M_s and the radius strongly depends on M_s. Otherwise, both the luminosity and the radius are very sensitive functions of M_s, depending upon how different are the rotation rates of the outer and the inner regions.

Since we have no information on the angular momentum flux throughout the stellar interior, to make the rotational velocity of the model at a given time agree with the observed (synchronized) velocity we directly modify the angular momentum of the outer region with M_s. To avoid the transitive course of a model reorganizing itself, it is better to make this modification in the pre-main-sequence (PMS) phase as soon as mass fraction of the receding convective envelope is less than M_s. Once, we determine the numerical derivatives of L and R of such a model, then, it is straightforward to calibrate it at any time to the observed values.

4. Double-lined eclipsing binaries

4.1. EK Cep

The fundamental properties of the components of EK Cep are well known (Andersen 1991): $M_A = 2.024 \pm 0.023\ M_\odot$, $R_A = 1.579 \pm 0.007\ R_\odot$, $\log(L_A/L_\odot) = 1.17 \pm 0.04$; $M_B = 1.121 \pm 0.012\ M_\odot$, $R_B = 1.315 \pm 0.006\ R_\odot$, and $\log(L_B/L_\odot) = 0.21 \pm 0.06$.

The most striking feature in this system is, which directs us toward differentially rotating models, that the ratio of the observed luminosities (L_A/L_B) of the components is smaller than the ratio found from the models (for example with solar compositions), in both the PMS and the MS phases (see Figure 2 in Yıldız (2003)). This is the case also for the ratio of the radii. These differences between the theoretical and the observational results cannot be removed by an interplay between hydrogen (X) and heavy element abundances (Z), and imply some more radical difference between the structures of the components than expected within the standard theory of stellar evolution (see also Marques $et\ al.$ (2004)).

Adopting the most probable case, that is, the secondary star is at the point where its luminosity is the maximum near the MS while the primary star is exactly at the ZAMS point, three equations (one for coevalness of the components and two for radius and luminosity of the secondary) can be written down with three unknowns $(X, Z$ and the mixing-length parameter for the secondary component; see Yıldız (2003)). The solution of the three equations yields the following values (SET E in Yıldız 2003):

$$X = 0.614, \qquad Z = 0.0395, \qquad \alpha = 1.30. \tag{4.1}$$

Nonrotating models with these values are in very good agreement with the observational results for the secondary star and the apsidal motion of the system, but are not in agreement with the observed primary star: the model expectations for the primary are brighter and greater than the observed values. These discrepancies can be eliminated by assuming a rapidly rotating core for this star which is a natural result of contraction plus the synchronization process due to tidal interactions (Zahn 1977).

The agreement between the theoretical and the observational results is obtained if the central regions of the rotating model of EK Cep A with SET E rotate 65 times faster than the surface, and the outer 48% of the star's mass rotates synchronously (surface equatorial velocity $= 22\,\mathrm{km\,s^{-1}}$).

Indeed, the rotation profile we found depends on the chemical composition (or, on our assumption about the position of EK Cep A relative to the ZAMS point). If the solar metallicity is assumed, then the core rotates 29 times faster than the surface and the synchronized outer mass is 10% of the total mass.

These two rotating models of EK Cep A (with SET E and B) have the same luminosity and radius (at different times; 26 My for SET E and 28.5 My for SET B), but the apsidal advance rates (AAR) computed from these models, using also the corresponding model of EK Cep B, are quite (but not exhaustively) different. While the AAR computed from the models with SET E is in perfect agreement with the observed AAR, the corresponding AAR for SET B is very close to the lower limit of the observed value. So, it is better to study a system which has more accurate data for the apsidal motion.

4.2. PV Cas

The fundamental properties of the components of the PV Cas binary system (Andersen 1991; Barembaum & Etzel 1995) are: $M_A = 2.82 \pm 0.05\ M_\odot$, $R_A = 2.297 \pm 0.021\ R_\odot$, $L_A/L_\odot = 51.24 \pm 5$; $M_B = 2.76 \pm 0.06\ M_\odot$, $R_B = 2.256 \pm 0.016\ R_\odot$, and $\log(L_B/L_\odot) = 49.17 \pm 5$. The uncertainty in luminosity is computed from the uncertainties in radius and temperature, assuming the uncertainty of the latter as 200 K.

This DLEB system is an excellent one in two respects: 1) It has a very short apsidal motion period $(91 \pm 2\ \mathrm{y})$. and 2) Its Ap-like variation is confirmed in its light curve. It is shown in Yıldız (2004) that there is no solution with nonrotating models and models rotating like a solid-body, in the sense that the models should give both the observed properties of the component stars and AAR of the system at the same time. As in the case of EK Cep, interplay between X and Z does not alter the case. Furthermore, overshooting, which is effective for the evolved models, worsens the situation.

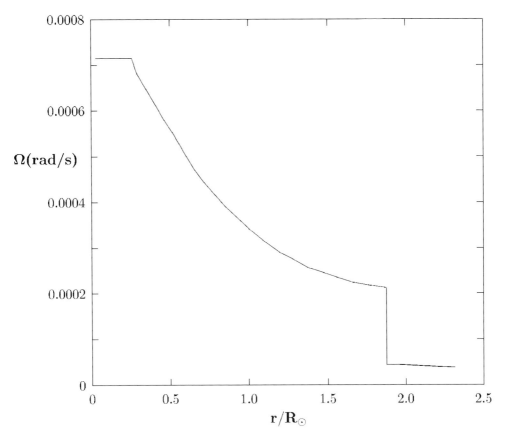

Figure 1. Internal rotation of model for PV Cas A as a function of radius. This model gives also the observed AAR.

Since the components of the binary system are almost identical ($M_A = 2.82 \pm 0.05\, M_\odot$, $M_B = 2.76 \pm 0.06\, M_\odot$), we have no chance to find the chemical composition and the age of the system from model computations. Therefore, initially, a solar composition is assumed (for simplicity we also assume that the two components are identical = PV Cas A).

For the age $t = 2.34 \times 10^8$ y, for example, after several tries we find that a model with $\Omega_o = 7.8\ 10^{-8}\,\mathrm{rad\,s^{-1}}$ ($\Omega_c = 8.5\ 10^{-4}\,\mathrm{rad\,s^{-1}}$) and $M_s = 0.016$ fits the observed luminosity, radius and observed rotational (=synchronized) velocity ($v_{eq} = 65\,\mathrm{km\,s^{-1}}$). This model does not gives the observed AAR at the corresponding time. The computed AAR is 23% greater than the observed value. Therefore, the amount of synchronized mass M_s should be decreased to reduce the theoretical AAR. Thus the solution we are seeking is at another time: At $t = 1.4 \times 10^8$, the model with $\Omega_o = 6.7\ 10^{-8}\,\mathrm{rad\,s^{-1}}$ ($\Omega_c = 7.1\ 10^{-4}\,\mathrm{rad\,s^{-1}}$) and $M_s = 0.000275$ is in perfect agreement with observations, including AAR. The angular velocity of this model is plotted in Figure 1 as a function of radius in solar units. The rotation rate is constant in the convective core and gradually decreases up to the base of the synchronized region ($r = 1.85\,R_\odot$) at which a sudden change occurs. This discontinuity is due to our simplification and there would be a smooth transition if the angular momentum transfer was computable in the physical conditions of stellar interiors.

If there is really such a sharp change in the rotational velocities of some stars, then it may somehow be confirmed by observations. The difference between effective

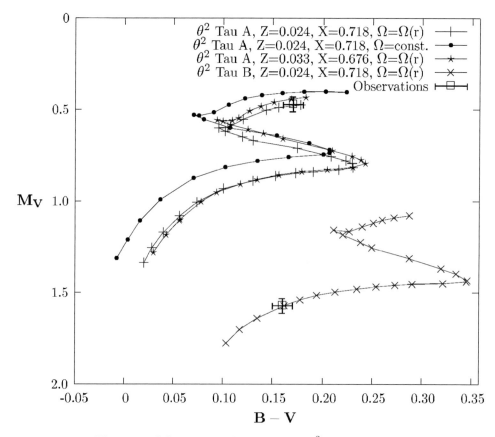

Figure 2. Color-magnitude diagram for θ^2 Tau A and B.

temperatures of the magnetic Ap stars and *normal* stars found by Hubrig *et al.* (2000) may be treated in this context, since the same amount of difference exists also between the effective temperatures of non-rotating models and models with such differential rotation (see Yıldız 2005).

5. The visual binaries θ^2 Tau and ϕ 342 in the Hyades open cluster

Although the masses of the component stars in θ^2 Tau found from the observations ($M_A = 2.42 \pm 0.30$, $M_B = 2.11 \pm 0.17$) are not as accurate as for the DLEBs discussed in the previous section, we can extract information about the evolutionary phase and the structure of these stars, by considering them together with the components of ϕ 342. While the calibration of latter gives X, for a given Z, we get the age of the cluster from calibration of θ^2 Tau A. However, the age, for a given chemical composition, depends on the rotational properties assigned. Therefore, regarding the observated rotational velocity of θ^2 Tau A ($v_{eq} \sim 100\,\mathrm{km\,s^{-1}}$, Torres *et al.* 1997a), we consider the rotational properties of the inner regions for two typical cases: 1) solid-body rotation, and 2) differential rotation as determined by contraction (no synchronization is assumed since the system is very wide).

In the latter case, for $Z = 0.024$, for example, $X = 0.718$ and $t = 721$ My. In Figure 2, the evolutionary tracks of models θ^2 Tau A (+) and B (×) with this chemical composition are plotted in the HRD. Both of the models are in very good agreement with the observed positions of the stars, but the agreement for θ^2 Tau B is obtained at a different time ($t = 440$ My; see below). On the other hand, the model of θ^2 Tau A with solid-body rotation (filled circle) is not in agreement with the observation. However, it is not so easy to rule out solid-body rotation in such an analysis by regarding the superiority of the differentially rotating model since the uncertainty in mass is not sufficiently small ($M_A = 2.42 \pm 0.30$). For comparison, another differentially rotating model for $Z = 0.033$, for which $X = 0.676$ and $t = 671$ My, is also plotted in Figure 2 ($\star$). It is also in good agreement with the observed values, but, not as good as the model with $Z = 0.024$.

As emphasized above both the differentially rotating models of θ^2 Tau A and B with $Z = 0.024$ are in agreement with the observations, but not at the same time. Although the observed rotational velocity of the secondary star is highly uncertain ($v \sin i = 90 - 170\,\mathrm{km\,s^{-1}}$; Torres *et al.* 1997a), it seems that maybe its rotation profile is somewhat different from that determined by contraction alone.

All the models of θ^2 Tau A are compatible with the subgiant phase in which hydrogen is burnt in the shell surrounding the He core. If θ^2 Tau A was an isolated star, we could neither be sure about its evolutionary aspect nor discuss its rotational properties by modelling.

6. Conclusion

The double-lined (eclipsing or visual) binaries, as unique sources of data for stellar mass, are essential systems for the determination of the physical properties and the evolutionary aspects of stars. We can assess the effect of internal rotation on the observable properties of the models for the components of these binaries, and then remove the discrepancy between the theoretical and observational results. In this respect, EK Cep, PV Cas and θ^2 Tau are very fruitful systems. To make the properties of differentially rotating models agree with the all available observational constraints, the internal rotational properties of components stars follow:

1) EK Cep A, a ZAMS star, has a very rapidly rotating core: For $Z = 0.04$, the core rotates 65 times faster than the surface and mass fraction of the synchronized outer mass is 48%. For $Z = 0.02$, the ratio of core rotation rate to that of the surface is 29 and the mass fraction is 10%. While the models for EK Cep A with the two different chemical compositions have the same observable properties (L and R), the theoretical AAR based on the models with a metal-rich composition is in better agreement than the models with solar metallicity. However, the latter is close to the lower limit of (but, still in) the observed range.

2) θ^2 Tau A, a subgiant near to the TAMS, may have a much steeper differential rotation. While properties of its model with differential rotation as determined by contraction are in very good agreement with its observed position in the HRD, the track of model with solid-body rotation never touches the observational box. But, for a definite conclusion, its mass should be measured more accurately.

3) PV Cas A & B seem to be slowly rotating stars, but the slow rotation is limited to a shallow region beneath the surface. For a solar composition, while only 0.03% of the outermost mass is synchronized, the rest is rotating very rapidly (see Figure 1). The confirmation of Ap-like variation in the light-curve of the system (Barembaum & Etzel 1995) leads us to seek a correlation between the chemical peculiarity and the internal rotation. Furthermore, the temperature difference found between the magnetic Ap stars and *normal* stars exists also between NR models and models with such differential rotation.

Acknowledgements

I thank Jean-Paul Zahn for his very stimulating discussions and suggestions, and Hiromoto Shibahashi for fruitful discussions during the symposium. Belinda Kalomeni is acknowledged for reading this manuscript.

References

Aerts, C., Thoul, A., Daszynska, J., *et al.* 2003, *Science* **300**, 1926
Andersen, J. 1991, *A&AR* **3**, 91
Barembaum, M.J., & Etzel, P.B. 1995, *AJ* 109, 2680
Claret, A. & Gimenez, A. 1993, *A&A* 277, 487
Dupret, M.-A., Thoul, A., Scuflaire, R., *et al.* 2004, *A&A* 415, 251
Goldreich, P. & Nicholson, P.D. 1989, *ApJ* 342, 1075
Guinan, E.F., Ribas, I., Fitzpatrick, E.L., *et al.* 2000, *ApJ* 544, 409
Harmanec, P. 1988, *BAICz* 39, 329
Hubrig, S., North, P., & Mathys, G. 2000, *ApJ* 539, 352
Lacy, C.H.S., Torres, G., Claret, A. & Sabby, J.A. 2003, *AJ* 126, 1905
Lastennet, E., & Valls-Gabaud, D. 2002, *A&A* 396, 551
Lebreton, Y., Fernandes, J., & Lejeune, T. 2001, *A&A* 374, 540
Marques, J.P., Fernandes, J., & Monteiro, M.J.P.F.G. 1997, *A&A* 239, 245
Pols, O.R., Tout, C.A., Schroder, K-P., Eggleton, P.P., Manners, J. 1997, *MNRAS* 289, 869
Popper, D. 1980, *ARA&A* 18, 115
Roxburgh, I.W. 1992, *A&A* 266, 291
Torres, G., Stefanik, R. P., & Latham, D. W. 1997a, *ApJ* 485, 167
Torres, G., Stefanik, R. P., & Latham, D. W. 1997b, *ApJ* 479, 268
Yıldız, M. 2003, *A&A* **409**, 689
Yıldız, M. 2004, *A&A*, submitted
Yıldız, M. 2005, *These Proceedings*, 89
Young, P.A., Mamajek, E.E., Arnett, D., & Liebert, J. 2001,*ApJ* 556, 230
Zahn, J.-P. 1977, *A&A* **57**, 383

Discussion

MOSS: It is a basic idea that nonaxisymmetric fields, as seen in Ap stars (especially the more rapid rotators), and strong differential rotation are incompatible - either the differential rotation destroys the nonaxisymmetric field, or the differential rotation is strongly reduced by the magnetic field.

How are your ideas compatible with this? How do you explain this problem in the context of your ideas?

YILDIZ: A very steep differential in depth is found near the surface for the components of PV Cas. But, in our models for the primary, the outermost 0.03% of the total mass (19% by radius) is rotating slowly. Maybe, this is the part where diffusion works and the magnetic structure has the same character as that of the observed surface.

The A-Star Puzzle
Proceedings IAU Symposium No. 224, 2004
J. Zverko, J. Žižňovský, S.J. Adelman, & W.W. Weiss, eds.
© 2004 International Astronomical Union
DOI: 10.1017/S1743921304004417

Effective mass: A new concept in stellar astrophysics based on the internal rotation, and its place in the A- and B-star puzzle

Mutlu Yıldız

Ege University, Dept. of Astronomy and Space Sciences, Bornova, 35100 Izmir, Turkey
email: yildiz@astronomy.sci.ege.edu.tr

Abstract. Putting their chemically peculiarities aside, Ap and Am stars have many common properties. The discriminating property between them could be the rotation of their interiors beneath their slowly rotating surface. In the PV Cas (HD 240208) binary system whose light curve shows Ap-like variation, the agreement between the theoretical and observed apsidal advance rate is satisfied only with differentially rotating models of the component stars in which rapid rotation is extended almost to the surface. Thus, it seems that there is a steep rotation rate gradient near the surface of the Ap stars. The conclusion we reach in the analysis of PV Cas system leads us to introduce the *effective mass* as a new conceptual tool in stellar astrophysics: $M_{\rm eff} = M_\star (1 - \bar{\Lambda})^{1.75}$, where $M_\star$ and $\bar{\Lambda}$ are mass of the star and average of ratio of centrifugal to the gravitational acceleration throughout the model, respectively. We find that the *effective mass* of PV Cas A whose true mass is $2.82 \pm 0.05\,M_\odot$ is $2.6\,M_\odot$ for both solar and metal rich compositions.

Keywords. Stars: rotation, stars: interiors, stars: evolution, stars: early-type, binaries: eclipsing, stars: individual (PV Cas)

1. Introduction

The conventional conceptual tools for our understanding of the stellar structure and the evolution fail to explain many important observed properties of A-stars. In this respect, the assumption of no or solid-body rotation may be the most important deficiency in the standard models. In fact, differential rotation with depth should be expected as a natural result of contraction, at least for some time, provided that the angular momentum transfer in the stellar interior is not a sudden process. The evolution of such a differential rotation depends on the details of angular momentum transfer layer by layer. Although we have no exact information about this transfer process so far, there are strong evidences that the internal regions of A- and B-type stars are rotating much more rapidly than their outer regions (see Yıldız 2005).

In this study, we further discuss the effect of rotation on the model properties and explain some basic irregularities arising from the rapid rotation of the inner regions, by introducing the *effective mass* concept. The main motivation for doing this comes from the comparison of the differentially rotating model of PV Cas A with the non-rotating model of $2.55\,M_\odot$. In Figure 1, these models (with metal rich composition) are plotted in the HR diagram: While the former is represented by $\star$, $\times$ marks the evolutionary track of the latter. For further comparison, the non-rotating model of $2.82\,M_\odot$ is also plotted in this figure ($\diamond$). Differential rotation of this kind makes the structure and evolution of PV Cas A with mass $2.82\,M_\odot$ very close to those of the non-rotating model of $2.55\,M_\odot$. Thus, a differentially rotating model of a certain mass is structured and evolves as though it has somewhat a reduced mass.

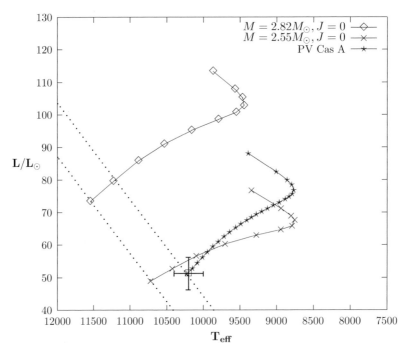

Figure 1. The observed luminosity of PV Cas A is sufficiently low that it has the same luminosity level as the non-rotating model of 2.55 $M_\odot$ ($\times$) rather than of 2.82 $M_\odot$ ($\diamond$). Differential rotation is required to fit model of PV Cas A ($\star$) to the observations. Two parallel dotted lines show the ZAMS lines for the non-rotating (lower) and differentially rotating models (upper).

Note also that the difference between the effective temperatures of the two ZAMS lines (one for the non-rotating models and the other for the differentially rotating models similar to that for PV Cas A), for a given luminosity, is the same as the difference found by Hubrig *et al.* (2000) between the effective temperatures of the magnetic Ap and *normal* stars.

In this study, the results are essentially based on a rich variety of rotating models constructed for the mass of the primary component PV Cas (Yıldız 2004). The rest of the paper is organized as follows: In Sect. 2, the basic effects of rotation are discussed and some relations between the quantities of non-rotating models and models with solid-body rotation are presented. An analytic expression for the homogeneous mass distribution is derived for *effective mass* in Sect. 3, while we obtain a similar expression for a more realistic case in Sect. 4. Finally, Sect. 5 contains some conclusions.

2. Basic effects of rotation

The mean effect of rotation on hydrostatic equilibrium in terms of rotation parameter Λ_r can be given as (Kippenhahn *et al.* 1970)

$$\frac{dP}{dr} = -\frac{GM_r\rho}{r^2}(1 - \Lambda_r) \tag{2.1}$$

where

$$\Lambda_r = \frac{2}{3}\frac{\Omega^2 r^3}{GM_r} \tag{2.2}$$

Due to the radial component of centrifugal acceleration, which is opposite to gravity, the central regions feel less mass than the true mass of the overlying regions. As a result the central temperature of a rotating model is always less than that of its non-rotating counterpart. Therefore, the former is always fainter than the latter. In the case of solid-body rotation, for example, we derive that luminosity of the rotating model can be given in terms of luminosity of the non-rotating model (L_o) and the value of the rotational parameter at the surface (Λ_s)

$$L = \frac{L_o}{(1 + \Lambda_s)^{0.25}}.\tag{2.3}$$

The effect of solid-body rotation on the radius, however, can be formulated from the model properties as

$$R = R_o(1 + \Lambda_s)^{0.45}.\tag{2.4}$$

Our aim is to find similar relations for the luminosities and the radii of differentially rotating models. But, the radius is a more complicated function of the rotational profile than the luminosity. Now, we shall see how the central physical conditions of a model depend on its total mass and rotational properties.

3. Effective mass for a homogeneous mass distribution

All the overlying layers compress the central regions. However, rotation acts to reduce this effect. Then, central physical conditions result as though there were less mass in the upper layers than the true mass. To find a simple relation between the physical conditions at the center and the total mass, and also how rotation affects this relation, we assume a constant mass distribution. Using the equation of continuity and integrating Eq. (2.1) from the center to the surface, we get

$$\frac{P_c}{c\rho_o^{4/3}} = \frac{\Re}{\mu} \frac{T_c}{c\rho_c^{1/3}} = \frac{3}{2} M_\star^{2/3}(1 - \bar{\Lambda})\tag{3.1}$$

where mean rotational parameter $\bar{\Lambda}$ is defined as

$$\bar{\Lambda} \equiv \frac{2}{3M_\star^{2/3}} \int_0^{M_\star} \frac{dM_r}{M_r^{1/3}} \Lambda_r.\tag{3.2}$$

For simplicity, in (3.1), we take only the gas pressure into consideration. Indeed, (3.1) describes implicitly the mass-luminosity relation in the case of rotation: The physical conditions at the center, where nuclear reactions are occurring, are determined by the total mass and $\bar{\Lambda}$, for a given chemical composition. Furthermore, these conditions could be very similar for two stars with very different masses, provided that their masses and mean rotational parameters give the same value for the right side of (3.1). We define *effective mass* in this context as

$$M_{\mathrm{NR}} \equiv M_{\mathrm{eff}}(\bar{\Lambda}) \equiv M_\star(1 - \bar{\Lambda})^{3/2}\tag{3.3}$$

where M_{NR} is the corresponding mass for no rotation. That is to say, a model of mass M with $\bar{\Lambda} = 0.63$ has the same physical conditions at its center as the non-rotating model of mass $M/2$.

4. The effective mass of PV Cas A

For a more realistic case than the homogeneous mass distribution, we construct models of mass $2.82\,M_\odot$ with a variety of rotational properties. In Figure 2, luminosities of these

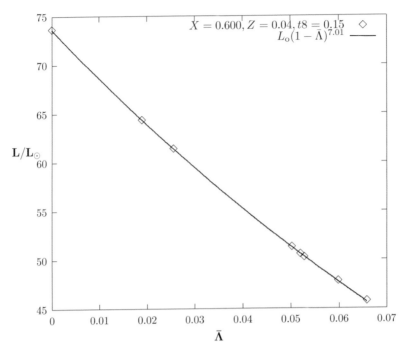

Figure 2. Luminosities of models of $2.82\,M_\odot$ with a variety of rotational properties as a function of the average of rotational parameter $\bar{\Lambda}$.

models at $t = 15$ My ($\Diamond$) are plotted with respect to their mean rotational parameter $\bar{\Lambda}$. The solid line represents the fitting formulae $L_o(1 - \bar{\Lambda})^{7.01}$, where L_o is luminosity of the non-rotating model. All points are on this curve.

To find an expression for the *effective mass*, we use the mass-luminosity relation derived from the non-rotating models of $2.82\,M_\odot$ and $2.55\,M_\odot$ with metal rich composition; $L = cM^4$. Then,

$$M_{\text{eff}}(\bar{\Lambda}) = M_\star(1 - \bar{\Lambda})^{7.01/4} \tag{4.1}$$

For the differentially rotating model with a metal rich composition given in Figure 1, using (4.1), we find that $M_{\text{eff}} = 2.58\,M_\odot$. Similarly, we can also compute the *effective mass* for the solar composition: $M_{\text{eff}} = 2.60\,M_\odot$. Thus these values for the *effective mass* are almost the same.

As it is emphasized above, the *effective mass* determines the evolution of stars as well as their structures. Comparison of the MS lifetimes of the models shows this very clearly: The MS lifetime of non-rotating model of $2.82\,M_\odot$ is 280 My, while the non-rotating model of $2.57\,M_\odot$ and the differentially rotating model of $2.82\,M_\odot$ have the same MS lifetime (370 My).

The use of *effective mass* may also explain why the observed mass-luminosity relation for the PV Cas binary system is so low (Barembaum & Etzel 1995):

$$\frac{\Delta \log L}{\Delta \log M} = 2.3. \tag{4.2}$$

The minimum value computed from the models of PV Cas A & B for this relation is 4. If we use the *effective masses* of PV Cas A & B instead of true masses we get 3.6, very close to the theoretical value (4).

5. Conclusions

There is strong evidence that the inner regions of some (if not all) early type stars are rotating very rapidly. We discussed the effects of such a rotation on the structure and the evolution of stars, by introducing *effective mass* concept. According to our findings concerning the calibrated model of PV Cas A, its *effective mass* is 10% less than its true mass, as a result of differential rotation. Another surprising result, pertaining to stellar evolution, is that the MS lifetime of a differentially rotating model is also determined by the *effective mass*.

Since the Ap-like variations are observed in the light curve of the PV Cas binary system, some or all of the Ap stars may have such a differential rotation. In that respect, maybe, ratio of the *effective mass* to the true mass is minimum for Ap stars with such rotational properties.

The *effective mass* concept also explains why the observed luminosity relation is so small for some eclipsing binaries in comparison with the minimum theoretical value. Furthermore, the *effective mass* may have also some cosmological implications.

Acknowledgements

D. Özlem Hürkal is thanked for reading the manuscript.

References

Barembaum, M.J., & Etzel, P.B. 1995, *AJ* **109**, 2680
Hubrig, S., North, P., & Mathys, G. 2000, *ApJ* **539**, 352
Kippenhahn, R., Meyer-Hofmeister, E., & Thomas, H.-C.i 1970, *A&A* **5**, 155
Yıldız, M. 2004, *A&A*, submitted
Yıldız, M. 2005, *These Proceedings*, 81

The A-star puzzle
Proceedings IAU Symposium No. 224, 2004   © 2004 International Astronomical Union
J. Zverko, J. Žižňovský, S.J. Adelman, & W.W. Weiss, eds.   DOI: 10.1017/S1743921304004429

First signatures of strong differential rotation in A-type stars

A. Reiners[1,2] and F. Royer[3,4]

[1] Astronomy Department, University of California, Berkeley, CA 94720, USA
email: areiners@astron.berkeley.edu

[2] Hamburger Sternwarte, Universität Hamburg, Gojenbergsweg 112,
21029 Hamburg, Germany

[3] Observatoire de Genève, 51 chemin des maillettes, 1290 Sauverny, Switzerland
email: frederic.royer@obs.unige.ch

[4] GEPI/CNRS UMR 8111, Observatoire de Paris–Meudon,
92195 Meudon cedex, France

Abstract. We reanalyzed high quality spectra of 158 stars of spectral types A0–F1 and $v \sin i$ values between 60 and 150 km s^{-1}. Using a least squares deconvolution technique we extracted high S/N broadening profiles and determined the loci of the Fourier transform zeros q_1 and q_2 where the S/N-ratio was high enough. For 78 stars q_2 could be determined and the ratio q_2/q_1 was used as a shape parameter sensitive to solar-like differential rotation (the equatorial velocity is faster than the near polar velocities). Seventy-four of the 78 stars have values of q_2/q_1 consistent with solid body rotation; in four of the 78 cases, values of q_2/q_1 are not consistent with rigid rotation. Although these stars may be binaries, none of their profiles shows any signatures of a companion. The Fourier transforms do not indicate any distortions and the broadening profiles can be considered due to single objects. One of these candidates may be an extremely rapid rotator seen pole-on, but for the other three stars of spectral types as early as A6, differential rotation seems to be the most plausible explanation for the peculiar profiles.

Keywords. Stars: rotation, stars: early-type, stars: activity, stars: individual (HD 6869, HD 44892, HD 60555, HD 109238)

---

1. Introduction

A substantial difference between the photospheres of solar-type stars and A-type stars is the existence of a convective envelope. Due to the ionization of hydrogen the cooler late-type stars harbour convective envelopes where turbulent motions of the photospheric plasma can occur. Stars of spectral types earlier than about F2 lack or have only very thin convective envelopes and the properties of granular flows change fundamentally.

The generally accepted activity paradigm places the stellar dynamo, believed to cause stellar activity, at the boundary between the convective envelope and the radiative core. Differential rotation drives the dynamo action by winding up and amplifying magnetic flux tubes. The interaction of magnetic fields, differential rotation and the convective envelopes is believed to be ultimately responsible for stellar activity.

Many investigators have searched for the onset of convection. It is generally accepted that the onset of stellar activity occurs between spectral types A7 and F5 depending on the observational strategy. Wolff *et al.* (1986) studied C II and He I emission and placed the onset of activity near $B - V = 0.28$, i.e., around spectral type F0. Schmitt (1997) concluded from X-ray data that coronal emission is universal in the spectral range A7 to G9 implying an onset of activity around spectral type A7. Hotter stars are expected to harbour shallow convective envelopes. These stars have higher convective velocities which peak at about spectral type A3 until convection disappears altogether at about spectral type A1 (Renzini *et al.* 1977). Gray & Nagel (1989) directly searched for the onset of convection by analyzing line bisectors of slowly rotating stars. In their targets the Doppler-shift distribution of the granulation dominates the broadening of spectral lines and a bisector reversal was found around spectral type F0. Stronger asymmetries were found in the stars at the hot side of the boundary indicating higher convective velocities.

Although stellar activity is not observed in early A-type stars, it is not clear whether differential rotation may take place in early-type stars. The absence of activity may simply reflect inefficient coupling of surface magnetic fields and the lack of an interface between the radiative core and the convective envelope. There is no reason to believe that rapidly rotating A-stars should rotate rigidly. For the later-type Sun the surface rotation law can be approximated by

$$\Omega(l) = \Omega_{\text{Equator}}(1 - \alpha \sin^2 l), \qquad (1.1)$$

with l the latitude and $\alpha_\odot \sim 0.2$ as derived from sunspots. Gray (1977) searched for differential rotation in line profiles of six A-stars and found none within his error bars.

Also using line profiles, Reiners & Schmitt (2003) found signatures of differential rotation in a sample of F-type stars. The earliest object in their sample indicating differential rotation is of spectral type F0IV/V. Applying the method used by Reiners & Schmitt (2003), we search for signatures of differential rotation in a large sample of A-star spectra. The results are presented in the following sections and have already been published by Reiners & Royer (2004a).

2. Observations and data analysis

The spectra were observed with the ECHELEC spectrograph (ESO/La Silla) and are part of a larger sample collected in the framework of an ESO Key Programme. The sample is described by Grenier *et al.* (1999) who measured the radial velocities and Royer *et al.* (2002, hereafter RGFG) who derived the rotational velocities from these spectra.

To search for the spectral signatures of stellar rotation laws, broadening profiles were derived by applying a Least Squares Deconvolution process (LSD). After constructing a δ-template comprising the strongest 150 lines taken from the Vienna Atomic Line Database (Kupka *et al.* 1999) and according to stellar temperature, a first-guess broadening profile was deconvolved using each pixel as a free parameter in the fit. Since the theoretical line depths match the observed ones poorly, the equivalent widths of the incorporated lines were optimized in a second step while leaving the broadening profile fixed. During a few iterations the broadening profile and the equivalent widths were optimized. Using this technique the spectral lines are effectively deblended, the information contained in every spectral line is used and the signal-to-noise ratio is significantly enhanced. Consistency of the fit is checked by comparing theoretical line depths to the derived ones.

Following Reiners & Schmitt (2002) we Fourier transformed the broadening functions and measured the position of the first and second zeros (q_1, q_2). The ratio q_2/q_1 is a robust observable for the shape of a rotational broadening function and a direct indicator for solar-like differential rotation with the equator rotating faster than the polar regions (Reiners & Schmitt 2002). We measured the ratio q_2/q_1 for all stars the LSD procedure yielded a stable and symmetric broadening function.

The spectral quality used in this analysis in principle was sufficient to follow the Fourier transformed broadening functions to the second zero q_2 in stars with projected rotational velocities in the range $60\,\text{km s}^{-1} < v \sin i < 150\,\text{km s}^{-1}$.

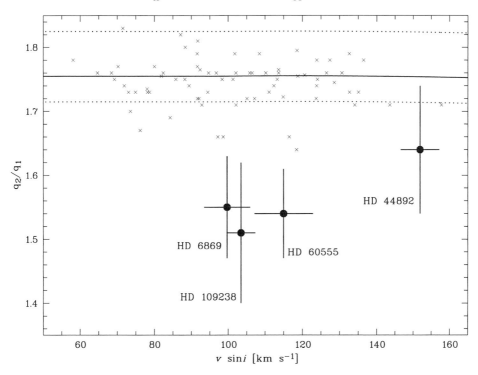

Figure 1. Derived values of q_2/q_1 plotted against $v \sin i$ as derived from the first zero of the Fourier transform (with 1-σ uncertainties). The region between dashed lines is consistent with solid body rotation for arbitrary limb darkening. For linear limb darkening with $\epsilon = 0.6$, $q_2/q_1 = 1.76$ is expected (solid line). Crosses indicate results consistent with solid body rotation, typical errors are of the order of $\Delta q_2/q_1 = 0.1$ (not plotted). Four stars not consistent with $q_2/q_1 = 1.76$ are indicated by solid circles, error bars are plotted for them.

3. Results

This method was applied to the spectra of 158 stars of spectral types A0–F1. The rotational velocity $v \sin i$ was derived from the first zero q_1 and is in good agreement with the results from RGFG. For 78 of our sample stars the ratio q_2/q_1 could be determined. For the discarded 80 stars, either the data quality was insufficient or the derived broadening function showed obvious peculiarities probably due to binarity.

We measured the second zeros of the Fourier transformed broadening profiles to calculate the ratios q_2/q_1. The results are plotted in Fig. 1. Typical errors are of the order of $\Delta q_2/q_1 \approx 0.1$. A rigid rotator is expected to yield a value of q_2/q_1 between 1.72 and 1.83 assuming a linear limb darkening law (indicated by the dashed lines in Fig. 1). For the stars of our sample linear limb darkening coefficients between 0.5 and 0.75 are expected during their time on the main sequence (Claret 1998). Assuming a limb darkening parameter of $\epsilon = 0.6$ rigid rotation would yield $q_2/q_1 = 1.76$ (solid line in Fig. 1). The results that are consistent with a value of $q_2/q_1 = 1.76$ within the error bars are indicated by small crosses in Fig. 1. For the sake of readability no errors are plotted for them. The second zero q_2 can only be determined in spectra where the signal exceeds the noise level beyond q_2, i.e., when a second sidelobe is detectable. Unfortunately the amplitude of the second sidelobe is in the noise level for many of our stars. These measurements of q_2 must be interpreted as lower limits; thus some of the measurements of q_2/q_1 plotted as crosses in Fig. 1 are essentially lower limits. For 74 of the 78 stars analyzed the broadening profiles are consistent with solid body rotation.

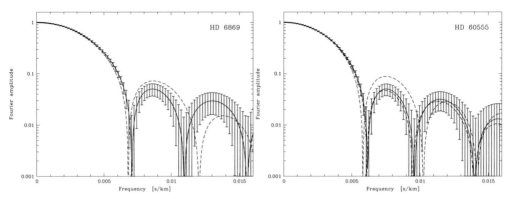

Figure 2. Fourier transforms of HD 6869 and HD 60555 plotted with error bars. These stars show extremely small values of q_2/q_1. The spectra of HD 154876 and HD 29920 have values of q_2/q_1 expected for rigid rotators and are plotted with dashed lines for comparison in the left and right panels, respectively.

Table 1. Derived values of $v \sin i$ and q_2/q_2 for the four stars with values of q_2/q_1 significantly smaller than 1.76. Also given are the strength of the differential rotation in terms of α (cp. Eq. 1.1) and the required values of the equatorial velocities $v_{\mathrm{e,rigid}}$ and inclination angles i if the value of q_2/q_1 is explained by rapid rotation seen pole-on (Sect. 3.2). Distances in pc and ROSAT X-ray luminosities are given in columns 8–9, respectively.

HD	Type	$v \sin i$ $(\mathrm{km\,s^{-1}})$	q_2/q_1	α	$v_{\mathrm{e,rigid}}$ $(\mathrm{km\,s^{-1}})$	i	d (pc)	L_X $(\mathrm{W\,m^{-2}})$
6869	A9V	100 ± 6	1.55 ± 0.08	0.28 ± 0.10	(460)	$(13°)$	87	570
60555	A6V	115 ± 7	1.54 ± 0.07	0.29 ± 0.08	(470)	$(14°)$	134	
109238	F0IV/V	103 ± 4	1.51 ± 0.11	0.32 ± 0.13	(500)	$(13°)$	133	
44892	A9/F0IV	152 ± 5	1.64 ± 0.10	0.16 ± 0.16	400	$22°$	160	

Four of our measurements are inconsistent with $q_2/q_1 = 1.76$. They are indicated by solid circles in Fig. 1 and errors bars are plotted for them. For three of those — HD 6869 (A9V), HD 60555 (A6V) and HD 109238 (F0IV/V) — the values of q_2/q_1 are significantly smaller than 1.7. The fourth star (HD 44892, A9/F0IV) has a value of q_2/q_1 marginally consistent with $q_2/q_1 > 1.7$ within its error bars. We will discuss this star in Sect. 3.2.

The Fourier transforms of our candidate stars are illustrated by the plots of HD 6869 and HD 60555 with error bars in Fig. 2. Overplotted are the Fourier transformed line profiles of stars with similar values of $v \sin i$ that are consistent with rigid rotation ($q_2/q_1 = 1.76$). While different velocity fields, e.g., turbulence, may influence the amplitudes of the sidelobes, the zeros of the Fourier transform arise from rotational broadening (Gray 1976). One mechanism known to change the ratio q_2/q_1 in the manner found in HD 6869, HD 60555 and HD 109238 is solar-like differential rotation. The strength of differential rotation in terms of the parameter α in Eq. 1.1 can be calculated from q_2/q_1 (Reiners & Schmitt 2003), and the respective values of α are given in Table 1 together with the spectral types and $v \sin i$ of the four suspected differential rotators.

3.1. *Binarity*

Using our deconvolution method, we find indications for double peaks in the broadening functions of 23 of our targets, but not in the spectra of these four stars. This does not mean that the 135 others are single stars since luminous A-type stars dominate spectra of, e.g., binaries consisting of an A-type and a G-type star. The G-type spectrum will easily be hidden in the light of the A-type star. To be complete we checked the shape of the correlation function by cross-correlating our template with the spectra and found no indications for binarity either.

For HD 44892 and HD 109238, the literature gives hints about a single star status. There is no evidence of binarity for HD 44892 neither in the HIPPARCOS data nor in Speckle observations (Mason *et al.* 2001). It can be considered as single with a high level of confidence, and its spectrum is surely not affected by any significant contamination. HD 109238 is part of the sample observed by Abt & Morrell (1995). Their MK classification is F0V with no suspicion of spectroscopic binarity.

We have two spectra each of both HD 6869 and HD 60555 in our data set. Individual observations of HD 6869 and HD 60555 are separated by 383 d and 767 d, respectively. Inspection of the broadening functions derived from the individual spectra yields no indication of variability due to the relative motions of binary components. Coadded spectra were used to derive the values of q_2/q_1 for both stars.

3.2. *Extremely fast rotation*

As an alternative to differential rotation, the shape of the broadening function and the value of q_2/q_1 can also be affected by very rapid rotation and gravity darkening possibly observed in pole-on stars (Reiners 2003). Flux is redistributed from the line's wings to the center when the equator becomes cooler due to gravity darkening. As far as the lines considered are not dominated by temperature and gravity variations over the stellar surface — which is the case, e.g, in the weak lines of early A-type stars as shown by Gulliver *et al.* (1994) — the value of q_2/q_1 is diminished by this effect. According to Reiners (2003), the ratio q_2/q_1 then only depends on the equatorial velocity v_e and on the gravity darkening law. We assume a linear gravity darkening law according to Claret (1998) and calculate the equatorial velocities $v_{e,rigid}$ required to produce the measured values of q_2/q_1 assuming solid body rotation for the four suspected differential rotators. The results and the respective inclination angles i are given in columns six and seven of Table 1.

For HD 6869, HD 60555 and HD 109238 the values of $v_{e,rigid}$ are larger than breakup velocity; for these stars rapid solid body rotation can be ruled out as the mechanism solely responsible for the diminished ratio q_2/q_1. In case of HD 44892 the rotational velocity required for the measured ratio $q_2/q_1 = 1.64$ is of the order of breakup velocity. Thus differential rotation as well as rapid solid body rotation are the two possible explanations for the measured profile shape of HD 44892.

4. Conclusions

We reanalyzed high quality data previously discussed by Grenier *et al.* (1999) and Royer *et al.* (2002) to search for differential rotation in early-type stars. With an iterative Least Squares Deconvolution method we obtained high quality broadening profiles of 158 stars with projected rotational velocities in the range $60 \, \text{km s}^{-1} < v \sin i < 150 \, \text{km s}^{-1}$. We discarded the profiles of 80 of them due to obvious asymmetries or multiplicity. For 78 stars the broadening profiles apparently reflect the rotational broadening law. Profile distortions were analyzed in terms of the ratio of the first two zeros of the Fourier transform q_2/q_1. Within the errors, 74 of the 78 measured profiles are consistent with the assumption of rigid rotation. Due to data quality many measurements must be considered lower limits and from this sample no conclusion can be drawn concerning values of q_2/q_1 possibly larger than 1.8.

For none of our sample stars interferometric measurements are avaliable, but Altair has been recently analyzed by Reiners & Royer (2004b) with this method. They find no signature of differential rotation and that Altair is seen with an inclination angle greater than $i = 68°$, improving the determination based on interferometric data.

Four stars are analyzed in detail. The profile of the A9/F0IV star HD 44892 is only marginally consistent with rigid rotation. It is likely that its profile is distorted either by differential rotation or by very rapid rotation seen pole-on; in the latter case HD 44892 would be the first star that directly shows signatures of gravity darkening in mean profile broadening as proposed by Reiners (2003). Comparison with interferometric results would be especially interesting for this star.

The broadening functions of the three stars HD 6869 (A9V), HD 60555 (A6V) and HD 109238 (F0IV/V) are not consistent with rigid — even very rapid — rotation since their equatorial velocities would be larger than breakup velocity. Although some authors suspect these stars are binaries, in our high quality spectra we find no indications of multiplicity neither in their data nor in Fourier space. Since contamination due to secondaries are easily visible in Fourier space — where no sharp zeros should occur in case of the profile being a sum of two — we consider their spectra single star spectra. Differential rotation seems to be the most plausible explanation for the observed profile distortions. For these three stars the equator is rotating about 30% faster than the polar regions. Thus we conclude that significant differential rotation seems to take place even in early-type stars not harbouring deep convection zones. The earliest object is the A6 dwarf HD 60555.

If differential rotation is the driving mechanism for stellar activity, these stars should be active, too. X-ray emission from HD 6869 was detected with the ROSAT mission and the other stars may also be X-ray sources, but were simply too far away for a detection. Whether differential rotation is a common phenomenon in these stars cannot be answered by this study since only very strong differential rotation is detectable with our method. The finding of strong differential rotation among A-type stars indicates that there is no abrupt change in the rotational laws of stars around the boundary where surface convection sets in.

References

Abt, H. A., & Morrell, N. I., 1995, *ApJS*, **99**, 135

Claret, A., 1998, A&AS, 131, 395

Gray D. F., 1976, The observation and analysis of stellar photospheres, Wiley, New York

Gray, D. F., 1977, *ApJ*, **211**, 198

Gray, D. F., & Nagel, T., 1989, ApJ, 341, 421

Grenier, S., Burnage, R., Faraggiana, R., *et al.* , 1999, A&AS, 135, 503

Gulliver, A. F., Hill, G., & Adelman, S. J., 1994, ApJ, 429, L81

Kupka, F., Piskunov, N.E., Ryabchikova, T.A., Stempels, H.C., & Weiss, W.W., 1999, A&AS, 138, 119

Mason, B. D., Hartkopf, W. I., Holdenried, E. R., & Rafferty, T. J., 2001, AJ, 121, 3224

Reiners, A. & Royer, F., 2004a, A&A, 415, 325

Reiners, A. & Royer, F., 2004b, A&A, in press [astro-ph/0408194]

Reiners, A., & Schmitt, J.H.M.M., 2002, A&A, 384, 155

Reiners, A., & Schmitt, J.H.M.M., 2003, A&A, 398, 647

Reiners, A., 2003, A&A, 408, 707

Renzini, A., Cacciari, C., Ulmschneider, P., & Schmitz, F., 1977, A&A, 61, 39

Royer, F., Gerbaldi, M., Faraggiana, R., & Gómez, A. E., 2002, A&A, 381, 105 (RGFG)

Schmitt, J.H.M.M., 1997, A&A, 318, 215

Wolff, S.C., Boesgaard, A.M., & Simon, T. 1986, *ApJ*, **310**, 360

Discussion

KHALACK: Could the anomalous low ratio q_2/q_1 be due to the nonspherical (ellipsoidal) shape of the aforementioned four stars?

ROYER: In the case of Altair (Reiners & Royer 2004b), whose oblateness has been measured by interferometry, the q_2/q_1 ratio is 1.77, higher than ratios obtained for our four differential rotator candidates. The effect of nonsphericity on $v \sin i$ is discussed in Reiners (2003).

MKRTICHIAN: The primaries of Algol-type semidetached close binary systems might show a strong differential rotation of the surface equatorial latitudes due to a gas stream impact on the surface of the accretor. Did you consider the possibility that a few stars from your sample exhibiting strong pole/equator differential rotation might be A-type mass-accreting primary components of noneclipsing ($i \approx 80°$) Algols? In such systems the contribution of low luminosity secondaries may be small and their spectral lines are not visible. For checking this hypothesis - have you any information about the radial velocity variabilty of your stars on time scales of several days?

ROYER: We used our broadening functions obtained by least square deconvolution as well as cross-correlation functions to find indications of binarity in our sample stars. We did discard 23 of our targets with this criterion. As detailed in Reiners & Royer (2004a) some of our candidate stars are indicated as binaries in the literature, HD 6869 for instance. The Fourier transforms of the derived broadening functions are probably not affected by secondary spectra, but the hypothesis of differential rotation caused by mass transfer is to be investigated. Our candidate stars should be monitored in terms of radial velocities, and the method we used to detect differential rotation could be applied to a known semi-detached system.

The A-Star Puzzle
Proceedings IAU Symposium No. 224, 2004
J. Zverko, J. Žižňovský, S.J. Adelman, & W.W. Weiss, eds.

© 2004 International Astronomical Union
DOI: 10.1017/S1743921304004430

Magnetorotational instability in Ap star envelopes

Reiner Arlt

Astrophysikalisches Institut Potsdam, An der Sternwarte 16, D-14482 Potsdam, Germany
email: rarlt@aip.de

Abstract. The rotational evolution of the radiative zone of magnetic Ap stars is investigated with numerical simulations. An angular-velocity profile decreasing with axis distance in combination with a magnetic field leads to a magnetorotational instability. The resulting flows efficiently transport angular momentum outwards. The corresponding decay of angular-velocity gradients in the radiative zone is estimated to take about 10–100 million years.

Keywords. MHD, instabilities, stars: rotation

1. The magnetorotational instability

Magnetic Ap stars rotate significantly slower than non-magnetic A stars. The Ap stars may either have magnetic fields because they are slow rotators, or they could rotate slowly because they possess magnetic fields. We are concerned with the latter of the two relations.

If the angular velocity of an object decreases with axis distance, the presence of a magnetic field causes an instability which will lead to flows reducing the gradient in angular speed. This local, linear instability was discovered by Velikhov (1959) in Taylor-Couette flows and was rediscovered and introduced to astrophysical flows by Balbus & Hawley (1991).

A few braking mechanisms acting on the star during its early life have been proposed such as magnetic star-disk locking and stellar winds. It is very important to note that these effects exert torques on the surface of the star and a coupling with the deeper layers of the star is needed to brake the entire interior. The microscopic viscosity in a radiative zone is much too small to provide an efficient coupling between the surface layers and the interior. This is why we have good reasons to assume that the interior of the star rotates faster than the surface.

The interesting fact about the magnetorotational instability (MRI) is that there is no lower limit for the magnetic field strength in ideal MHD. If there is a finite conductivity, there is of course a lower limit for the field strength, but since the microscopic conductivity of the stellar plasma in the radiative zone is extremely high, the lower limit for the MRI is below 1 G. As the magnetic field strength is decreased, the wavelength of the most unstable mode becomes shorter, too. The diffusive damping rate increases thereby. Since the diffusive decay and MRI growth will balance at a certain magnetic field strength, we consider this the minimum field strength. The minimum is plotted in Fig. 1 for various densities and differential rotations based on an A0 star. Various densities mean various loci in the star.

In the context of Ap stars, the upper limit will be more relevant. Again, a given magnetic field strength corresponds to a certain wavelength of perturbation which will grow the fastest. As the field strength is increased, this wavelength gets larger and may eventually exceed the size of the object, i.e., the stellar radius. The MRI will be much

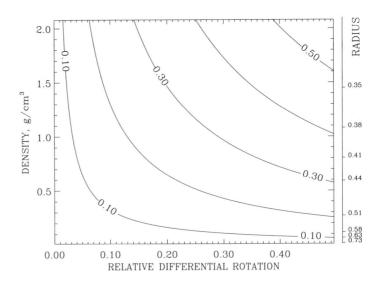

Figure 1. Minimum magnetic field strength in G for an A0 star envelope depending on differential rotation and density. At smaller field strengths, diffusion of the most unstable MRI mode will be faster than its growth rate in the nondiffusive case.

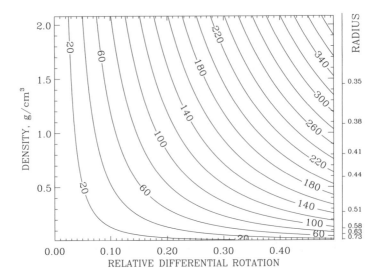

Figure 2. Maximum magnetic field strength in kG for an A0 star envelope depending on differential rotation and density. With stronger fields, the MRI will develop at highly reduced growth rates.

weaker beyond this field strength. We plotted this maximum field strength again for an A0 star as a function of differential rotation and density (or radius fraction) in Fig. 2. If the differential rotation is only about 10%, we can expect the MRI to occur up to field strengths of several tens of kG. Typically the surface fields of Ap stars do not exceed 3.5 kG (Bychkov *et al.* 2003). Assuming a dipole as the simplest field geometry, internal fields at $0.5R_*$ will then be 20–30 kG.

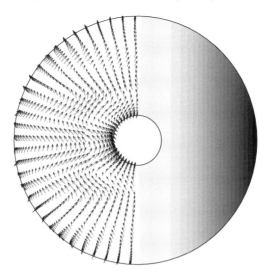

Figure 3. A vertical cross-section through the initial configuration with the magnetic field perturbation on the left side, and the angular velocity contours on the right side.

2. The numerical setup

A spherical, spectral numerical code developed by Hollerbach (2000) is applied to solve the MHD equations in the Boussinesq approximation,

$$\left(\frac{\partial}{\partial t} + \boldsymbol{u} \cdot \nabla\right)\boldsymbol{u} = -\nabla p + \mathrm{Pm}\nabla^2\boldsymbol{u} + (\nabla \times \boldsymbol{B}) \times \boldsymbol{B} \tag{2.1}$$

$$\frac{\partial \boldsymbol{B}}{\partial t} = \nabla^2\boldsymbol{B} + \nabla \times (\boldsymbol{u} \times \boldsymbol{B}) \tag{2.2}$$

with the usual meanings of $\boldsymbol{u}$, $\boldsymbol{B}$, and p as the velocity, magnetic field, and pressure which is not explicitly calculated in this model, but eliminated by applying the curl-operator to Eq. (2.1). Additionally, $\nabla \cdot \boldsymbol{u} = 0$ and $\nabla \cdot \boldsymbol{B} = 0$ hold, which are exactly preserved because potentials for velocity and magnetic field are used.

Normalizations are introduced using the outer radius R_*, the magnetic diffusivity η and an average density ρ, which lead to the non-dimensional quantities $\mathrm{Rm} = R_*^2\Omega_0/\eta$ – the magnetic Reynolds number, and $\mathrm{Pm} = \nu/\eta$ – the magnetic Prandtl number with ν being the kinematic viscosity. The normalized inner radius was $r_\mathrm{i} = 0.2$ and the outer radius was $r_\mathrm{o} = 1$.

The initial rotation profile depends on the axis distance, $s = r\sin\theta$ only. Such a profile fulfills the Taylor-Proudman theorem ensuring a minimum of purely hydrodynamic meridional circulations. The actual initial condition is

$$\Omega(s) = \frac{\mathrm{Rm}}{\sqrt{1 + (2s)^{2q}}}. \tag{2.3}$$

The exponent q is always set to 2; it will later be used as a fitting parameter to obtain a measure for the steepness of the rotation profile. The initial magnetic field is composed of a homogenous vertical field and a perturbation in a plane, with a wave vector parallel to the rotation axis,

$$\boldsymbol{B} = B_0[\hat{\boldsymbol{z}} + \epsilon\sin(kz + \pi/4)\hat{\boldsymbol{x}}], \tag{2.4}$$

where $\hat{\boldsymbol{z}}$ is the unit vector in the direction of the rotation axis and $\hat{\boldsymbol{x}}$ is a unit vector in

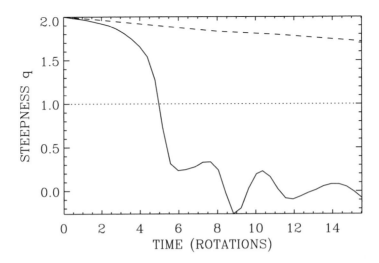

Figure 4. Decay of differential rotation measured by a steepness parameter q for Rm $= 10^4$, taken from Arlt, Rüdiger & Hollerbach (2003). The dashed line is a hydrodynamic run for comparison.

the equatorial plane. The wave number of the perturbation is $k = 4\pi$. We added $\pi/4$ to the second term in (2.4) to provide mixed parity to the system. Equatorial and axial symmetry are thus broken allowing the nonlinear system to develop flows and fields in all modes. The initial configuration of magnetic field and angular velocity is shown in Fig. 3. The spectral truncation was set to 50 Chebyshev polynomials for the radial, 100 Legendre polynomials for the latitudinal and 30 Fourier modes for the azimuthal structure.

The MHD run quickly shows strong flows redistributing angular momentum in the radiative zone. Now we need a measure for the strength of differential rotation to evaluate the decay. The profile (2.3) is used with Rm and q now being free parameters. From each snap-shot of the simulation, we derive $\Omega(s)$ in the equatorial plane of the 'star'. This function is an average over all azimuths, since we are interested in the global rotation profile. A fit of (2.3) delivers a q for a number of time-steps. This steepness q is plotted versus time and shown in Fig. 4 for Rm $= 10^4$. The dashed line shows the decay of differential rotation by pure viscosity. The decay time is much longer than the MHD decay. In a real star, the viscosity in the radiative zone is so small that the time-scale of viscous differential-rotation decay is of the order of 10^{13} yr.

Stellar Reynolds numbers are much larger than 10^4. Several runs were performed at various magnetic Reynolds numbers to get an extrapolated estimate for stellar conditions. As a result, the decay time amounts to roughly 10^7–10^8 yr. Since the age of the Sun is $5 \cdot 10^9$ yr, the process was fast enough to equalize differential rotation in the radiative core of the Sun. The more massive Ap stars have MS life-times of 10^9 yr or less. It is argued that the magnetorotational instability may be an ongoing process in the radiative envelopes of many of these stars.

Since radial motions will be strongly suppressed by buoyancy in radiative zones which are stably stratified, the simulation setup was improved by adding the effect of buoyancy. The MHD equations thus read

$$\left(\frac{\partial}{\partial t} + \boldsymbol{u} \cdot \nabla\right)\boldsymbol{u} = -\nabla p + \mathrm{Pm}\nabla^2 \boldsymbol{u} + (\nabla \times \boldsymbol{B}) \times \boldsymbol{B} + \mathrm{Ra}_\eta \theta \boldsymbol{r} \qquad (2.5)$$

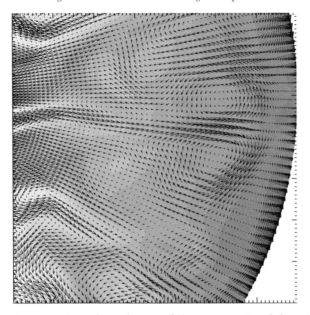

Figure 5. A vertical cross-section through part of the computational domain. The arrows are the meridional flow with the grey shading represents the angular velocity.

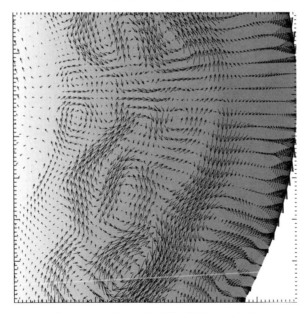

Figure 6. The same vertical cross-section as in Fig. 5, but including the stabilizing buoyancy force in the radiative zone.

$$\left(\frac{\partial}{\partial t} + \boldsymbol{u} \cdot \nabla\right)\theta = -u_r \frac{\partial T}{\partial r} + \frac{\text{Pm}}{\text{Pr}}\nabla^2\theta, \tag{2.6}$$

with the unchanged induction equation (2.2). T and θ now denote an adiabatic background temperature profile and deviations from that, resp. Pr is the 'usual' Prandtl number defined by the viscosity and thermal diffusivity, ν/κ. The strength of the subadiabaticity is controlled by the dimensionless, modified Rayleigh number Ra_η which is based on the magnetic diffusivity rather than the viscosity.

A cross-section through the velocity field at Ra $= -10^8$ is shown in Fig. 6 at about the same time when the snap-shot in Fig. 5 was taken. Flows are much more horizontal. Yet, sufficient transport of angular momentum is found, leading to nearly the same decay time of 6 rotations (compared to 5 in the earlier model free of negative buoyancy). The angular-momentum transport will be highly suppressed at Rayleigh numbers Ra $> 10^{10}$. Note again that these times from the numerical simulations will translate to much longer times using stellar parameters. Mixing in the outermost stellar layers by the MRI is less likely, because Ap star fields may exceed the maximum field there according to Fig. 2.

Acknowledgements

I am very grateful to R. Hollerbach for his help and making his MHD code available.

References

Arlt, R., Hollerbach, R. & Rüdiger, G. 2003, *A &A*, 401, 1087
Balbus S.A. & Hawley J.F. 1991, *ApJ*, 376, 214
Bychkov V.D., Bychkova L.V. & Madej J. 2003, *A &A*, 407, 631
Hollerbach, R. 2000, *Int. J. Numer. Meth. Fluids*, 32, 773
Velikhov, E.P 1959, *Sov. Phys. JETP*, 9, 995

Discussion

MOSS: If this works to produce Ap stars as you suggest, why are not all A stars magnetic? A small seed-sized magnetic field could be expected to be present in any newly formed star. With any differential rotation (also likely to be present), the magnetorotational instability should be universal. Thus all A stars would be observably magnetic!

ARLT: The mechanism is not meant to *produce* Ap stars. Up to the moment when diffusion dominates the simulation, field amplifications of a factor of 10 are seen in the poloidal field. Starting with about 1 G, the resulting 10 G field will not easily be seen in A stars. Toroidal fields are amplified by a factor of $\sim$ Rm, but are hidden.

STÜTZ: To which extent do you think it is now possible to include more realistic microphysics and radiative transfe to get more realistic models?

ARLT: A full MHD simulation of a radiative envelope would require extensive radiative transfer computations at each MHD time-step. At present, these cannot be achieved in combination with the dynamics.

NOELS: If the differential rotation is destroyed during the lifetime of an A star, is not there a danger to destroy the magnetic field at the same time? If this is the case, could it be that the field is smaller near the end of MS than at the ZAMS?

ARLT: The decay of differential rotation will have only a mediocre effect on the poloidal magnetic field which is observed. The magnetic field can actually be amplified using the differential rotation as an energy source. The following diffusive decay is very slow. Since the amplification of the (visible) poloidal field is weak, an observable evolutionary effect is not expected.

MATHYS: The question of the evolution of the magnetic field during the MS lifetime has been looked into indeed: there is no evidence for a significant relation between the magnetic field intensity and the fraction of MS lifetime completed by the star.

The A-Star Puzzle
Proceedings IAU Symposium No. 224, 2004   © 2004 International Astronomical Union
J. Zverko, J. Žižňovský, S.J. Adelman, & W.W. Weiss, eds.   DOI: 10.1017/S1743921304004442

Rotational velocity distributions of A-type stars

F. Royer[1,2], J. Zorec[3] and A.E. Gómez[2]

[1]Observatoire de Genève, 51 chemin des maillettes, 1290 Sauverny, Switzerland
email: frederic.royer@obs.unige.ch

[2]GEPI/CNRS UMR 8111, Observatoire de Paris–Meudon,
92195 Meudon cedex, France

[3]Institut d'Astrophysique de Paris, 98bis boulevard Arago, 75014 Paris, France

Abstract. Using an homogeneous sample of $v \sin i$ values for A-type main sequence stars (Royer *et al.* 2002), the equatorial velocity (v) distributions are determined as function of spectral class, from B9 to F2. The chemically peculiar and binary stars are discarded. These distributions of "normal" stars are discussed in terms of stellar formation and evolution, in particular the remaining bimodality observed for the earliest spectral types of the sample. We show that late B and early A-type main-sequence stars have genuine bimodal distributions of true equatorial rotational velocities probably due to the phenomena of angular momentum loss and of redistribution the star underwent before reaching the main sequence. A striking lack of slow rotators is noticed among intermediate and late A-type stars. The bimodal-like shape of their true equatorial rotational velocity distributions could be due to evolutionary effects.

Keywords. Stars: early-type, stars: rotation

1. Introduction

Thirty years ago investigations by Deutsch (1970) and Dworetsky (1974) already focused on the rotational velocity distributions of A-type stars and the bimodality observed around A0 in particular. Guthrie (1982) found that cluster late B-type stars have bimodal rotational velocity distributions, while for the same class of field stars they are unimodal. Bimodality was also observed in rotation of solar mass stars in Orion by Attridge & Herbst (1992), Choi & Herbst (1996) and Herbst *et al.* (2001).

Thus since bimodal distributions were found for cluster late B-type stars and for solar mass stars, it is important to review the distributions of rotational velocities of field late B-type stars in the early MS phases, to see whether it is possible to detect signatures of bimodality using new and highly homogeneous data. In the same way, it is also significant to know whether distributions of rotational velocities of stars in the spectral range from A0 to F0 are bimodal. This spectral range is of particular interest because it represents the transition from objects with radiative to those with convective envelopes.

Apart from the distinct type of stellar formation evoked by Guthrie (1982) to account for the velocity distribution aspects, fragmentation of large clouds for the bimodal star sample and through turbulence in small clouds in loose clusters for the unimodally distribution, differences could be also searched in circumstances that could lead in each case to a distinct internal angular momentum redistribution. A higher concentration of the angular momentum towards the stellar core might trigger phenomena with surface proxies noted as "stellar peculiarities". In this context can be considered the study of Abt & Morrell (1995) who tried to associate the Am phenomenon to the slow-rotation excess in the velocity distributions of A-type stars.

Thus, according to Abt & Morrell's (1995) attempt and the comments above, the goals of the paper are **a)** to see whether the chemically peculiar characteristics and the excess of slow rotators among the more massive stars of our sample, late type B and early A-type stars, are necessarily concomitant phenomena and **b)** to investigate whether the rotational velocity distributions of intermediate and late A-type stars bear signatures that could be related to the complex structure of their envelopes.

The full study will be published in a forthcoming paper (Royer *et al.* 2004).

2. Rotational velocities data

This study is based on the catalog of Royer *et al.* (2002), which compiles homogeneous $v \sin i$ data for B8 to F2-type stars. The luminosity class range V to IV is used as synonymous for "dwarfs" representing the first MS evolutionary phases. The first selection is then based on the luminosity class of the stars, and only the main sequence stars (classes V, IV-V and IV) are retained.

In this study, all known "close" binaries and chemical peculiar stars in the studied spectral range have been removed. Binaries were eliminated because of possible tidal braking mechanisms which introduce deviations in the distributions that cannot be ascribed to initial MS conditions related to single stars. Following the same line of thought, Am stars were taken out of the studied sample, as they are considered to be close binaries (Debernardi 2000). Moreover, they are found among the slow rotators (Abt & Morrell 1995).

Evidence exists that the Ap phenomenon appears when the star has already completed at least the first third of its MS life span (Hubrig *et al.* 2000a). Ap stars represent then a population which is systematically separated from what it could be meant as the first MS stages. No indications were found for significant magnetic braking in the stellar surface (Hubrig *et al.* 2000b). Be and A-type stars with shell spectra were also removed. A-shell stars seem to be the A-type counterparts of early-type Be stars (Abt & Moyd 1973). The Be phenomenon is associated with fast rotation and is age-dependent (Zorec 2004).

CP stars (chemically peculiar stars): The catalogs of Ap and Am stars from Renson *et al.* (1991) are used to identify peculiar stars, as well as the spectral classifications of Abt & Morrell (1995) and Abt *et al.* (2002).

CB stars ("close" binary stars): Tidal effects in a multiple system tend to synchronize the axial rotation period with the orbital period. This category of stars are selected using criteria based on HIPPARCOS data and spectroscopic data. Except for a few, the selected stars are in the HIPPARCOS catalog (ESA 1997). The binaries detected by this satellite, with $\Delta Hp < 4$ mag are flagged as CB stars. The compilation by Pedoussaut et al. (1985) of spectroscopic binaries is used to complete the identification, together with the Eighth Catalog of the Orbital Elements of Spectroscopic Binaries (Batten et al. 1989). Stars with both CB and CP criteria are classified as CP.

The sample is divided according to spectral type, chosen so as to warrant in each statistical significance and mass resolution, to allow for the detection of possible mass dependencies of the velocity distributions. The six resulting subsamples are listed in Table 1. Their respective $v \sin i$ distributions are displayed in Fig. 1 as hatched histograms.

3. Distributions of equatorial rotational velocities

3.1. *Rectified distributions*

The observed $v \sin i$ parameter is the projection of the equatorial velocity v of the star on the line of sight, i being the inclination between the stellar rotation axis and the line of

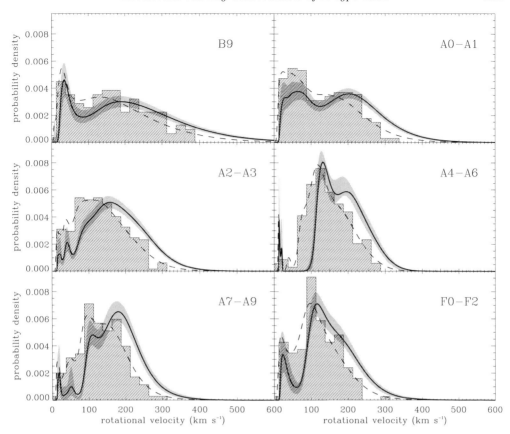

Figure 1. Rotational velocity distributions: observed $v \sin i$ (histograms), fitted $v \sin i$ distributions (dashed lines), deconvolved v distributions (solid lines). The gray strips around the solid lines are the variability bands which shows the significance of the modes in the distribution.

Table 1. Moments of the $v \sin i$ distributions. For each subsample, the number of stars, the median, the mean and the dispersion of the $v \sin i$ distribution are indicated

Subsample	#	Center ($\mathrm{km\,s^{-1}}$) median	mean	Disp. ($\mathrm{km\,s^{-1}}$)
B9	125	159 ±11.0	161 ±8.7	98 ±6.2
A0–A1	271	128 ±5.9	131 ±4.7	78 ±3.4
A2–A3	258	128 ±5.2	133 ±4.2	67 ±3.0
A4–A6	137	138 ±6.0	144 ±4.8	56 ±3.4
A7–A9	141	133 ±6.3	134 ±5.0	60 ±3.6
F0–F2	150	110 ±5.9	114 ±4.7	57 ±3.3

sight. The assumption that stellar rotation axes are randomly oriented is adopted. This hypothesis has been tested many times (Gray 1992, Gaigé 1993) and is still considered to be valid.

The Probability Density Function (hereafter PDF) of the $v \sin i$ values is thus the result of the convolution between the distribution of "true" equatorial velocities v, the distribution of inclination angles i, and the observational error law. It is estimated from the observed histograms, using a Kernel method. The smoothed $v \sin i$ distributions are displayed in Fig. 1 (dashed lines). The PDF of $v \sin i$ values is first deconvolved by the

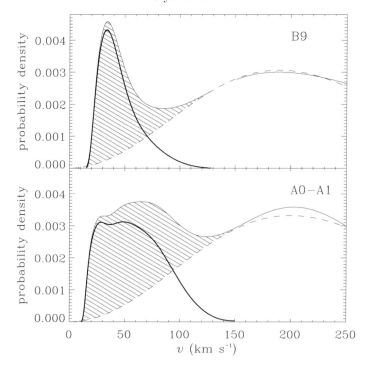

Figure 2. Slow rotators in B9 and A0–A1 stars. The fast rotator modes are fitted by Maxwellian distributions (dashed line). The excesses of slow rotators is represented by the hatched areas. The thick solid lines are the differences between the PDF of the sample and the Maxwellian distribution.

distribution of errors (chosen as log-normal) and then by the distribution of inclinations (axes randomly oriented). Both processes are carried out using the Lucy (1974) iterative method.

The resulting v distributions are displayed for the six subsamples in Fig. 1.

3.2. *Description of the v distributions*

The distributions of true equatorial velocities imply the following :
- The bimodal character of distributions corresponding to B9 and A0–A1 field dwarf stars is clearly present, even for the samples cleaned of all known Am, Ap and close binary stars. For B9-type stars: the mode of slow rotators is ~ 35 km s^{-1}, and for fast rotators ~ 190 km s^{-1}. For the A0–A1 subsample: they are ~ 60 km s^{-1} and ~ 200 km s^{-1}, respectively. The modes of fast rotators ($v \gtrsim 150$ km s^{-1}), for B9 and A0–A1 type stars, are fitted by Maxwellian PDFs (Fig. 2). The proportion of slow rotators ($v \lesssim 150$ km s^{-1}) is taken as the excess compared to the Maxwellian fast rotator distribution. The slow rotator peak corresponds to some 19 B9-type stars, and about 66 A0–A1-type stars.
The distribution of the rotation velocities of B9 to A1-type "normal" field dwarf stars are genuinely bimodal. This finding is contrary to the results obtained by Guthrie (1982), Abt & Morrell (1995), and Abt (2000). The presence of modes around 50 and 200 km s^{-1} may be due to formation processes and phenomena of AM loss and redistribution undergone during PMS phases.
- In the A2 to A9 spectral type groups, the small wiggles of the distributions in the velocity interval $0 \lesssim v \lesssim 70$ km s^{-1} concern only a negligible fraction of the stars. Moreover the variability bands associated with the distributions argue in the sense that

the presence of these slow rotators is not significant. These objects are probably unknown synchronized binaries or chemically peculiar stars that pollute the sample of "normal" stars. There is a net lack of rotators with $v \lesssim 70$ km s^{-1}. An absolute minimum of this fraction is seen in the A4–A6 group.

• An excess of slow rotators with velocities $0 \lesssim v \lesssim 60$ km s^{-1} appears in the F0–F2 group. This mode of slow rotators, around 20 km s^{-1}, in the distribution for early-F stars, is significant. The $B - V$ distribution for all the "normal" F0–F2 stars, ranging from 0.2 to 0.4 mag, shows that all (but one) stars with $v \sin i < 75$ km s^{-1} have color indices $B - V \gtrsim 0.3$ mag. It is known that the Böhm-Vitense gap occurs at $0.33 \lesssim B - V \lesssim 0.38$ mag (D'Antonna *et al.* 2002) and is a signature of the transition from radiative to convective atmospheres. The group of slow rotators for F0–F2 stars is then probably composed of stars that undergone a braking due to their convective atmosphere.

• The high-velocity side of the distributions of stars with spectral types ranging from A2 to F2 have a double-like structure. A well defined peak can be noticed at $v \simeq 200$ km s^{-1} in all groups, except for the F0–F2 stars for which at $v \simeq 200$ km s^{-1} there is only an inflection. The maximum of fast rotators in the B9 and A0–A1 groups is also situated at $v = 200$ km s^{-1}, although the corresponding distributions do not show any noticeable double structure.

The double structure might be due to an evolution-fast rotation interplay effect.

References

Abt, H. A. 2000, *ApJ*, 544, 933

Abt, H. A., Levato, H. & Grosso, M. 2002, *ApJ*, 573, 359

Abt, H. A. & Morrell, N. I. 1995, *ApJS*, 99, 135

Abt, H. A. & Moyd, K. I. 1973, *ApJ*, 182, 809

Attridge, J. M. & Herbst, W. 1992, *ApJ Lett.*, 398, L61

Batten, A. H., Fletcher, J. M. & MacCarthy, D. G. 1989, *8th Catalogue of the orbital elements of spectroscopic binary systems* (Victoria: Dominion Astrophysical Observatory)

Choi, P. I. & Herbst, W. 1996, *AJ*, 111, 283

D'Antona, F., Montalbán, J., Kupka, F. & Heiter, U. 2002, *ApJ Lett.*, 564, L93

Debernardi, Y. 2000, in *Birth and Evolution of Binary Stars*, IAU Symp. 200, ed. B. Reipurth & H. Zinnecker, 161

Deutsch, A. J. 1970, in *Stellar rotation*, IAU Colloquium No.4, ed. A. Slettebak, 207

Dworetsky, M. M. 1974, *ApJS* 256, 101

ESA. 1997, *The Hipparcos and Tycho Catalogues*, ESA-SP 1200

Gaigé, Y. 1993, *A&A*, 269, 267

Gray, D. F. 1992, *The observation and analysis of stellar photospheres*, 2nd edn. (Cambridge University Press)

Guthrie, B. N. G. 1982, *MNRAS*, 198, 795

Herbst, W., Bailer-Jones, C. A. L., & Mundt, R. 2001, *ApJ Lett.*, 554, L197

Hubrig, S., North, P. & Mathys, G. 2000a, *ApJ*, 539, 352

Hubrig, S., North, P. & Medici, A. 2000b, *A&A*, 359, 306

Lucy, P. 1974, *AJ*, 79, 745

Pedoussaut, A., Capdeville, A., Ginestet, N. & Carquillat, J. M. 1985, *List of spectroscopic binaries from the Toulouse general catalogue*, Observatoire de Toulouse

Renson, P., Gerbaldi, M., & Catalano, F. A. 1991, *A&AS*, 89, 429

Royer, F., Grenier, S., Baylac, M.-O., Gómez, A. E. & Zorec, J. 2002, *A&A*, 393, 897

Royer, F., Zorec, J. & Gómez, A. E. 2004, submitted to *A&A*

Zorec, J. 2004, in *Stellar Rotation*, IAU Symp. 215, ed. A. Maeder & P. Eenens, in press

Discussion

DWORETSKY: I saw something similar in A0 rotational velocities many years ago (ApJS, 1994). There is a "double-peaked" distribution of v_e even with all known peculiar and binary stars excluded. Helmut Abt saw something similar. Are any of the "genuine" slow rotators actually chemically peculiar, but we do not know it?

ROYER: The proportion of chemically peculiar stars in my sample is fairly constant at $\approx 15\%$ for stars brighter than $V = 6.5$ mag. This magnitude corresponds to the faint limit of the catalogues of Renson *et al.* (1991) and Abt & Morell (1995). Making the hypthesis that this prportion remains constant for stars fainter than $V = 6.5$, the estimated number of undetected chemically peculiar stars in my sample is about 20 stars, much smaller than the observed excess for slow rotators.

The A-Star Puzzle
Proceedings IAU Symposium No. 224, 2004   © 2004 International Astronomical Union
J. Zverko, J. Žižňovský, S.J. Adelman, & W.W. Weiss, eds.   DOI: 10.1017/S1743921304004454

Panel discussion section B

Chair: **N. Piskunov**

Section organizer & key-note speaker: A. Noels, J. Montalbán, C. Maceroni
Invited speakers: M. Marconi, M. Yıldız,
Contribution speakers: M. Yıldız, F. Royer, R. Arlt

Discussion

Piskunov: I would like to start this discussion with evolutionary aspects. There are a few recent developments both observational and theoretical. So, some might disagree and others might agree that there are many things happening concerning the pre-Main sequence evolution of early A type and Ap stars and that is what makes them different. Would anybody like to comment?

Mathys: A result that has raised controversy in the past few years is the claim by Hubrig *et al.* (2000, ApJ 539, 352) that Ap stars with mass below $3\,M_\odot$ become observably magnetic only once they have completed approximately 30% of their Main sequence lifetime. However, this study was restricted to slowly rotating stars (mostly with $P > 10\,\mathrm{d}$.), and the validity of its results for Ap stars with typical rotation periods of 2 - 4 d. could be questioned. Swetlana Hubrig has now addressed this point with new observations. The first results that she has recently presented at another conference are fully consistent with the conclusions of the original work. The recent discovery of a strong magnetic field in HD 66318 by Bagnulo *et al.* (2003, A&A 403, 645), a $2\,M_\odot$ star which, based on its membership in an open cluster, has completed only 16% of its MS lifetime, suggests that the 30% limit is not strict and may vary from star to star, e.g., according to field strength. But it does not by itself question the deficiency of magnetic stars close to the ZAMS.

Cowley: This is a question to those experts in the structure of stars. Upper MS stars have convective cores and radiative envelopes. Lower MS stars have the opposite structure. Is there ever a time or range of stellar mass where a star could be fully convective or fully radiative, or perhaps could the star oscillate from one structure to the other? Finally, are there uncertainties, perhaps in the stellar opacities, that could modify your answer?

Noels: You are talking about the models at the transition between the pp chain and the CNO cycle, which occurs at about $1.1\,M_\odot$. Low mass stars burn hydrogen through pp chain while more massive stars undergo CNO cycle reactions on the MS. The pp chain reactions are much less temperature sensitive than the CNO cycle reactions so low mass stars have no convective cores on the MS but they have a convective envelope. On the other hand, more massive stars have a convective core, but no convective envelope, due to a lower opacity in the external layers. In other words, increasing the mass favours the appearance of a convective core at the same time that it lessens the extent of the convective envelope. These two effects compete with each other so there are no fully radiative nor fully convective stars on the MS. The only fully convective structures are found high on the Hayashi track during the PMS phase.

The transition is very sensitive to the physical parameters and your question is very interesting indeed. This is especially true for the opacities and the chemical composition.

For α Cen for example, whose mass is $1.1\,M_\odot$, changing by a small amount the metallicity, can change the structure, from no convective core (Z high enough for the luminosity to be sufficiently small for pp chain) to a small convective core (lower Z meaning a higher luminosity and the CNO cycle).

MATHYS: I wish to call attention to a result that may be important to understanding rotation in Ap stars. While the general deficiency of short-period binaries ($P_{\mathrm{orb}} < 2.5\,\mathrm{d}$) among Ap stars is well known, it is less known that among the dozen Ap stars with a rotation period longer than 30 days, known to be SB1 stars, there is at most one that may have an orbital period shorter than 100 days. In other words it seems that very long rotation periods in binary systems occur only if the orbital period is very long too.

KUBÁT: How could the stellar oblateness of rapidly rotating stars affect the processes in stellar envelopes?

TALON: The oblateness becomes significant (of order 10%) for stars rotating faster than about $150\,\mathrm{km\,s^{-1}}$. In that case, the temperature of the atmosphere will vary according to latitude.

ARLT: Can you comment on the interaction between magnetic fields and meridional circulation?

MOSS: The problem of meridional circulation and magnetic fields was studied intensively about 15 to 20 years ago. The numerical results agreed with earlier analytical work. For axisymmetric fields, a state of approximate uniform rotation along joint field/streamlines is obtained (very locally the angle between the $\mathbf{B}$ and $\mathbf{v}$ vectors may not be near zero). With nonaxisymmetric fields, the probable end state is near-uniform global rotation. Given a strong, continually present, field, the magnetorotational instability will be inhibited. The MRI works well with a weak vertical field (for the upper limit, see Arlt, this proceedings, p. 103).

TALON: There was some recent numerical simulation by Pascale Garaud about the interaction of some weak magnetic field with meridional circulation. In that case, the end state is not solid body rotation, and some circulation remains, although its shape is strongly modified.

KRTIČKA: There are new measurements of the CNO cycle cross sections. How does this influence the properties of hot stars?

NOELS: We have tested the effect of the cross section of $^{14}\mathrm{N}(\mathrm{p},\gamma)^{15}\mathrm{O}$ in an A star. With a smaller cross section, the time needed to reach equilibrium of the CNO abundance in the CNO cycle is larger. Remember that this reaction is the slowest in the CNO cycle. So the "second loop" in the approach to the MS is longer and the ZAMS has a slightly higher T_{eff}. The MS itself as to its duration or its HR track is not significantly affected. However, it has been shown that in globular clusters the effect is quite important in lowering the luminosity of the turn-off. The ages determined by fitting of isochrones are affected by about 0.7×10^9 years.

YILDIZ: Before the opacity projects, there were systematic differences between the models and observations. Now, it seems that there is no such systematics in the early type stars. The problem is probably not with the opacity or the nuclear reactions since uncertainties in them would cause systematic differences. However, a rich variety of rotational properties of A type stars may be the dominant factor for the differences between observational and theoretical results pertaining to the stellar structure.

NOELS: You said that OPAL opacities have solved many problems, which is true, especially for Cepheids. However, the opacities have not had their last word! With the new CNO solar abundances, recently obtained by Asplund *et al.* (2004, A&A, 417, 751), the solar convective envelope is too shallow to reproduce the helioseismic data. To get a better agreement with the helioseismological (inversion) model, an increase of opacity is needed. Near the bottom of the convective envelope, Mike Seaton (OP Project) indeed finds an increase of about of 7% in the values he obtains versus the OPAL opacities.

TALON: There is another example of a case where some opacities were not properly taken into account but which had no real impact on models except in the determination of the beryllium abundance; this is the missing UV opacity.

GREVESSE: Comment on the missing "near UV" opacity In the Sun, using 1D photospheric models and classical opacity sources, the predicted flux below about 400 nm is always greater than the observed one. The abundance of Be derived from the two Be II lines around 313 nm is smaller than the meteoritic value. A few years ago, Balachandran & Bell (1998, Nature, 392, 791) and more recently M. Asplund (but using new 3D models) have shown that when an extra opacity source is taken into account in this spectral region the disagreement disappears and the Solar Be abundance is NOW in perfect agreement with the meteoritic value! The origin of this missing opacity in the blue and UV could be found in the filling by quite a large number of very faint atomic lines due to Iron group elements which simulates a quasicontinuous opacity which adds to the well known continuous opacity sources (H, H^-) below 400 nm.

MATHYS: A semantic point in this session: differential rotation has been used to refer to two different situations. On the one hand, the rotation velocity difference between stellar interior and atmosphere and on the other hand, difference of (surface) velocity between pole and equator. This is potentially confusing.

The A-Star Puzzle
Proceedings IAU Symposium No. 224, 2004   © 2004 International Astronomical Union
J. Zverko, J. Žižňovský, S.J. Adelman, & W.W. Weiss, eds.   DOI: 10.1017/S1743921304004466

Convection in stars

F. Kupka

Max-Planck-Institute for Astrophysics, Karl-Schwarzschild-Strasse 1, 85741 Garching,
Germany email: fk@mpa-garching.mpg.de

Abstract. Convection is one of the most intricate processes studied in stellar astrophysics and has challenged both theorists and observers since the beginnings of astrophysics. But during the last two decades observational data of unprecedented resolution and accuracy have been collected in solar and stellar research which permit a new look at the field. An enormous increase of computer speed now permits solving more complete model equations with more accurate numerical approximations. Modelling and theoretical understanding of convection, however, are lagging behind observational progress and are still wanting.

As a background to the contributions to this session on convection, I first provide an overview on its basic physics and its observational evidence. I point out why astrophysicists have a general interest in improvements of our understanding of stellar convection and then focus on convection in A-stars with their unique combination of convection zones. I summarise how this richness of different manifestations can arise in A-stars, such as convection zones near the surface and in the core, several on top of each other, or some of them depleted by diffusion processes, suppressed by or even creating magnetic fields, suspected to create a chromosphere in some of them, or influenced by binaries, to name just a few. In the last part I will present a few recent results on modelling of convection in A-stars.

Keywords. Convection, hydrodynamics, turbulence, stars: atmospheres, stars: interior

---

1. Introduction

Conferences on stellar astrophysics frequently devote an entire session to the topic of convection. Why is there such an ongoing interest in that particular field? Back in the 1920s the existence of a convection zone near the solar surface and in stars in general had been intensively debated until Unsöld (1931) proved that the layers at and below the bottom of the solar photosphere have to be convective due to partial ionisation of hydrogen. The latter lowers the adiabatic temperature gradient to such an extent that the plasma is unstable to buoyancy and hence fulfills Schwarzschild's criterion of convective instability (Schwarzschild 1905). Although a simple model for this process was proposed a year later (Biermann 1932), the well-known mixing length theory (MLT), it turned out to be extremely difficult to make accurate predictions of heat transport, mixing, and other properties, and not only just the rough order of magnitude estimates which can be obtained from this early model and its siblings. Indeed, during the last three decades convection has become the largest single factor of uncertainty in many problems of stellar evolution, stellar structure, pulsational stability, or model stellar atmospheres, to name just a few.

As a response to this demand, research has been performed and considerable progress has been achieved in our understanding of solar and stellar convection through new types of observations. Examples include high resolution spectra and time series of spectra of solar granules, as well as accurate measurements of the depth of the solar convection zone by means of helioseismology. At the same time, numerical simulations of solar and stellar convection have become more and more powerful because of enhanced numerical resolution and more realistic microphysics.

Despite such success, actual calculations of stellar structure, pulsation, and evolution are still based on the classical MLT in a variation similar to that given by Böhm-Vitense (1958). One reason is that numerical simulations are computationally too expensive to be directly linked to those calculations. For those convection models which are in widespread use, "local tuning" of a model parameter could be performed with numerical simulations so as to match a particular physical quantity. But in that case, a suitable numerical simulation has to be available for close enough physical parameters (i.e., effective temperature, surface gravity, etc.), because as always in uncharted territories, extrapolations can be dangerous. The contributions to this session have once more shown what kind of unexpected phenomena may actually occur in such seemingly simple physical systems as atmospheres and envelopes of A-stars.

2. Physics of convection

The basic physical scenario underlying stellar convection is simple. For a fluid with a density ρ stratified as a function of depth by gravitational forces such that $\rho_{\rm top} < \rho_{\rm bottom}$, a temperature stratification in which cold fluid is situated on top of hot one ($T_{\rm top} < T_{\rm bottom}$) can be unstable. This will indeed be the case, if the temperature gradient is steep enough. Then hot fluid which moves upwards will expand adiabatically by an amount sufficiently large for its density to become smaller than the density of the surrounding, colder fluid at the new location further upwards and a net buoyancy will prevail. Of course, this requires other physical processes such as radiative cooling, viscous friction, or "external forces" (rotation, magnetic fields, etc.) not to interfer too much so as to suppress the process of moving upwards while expanding. This *buoyancy driven instability*, which may occur in the Sun and in stars, was first described by Schwarzschild (1905) and requires, in modern notation, the local temperature gradient to be steeper than the adiabatic one, $\nabla > \nabla_{\rm ad}$.

In the case of the Sun there are actually two effects which together are responsible for the convective instability. The first is caused by partial ionisation of hydrogen lowering $\nabla_{\rm ad}$ such that $\nabla > \nabla_{\rm ad}$ (Unsöld 1931). The same can also occur as a result of the single or the double ionisation of He. Other species are usually not sufficiently abundant to provide a convective instability through this mechanism. A second possible source of convective instability generally also coincides with regions of partial ionisation: high opacity. Instead of decreasing $\nabla_{\rm ad}$, a high opacity directly increases ∇. This is readily understood by looking at the dimensionless temperature gradient for purely radiative energy transport in the diffusion approximation, $\nabla_{\rm rad} = (3\kappa_{\rm ross}PL_r)/(16\pi\,{\rm a\,c\,G}\,T^4 M_r)$. Here, P is the pressure, L_r is the luminosity at radius r, T is the temperature, M_r is the mass inside r, $\kappa_{\rm ross}$ is the Rosseland opacity, and the other symbols have their usual meaning as physical and mathematical constants, including the radiative constant a. For large $\kappa_{\rm ross}$, $\nabla_{\rm rad} > \nabla_{\rm ad}$ and hence convection can set in such that finally $\nabla_{\rm rad} > \nabla > \nabla_{\rm ad}$ within and around an opacity peak. The third reason for convective instability, high luminosity, can also be understood from considering $\nabla_{\rm rad}$: if the energy production ε_c in the core of a star is large, which occurs for instance in sufficiently massive stars, then because of $\varepsilon_c = {\rm d}L_r/{\rm d}M_r \approx L_r/M_r$, the ratio L_r/M_r and thus $\nabla_{\rm rad}$ are large as well.

The most direct observational evidence for convection in stars is provided by the solar granules. Observations of sufficient quality to undoubtfully reveal their shapes and sizes were first made by M. Schwarzschild (1959) (see also Leighton 1963) with a stratoscope. Previous observations from the ground were resolution limited by atmospheric seeing effects. Naturally, similar observational evidence cannot be given for the case of convection in the cores of stars. The lower mass limit for core convection to occur throughout the Main Sequence lifetime is found to be $\sim 1.2\ M_\odot$ using state-of-the-art microphysics and assuming solar element abundances. But the presence of convection in the cores of more massive stars is a very solid theoretical result known since the beginning of stellar evolution calculations (Ledoux 1947 and references therein). Massive stars are commonly considered to have no convection zones in their envelopes, as hydrogen and helium are essentially fully ionised. A revision of opacity sources has changed this picture: stars with at least $7\ M_\odot$ and a sufficiently high (i.e., solar) metallicity are now also predicted to have one or even two convection zones within their envelope due to opacity peaks from higher ionisation stages of the Fe-group elements (cf. Stothers 2000). The upper of these two zones is also important for A-stars (see Sect. 4).

Models of stellar convection have to account for physical effects interacting with the basic instability. Among them are radiative losses which go along with a comparatively very low viscosity. Forces acting on large scales (say on the scales of granules or larger), including those originating from magnetic fields or a global mean flow caused by rotation or pulsation, may inhibit or modify convection. Finally, there is a direct competitor which also causes buoyancy driven instabilities. Heavy fluid on top of lighter one is unstable to small fluctuations, the Rayleigh-Taylor instability. Such conditions can occur because of a gradient ∇_μ in the mean molecular weight μ. The competition between instabilities driven by either a temperature gradient or a gradient of mean molecular weight was first studied in detail by Ledoux (1947). He derived that $\nabla - \nabla_{\rm ad} > \nabla_\mu$ for a stratification to be unstable. Schwarzschild & Härm (1958) noticed difficulties when calculating convective cores of massive stars, where $\nabla_\mu > 0$ (i.e., stable) while $\nabla > \nabla_{\rm ad}$ (i.e., unstable according to Schwarzschild 1905) in some region just outside the core. This scenario is known as semi-convection or diffusive convection. An accurate stability criterion for this situation has to account for the differences in the diffusivities of heat, $K_{\rm h}$, and molecular weight (or concentration), $K_{\rm c}$. Only if $\nabla - \nabla_{\rm ad} > (K_{\rm c}/K_{\rm h})\nabla_\mu$, semi-convection will actually occur (cf. Canuto 1999). The opposite case in which $\nabla < \nabla_{\rm ad}$ (i.e., stable) while $\nabla_\mu < 0$ (i.e., unstable) was first studied by Stern (1960) in a geophysical context (salt-fingers) and later by Stothers & Simon (1969) and Ulrich (1972) for astrophysical problems. It is also known as inverse μ-gradient effect or thermohaline convection and may be important in nuclear shell burning of stars or when there is some form of mass transfer onto the stellar surface. For this type of "convective instability" to occur, $|\nabla_\mu| > (K_{\rm h}/K_{\rm c})(\nabla_{\rm ad} - \nabla)$ (see also Canuto 1999). This problem and its relation to A-stars have been discussed by Vauclair (2005).

3. Astrophysical interest in convection

The main physical consequences of convection are heat transport, mixing, and interactions with mean flow and magnetic fields and the dynamo generation of magnetic fields. Convective heat transfer changes (usually reduces) temperature gradients and causes horizontal temperature inhomogeneities. These in turn change the emitted radiation of stellar atmospheres compared to a purely radiative environment. Convection can thus be observed through its effects on photometric colours and spectral line profiles. As a consequence thereof, it contributes to the uncertainties in secondary distance indicators based on photometric indices (Smalley *et al.* 2002 and references therein). Provided sufficient mechanical energy is transfered through the photosphere, it can also cause chromospheric activity indicators such as UV emission lines to appear (Sect. 4).

Through changing the vertical temperature gradient in stars convection affects stellar structure and evolution. This is particularly evident for evolutionary tracks computed for the Pre-Main Sequence (PMS) phases or for any evolutionary stages beginning with the lower tip of the red giant branch. Even the location of the Main Sequence within the HR diagram is influenced through the dependence of the radii of stars with deep convective envelopes, such as the Sun, on the efficency of convection just underneath the surface. This causes one of the main uncertainties in mass determinations based on stellar model calculations and complicates the interpretation of the observed HR diagram. An extensive comparison of different models of convection on the PMS evolution was recently given by Montalbán *et al.* (2004). Particularly instructive are 1 $M_\odot$ models of solar metallicity which are shown therein and which all match the solar effective temperature and luminosity for the present age of the Sun. Due to the effects of convection alone their evolutionary tracks differ by up to several 100 K at 10× solar luminosity in both earlier and later stages of evolution. Other calculations demonstrate how the change of assumed convective efficiency of MLT models for the upper envelope of stars with less than 1.5 $M_\odot$ changes the location of (or radii at) the Zero-Age Main Sequence, the shapes of the Pre-Main Sequence tracks, and masses determined from such model calculations.

Convection zones are usually well mixed, provided the velocities are large enough to overcome molecular diffusion (segregation or levitation). Accurate predictions of mixing efficiency become a lot more difficult for layers which are stably stratified, when overshooting from an adjacent convection zone can occur. This affects the evolution of convective cores and consequently the stellar evolution at stages from the turn-off point of the Main Sequence onwards and particularly on the asymptotic giant branch (AGB, see, e.g., contributions to the Granada workshop in Gimenez *et al.* 1999). The final results of these processes become visible in differences in the terminal phases of a star including post-AGB development (see also the comments and references in Kupka 2003) and supernovae (see contributions in van der Hucht *et al.* 2003). The effects of convective mixing can also be observed at the Sun and similar stars through the amount of Li depletion stemming from PMS (and Main Sequence) evolution. The ^{7}Li isotope is destroyed at temperatures around 2.5×10^6 K and a deep convective envelope ranging from the photosphere to the burning temperatures of Li during PMS phases or also later on will remove some fraction of this species from the observable and much cooler surface layers (see contributions in da Silva *et al.* 2000).

Convection is thought to drive p-mode oscillations in solar-like stars through stochastic excitation (Kumar & Goldreich 1989). In other types of stars such as RR Lyrae stars, Cepheids, or white dwarfs convection is expected to modulate the pulsation and introduce nonlinearities which are observable in the light curves of these objects (see Feuchtinger 1999 for the RR Lyrae stars). A different example of interaction of convection with a large scale velocity field is the transport of angular momentum (which is also discussed by Talon 2005). The redistribution of angular momentum in the solar envelope can be derived from helioseismology (see Gilman (2000) for a review). Numerical simulations of rotating convective shells, particularly those presented by Miesch *et al.* (2000), have made considerable steps towards recovering the averaged longitudinal angular velocity profile in the equatorial and mid-latitude regions of the Sun which in turn has been reconstructed from helioseismological inversions.

A closely related effect of the convection-rotation interaction is the creation of magnetic dynamos (see Gilman (2000) for further references). Sunspots and the solar activity cycle are the visible results of a process powered by convection. Interestingly, sunspots were also studied as the first astrophysical examples of the inhibition of convection through a magnetic field (Biermann & Cowling 1938, Biermann 1941, Cowling 1953). While sunspot models have evolved considerably (Solanki 2003), they have not led to a general answer to the question: under which conditions do magnetic fields inhibit convection? Solar magnetograms by Domínguez Cerdeña *et al.* (2003) have shown there are considerable fields of 50 to 150 G in the intergranular lanes of quiet solar regions. These fields can only modify convection, but not suppress it. Gough & Tayler (1966) have derived analytical stability results for several configurations with a vertical field component and found the necessary field strengths for suppression of convection to be several kG for conditions typical for sun spots. Cool magnetic Ap stars hence range an interesting transition region, but a thorough study of magnetoconvection in these objects has not yet been made.

4. Convection in A-stars

The existence of convection zones in the photospheres of A-stars was first predicted by Siedentopf (1933). He extended Unsöld's argument for the lowering of ∇_{ad} as the cause of solar convection to all stars which have a region of partial ionisation of hydrogen reaching the photosphere from below (as with the Sun the role of opacity became clear only later on when more accurate microphysics data were available).

Evidences from spectroscopy for the presence of convection in A-stars include the peculiar behaviour of Balmer lines as a function of effective temperature T_{eff} when trying to match theoretical line profiles with the observed data (see Smalley 2005). Another indication are the shapes of line bisectors and detailed high resolution line profiles (see Kupka *et al.* 2005) for A-stars with low projected rotational velocity ($v \sin i$) as well as the necessity to introduce a large "microturbulent velocity" ξ_t of several km s^{-1} when comparing spectral lines from the same ion but having different strengths with observations. Remarkably, ξ_t appears to be (roughly) a function of T_{eff} (see also Smalley 2005). Likewise, chromospheric activity indicators such as UV emission lines disappear only for A-stars with T_{eff} greater than about 8300 K (Simon *et al.* 2002). The main conclusions from these observations are that convective velocity fields of several km s^{-1} are present in A-stars, appear to have a filamentary topology (columns of fast upflow embedded in slow downflows), while the temperature gradient must be close to the radiative one except for the coolest A-stars with T_{eff} not much greater than 7000 K.

All these observational evidences deal with the photospheric convection zone caused by partial ionisation of hydrogen which stems from both a peak in opacity and a minimum in ∇_{ad} for near surface temperatures and densities. Both disappear for the late B-stars. Further inside the envelopes of A-stars, partial ionisation of He I and He II can extend the upper convection zone and cause a second convection zone (due to He II) to appear. In both cases the instability is primarily caused by the lowering of ∇_{ad}. Layers in between the two zones are usually found to be subadiabatic ($\nabla < \nabla_{ad}$). Both zones can disappear due to diffusion of He from the bottom of the mixed (convective or overshooting) zone further down into the envelope. Diffusion is also responsible for further convection zones which may appear deeper inside the envelope of A-stars: radiative levitation can accumulate ions of Fe-group elements in layers with T around 200000 K (as has been discussed by Michaud 2005). The traditional claim that envelope convection is not important for the evolution of A-stars is based on the notion that such "thin" convection zones barely alter the stellar radius in comparison with a purely radiative envelope model. But if the chemical composition of A-star envelopes is to be modelled, diffusion has to be taken into account in its two forms of gravitational settling and of radiative levitation and at that point envelope convection cannot be ignored any more. Contrary to the case of surface convection, observational evidence about the properties of the deeper envelope convection zones can only be obtained indirectly, from comparisons of abundance peculiarities and perhaps at some point from asteroseismology.

Compared to their envelope convection zones the convective cores of A-stars have received a more general attention. The most important problems related to them are the extent of well mixed stably stratified layers above the convective cores due to over-shooting, the influence of stellar rotation on that mixing process, and the possibility of convective core dynamos. Convective mixing of stably stratified layers is thought to be observable through the colour distributions of open clusters, the exact position of binary pairs near the turn-off of the Main Sequence, and through changes of the internal com-position by nuclear reactions in later stages of stellar evolution. Convective overshooting changes the local temperature gradient by making it more flat close to the convection zone and steeper a little further away compared to what would happen for purely radia-tive transport of energy. A much more important consequence of overshooting is the extra mixing provided by (He rich) material flowing into the envelope and (H rich) material being drained into the core which changes the lifetime of the star and its final compo-sition at the end of the Main Sequence. The extent of mixing is frequently measured as a fraction d of the pressure scale height at the boundary of the convective core. Many simple models for overshooting use this measure and the application of such models has been criticized by Yıldız (2005) at this conference. These prescriptions also fail because the definition of d loses its meaning for smaller convective cores: the pressure scale height diverges in the centre of a star and to expect d to be a "constant" in this mass range (or any mass range) is hopeless and misleading. More meaningful measures could be based on the amount of core mass mixed by overshooting, as has been suggested by several researchers. Indeed, models similar to that one discussed in Sect. 6 provide the mixed mass fraction of stably stratified layers as an output result instead of requiring it to be an input parameter. Further progress in our understanding of core convection in A-stars can also be expected from numerical simulations such as those discussed in Browning *et al.* (2005), particularly for questions concerning the influence of rotation and possible dynamo mechanisms in the convective core of A-stars.

The influence of rotation and tidal forces through close binaries on the convection zones near the surface pose further unresolved problems in the physics of A-stars. Both might be needed to fully understand the observed data such as line profiles of A-stars. How shall theoreticians match all these challenges?

5. Simulations and models

The main problem of constructing convective stellar models is that the underlying equations describing the structure and dynamics of fluids (the fully compressible Navier-Stokes equations coupled to equations for radiative transfer, magnetic fields, etc.) are highly nonlinear and have to be solved numerically. Because of the extremely high strat-ification, density and pressure in a star change by many orders of magnitude from the core to the photosphere. As a consequence, a vast range of time scales is encountered in stars from radiative cooling of fluid at the surface (on the order of seconds) to the ther-mal time scale of gravitational contraction (on the order of 10^5 to 10^7 years). Likewise, the vast difference between buoyancy or inertial forces and their inherent time scales on the one hand and the time scales of viscous dissipation and radiative conduction in the stellar interior on the other allows for an enormous range of spatial scales on which dynamical processes can occur (up to 9 orders of magnitude). The way taken around this is to explicitly account for the length and the time scales of the most relevant to the physical problem. In the case of convection these are the spatial scales of up- and down-flow patterns (and corresponding horizontal flows), such as granules and plumes, and the time scales of the flow (speed of sound or flow motion) as well as the time required to achieve a quasi-stationary equilibrium state. The latter is set by the boundary condi-tions or radiatively cooled layers. This approach is also called "large eddy simulation" and is underlying all the numerical simulations of A-stars presented in this session. The physical picture behind this technique is that of volume averages performed within a computational domain.

Convection models are usually developed to predict averages over horizontal areas which allows a dimensional reduction of the problem. The physical idea underlying these models in general is that of an ensemble average. In this case equations are derived from the underlying dynamical equations with some additional approximations or heuristic assumptions to obtain a closed set for the prediction of equilibrium ensemble quantities such as the convective enthalpy (heat) flux or the (turbulent) pressure generated by the flow. Such ensembles quantities can also be computed from numerical flow simulations by considering a sufficiently large number of "states" generated through time integration which are averaged later on. This procedure is computationally much more expensive. However, no rigorous theory exists for the construction of more affordable ensemble averaged equations.

In a direct comparison the advantage of numerical simulations is that they explicitly account for the nonlinearities of the most important, energy carrying scales and the spatially inhomogeneous nature of convection. High computational costs are their most important drawback, if integral properties of a large numbers of objects have to be computed, for instance during the automated analysis of millions of spectra expected from the GAIA mission. Similar limitations hold for the modelling of complete stars or groups of stars and their time evolution. Possible caveats are the required independence of the results on specific (artificial) boundary conditions and properties of unresolved scales which both have to be excluded carefully. Convection models, on the other hand, are designed to be computationally affordable. However, their range of validity cannot be determined from first principles or from studying related physical scenarios such as geophysical or laboratory flows alone. Most convection models are derived from methods developed to describe homogeneous turbulence while stellar convection is demonstrably inhomogeneous, as shown by the solar granules. Hence, for the time being models always have to be carefully tested with observations (and suitably tested numerical simulations) to corroborate their applicability.

6. Modelling convection in A-star envelopes

For a long time MLT remained the standard model of convection for A-stars. Xiong (1990) first suggested a nonlocal model for A-star envelopes. This model describes the transport of velocity and temperature fluctuations created by convection as a diffusion process. It had previously been used to model convection in the cores of massive stars by Xiong (1985). Canuto (1992), Canuto (1993), and Canuto & Dubovikov (1998) suggested a new model which abandoned the diffusion ("down-gradient") approximation for nonlocal transport and also avoided the use of a mixing length to compute the dissipation rate of turbulent kinetic energy. An improvement of that model of nonlocal transport was suggested by Canuto et al. (2001) and in this form the convection model was adopted by Kupka & Montgomery (2002) to compute envelope models for A-stars. These computations ranged from the mid-photosphere down into layers where $T \sim 100000$ K. Figure 1 shows a comparison of the convective enthalpy (heat) flux for a mid A-type Main Sequence star with a numerical simulation by Freytag (1995) and with an MLT model tuned to match the peak flux of the nonlocal model. The MLT model would require a mixing length four times as large to obtain the convective flux found in both the nonlocal model and the numerical simulation for the lower lying (He II) convection zone. Compared to that a modification of the original model of Canuto et al. (2001) for nonlocal

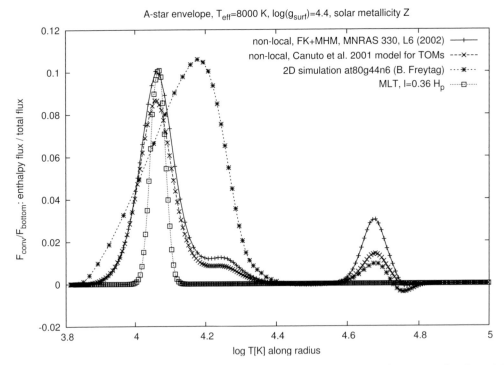

Figure 1. Relative convective heat flux for a mid A-star as obtained from a nonlocal model with two different closure approximations compared to averages from a 2D numerical simulation (courtesy B. Freytag) and an MLT model adjusted to match the peak flux.

transport as done by Kupka & Montgomery (2002) results only in a small change. The simulations shown by Freytag (2005) at this conference improve over their predecessors in Freytag (1995) also by replacing the 2D geometry with a 3D one and by using updated microphysics. Still, the MLT model remains to be by far the one most discrepant from the others.

That discrepancy is even more evident when considering the velocity fields. A comparison for the vertical root mean square convective velocity for the same nonlocal model, the MLT model, and the numerical simulations by Freytag (1995) has already been shown by Kupka (2003). For the MLT model the flow remains localised to the middle of the partial ionisation zones, predicts too small velocities for the lower lying convection zone, and neglects overshooting. The nonlocal model compares well to the simulation also quantitatively, although the surface velocities are clearly lower. The horizontal root mean square velocities allow a similar conclusion, as can be seen in Fig. 2. A local model such as MLT provides no framework to compute horizontal flow velocities and the extent of the convectively mixed region would again be grossly underestimated.

The performance of the nonlocal model used by Kupka & Montgomery (2002) for A-star envelopes was recently corroborated when Montgomery & Kupka (2004) computed models for envelopes of DA white dwarfs with the same code. Interestingly, much larger values of the mixing length parameter are required for an MLT model to match the peak fluxes of the nonlocal model, while the quantitative agreement of the latter with 2D numerical simulations of Freytag (1995) is comparable to that one found for A-stars. Limitations of the current nonlocal model become clear when computations for deep

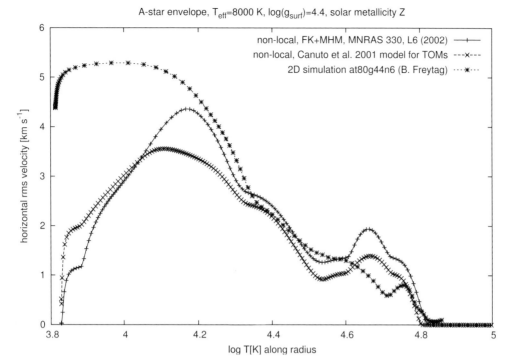

Figure 2. Horizontal root mean square convective velocity for a mid A-star as obtained from a nonlocal model with two different closure approximations (for TOMs – third order moments) compared to averages from a 2D numerical simulation (courtesy B. Freytag).

convection zones (similar to the case discussed by Trampedach 2005) are performed. Further improvements to the nonlocal transport model such as that one suggested by Gryanik & Hartmann (2002) are thus now investigated. Preliminary studies based on numerical simulations for solar granulation by F. Robinson appear encouraging.

7. Conclusions

Observations of solar and stellar convection have demonstrated the numerous shortcomings of MLT like convection models, but have not provided a way out through simple parameter calibrations. Nonlocal models appear to be a valuable alternative at least for the case of A-stars. Recently, numerical simulations of A-stars in 3D have been started and some first (surprising !) results have been presented here, together with a host of new observational data. The field thus provides further challenges to anyone interested in a theoretical understanding of stellar convection, and such challenges are usually fun for observers, too. The study of solar and stellar convection hence is neither in its infancy, nor is it a "mature field", nor stagnant. It is in fact right in the middle of new developments which have long been waited for by researchers in several fields of astrophysics.

Acknowledgements
I would like to acknowledge support by an IAU grant for parts of the conference fee.

References

Biermann, L. 1932, *Z. Astrophys.* 5, 117

Biermann, L. & Cowling, T.G. 1938, *private letter exchange*

Biermann, L. 1941, *Vierteljahresschr. der AG* 76, 194

Böhm-Vitense, E. 1958, *Z. Astrophys.* 46, 108

Browning, M.K., Brun, A.S. & Toomre, J. 2005, *These Proceedings*, 149

Canuto, V.M. 1992, *ApJ* 392, 218

Canuto, V.M. 1993, *ApJ* 416, 331

Canuto, V.M. & Dubovikov, M. 1998, *ApJ* 493, 834

Canuto, V.M. 1999, *ApJ* 524, 311

Canuto, V.M., Cheng, Y., & Howard, A. 2001, *J. Atm. Sci.* 58, 1169

Cowling, T.G. 1953, in: G.P. Kuiper (ed.) *The Sun*, Chicago Univ. Press

D'Antona, F. & Montalbán, J. 2003, *A&A* 412, 213

L. da Silva, M. Spite & J.R. de Medeiros (eds.) 2000, *The light elements and their evolution*, IAU Symp. 198

Domínguez Cerdeña, I., Sánchez Almeida, J. & Kneer, F. 2003, *A&A* 407, 741

Feuchtinger, M.U. 1999, *A&A* 351, 103

Freytag, B. 1995, PhD thesis, University of Kiel

Freytag, B. 2005, *These Proceedings*, 139

Gilman, P.A. 2000, *Sol. Phys.* 192, 27

A. Gimenez, E.F. Guinan & B. Montesinos (eds.) 1999, *Theory and Tests of Convection in Stellar Structure*, ASP Conf. Ser. (San Francisco), vol. 173

Gough, D.O. & Tayler, R.J. 1966, *MNRAS* 133, 85

Gryanik, V.M., & Hartmann, J. 2002, *J. Atm. Sci.* 59, 2729

Kumar, P. & Goldreich, P. 1989, *ApJ* 342, 558

Kupka, F. & Montgomery, M. H. 2002, *MNRAS* 330, L6

Kupka, F. 2003, in: L.A. Balona, H.F. Henrichs & R. Medupe (eds.), *International Conference on magnetic fields in O, B and A stars*, ASP Conf. Ser. (San Francisco), vol. 305, 190

Kupka, F., Landstreet, J.D., Sigut, A., Bildfell, C., Ford., A., Officer, T., Silaj, J. & Townshend, A. 2005, *These Proceedings*, CP2

Miesch, M.S., Elliott, J.R., Toomre, J., Clune, T.L., Glatzmeier, G.A. & Gilman, P.A. 2000, *ApJ* 532, 593

Michaud, G. 2005, *These Proceedings*, 173

Landstreet, J. D. 1998, *A&A* 338, 1041

Ledoux, P. 1947, *ApJ* 105, 305

Leighton, R.B. 1963, *ARA&A* 1, 19

Montalbán, J., D'Antona, F., Kupka, F. & Heiter, U. 2004, *A&A* 416, 1081

Montgomery, M.H. & Kupka F. 2004, *MNRAS* 350, 267

Schwarzschild, K. 1905, *Göttinger Nachrichten* 1906, 41

Schwarzschild, M. & Härm, R. 1958, *ApJ* 128, 348

Schwarzschild, M. 1959, *ApJ* 130, 345

Siedentopf, H. 1933, *Astron. Nachr.* 247, 297

Smalley, B., Gardiner, R.B., Kupka, F. & Bessell, M.S. 2002, *A&A* 395, 601

Smalley, B. 2005, *These Proceedings*, 131

Solanki, S.K. 2003, *A&A Rev.*, 11, 153

Stern, M.E. 1960, *Tellus* 12, 172

Stothers, R.B. & Simon, R. 1969, *ApJ* 157, 673

Stothers, R.B. 2000, *ApJ* 530, L103

Talon, S. 2005, *These Proceedings*, 59

Trampedach, R. 2005, *These Proceedings*, 155

Ulrich, R. 1972, *ApJ* 172, 165

Unsöld, A. 1931, *Z. Astrophys.* 1, 138

K.A. van der Hucht, A. Herrero & C. Esteban (eds.) 2003, *A massive star odyssey: from main sequence to supernova*, IAU Symp. 212

Vauclair, S. 2005, *These Proceedings*, 161
Xiong, D.R. 1985, *A&A* 150, 133
Xiong, D.R. 1990, *A&A* 232, 31
Yıldız, M. 2005, *These Proceedings*, 81

Discussion

MOSS: I would assume that in the surface region of A stars the simulations of convection are relatively easier because of the shallow depth-reduced stratification. A quite successful model exists for photospheric solar convection in the presence of large magnetic fields. So, can/will simulations soon be able to tell us something useful about near-surface convection in A stars in the presence of global-scale kG-strength magnetic fields?

KUPKA: Yes, I think that this is feasible. At least if one does not aim at including a global field distribution and the up-/downflow structure of convection at the same time, but rather restricts oneself to a study based on a series of box type convection simulations with a large scale field entering the simulation domain. That should be sufficient to tell us about convection suppression in Ap stars and its relation to the geometry of the global scale field.

VAUCLAIR: A comment: we now have a powerful tool for studying convection in stars and obtaining constraints on it: stellar seismology. Due to helioseismology we know the depth of the solar convection zone with a precision of 0.1%. In the future, we hope to be able to use asteroseismology to obtain constraints on stellar convective zones.

The A-Star Puzzle
Proceedings IAU Symposium No. 224, 2004　　　　　ⓒ 2004 International Astronomical Union
J. Zverko, J. Žižňovský, S.J. Adelman, & W.W. Weiss, eds.　　　DOI: 10.1017/S1743921304004478

Observations of convection in A-type stars

Barry Smalley

Astrophysics Group, School of Chemistry & Physics, Keele University, Staffordshire ST5 5BG,
United Kingdom

Abstract. Convection and turbulence in stellar atmospheres have a significant effect on the emergent flux from A-type stars. The recent theoretical advancements in convection modelling have proven to be a challenge to the observers to obtain measurements with sufficient precision and accuracy to permit discrimination between the various predictions.

A discussion of the current observational techniques used to evaluate the various convection theories is presented. These include filter photometry, spectrophotometry, hydrogen lines, and metal lines. The results from these techniques are given, along with the successes and limitations.

Keywords. Convection, turbulence, techniques: photometric

1. Introduction

The gross properties of a star, such as broad-band colours and flux distributions, are significantly affected by the effects of convection in stars later than mid A-type. Consequently, our modelling of convection in stellar atmosphere models can significantly alter our interpretation of observed phenomena.

Convection in stellar atmospheres is usually based on mixing-length theory (MLT) of Böhm-Vitense (1958). In their discussion of the Kurucz (1979) ATLAS6 models Relyea & Kurucz (1978) found discrepancies between theoretical and observed *uvby* colours which might be the result of an inappropriate treatment of convection within the models. Subsequently, several attempts have been undertaken to improve the situation. Lester *et al.* (1982), for example, introduced "horizontally averaged opacity" and a "variable mixing length" which improved the match with observed *uvby* colours, but did not remove all the discrepancies. The ATLAS9 models (Kurucz 1993) introduced an "approximate overshooting" which has not been without its critics (see Castelli *et al.* 1997 for details).

Canuto & Mazzitelli (1991, 1992) proposed a turbulent model of convection to overcome one of the most basic short-comings of MLT, namely that a single convective element (or "bubble" or "eddy") responsible for the transport of all the energy due to convection. This new model accounts for eddies of various sizes that interact with each other. The CM convection model was implemented in the ATLAS9 code by Kupka (1996).

Theoretical studies of convection in A-type stars have suggested that convection is inefficient (Freytag *et al.* 1996), and thus has only a very small influence on atmospheric structure (Heiter *et al.* 2002). However, these effects are nonetheless still important and in need of observational confirmation. For the early A-type stars (hotter than ∼8500 K) convective flux is negligible. It is only from mid A-type and later when convection becomes important, as an extensive convection zone develops in the photosphere and below (e.g. Weiss & Kupka 1999).

Convection in A-type stars poses a challenge to both theorists and observers. In this review I will concentrate on the observational effects of convection and what can be deduced from various types of observations.

2. Colours

Photometric indices are a fast and efficient method for determining approximate atmospheric parameters of stars. For the commonly-used Strömgren *uvby* system a vast body of observational data exists (e.g., Hauck & Mermilliod 1998) which can be used to estimate parameters using calibrated model grids (e.g., Moon & Dworetsky 1985, Lester *et al.* 1986, Smalley & Dworetsky 1995). Conversely, knowing atmospheric parameters from other methods, allows observed colours to be compared to model predictions. This method has been used to compare various treatments of stellar convection.

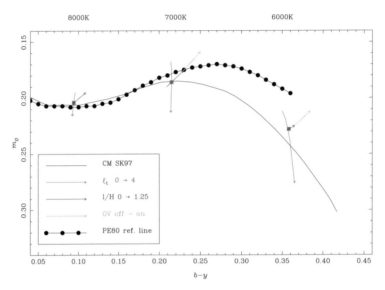

Figure 1. The variation of m_0 index with $b - y$ showing the sensitivity to microturbulence (ξ_t), mixing-length (l/H) and "approximate overshooting". At each temperature the model with $\log g = 4.0$, $\xi_t = 2$ km s^{-1} and $l/H = 0.5$ is denoted by a square. The arrows indicate the effect of varying ξ_t, l/H and including overshooting (for $l/H = 1.25$). The Philip & Egret (1980) main-sequence and the Smalley & Kupka (1997) CM relationships are included for reference. Note that the squares do not lie on the CM relationship due to differing $\log g$ values.

Smalley & Kupka (1997) compared the predicted *uvby* colours for the CM model with that from the standard Kurucz (1993) MLT models with and without "approximate overshooting". Comparison against fundamental $T_{\rm eff}$ and $\log g$ stars revealed that the CM models gave better agreement than MLT without overshooting. Models with overshooting were clearly discrepant. This result was further supported by stars with $T_{\rm eff}$ values obtained from the Infrared Flux Method (IRFM) and $\log g$ values from stellar evolutionary models. However, some discrepancies still remained, including a "bump" around 6500 K in the $\log g$ values obtained for the Hyades and continued problems with the Strömgen m_0 index. Similar conclusions were found by Schmidt (1999) using Geneva photometry.

The m_0 index is sensitive to metallicity and microturbulence, but also to convection efficiency as discussed by Relyea & Kurucz (1978) and Smalley & Kupka (1997 Fig. 6). Inefficient convection (CM and MLT $l/H \sim 0.5$) clearly works in the domain of the A stars down to $b - y \sim 0.20$ (approx 7000 K, F0). For cooler stars, convection becomes more efficient and substantive within the atmosphere and higher values of mixing-length and lower microturbulent velocities would be required to fit the observed m_0 indices of the main sequence (see Fig. 1). The 2d numerical radiation hydrodynamics calculations of Ludwig *et al.* (1999) are useful in this regard. They indicate a rise in mixing-length from $l/H \sim 1.3$ at 7000 K to $l/H = 1.6$ for the Sun (5777 K), while a much lower l/H ~ 0.5 was found for models at 8000 K (Freytag 1995). This is in agreement with that implied by m_0 index. However, around 6000 K there still remains a significant discrepancy, which could only be reduced by invoking the approximate overshooting option (Fig. 1). Thus, none of the convection models used in classical model atmospheres allows for the reproduction of the m_0 index, unless two parameters, i.e., l/H and the amount of "approximate overshooting", are varied over the H-R Diagram.

3. Fluxes

The observed stellar flux distribution is influenced by the effects of convection on the atmospheric structure of the star. As we have seen with photometric colours, these effects have a clearly observable signature. Hence, high precision stellar flux measurements will provide significant and useful information on convection.

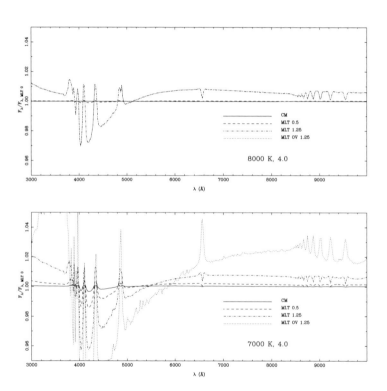

Figure 2. Fluxes for CM models and MLT models, with $l/H = 0.5$ and 1.25, compared to that for a model with zero convection. At 8000 K the CM and MLT $l/H = 0.5$ models give essentially the same fluxes as zero convection. Note that the fluxes for MLT $l/H = 1.25$ with and without overshooting are almost identical. At 7000 K the differences are more noticeable, especially in the region $4000 \sim 5000$ Å, and the effect of overshooting is now considerable.

Lester *et al.* (1982) presented a study of convective model stellar atmospheres using a modified mixing-length theory. They found small, systematic differences in the optical fluxes. Their figures demonstrate that convection can have a measurable effect on stellar fluxes.

Figure 2 shows the effects of changing mixing length from 0, through 0.5 to 1.25 on the emergent flux for solar-composition models with 8000 K and 7000 K ($\log g = 4.0$ and

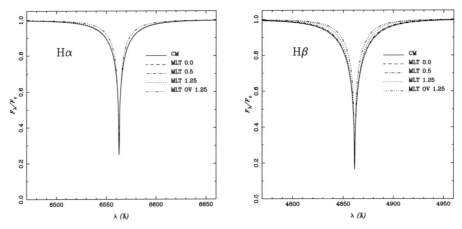

Figure 3. The effects of convection on the predicted shape of Balmer profiles for models with $T_{\mathrm{eff}} = 7000$ K, $\log g = 4.0$, $[\mathrm{M/H}] = 0.0$ and $\xi_t = 2$ km s^{-1}. Hα is unaffected by the values of l/H, but sensitive to "approximate overshooting", while Hβ is sensitive to both.

$\xi_t = 2$ km s^{-1}). At 8000 K both the CM and MLT $l/H = 0.5$ models give essentially the same fluxes as zero convection. At 7000 K, however, the differences are more noticeable, with the effect of overshooting being considerable.

In the ultraviolet the effects are even more significant, but comparison with observations would be complicated by the significant amount of metal line blanketing in this region. In contrast, in the infrared the effects are generally less compared to the optical. Hence, a combination of optical and infrared fluxes should provide a good basis for fixing the effective temperature of the star as well as testing the convective nature of the stellar atmosphere. In fact, the IRFM (Blackwell & Shallis 1977, Blackwell & Lynas-Gray 1994) is a very useful method for determining nearly model-independent T_{eff} for stars and has proved invaluable in our quest for accurate atmospheric parameters.

Unfortunately, very little high-precision stellar spectrophotometry exists. This situation will be rectified, once the ASTRA Spectrophotometer (Adelman *et al.* 2003, 2005) begins operation. This will allow spectrophotometry to be added to our observational diagnostic toolkit.

4. Balmer profiles

The temperature sensitivity of Balmer lines makes them an excellent diagnostic tool for late A-type stars and cooler (Gardiner 2000). However, as emphasised by van't Veer & Mégessier (1996), the Hα and Hβ profiles behave differently due to convection: Hα is significantly less sensitive to mixing-length than Hβ. Both profiles are affected by the presence of overshooting, with Hβ being more influenced than Hα (see Fig. 3). Since Hα is formed higher in the atmosphere than Hβ, Balmer lines profiles are a very good depth probe of stellar atmospheres. Naturally, Balmer profiles are also affected by microturbulence, metallicity and, for the hotter stars, surface gravity (Heiter *et al.* 2002).

In their comparison of Balmer line profiles, Gardiner *et al.* (1999) found that both CM and MLT without overshooting gave satisfactory agreement with fundamental stars. Overshooting was again found to be discrepant. Intriguingly, they found that while $l/H = 0.5$ was generally preferred, the region between 6000 K and 7000 K required a higher value ($l/H \geqslant 1.25$). This corresponds to the "bump" region found for Hyades stars using

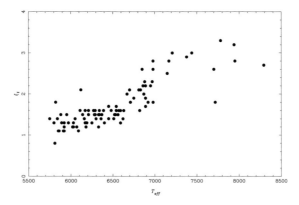

Figure 4. Variation of microturbulence with effective temperature. Based on results of Gray (2001) for stars with $\log g > 4.0$. Note the apparent relatively abrupt change in behaviour around 6600 K

uvby. However, this is not supported by either Fuhrmann *et al.* (1993) or van't Veer & Mégessier (1996).

Gardiner *et al.* (1999) also found evidence for significant disagreement between all treatments of convection for stars with $T_{\rm eff}$ around $8000 \sim 9000$ K. Smalley *et al.* (2002) reviewed this region using binary systems with known $\log g$ values and their revised fundamental $T_{\rm eff}$ values of the component stars. They found that the discrepancy found by Gardiner *et al.* (1999) was no longer as evident. However, this region was relatively devoid of stars with fundamental values of both $T_{\rm eff}$ and $\log g$. Further fundamental stars are clearly required in this region.

One potential difficulty with Balmer profiles is their great width, which can pose problems during the extraction and rectification of observations. Locating the 'true' continuum is non-trivial and a small error can lead to significant errors in any temperature obtained from fitting the profile (Smith & van't Veer 1988).

5. Microturbulence

Microturbulence (long used as a free parameter in abundance analyses) does appear to vary with effective temperature. Chaffee (1970) found that microturbulent velocity (ξ_t) rose from 2 km s^{-1} for early A-type stars up to 4 km s^{-1} for late A-type stars, and then back to 2 km s^{-1} for mid F-type stars and back up to 4 km s^{-1} for solar-type stars. He also found that microturbulence correlated weakly with the Strömgren m_0 index. Recently, Gray (2001) found that ξ_t varied from around 3 km s^{-1} for mid-A type down to around 1 km s^{-1} for solar-type stars, confirming the results of Coupry & Burkhart (1992) who found a variation from around 0 km s^{-1} for late B-type, up to around 3 km s^{-1} for mid A-type type, down to around 2 km s^{-1} for early F-type stars. Nissen (1981) gave a regression fit to both $T_{\rm eff}$ and $\log g$ for near-main-sequence solar-composition stars, from 2 km s^{-1} for F-type down to 1.5 km s^{-1} for solar-type stars. Fig. 4 shows the variation of ξ_t with $T_{\rm eff}$ for near main-sequence stars ($\log g > 4.0$) based on the results given by Gray (2001).

Velocity fields are present in stellar atmospheres which can be measured using line bisectors (Dravins 1987, Gray 1992). Compared to solar-type stars, the line bisectors are reversed, indicating small rising columns of hot gas and larger cooler downdrafts in A-type stars (Landstreet 1998). It is these motions that are thought to be responsible, at

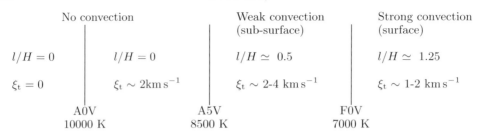

Figure 5. Schematic summary of typical values of mixing-length and microturbulence found in the literature for various spectral types.

least in part, for the existence of microturbulence. In fact, 3-D numerical simulations of solar granulation can account for observed line profiles without the need for any microturbulence (Asplund *et al.* 2000). Similar results have been found for Procyon (Gray 1985, Allende Prieto 2002), which is also a star with well-known physical parameters (e.g., Kervella *et al.* 2004). Unfortunately, current numerical simulations for A-stars predict bisectors in the opposite direction to that observed (Freytag & Steffen 2005).

6. Böhm-Vitense gap

Böhm-Vitense (1970) found a scarcity of stars around 7000 K which were attributed to the abrupt onset of convection. Observationally, as discussed above, this region corresponds to the *uvby* "bump" which is a mismatch between the $\log g$ from evolutionary models and model atmosphere calculations, an apparent change in behaviour of microturbulence around 6600K, and the suggestions that Balmer profiles require a higher value for mixing-length than other regions. In addition, there is a dip in chromospheric activity in this region (Böhm-Vitense 1995). D'Antona *et al.* (2002) discussed the Böhm-Vitense gaps and the role of turbulent convection, concluding that a gap in the region just below T_{eff} of 7000 K is a T_{eff} rather than just a colour gap. This is an interesting area of the H-R Diagram where the convective core shrinks to become radiative, while the envelope and the atmosphere change from radiative to convective (de Bruijne *et al.* 2001). Chromospheric activity indicators are another useful diagnostic, with another transition regime around 8300 K (Simon *et al.* 2002). All in all, this in an observationally challenging region of the H-R Diagram, since there are both atmospheric and internal effects which we need to differentiate between.

7. Conclusions

Fig. 5 summarizes the typical results of the various observational tests. While there are some contradictions and anomalies the Figure does illustrate the current state of our observational understanding of stellar convection. For stars hotter than A0 there is no convection or significant microturbulence. For the early A-type stars there is essentially no convection within the atmosphere, since the temperature gradient is radiative, but there are velocity fields as indicated by the modest microturbulence values. Velocity fields increase as we go through mid to late A-type stars, and inefficient convection is required within the atmosphere. Once convection becomes efficient (F-type and later) the value of microturbulence is found to drop, while the mixing-length is required to increase (although some evidence points against this, such as the Balmer-line results discussed earlier).

Overshooting is still an issue to be resolved, since there are clearly velocity fields above the convection zone. While the "approximate overshooting" of Kurucz appears to have been discounted by observational tests, there is clearly the need for some sort of overshooting to be incorporated within model atmosphere calculations.

The effects of convection on the stellar atmospheric structure can be successfully probed using a variety of observational diagnostics. The combination of photometric colours and Balmer-line profiles has given us a valuable insight into the nature of convection in A-type stars. High quality observations that are currently available and those that will be in the near future, will enable further refinements in our theoretical models of convection and turbulence in stellar atmospheres.

Acknowledgements

The author gratefully acknowledges the support of PPARC, the Royal Society and the IAU. This research has made use of model atmospheres calculated with ATLAS9 at the Department of Astronomy of the University of Vienna, Austria, available from http://ams.astro.univie.ac.at/nemo/.

References

Adelman S.J., Gulliver A.F., Smalley B., Pazder J.S., Younger P.F., Boyd L. & Epand D., 2003 In: Adelman S.J., Kupka F., Weiss W.W. (eds.) *Model Atmospheres and Spectrum Synthesis.* A.S.P. Conf. Proc. 44, Poster E55.

Adelman S.J., Gulliver A.F., Smalley B., Pazder J.S., Younger P.F., Boyd L. & Epand D., 2005, *These Proceedings*, JP2

Allende Prieto C., Asplund M., López, R.J.G., Lambert D.L., 2002, *ApJ* 567, 544

Asplund M., Nordlund Å., Trampedach R., Allende Prieto C., Stein R.F., 2000, *A&A* 359, 729

Blackwell D.E., Lynas-Gray A.E., 1994, *A&A* 282, 899

Blackwell D.E., Shallis M.J., 1977, *MNRAS* 180, 177

Böhm-Vitense E., 1958, *ZfA* 46, 108

Böhm-Vitense E., 1970, *A&A* 8, 283

Böhm-Vitense E., 1995, *A&A* 297, L25

Canuto V.M., Mazzitelli I., 1991, *ApJ* 370, 295

Canuto V.M., Mazzitelli I., 1992, *ApJ* 389, 724

Canuto V.M., Goldman I., Mazzitelli I., 1996, *ApJ* 473, 550

Castelli F., Gratton R.G., Kurucz R.L., 1997, *A&A* 318, 841

Chaffee F.H., 1970, *A&A* 4, 291

Coupry M.F., Burkhart C., 1992, *A&AS* 95, 41

D'Antona F., Montalbán J., Kupka F., Heiter U., 2002, *ApJ* 564, L93

de Bruijne J.H.J., Hoogerwerf R., de Zeeuw P.T., 2001, *A&A* 367, 111

Dravins D., 1987, *A&A* 172, 200

Freytag B., 1995, *Ph.D. Thesis*, Universität Kiel

Freytag B., Ludwig H.-G., Steffen M., 1996, *A&A* 313, 497

Freytag B., Steffen, M. 2005, *These Proceedings*, 139

Fuhrmann K., Axer M., Gehren T., 1993, *A&A* 271, 451

Gardiner R.B., 2000, *Ph.D. Thesis*, University of Keele

Gardiner R.B., Kupka F., Smalley B., 1999, *A&A* 347, 876

Gray D.F., 1985, *ApJ* 255, 200

Gray D.F., 1992, *The observation and analysis of stellar photospheres*, (2nd ed.), Cambridge University Press

Gray R.O., Graham P.W., Hoyt S.R., 2001, *AJ* 121, 2159

Hauck B., Mermilliod M., 1998, *A&AS* 129, 431

Heiter U., Kupka F., van't Veer-Menneret C., Barban C., Weiss W.W., Goupil M.-J., Schmidt W., Katz D., Garrido R., 2002, *A&A* 392, 619

Kervella P., Thévenin F., Morel P., Berthomieu G., Bordé P., Provost J., 2004, *A&A* 413, 251

Kupka F., 1996. In: Adelman S.J., Kupka F., Weiss W.W. (eds.) *Model Atmospheres and Spectrum Synthesis. ASP. Conf. Proc.* 44, p. 356

Kurucz R.L., 1979, *ApJS* 40, 1

Kurucz R.L., 1993, Kurucz CD-ROM 13: ATLAS9, SAO, Cambridge, USA

Landstreet J.D., 1998, *A&A* 338, 1041

Lester J.B., Gray R.O., Kurucz R.L., 1986, *ApJS* 61, 509

Lester J.B., Lane M.C., Kurucz R.L., 1982, *ApJ* 260, 272

Ludwig H.-G., Freytag B., Steffen M., 1999, *A&A* 346, 111

Moon T.T., Dworetsky M.M., 1985, *MNRAS* 217, 305

Nissen P.E., 1981, *A&A* 97, 145

Philip A.G.D., Egret D., 1980, *A&AS* 40, 199

Relyea L.J., Kurucz R.L., 1978, *ApJS* 37, 45

Schmidt W., 1999, *Master's Thesis*, Johannes Kepler University Linz, Austria

Simon T., Ayres T.R., Redfield S., Linsky J.L., 2002, *ApJ* 579, 800

Smalley B., Dworetsky M.M., 1995, *A&A* 293, 446

Smalley B., Kupka F., 1997, *A&A* 328, 439

Smalley B., Gardiner R.B., Kupka F., Bessell M.S., 2002, *A&A* 395, 601

Smith K.C., van't Veer C., 1988 in *Elemental Abundance Analyses*, Adelman S.J., Lanz T. (eds.), Institut d'Astronomie de l'Universite de Lausanne, p. 133

van't Veer C., Megessier C., 1996, *A&A* 309, 879

Weiss W.W., Kupka F., 1999 In: Giménez Á., Guinan E.F., Montesinos B. (eds.), *Theory and Tests of Convection in Stellar Structure, ASP Conf. Proc.* 173, p. 21

Discussion

GREVESSE: Turbulence in the Sun not only shows up in the asymmetries of the lines through the shapes of the bisectors but also in a systematic but small blue shift of the centre of the line which is related to the depth of formation of the line.

KUPKA: In stellar evolution, "alpha" is tuned to match the radius of the present Sun. Now it may be calibrated by numerical simulations. But although such an MLT calibrated model matches the right entropy jump, its temperature gradient is likely to be wrong (and it will actually continue to fail in helioseismological tests), as will be the radiation field, and the colours computed from it. Thus, just tuning the alpha value with simulations will not help. You can only get one quantity at and perhaps near the calibration point. It does not prevent us from finding a better model, and the alpha to match a particular quantity may be whatever in a particular case.

SMALLEY: Indeed, from a purely observational viewpoint we could use a negative mixing-length if that fitted the observational data better!

TRAMPEDACH: Overshooting is not contentious, but it does transport a negative convective flux. Kurucz' s 'overshoot" is a flux-smoothing scheme, that results in a positive convective flux in the stable layers adjacent to a convection zone.

KUPKA ON TRAMPEDACH'S REMARK: On the overshooting: Yes, I agree that is what is found in numerical simulations, in refined non-local models, and also in geophysical measurements of convection zones in geophysical convection zones. But they all usually also find a small zone where $\nabla - \nabla_{ad} < 0$ while $F_c > 0$ which is what has been smoothed out by Kurucz. For 1D models it was just a fitting parameter and having a negative flux instead does not resolve the problems one finds for these models when comparing them to observations. In this sense one still has the problem of how to carry over these findings to single models and reproduce the onbservations without further tuning.

The A-Star Puzzle
Proceedings IAU Symposium No. 224, 2004   © 2004 International Astronomical Union
J. Zverko, J. Žižňovský, S.J. Adelman, & W.W. Weiss, eds.   DOI: 10.1017/S174392130400448X

Numerical simulations of convection in A-stars

Bernd Freytag[1] and Matthias Steffen[2]

[1]GRAAL, Université de Montpellier II, F-34095 Montpellier, France
e-mail: Bernd.Freytag@graal.univ-montp2.fr

[2]Astrophysikalisches Institut Potsdam, 14482 Potsdam, Germany

Abstract. Radiation hydrodynamics simulations have been used to produce numerical models of the convective surface layers of a number of stars, including the Sun and other stars on or above the main-sequence, white dwarfs of type DA, and red supergiants.

While granulation of main-sequence solar-type stars resembles that of the Sun, the convective velocity fields of F-type stars are much more violent and accompanied by strong pulsations. The properties of the thin convection zone(s) of A-type stars differ again (see Fig. 1). In this contribution, the pattern and dynamics of their surface granulation, the photospheric velocity fields and their effect on line profiles are investigated, based on new 3-D models of surface convection in main-sequence A-type stars with T_{eff}=8500 K and 8000 K. Furthermore, we will look below the surface to study overshoot and the interaction of the surface convection zone and the deeper helium II convection zone.

Keywords. Convection, hydrodynamics, radiative transfer, methods: numerical, stars: atmospheres

1. Introduction

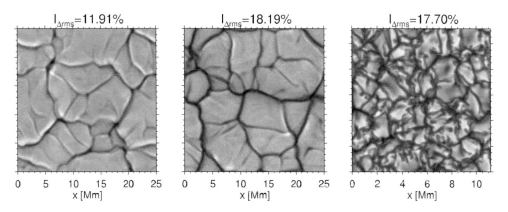

Figure 1. Grey surface intensity maps for model at85g44n16 (left: T_{eff}=8500 K, $\log g$=4.40); model at80g44n10 (center: T_{eff}=8000 K, $\log g$=4.40); model gt57g44n67 (right: T_{eff}=5770 K, $\log g$=4.44), cf. Table 1. The contrast is shown on top of each image. The granules on an A-star are significantly larger than on the Sun both absolutely and as measured, e.g., in pressure scale heights.

1.1. Number of convection zones in A-type stars

In the star modelling business it is common wisdom that surface convection in A-type stars reaches to a much smaller depth than in our Sun. However, the opinions about the exact extent vary significantly: modellers of the internal stellar structure sometimes claim that there is no convection at all, based on the fact that it is too weak to affect the stellar radius. Similar arguments can be heard about atmosphere models: the stratification is essentially in radiative equilibrium and convection can be ignored. However, if diffusion processes are included in the structure equations, one convectively mixed zone is required to explain observed abundance patterns. Mixing-length theory, on the other hand, predicts two separate shallow convectively unstable zones.

Can these views be reconciled with the help of radiation hydrodynamics simulations?

1.2. Numerical simulations with CO5BOLD

Earlier numerical models of the helium II convection zone in A-type stars by Sofia & Chan (1984) or of both convection zones by Freytag et al. (1996) were restricted to two dimensions.

During this conference, the first 3-D radiation hydrodynamics simulations of surface convection in A-type stars performed with CO5BOLD are presented (see Table 1). The equations of hydrodynamics are numerically integrated on a Cartesian grid (with varying grid size in vertical direction) using an approximate Riemann solver of the Roe type, modified to account for an external (here constant) gravity field. Ionization is treated by means of a tabulated equation of state. Non-local radiative energy transfer is computed with a long-characteristics scheme for a number of representative rays. The code takes tabulated opacities (grey for the A-star models) as input. Strict LTE is assumed and scattering is not taken into account at present. Radiation pressure is ignored. The lateral boundaries are periodic. Further information can be found in Wedemeyer et al. (2004) or Freytag et al. (2004).

Table 1. Basic model parameters: model name, grid points in two horizontal and one vertical direction, geometrical extent horizontally and vertically, effective temperature, gravity, and a short description.

model	grid	extent Mm^3	T_{eff} K	$\log g$ (cgs)	comment
gt57g44n67	$400^2 \times 150$	$12^2 \times 3$	5770	4.44	reference solar model
at80g44n10	$180^2 \times 90$	$25^2 \times 12$	8000	4.40	cooler A-star, vertically not very extended
at85g44n13	$180^2 \times 90$	$25^2 \times 14$	8500	4.40	hotter A-star model, transition 2-D to 3-D
at85g44n16	$180^2 \times 110$	$25^2 \times 15$	8500	4.40	hotter A-star, large overshoot region

2. Results of the simulations

2.1. Radiative versus hydrodynamic time-scales

Efficient radiative energy exchange in A-type stellar atmospheres causes radiation to be the dominant form of energy transport. Furthermore, the radiative relaxation time-scales are extremely short, much smaller than, e.g.. in the Sun (compare the continuous lines in the top right panels in Figs. 2, 3, and 4). The time-step in CO5BOLD is bound by these time-scales due to the explicit nature of its solvers and is below 0.2 sec for model at85g44n16, even with multiple radiation transport steps per hydrodynamics step. Therefore, the computational effort for an A-type model is typically a factor 10 to 100 larger than for a solar model with comparable resolution.

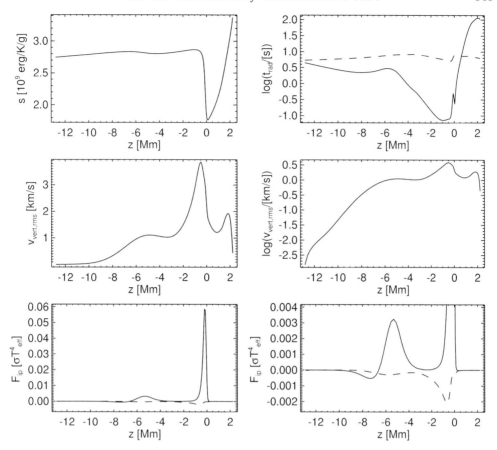

Figure 2. Various horizontally and temporally averaged quantities over height for model at85g44n16 (T_{eff}=8500 K): top left: specific entropy; top right: decadic logarithm of an estimate of the radiative relaxation time scale (continuous line) and grid cell sound crossing time (dashed line); middle left: root-mean-square of vertical velocity; middle right: decadic logarithm of root-mean-square of vertical velocity; bottom left and right: enthalpy flux (continuous line) and flux of kinetic energy (dashed line). The two small regions with negative entropy gradient indicate convective instability. Convection is inefficient in terms of energy transport, but can easily mix large regions.

2.2. *Granules on the Sun and A-type stars*

In Fig. 1 three snapshots showing the grey surface intensities of two A-type star models and a solar model are put side by side. The general appearance is somewhat different, although not dramatically. The intensity contrast is similar. However, it would decrease drastically for even higher effective temperatures. On the other hand, the granules are significantly larger in the A-type models than in the solar model, an effect that cannot be explained by the slight increase in the pressure scale height (see also, Freytag *et al.* (1997)). Instead, the faster moving intergranular lanes mean that granules are able to grow faster. Further the efficient radiation transport slows down the generation and growth of new downdrafts, causing granules to split at larger diameters (see Fig. 5).

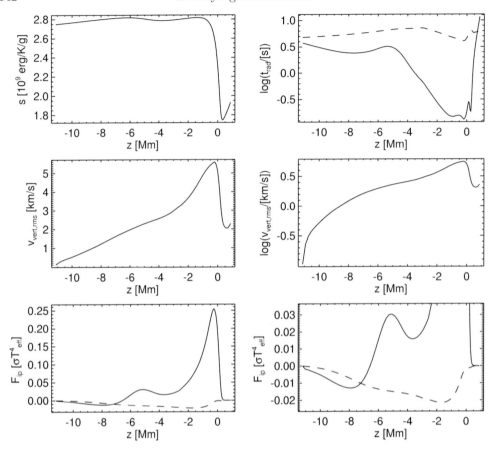

Figure 3. Various horizontally and temporally averaged quantities over height for model at80g44n10 ($T_{\rm eff}$=8000 K). See Fig. 2 for further details. This model shows two separate instability zones in a similar way as the hotter model displayed in Fig. 2. However, convection is more efficient (larger fluxes and velocities) and overshoot is even stronger.

2.3. Line profiles and bisectors

As expected from the rather ordinary granular cell pattern in Fig. 1, the line profiles in Fig. 6 do not look peculiar at all. Their bisectors do not differ much from the solar ones and show the common C-shape.

This is at odds with the observations by Landstreet (1998) who found strange looking line profiles with wide wings and an "inverted C-shape" bisector.

This might be an observational problem, because the very few A-stars, which have a low enough $v \sin i$ to allow the detection of the photospheric velocity field, might also be peculiar otherwise. In fact, both of the A stars with $T_{\rm eff} \sim 8000$ K listed by Landstreet (1998) are classified as peculiar (Am type) and are known to be spectroscopic binaries.

Or, there might be shortcomings in the simulations and assumptions valid for solar granulation models might not be appropriate: geometrically further extended models might show larger structures after a longer time-span than covered so far. A higher numerical resolution especially of layers near the steep subphotospheric temperature jump might alter the flow somewhat. The observed A-stars have slightly lower gravity than the presented models, which would cause an even steeper subphotospheric temperature jump.

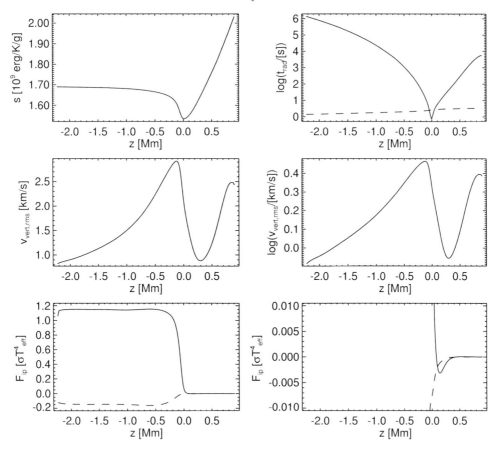

Figure 4. Various horizontally and temporally averaged quantities over height for model gt57g44n67 (T_{eff}=5770 K). See Fig. 2 for further details. The entropy profile shows that only the top of the very deep solar convection zone (and the stable photosphere) are included in the model. In the deeper layers of the model convection provides virtually the entire energy flux.

Non-grey opacities will affect the mean photospheric temperature structure. Effects due to magnetic fields or rotation (thanks to Rainer Arlt for the hint) are not included at all, yet.

One possibility is, that the bright borders of some granules in Figs. 1 and 5 might become even more prominent, enough to change the shape of the line profiles.

However, based on the available models and observations no firm conclusions can be drawn at present.

2.4. Overshoot

The two small regions with negative entropy gradient in Figs. 2 and 3 (top left) indicate convective instability and correspond closely to predictions of the standard mixing-length theory. Significant velocities are found in and near these unstable zones, which mix the entire computational domain efficiently. The large central velocity peak in Fig. 2 (middle left) marks the hydrogen convection zone, the lower one the helium II convection zone. The photospheric peak close to the top at the right is caused by waves emitted by the instationary convective flow (have a look at the velocities in the upper photosphere in Fig. 7). Below the helium II convection zone the velocities decay approximately exponentially

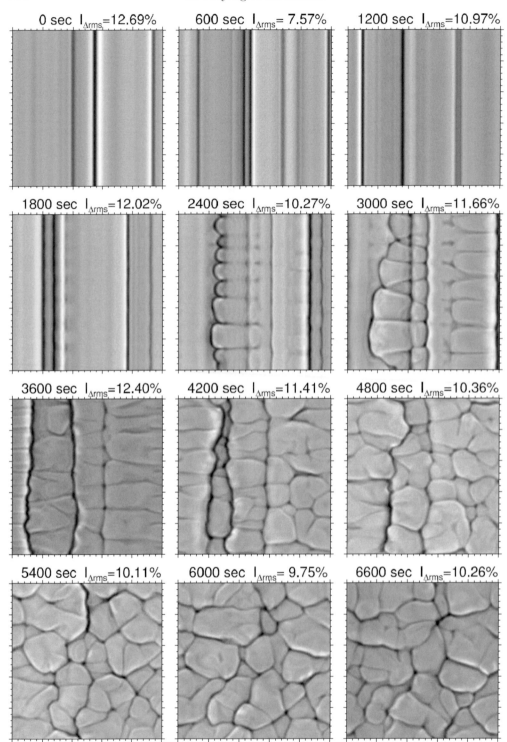

Figure 5. This sequence of grey intensity snapshots for model at85g44n13 (T_{eff}=8500 K) shows the transition from a flow with 2-D topology (inherited from the 2-D start model) to fully 3-D convection. The later frames show large granules with fast moving intergranular lanes. The model covers 25×25 Mm². Time and contrast are shown in the title.

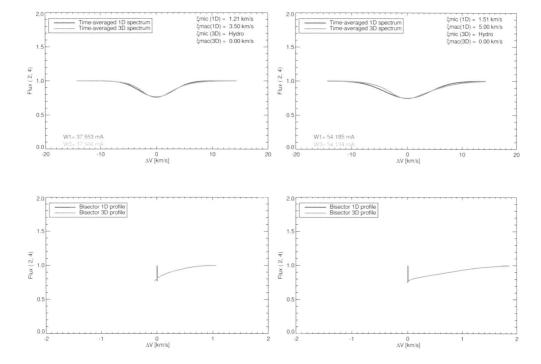

Figure 6. Flux line profiles and bisectors for model at85g44n16 (left: T_{eff}=8500 K), and model at80g44n10 (right: T_{eff}=8000 K). The dark (or red) symmetric profiles in the top row are computed from 1-D models generated from the the 3-D models by horizontal averaging with an appropriate value for the microturbulence. The lighter (or green) slightly asymmetric profiles are synthesized directly from the 3-D models taking into account the local velocities (and averaging afterwards over all columns in a model and a number of representative inclinations). The lower row shows the corresponding bisectors.

with depth, until they "hit" the closed bottom boundary. This happens, if the model is deep enough (Fig. 2, middle right). Otherwise, the computational domain is too shallow to allow the appearance of the exponential overshoot tail of the velocity (as in Fig. 3). For this cooler model, the lower convection zone is hardly discernible in the vertical velocity profile. However, in the flux profiles both zones can be distinguished. The energy fluxes show a significant overshoot beyond the unstable layers, too. The flux of kinetic energy is always non-positive. For even hotter models both instability zones will eventually merge to form one single convection zone.

3. Conclusions

First 3-D radiation hydrodynamics simulations of the photosphere and subphotospheric convection zone(s) of A-type stars have been presented.

They show, that there is no fundamental contradiction between the views presented in the introduction: The deviations from a radiative equilibrium stratification are small for the cooler A-type stars and vanish for the hotter ones, as predicted by the mixing-length theory. There are two separate instability zones. However, the regions with non-vanishing convective flux overlap for the cooler models. The convective overshoot above and below the unstable layers is so efficient that the velocity fields merge leading to complete mixing

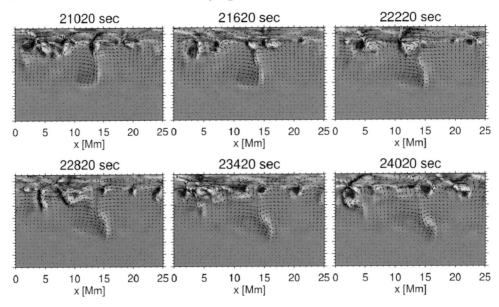

Figure 7. This sequence of vorticity slices (grey: small velocities; dark: clockwise motion; bright: counter-clockwise motion) from the 3-D model at85g44n16 shows the convective signature of the flow. The stellar surface ($\tau=1$) is approximately 2 Mm below the top of the box and the steep temperature (and density) jump has an imprint on the vorticity. Downdrafts usually do not reach very deep and tend to form horizontally moving surface eddies. However, occasionally a downdraft can reach the lower instability zone and generate a longer-lived structure (close to the center of the frames).

in the photosphere, between the instability zones, and in a layer of considerable extension below the helium II convection zone.

The granular flow in A-type stars organizes into larger cells than on the Sun. However, it shows normal solar-type topology and synthetic spectra display a solar-type bisector shape.

Acknowledgements

The numerical simulations have been performed on SGI machines at CINES in Montpellier and on an Hitachi SR8000 at the AIP in Potsdam.

References

Freytag, B. 1996, in *ASP Conf. Ser. 108: M.A.S.S., "Model Atmospheres and Spectrum Synthesis", ASP Conference Series; Vol. 108; 1996; ed. Saul J. Adelman; Friedrich Kupka; Werner W. Weiss*, p. 93

Freytag, B., Ludwig, H.-G., & Steffen, M. 1996, *A&A*, 313, 497

Freytag, B. and Holweger, H. and Steffen, M. and Ludwig, H.-G. 1997, in *"Science with the VLT Interferometer", Proceedings of the ESO workshop, held at Garching, Germany, 18-21 June 1996, Publisher: Berlin, New York: Springer-Verlag*, p. 316

Freytag, B., Steffen, M., Wedemeyer-Böhm, S., & Ludwig, H.-G. 2004, *CO5BOLD User Manual*, http://www.astro.uu.se/~bf/co5bold_main.html

Landstreet, J.D. 1998, *A&A*, 338, 1041

Sofia, S. & Chan, K.L. 1984, *ApJ*, 282, 550

Wedemeyer, S., Freytag, B., Steffen, M., Ludwig, H.-G., & Holweger, H. 2004, *A&A*, 414, 1121

Discussion

PISKUNOV: What causes the sharp horizontal strip in the pressure and velocity movies?

FREYTAG: The sharp opacity peak just below the photosphere causes a steep inward increase in temperature and corresponding drop in density. This jump shows up in many other quantities. For example, the density change is reflected in the velocity and is prominent in all gradient–like quantities like the velocity. The movie of the pressure fluctuations show the change of log(*pressure*) in time, scaled by a power of the temperature to enhance the tiny fluctuations in deeper layers relative to the larger photospheric value. This scaling also magnifies the fluctuations caused by overturning convective motions or waves "on top" of the subphotospheric temperature jump.

SHIBAHASHI: Your simulation reminds me of the stochastic excitation of p–modes in the Sun. Do you think that convection in A-type stars may induce the p–modes in these stars? How about the possibility of stochastic excitation in δ Sct stars?

FREYTAG: Actually, all simulations of convective zones on or above the main sequence show the stochastic excitation of p–modes in the computational box. However, the observable amplitude depends on the excitation, the damping, and the mass distribution. Stochastically excited p–modes in A stars would have very high order and a small amplitude, different from the observed radial or low–order modes.

KOCHUKHOV: (a) Do you expect that atmospheres of nonmagnetic A stars are well mixed due to convection and should not show vertical abundance stratification?
(b) How would the possible presence of a magnetic field change the picture?

FREYTAG: (a) All presented models do not include magnetic fields. The photopheric velocities are of the order of a few km s^{-1}. In the lower atmpsphere they are due to overturning convective motions (overshoot). In the upper atmosphere they are due to pressure waves generated by nonstationary convective motions. A nonmagnetic atmosphere should be mixed on a time scale of a few hours.
(b) The weak convection in A-type stars is very sensitive to all "obstacles". A magnetic field could easily suppress convection and its mixing of the atmosphere.

KUPKA: Have I understood your figure correctly that the bisectors obtained from your simulations are looking redwards? Because the observed ones are looking bluewards.

FREYTAG: Yes.

KUPKA: Could it be that this result changes if you evolve the simulations for longer time scales? You have started from a 2-D initial condition and the features in the He II ionisation region evolve on a much longer time scale.

ARLT: Is it possible to estimate the influence of the rotation of A stars on the results, since considerable Coriolis forces will be involved?

FREYTAG: I have done neither numerical simulations nor analytical calculations to estimate this influence. However, a strong Coriolis force should bend the downdrafts, favoring the development of fast–moving roll–like intergranular lanes. The depth of the region of material mixed by convective overshoot might be smaller.

The A-Star Puzzle
Proceedings IAU Symposium No. 224, 2004
J. Zverko, J. Žižňovský, S.J. Adelman, & W.W. Weiss, eds.
© 2004 International Astronomical Union
DOI: 10.1017/S1743921304004491

Simulations of core convection and resulting dynamo action in rotating A-type stars

Matthew K. Browning[1], Allan S. Brun[2,1] and Juri Toomre[1]

[1]JILA, University of Colorado at Boulder, 440 UCB, Boulder CO, USA 80309-0440
email: matthew.browning@colorado.edu

[2]SAp, CEA-Saclay, 91191 Gif-sur-Yvette, France

Abstract. We present the results of 3–D nonlinear simulations of magnetic dynamo action by core convection within A-type stars of $2\,M_\odot$ with a range of rotation rates. We consider the inner 30% by radius of such stars, with the spherical domain thereby encompassing the convective core and a portion of the surrounding radiative envelope. The compressible Navier-Stokes equations, subject to the anelastic approximation, are solved to examine highly nonlinear flows that span multiple scale heights, exhibit intricate time dependence, and admit magnetic dynamo action. Small initial seed magnetic fields are found to be amplified greatly by the convective and zonal flows. The central columns of strikingly slow rotation found in some of our progenitor hydrodynamic simulations continue to be realized in some simulations to a lesser degree, with such differential rotation arising from the redistribution of angular momentum by the nonlinear convection and magnetic fields. We assess the properties of the magnetic fields thus generated, the extent of the convective penetration, the magnitude of the differential rotation, and the excitation of gravity waves within the radiative envelope.

Keywords. Convection, magnetic fields, turbulence, stars: interiors

---

1. Introduction

The observational pathologies of A-type stars have attracted scrutiny for more than a century (e.g., Maury 1897). The striking surface features seen in some of these stars are varied and extensive: abundance anomalies, strong global magnetic fields, rapid acoustic oscillations. Though many of these may be driven primarily by phenomena near the stellar surface, some could be influenced by the core convection occurring deep within the interiors of A-type stars. In particular, the long-standing puzzle regarding the origins of the surface magnetism in Ap stars raises questions about whether that core convection can sustain dynamo action, and if so, whether such interior fields could have any influences at the stellar surface or are instead buried from view.

Convection within the core clearly impacts the structure and the evolution of massive stars. Convective motions can overshoot beyond the region of superadiabatic stratification, bringing fresh fuel into the nuclear-burning core and prolonging the star's Main-Sequence lifetime. Observational hints at the extent of these overshooting motions have been provided through studies of best-fit isochrones to clusters (e.g., Meynet *et al.* 1993). Likewise some theoretical estimates have served to constrain the penetrative properties of the convection (Roxburgh 1978, 1989, 1992). Yet a detailed model of the overshooting realized from convective cores, one that gives its extent and its variation with latitude, remains lacking.

Motivated in part by these observational and theoretical challenges, we have conducted 3-D simulations of core convection deep within A-type stars. We begin by discussing

first some properties of such nonmagnetic modelling from Browning *et al.* (2004), and
then turn to our most recent MHD simulations. Among our aims here are to assess the
penetration and overshooting achieved from convective cores, and the nature of the convective flows and the differential rotation that are established. In the calculations with
magnetism, we explore whether dynamo action is realized, and if so what are the gross
properties of the magnetism, its morphology, its intensity, its variations with time. Although the wealth of A-star observations lends vibrancy to our theoretical study, our
work has little to say about the surface conundrums at this stage. The models here simply provide some glimpses into the complicated dynamics happening within the central
regions of these stars.

2. Approach

Our simulations here examine the inner 30% by radius of A-type stars of $2 \, M_\odot$, rotating at the solar rate (case A) and fourfold faster (case B). The spherical domain thus
encompasses the entire convective core (central 15% of star), but only a portion of the
surrounding radiative envelope. We exclude the innermost 2% of the star from our computations for numerical reasons. Thus we consider turbulent convection interacting with
rotation and magnetism mainly within the cores of such stars. We employ our Anelastic Spherical Harmonic (ASH) code, which solves the full 3-D anelastic MHD equations
(Brun *et al.* 2004). The radial stratification is consistent with a 1-D stellar structure
model. The MHD simulations were begun by introducing small seed dipole magnetic
fields into the progenitor mature nonmagnetic simulations (Browning *et al.* 2004). We
adopt a Prandtl number P_r of 0.25, a magnetic Prandtl number P_m of 5, and consider
stratifications within the radiative envelope that are somewhat less stable (subadiabatic)
than those of real stars. We adopt stress-free and impenetrable boundary conditions at
the top and bottom of the domain, and require the magnetic field to be purely radial
there. The intricate spatial variation of flows and magnetic fields in our simulations is
captured by expanding both in spherical harmonics in the horizontal directions (including all degrees up to $\ell = 170$) and in Chebyshev polynomials in the radial (employing
two stacked expansion domains with $N_r = 82$ colocation points). The high-resolution
modelling of the evolving turbulent convection and intricate magnetism requires the use
of massively parallel supercomputers.

3. Results

Our simulations without magnetism (Browning *et al.* 2004) revealed that the convective flows within the core are vigorous and time dependent. The global connectivity allowed in the spherical domain means that the motions can couple widely separated sites,
extend radially through much of the core, and span large fractions of a hemisphere. The
flows penetrated into the surrounding radiative envelope, establishing a nearly-adiabatic
penetrative region that was prolate in shape, having greater spatial extent near the poles
than at the equator. The farther region of overshooting, where flows persisted but did not
appreciably modify the prevailing stratification, had a basically spherical outer boundary.
By pummelling the lower boundary of the radiative envelope, the penetrative convective
plumes excited internal gravity waves within the radiative envelope.

The present MHD simulations also reveal that sustained dynamo action is realized,
with the vigorous core convection serving to amplify initial seed fields by many orders of
magnitude (Fig. 1). The magnetic fields that result ultimately possess energy densities
comparable to that in the flows relative to the rotating frame. The magnetism is then

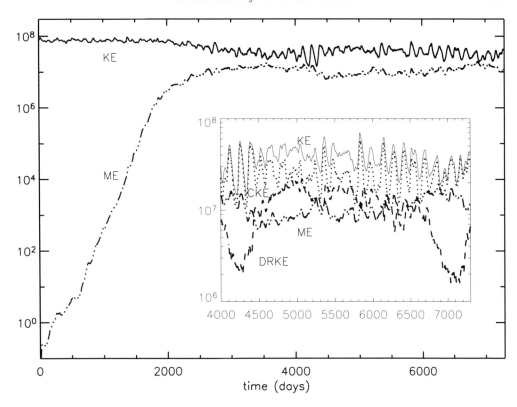

Figure 1. Temporal evolution of volume-averaged magnetic energy density (ME) from an initial seed dipole field, together with total kinetic energy density (KE). *Inset*: vigorous time dependence of the energy densities in the convection (CKE) and differential rotation (DRKE), with pronounced successive minima in DRKE.

sustained against decay for the multiple ohmic diffusion times that we have studied. Both simulations exhibit a rich time dependence, as magnetic fields are generated by the convective and the zonal flows and in turn feed back upon the motions, yielding the pronounced oscillatory behavior in the timetrace of energy densities (Fig. 1).

The morphologies of the flows and magnetic fields within the core are distinctive (Fig. 2): the radial velocity (v_r) shows columnar alignment reminiscent of Taylor columns, whereas the magnetism is both fibril (B_r) and stretched into ribbon-like structures that extend around much of the core (B_ϕ). The differing morphologies of v_r and B_r arise partly from the smaller value of magnetic diffusivity relative to the viscous diffusivity. The organized bands of the toroidal field may be attributable to stretching by angular velocity gradients, described in mean-field dynamo theory as the ω-effect, near the boundary between the core and the radiative envelope. Such stretching and amplification of the field by differential rotation also means that B_ϕ near the core boundary is only slightly diminished from its interior strength. In contrast, v_r and B_r decline rapidly in amplitude outside the convective core (Fig. 2a,b). Within the core, B_r and B_ϕ possess comparable strengths, but in the region of overshooting, where convective motions have waned, B_r has decreased by about a factor of 40 whereas B_ϕ is only roughly a factor of 3 smaller than in the core.

The magnetic fields realized within the core are predominantly fluctuating, accompanied by weak mean fields. Throughout the core, the energy associated with fluctuating

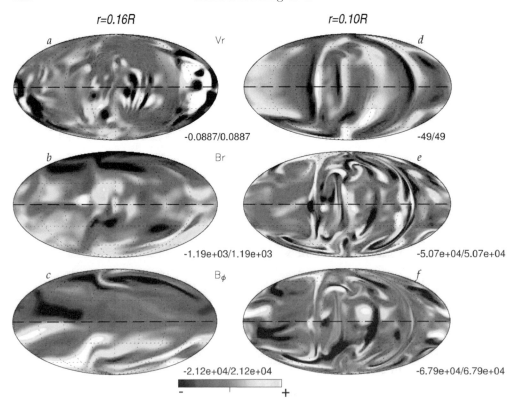

Figure 2. Instantaneous global views for case A of the fields v_r, B_r, and B_ϕ on spherical surfaces (rendered as Mollweide projections) at mid-core ($r = 0.10R$, right), and in the region of overshooting ($r = 0.16R$, left). Dashed line denote the equator. Positive values are shown in bright tones, negative values in dark tones, and the ranges rendered are indicated in $\mathrm{m\,s}^{-1}$ and in G. Within the core, intricate magnetic fields are well-correlated with the broad convective flows, whereas outside, differential rotation stretches B_ϕ into larger-scale features.

fields accounts for more than 90% of the total magnetic energy, split in roughly equal measure between toroidal and poloidal fluctuating components. In contrast, the comparatively weak mean (axisymmetric) fields are divided unevenly between toroidal and poloidal components, with B_t stronger than B_p by about a factor of two. Outside the core, the mean toroidal field becomes the dominant component of the plummeting magnetic energy. This likely arises because the generation of fields by helical convection largely vanishes, but the stretching and amplification of toroidal fields by angular velocity contrasts persists.

The strong differential rotation established in simulations without magnetic fields by the Reynolds stresses is diminished by the presence of magnetism (Fig. 3) due to the poleward transport of angular momentum by Maxwell stresses. In case A, with ME about 40% of KE, prominent differential rotation continues to be realized. However, in the more rapidly rotating case B with nearly equipartition fields, contrasts in angular velocity are greatly weakened. Case A exhibits major variations in differential rotation, with brief intervals during which the angular velocity contrasts become quite small. These grand minima in the differential rotation are visible in Figure 1 (*inset*) as broad dips in DKRE. The onsets of these intervals of small DRKE coincide with times when ME has grown larger than about 40% of KE, suggesting a complex interplay between strengthening magnetic fields and weakening differential rotation.

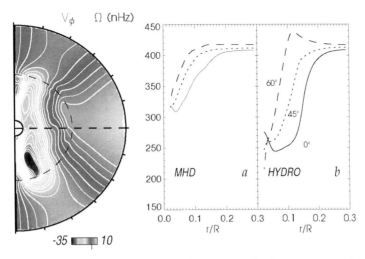

Figure 3. Differential rotation realized in case A and its hydrodynamic progenitor. (*left*) Contour plot in radius and latitude of longitudinal (zonal) velocity $\hat{v}_\phi$, with gray (color) bar and ranges (in $\mathrm{m\,s^{-1}}$) indicated. (*right*) Angular velocity $\hat{\Omega}$ with radius for indicated latitudinal cuts in both (*a*) case A and (*b*) its progenitor. The magnetism from dynamo action weakens the strong angular velocity contrasts seen in the hydrodynamic simulations.

4. Reflections

Our simulations have revealed some striking dynamical properties of core convection deep within prototypical A-type stars. The vigorous flows penetrate into the surrounding radiative envelope, thereby exciting gravity waves, and can establish differential rotation within the core. Our MHD models further show that the flows admit magnetic dynamo action, with initial seed fields amplified by many orders of magnitude and sustained against ohmic decay. The resulting magnetic fields are mostly nonaxisymmetric and intermittent in nature, and attain strengths of up to several hundred kG.

The generation of magnetic fields in our simulations is a complex process. Both helical convective motions and differential rotation appear to play roles in giving rise to the mean fields within the core. Thus the dynamo operating there may loosely be classified in the language of mean-field theory as being of the $\alpha^2 - \omega$ type. The generation of the far stronger fluctuating fields is less readily parameterized in terms of such concepts, but there, too, convection and differential rotation are likely both important. Whether such differential rotation is indeed an essential element in the operation of the dynamo, and how the balance between it and the magnetism is set, remains unclear at this stage.

How the magnetic fields and flows realized in our simulations are affected by rotation, and how they are influenced by our simulation parameters, are both likely to be sensitive matters. Future work is required to assess whether the dynamos operating here are realized at all rotation rates, and if so, how the strength of the fields may scale with rotation. Also unaddressed by the present simulations is the question of whether the dynamo-generated fields can migrate to the surface where they might be observed. Magnetic buoyancy instabilities could potentially lead to such field emergence (e.g., MacGregor & Cassinelli 2003) given strong enough magnetism, but MacDonald & Mullan (2004) indicate difficulties with this process for realistic stratifications. The highly fibril interior fields needed for such instabilities to arise are also somewhat at odds with the large-scale magnetism observed at the surface, although the detailed morphology of fields that have risen from the core through the radiative envelope is difficult to predict.

Acknowledgements

We thank Douglas Gough for helpful discussions, NASA for providing partial funding through SEC Theory Program grant NAG5-11879 and through the Graduate Student Researchers Program (NGT5-50416), and NSF PACI support for providing supercomputing resources at PSC, NCSA, and SDSC. Browning's travel was supported by the AAS and NSF through an International Travel Grant.

References

Browning, M.K., Brun, A.S. & Toomre, J. 2004, *ApJ*, 610, 512

Brun, A.S., Miesch, M.S. & Toomre, J. 2004, *ApJ*, in press

Cunha, M.S. 2002, in: C. Aerts, T.R. Bedding & J. Christensen-Dalsgaard (eds.), *Radial and Nonradial Pulsations as Probes of Stellar Physics* (San Francisco: ASP), p. 272

Kochukhov, O., Bagnulo, S., Wade, G., Sangalli, L., Piskunov, N., Landstreet, J.D., Petit, P., Sigut, T.A.A. 2004, *A&A*, 414, 613

MacDonald, J. & Mullan, D.J. 2004, *MNRAS*, 348, 702

MacGregor, K.B. & Cassinelli, J.P. 2003, *ApJ*, 586, 480

Matthews, J. 1991, *PASP*, 103, 5

Maury, A.C. 1897, *Harvard Ann.*, 28, 96

Stibbs, D. 1950, *MNRAS*, 110, 395

Wolff, S.C. 1980, *The A-Stars: Problems and Perspectives*, (Washington, D.C.: NASA)

Discussion

DWORETSKY: For many years the generally accepted model for Ap star fields has been the "frozen fossil field." Does this work imply that the dynamo model for Ap stars is staging a comeback? Do I need to change my standard "Ap star" lecture slides?

BROWNING: No, I am not advocating a change in worldview regarding the magnetism of Ap stars. We have shown that strong fields are produced within the core, but getting those fields to the surface is another matter entirely. At this stage, we simply do not know for sure whether that is possible; modelling by MacGregor & Cassinelli, and more recently MacDonald & Mullan, though, suggests that it is very difficult to get the fields out.

PISKUNOV: Do you see a magnetic field strength maximimum at some spherical harmonics and if yes, what order of spherical harmonic is it?

BROWNING: The power spectra are fairly flat until about $\ell = 15$, after which they fall off rapidly. There's a modest peak at around $\ell = 7$.

MOSS: First, a comment to Dworetsky's question: Fossil field advocates have always recognized that the cores of A stars will host dynamos! What this work does is to provide a detailed picture which, comfortingly, agrees reasonably well with prior estimates. As the speaker said, the question is whether the field can manifest itself at the surface. So there is no need for a change in paradigm. Question: I assume that there is not a large dipole moment of the dynamo field?

BROWNING: The maximum amplitude of the dipole component is typically about 5% of the maximum amplitude of the total radial magnetic field. So yes, the dipole is fairly weak.

The A-Star Puzzle
Proceedings IAU Symposium No. 224, 2004   © 2004 International Astronomical Union
J. Zverko, J. Žižňovský, S.J. Adelman, & W.W. Weiss, eds.   DOI: 10.1017/S1743921304004508

3D-simulation of the outer convection-zone of an A-star

Regner Trampedach

Research School of Astronomy and Astrophysics, Mt. Stromlo Observatory, Cotter Road,
Weston ACT 2611, Australia
email: art@mso.anu.edu.au

Abstract. The convection code of Nordlund & Stein has been used to evaluate the 3D, radiation-coupled convection in a stellar atmosphere with $T_{\rm eff} = 7300$ K, $\log g = 4.3$ and [Fe/H]= 0.0, corresponding to a main-sequence A9 star. I present preliminary comparisons between the 3D-simulation and a conventional 1D stellar structure calculation, and elaborate on the consequences of the differences.

Keywords. Convection, stars: atmospheres, stars: early-type

1. Introduction

From 3 dimensional simulations of convection, it has been known for the last two decades that convection grows stronger with increasing effective temperature and decreasing gravity. By stronger, is here meant larger Mach-numbers, larger turbulent- to total-pressure ratios and larger convective fluctuations in temperature and density. The stronger convection has also been accompanied by increasing departures from 1D stellar models that fail to predict the extensive overshoot into the high atmosphere, the turbulent pressure and its effect on the hydrostatic equilibrium, the temperature fluctuations and the coupling with the highly nonlinear opacity. The latter has the effect of heating the layers below the photosphere, thereby expanding the atmosphere, as also done by the turbulent pressure. The various 1D convection theories/formulations, *e.g.*, classical mixing-length (Böhm-Vitense 1958), non-local extensions to it (Gough 1977) or an independent formulation based on turbulence (Canuto & Mazzitelli 1992), all have similar shortcomings with respect to the simulations. Their predictive power is further limited by the free parameters involved.

Going towards earlier type stars, apart from stronger convection, also means a more shallow outer convection zone. This combination is rather unpredictable and is the main motivation for the work presented here. Classical predictions call for the outer convection zone to disappear close to the transition between A and F stars, but details about where and how this transition occurs can only be gained from realistic, 3D simulations, as outlined below.

2. The simulations

The simulation presented here was carried out using the code of Nordlund & Stein (1990) and is further described in Nordlund (1982), Stein (1989) and Stein & Nordlund (2003).

The Navier-Stokes equations are the basis of hydrodynamics. The code employs the *conservative* or *divergence* form

$$\frac{\partial \varrho}{\partial t} = -\nabla \cdot (\varrho \boldsymbol{u}) \tag{2.1}$$

$$\frac{\partial \varrho \boldsymbol{u}}{\partial t} = -\nabla \cdot (\varrho \boldsymbol{u}\boldsymbol{u}) - \nabla P_{\mathrm{g}} + \varrho \boldsymbol{g} \tag{2.2}$$

$$\frac{\partial \varrho \varepsilon}{\partial t} = -\nabla \cdot (\varrho \varepsilon \boldsymbol{u}) - P_{\mathrm{g}}\nabla \cdot \boldsymbol{u} + \varrho(Q_{\mathrm{rad}} + Q_{\mathrm{visc}}) \,, \tag{2.3}$$

where ϱ is the density, P_{g} is the gas pressure, ε is the specific internal energy, $\boldsymbol{u}$ is the velocity field, $\boldsymbol{g}$ is the gravitational acceleration and Q_{rad} and Q_{visc} are the radiative and viscous heating, respectively, the latter arising from the numerical diffusion applied.

Equations (2.1)–(2.3) describe the conservation of mass, momentum and energy, respectively, with sources and sinks on the right-hand-side. For the convection code the equations are preconditioned, by dividing by ϱ, to improve the handling of the large density contrast between the top and bottom of the simulations.

The vertical component of the momentum equation, Eq. (2.2), can be written

$$F_z = -\frac{\partial(\varrho u_z^2 + P_{\mathrm{g}})}{\partial z} + \varrho g \,, \tag{2.4}$$

as we have chosen $\boldsymbol{g}$ to be in the z-direction. With $F_z = 0$ this equation describes hydrostatic equilibrium, where the gas pressure and the turbulent pressure, $P_{\mathrm{turb}} = \varrho u_z^2$, provide support against gravity.

The gas pressure, $P_{\mathrm{g}}(\varrho, \varepsilon)$, and the opacities going into the computation of the radiative heating, Q_{rad}, form the atomic physics basis for the simulations. The continuous opacities are calculated from the MARCS-package (Gustafsson 1973) and subsequent updates as detailed in Trampedach (1997), the line opacity is in the form of opacity distribution functions (ODFs) (Kurucz 1992), and the equation of state accounts explicitly for all ionization stages of the 15 most abundant elements (Hummer & Mihalas 1988; Däppen et al. 1988).

The present simulation is performed on a $100 \times 100 \times 82$-point grid, has $T_{\mathrm{eff}} = 7300\,\mathrm{K}$, $\log g = 4.3$ and [Fe/H]= 0.0, and therefore corresponds to a A9 dwarf on the main-sequence. The computational domain is 11.5 Mm on each side, and 13.1 Mm deep, of which 1.5 Mm is above the photosphere.

So far the convection code has only been used for stars that were convective at the bottom boundary, so to accommodate this simulation, the boundary condition was changed and evaluation of radiative heating in optically thick layers was included.

The simulation was carried out in the plane-parallel approximation and includes no rotation or magnetic fields, and has a simple Solar abundance (Anders & Grevesse 1989). Both thermodynamics and radiative transfer is performed in strict LTE. The velocity-field is too large, throughout the simulation domain, to support segregation of elements, rendering diffusion and radiative levitation of individual species, comfortably irrelevant.

3. Comparison with 1D models

In Fig. 1 the temporal and horizontal averages of the simulation are compared with a corresponding 1D stellar model. From the solid black curve in panel **a)** we see that this simulation has an impressive turbulent pressure, making up almost 35% of the total pressure about 700 km below the photosphere (it is about 13% in Solar simulations). The mach-numbers reaching Mach 0.7 as shown with the dashed black line, is equally

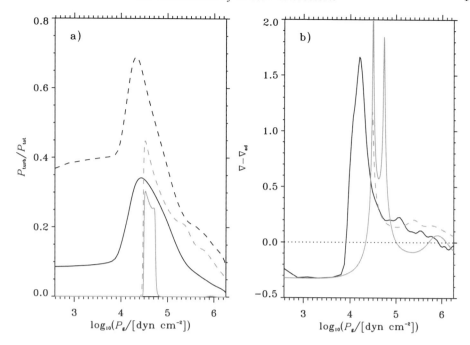

Figure 1. Comparisons between the simulation and two 1D, mixing-length models. In both panels the black solid line shows the temporal- and horizontally averaged simulation, and the gray solid line shows the same quantity for a 1D model with $\alpha = 1.0$ and the dashed gray curve is for $\alpha = 2.0$. Panel **a)** shows the turbulent- to total-pressure ratios, and the dashed black line shows the RMS Mach-numbers. Panel **b)** shows the super-adiabatic gradient, with the dotted zero-line aiding the location of the borders of the convection zones.

impressive. The convection is much less efficient compared to the Solar case, and invokes both higher velocities, as seen in panel a), and a higher super-adiabatic gradient, as seen from the black curve in Panel b) of Fig. 1. The two gray curves show the $P_{\mathrm{turb}}/P_{\mathrm{tot}}$-ratio and $\nabla - \nabla_{\mathrm{ad}}$ for two 1D envelope models using the standard mixing-length formulation of convection (Böhm-Vitense 1958). The solid line is for $\alpha = 1.0$ and the dashed line for $\alpha = 2.0$, in an attempt to bracket the behavior of the simulation. One glance at Fig. 1 makes it clear that no mixing-length model can reproduce the outer few percent of an A9 dwarf. The convection zone extends further up into the atmosphere and has much broader features than are possible with mixing-length models. We also see that overshoot from the convection zone supports an appreciable velocity-field, with velocities of 2–$3\,\mathrm{km\,s^{-1}}$, and has obvious consequences for spectral line shapes.

The large P_{turb} results in smaller temperatures at the same hydrostatic pressure, when compared to a 1D model. This can be translated into higher pressures and densities on an optical depth scale. Consequently the derived population of line-producing states in atoms, ions and molecules will be misleading and result in erroneous abundance analysis.

The mismatch between the simulation and the 1D models in the atmosphere is actually so profound that no mixing-length model with T_{eff} and g_{surf}, consistent with the simulation, can be found to match the simulation (*i.e.*, P, ϱ and T) at the bottom. This means that stellar structure and evolution calculations are misplaced in the HR-diagram, with implications for age-determinations of globular clusters and our general knowledge of the interior and the evolution of A-stars.

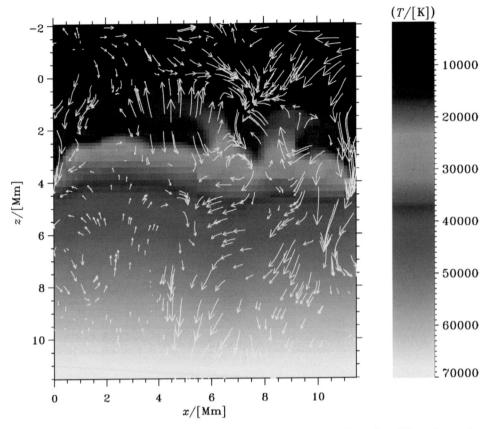

Figure 2. A vertical slice of a snapshot showing a temperature inversion. The color-scale shows temperature, and the arrows indicate the velocity-field.

4. Temperature inversion

This simulation, which is the hottest performed with this code, displays some large and persistent temperature inversions, 1.5-2.5 Mm below the photosphere. They have an amplitude of up to 8 000 K and are about 0.5 Mm deep. During the meeting, Noels *et al.* (2005) suggested that there might be a correlation between the turbulent pressure and the low temperatures; keeping $P_{\rm tot}$ and ϱ constant to ensure hydrostatic equilibrium, a large $P_{\rm turb}$ would force a low temperature. This is, however, not observed in the simulations. There is no correlation between the temperature inversion and $P_{\rm turb}$ nor are there correlations with vertical or horizontal velocities or with the density contrast. There is, however, a strong correlation with the vertical force (see Eq. 2.4), as the force on the plasma in the temperature inversion is always outwards (the $-z$-direction). Furthermore, the vorticity is clearly skewed towards larger values in the inversion layer, especially when compared to the upflow, but also with respect to the down-drafts.

Fig. 2 shows a vertical snapshot in temperature, displaying a typical temperature inversion. The arrows indicate the velocity field, with the maximum length corresponding to 2.9 km s^{-1}. The temperature inversion extends some 6 Mm across the left side of the plot, at a height of 2 Mm. It is connected to the downdraft at the edge, but it never connects with the downdraft to the right of the center. The velocity field there is connected, though, but the colder plasma from the inversion is compressed on the way to the downdraft, and, therefore, heated. The velocity field is diverging in the z-direction, neutral in the x-direction and convergent in the y-direction, perpendicular to the plane of the Figure the net-flow is convergent. The features observed in this simulation are cylindrical in the horizontal direction, with width and depth being similar and the length being 5–10 times larger.

The temperature inversions seem to develop when the local photosphere has subsided to about 2 Mm below the average photosphere, at the edge of a granule. The layers above, closer to the height of the average photosphere, then heats up, leaving a cooler area in between, the temperature inversion. In the white-light surface intensity, this now looks like the edge of a normal granule. The inversion immediately starts eroding from the newly heated region on top moving down, and presumably from heating by the surroundings. The temperature profile after this sequence, looks like that of a downdraft, and from the surface, what looked like the edge of a normal granule collapses as the temperature inversion below disappear. The whole cycle takes about 5 min.

The reason for this behavior is still under investigation, but it might be connected to the local minimum in ∇_s, as seen in Fig. 1 around $\log_{10} P_g \simeq 4.9$ corresponding to $z \simeq 1.9$ Mm.

5. Summary

A realistic 3D simulation of convection in the surface layers of a A9 dwarf reveals profound differences with a conventional 1D stellar structure model, as indicated in Sect. 3. In general, the simulations have much smoother and broader convection-zone features, compared to mixing-length models.

We are also treated to a new phenomena, as the simulations display repeated local temperature inversions just below the photosphere (*cf.*, Sect. 4). The mechanism responsible for these large inversions, has not been uncovered yet, but it being an effect of large fluctuations in the turbulent pressure, has been ruled out.

This is still work in progress, and in the future, this and other simulations will be used for the evaluation of spectral lines, limb-darkening, broad-band colors, granulation-spectra and p-mode spectra.

References

Anders, E., Grevesse, N. 1989, *Geochim. Cosmochim. Acta* 53(1), 197
Böhm-Vitense, E. 1958, *Zs. f. Astroph.* 46, 108
Canuto, V. M., Mazzitelli, I. 1992, *ApJ* 389, 724
Däppen, W., Mihalas, D., Hummer, D. G., Mihalas, B. W. 1988, *ApJ* 332, 261
Freytag, B., Steffen, M. 2005, *These Proceedings*, 139
Gough, D. O. 1977, *ApJ* 214, 196
Gustafsson, B. 1973, *Upps. Astr. Obs. Ann.* 5(6)
Hummer, D. G., Mihalas, D. 1988, *ApJ* 331, 794
Kurucz, R. L. 1992, *Rev. Mex. Astron. Astrofis.* 23, 45
Noels, A., Montalbán, J., Maceroni, C. 2005, *These Proceedings*, 47
Nordlund, Å. 1982, *A&A* 107, 1
Nordlund, Å., Stein, R. F. 1990, *Comput. Phys. Commun.* 59, 119
Stein, R. F. 1989. In: R. J. Rutten, R. J. and G. Severino, G. (eds.), "*Solar and stellar granulation*", Kluwer Academic Publishers, 381
Stein, R. F., Nordlund, Å. 2003. In: I. Hubeny, D. Mihalas, and K. Werner, (eds.), "*Stellar atmosphere modeling*", Vol. 288 of Conf. Ser., ASP, ASP, San Francisco, 519
Trampedach, R. 1997. *Master's thesis*, Aarhus University, Århus, Denmark

Discussion

COWLEY: Did you get a higher contrast between hot and cold regions than in simulations of cooler stars? At one time I though the ionization imbalance that we observe in some cool Ap stars might be explained by such temperature inhomogeneities. But I did some test calculations and found this could not explain the ionization, at least not with the numerical models I had.

TRAMPEDACH: The RMS temperature fluctuations at $\tau = 1$ on a Rosseland optical depth scale, are about twice as large for the A9 simulation ($\sim 610\,\mathrm{K}$) as for the Solar simulation ($\sim 300\,\mathrm{K}$). For the latter, the RMS fluctuations stay below $750\,\mathrm{K}$, peaking at $\tau = 10$, whereas in the A9 simulation they grow almost exponentially to reach $5000\,\mathrm{K}$ at $\tau = 100$.

I do not know what kind of numerical models you have used, but a decisive factor in the emergent spectra from the convection simulations is the asymmetry of the distributions of temperatures, densities and velocities, as well as their correlations. These characteristics are hard to predict without realistic simulations.

PISKUNOV: What is the height of the temperature inversion and the filling factor of the effect? Do you see significant changes of the filling factor?

TRAMPEDACH: The temperature inversions are about 0.5–$1.0\,\mathrm{Mm}$ deep and peak at a depths of 1.5–$2.5\,\mathrm{Mm}$. They involves from 2% to 8% of the horizontal area, with an average of about 5%.

NOELS: Do you find a high turbulent pressure in the region of temperature inversion? If so, I would suggest you investigate this point relative to the origin of the temperature inversion.

TRAMPEDACH: There is no obvious correlation between the temperature inversion and the local turbulent pressure. I refer you to Sect. 4 for the present extent of my analysis.

GREVESSE: Microturbulence and macroturbulence needed with 1D stellar photosphere models are not needed anymore with 3D models. Why is it not the same when 3D models are used, instead of 1D models, for A stars?

TRAMPEDACH: I do not think that has changed, going from solar-like stars to A stars. I think what you are alluding to, is the lack of reversal of the bisector shape in the A star simulations presented by Freytag & Steffen (2005). I do not know the reason for this and the only major thing missing from those simulations is line-blanketing. Whether that will make a difference is unclear. We will have to await further simulations. It is important to understand that theoretical bisectors from 1D models are straight, no matter how much micro- or macroturbulence is applied. The shape of bisectors is a higher order problem, that cannot be addressed with 1D models.

The A-Star Puzzle
Proceedings IAU Symposium No. 224, 2004   © 2004 International Astronomical Union
J. Zverko, J. Žižňovský, S.J. Adelman, & W.W. Weiss, eds.   DOI: 10.1017/S174392130400451X

Thermohaline convection and metallic fingers in polluted stars

Sylvie Vauclair

Laboratoire d'Astrophysique, Observatoire Midi-Pyrénées, 14 avenue Edouard Belin, 31400
Toulouse, France
email:sylvie.vauclair@obs-mip.fr

Abstract. When a layer of heavy matter is above lighter matter in a star, the inverse μ-gradient may lead to thermohaline (or double-diffusive) convection. This has been studied in the past for helium-rich atmospheres, but it may also occur for metal-rich layers. It has recently been studied for the accretion of hydrogen poor material onto the host stars of exoplanets. These stars present a metallicity excess compared to stars in which no planets have been detected. However, the reason for this excess is still a subject of debate. It may be primordial or the result of accretion, or both. In this last case, thermohaline convection may lead to "metallic fingers" which partially dilute the accreted matter inside the star. Such an effect can also be important in the chemically peculiar A stars in which metals accumulate in the atmospheric layers.

Keywords. Accretion, convection, diffusion, stars: abundances

1. Introduction

Spectroscopic observations of stars sometimes give evidence of overmetallicity compared to "normal" stars. In the chemically peculiar A stars, heavy elements are supposed to accumulate in the atmosphere due to radiative support. Solar type stars around which planets have been detected, compared with stars without planets clearly show a metallicity excess of a factor two on the average, while the individual [Fe/H] values lie between -0.3 and +0.4 (Santos *et al.* 2003 and references therein). The proposed possible explanations for this behavior are either that the star and its planets formed out of an already metal-rich cloud (primordial hypothesis) or that a non-negligible number of planets fell into the star during the formation process (accretion hypothesis). In the last case, overmetallic matter would be above matter with a normal composition, as in peculiar A stars.

Vauclair (2004) showed that in this case we must account for the convective instability induced by inverse μ-gradients, or thermohaline convection (also called "double-diffusive convection"). If metals accumulate in stellar outer layers, they will not stay there but will be partially diluted downwards in "metallic fingers" similar to the "salt fingers" observed in the ocean. The exact abundances of the metals which may remain in the stellar outer layers depend on parameters like the size and depth of the metallic fingers, which cannot be precisely constrained in the framework of our present knowledge. Here we show the importance of this process, which should be taken into account in the computations, and studied more precisely with the help of numerical simulations.

2. Thermohaline convection

Thermohaline convection is a well known process in oceanography. Warm salty layers on the top of cool unsalted ones rapidly diffuse downwards even in the presence of stabilizing temperature gradients. When a blob is displaced downwards, it continues moving

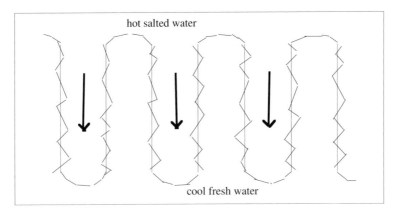

Figure 1. Schematic drawing of salt fingers: for hot salted water on top of cool fresh water, the salinity gradient is destabilizing while the temperature gradient is stabilizing; the medium should be stable except that heat diffuses more quickly than salt between the fingers and the interfinger medium. Because of the difference in diffusivities, a blob which begins to fall continues this motion until it mixes completely with the surroundings. This creates a special kind of convection which is well known in oceanography. A similar case occurs in stars for layers with inverse μ-gradients when the temperature gradients are stabilizing. The broken lines at the edges of the fingers are drawn as symbols of the partial mixing which occurs there, due to shear flow instabilities. This mixing is difficult to evaluate, which is the basic reason for the uncertainties in the computations of the final abundances.

downward due to its excess density compared with the surroundings, but at the same time that it is hotter contradicts this tendency. When the salt gradient is large compared to the thermal gradient, salted water normally mixes down until the two effects compensate. Then thermohaline convection begins. While the medium is marginally stable, salted blobs fall like fingers while the unsalted matter goes upward. This process is commonly known as "salt fingers" (Stern 1960, Kato 1966, Veronis 1965, Turner 1973, Turner & Veronis 2000, Gargett & Ruddick 2003). The reason why the medium is still unstable is due to the different diffusivities of heat and salt (for this reason it is also called "double-diffusive convection"). A warm salted blob falling in cool fresh water sees its temperature decrease before the salt has time to diffuse out. The blob goes on falling due to its density until it mixes with the surroundings (Figure 1).

The condition for the salt fingers to develop is related to the density variations induced by the temperature and the salinity perturbations. Two important characteristic numbers are defined :
- the density anomaly ratio

$$R_\rho = \alpha \nabla T / \beta \nabla S \qquad (2.1)$$

where $\alpha = -(\frac{1}{\rho}\frac{\partial \rho}{\partial T})_{S,P}$ and $\beta = (\frac{1}{\rho}\frac{\partial \rho}{\partial S})_{T,P}$ while ∇T and ∇S are the average temperature and salinity gradients in the considered zone
- the so-called "Lewis number"

$$\tau = \kappa_S / \kappa_T = \tau_T / \tau_S \qquad (2.2)$$

where κ_S and κ_T are the saline and thermal diffusivities while τ_S and τ_T are the saline and thermal diffusion time scales.

The density gradient is unstable and overturns into dynamical convection for $R_\rho < 1$ while the salt fingers grow for $R_\rho \geqslant 1$. On the other hand they cannot form if R_ρ is larger than the ratio of the thermal to saline diffusivities τ^{-1} as in this case the salinity difference between the blobs and the surroundings is not large enough to overcome buoyancy (Huppert & Manins 1973, Gough & Toomre 1982, Kunze 2003). Salt fingers can grow if the following condition is satisfied :

$$1 \leqslant R_\rho \leqslant \tau^{-1} \tag{2.3}$$

3. The stellar case

Thermohaline convection may occur in stellar radiative zones when a layer with a larger mean molecular weight sits on top of layers with smaller ones (Kato 1966, Spiegel 1969, Ulrich 1972, Kippenhahn *et al.* 1980). In this case $\nabla_\mu = \mathrm{d}\ln\mu/\mathrm{d}\ln P$ plays the role of the salinity gradient while the difference $\nabla_{\mathrm{ad}} - \nabla$ (where ∇_{ad} and ∇ are the usual adiabatic and local (radiative) gradients $\mathrm{d}\ln T/\mathrm{d}\ln P$) plays the role of the temperature gradient. Different situations may occur according to the respective values of these gradients. When ∇_{ad} is smaller than ∇_{rad}, the temperature gradient is unstable against convection (Schwarzschild criterion) which corresponds to warm water below cool water in oceanography. In the opposite case the temperature gradient is stable but the medium can become convectively unstable for inverse μ gradients (heavy matter above lighter one).

When the destabilizing effect of the μ gradient is larger than the stabilizing effect of the temperature gradients, the medium is unstable against dynamical convection: this is the so-called "Ledoux criterium". Namely the medium is unstable if the following condition is satisfied : :

$$\nabla_{\mathrm{crit}} = \frac{\phi}{\delta}\nabla_\mu + \nabla_{\mathrm{ad}} - \nabla < 0 \tag{3.1}$$

where $\phi = (\partial\ln\rho/\partial\ln\mu)$ and $\delta = (\partial\ln\rho/\partial\ln T)$.

When this situation occurs, convection first takes place on a dynamical time scale and the μ enriched matter mixes down with the surroundings until ∇_{crit} vanishes.

When the stability condition is achieved, we could suppose that the medium remains stable. When a blob begins to fall , it is heavier than its surroundings and would like to go on falling, but in this situation the temperature gradients act in the inverse direction and should be able to prevent it from falling. However, this is when thermohaline convection begins because, as for the salted case in the ocean, the heat diffuses more rapidly out of the fingers than the particles (Figure 1). In the process of the falling blob, heat diffuses between the blob and its surroundings so that it becomes less efficient to support it against its extra-weight. This leads to thermohaline or double-diffusive convection which mixes matter as a "secular process", namely on a thermal time scale (short compared to the stellar lifetime!).

Such an effect has previously been studied for stars with a helium-rich accreted layer (Kippenhahn et al. 1980). It was also invoked for helium-rich stars in which helium is supposed to accumulate due to diffusion in a stellar wind (as proposed by Vauclair 1975) and for roAp stars in case some helium accumulation occurs (Vauclair *et al.* 1991).

The study of thermohaline mixing in stars is far from trivial. Detailed comparisons of numerical simulations and laboratory experiments in the water case have recently been published (Gargett & Ruddick 2003), but the stellar case may be different as mixing then occurs in a compressible stratified fluid. Comparing the stellar case with the water case, we can guess that metallic fingers will form if the following condition is verified :

$$1 \leqslant \left|\frac{\delta(\nabla_{\mathrm{ad}} - \nabla)}{\phi(\nabla_\mu)}\right| \leqslant \tau^{-1} \tag{3.2}$$

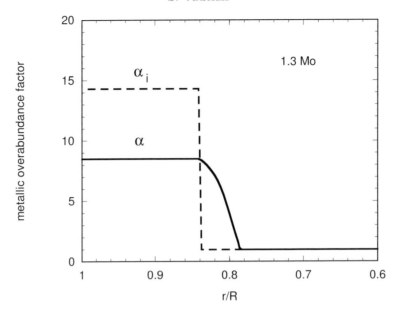

Figure 2. This figure presents the metallicity profiles in a 1.3 $M_\odot$ star, in the case of accretion, before and after dynamical convection. The ordinate α stands for the overmetallic ratio, i.e., the ratio of the global abundances of metals compared to the initial (solar) one. Thermohaline convection begins after this phase.

with $\tau = D_\mu/D_T = \tau_T/\tau_\mu$ where D_T and D_μ are the thermal and molecular diffusion coefficients while τ_T and τ_μ are the corresponding time scales.

As will be discussed below, the most important unknown about the fate of the fingers concerns the local mixing which occurs between the fingers and the interfinger regions. This diffusion process is basically due to shear flow instabilities at the boundary of the fingers (Figure 1) and depends on their velocity, which in turn depends on the diffusion process. Moreover, the whole process may be perturbated in case of other kinds of turbulence, like rotation-induced mixing. Only numerical simulations can help these diffculties.

4. The fate of overmetallic layers in stars

Vauclair (2004) studied Main-Sequence solar type stars which would have accreted hydrogen-poor material at the beginning of their lifetime. I assumed, for simplicity, that the accretion occurred in a very short time scale compared to stellar evolution. I then studied the fate of the accreted metals and chose the examples of 1.1 $M_\odot$ and 1.3 $M_\odot$ stars. I first computed the depth at which metal-enriched material is diluted and the actual overabundance ratio in the convective zone, when it reaches the marginal equilibrium phase, which is obtained when $\nabla_{\rm crit}$ vanishes (Equation 3.1) Then I used the formalism proposed by Kippenhahn *et al.* (1980) to evaluate the effect of thermohaline convection.

The whole picture may be described as follows: blobs of metal enriched matter begin to fall from the convective zone and exchange heat and heavy elements with their surroundings, Chemicals diffuse more slowly than heat, so that the blobs continue falling until they are completely disrupted, thereby creating finger shapes. The most efficient process for element diffusion out of the blobs is the shear flow instabilities at the edge of the fingers. As the falling matter undergoes friction with the rising matter, turbulence

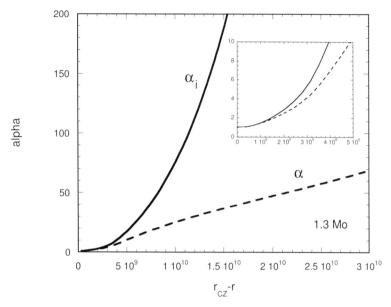

Figure 3. This graph displays, for a 1.3 $M_\odot$ star, the overmetallic ratio at the end of the dynamical phase, according to the original one. It also shows the depth at which the overmetallic matter is mixed after this phase. Namely, for a given initial overmetallic ratio α_i, the depth (below the convective zone) at which it will be mixed is obtained in abscissa from the α_i curve, and the resulting overmetallic ratio is then obtained from the α curve. Thermohaline convection occurs after this phase, and the final overmetallic ratio is difficult to evaluate.

occurs and mixes part of the fingers with their surroundings, on a horizontal length scale which is a fraction ϵ of the horizontal size of the blobs. The blobs disappear when they have travelled a distance long enough for this mixing to disrupt them completely. The remaining metal abundances in the convective zone are related to the depth of the fingers, which, as we have seen, is difficult to evaluate.

Although approximative, the computations show that metallic matter in the stellar outer layers should not stay there. It is expected to be partially mixed due to the dynamical convection and then should diffuse due thermohaline convection.

In chemically peculiar A stars, the destabilizing effect due to metal accumulation in the outer layers may be compensated by the stabilizing effect of the helium-gradient induced by downward diffusion. It should, however, be studied in details as it may modify in a non-negligible way the final abundance anomalies.

This type of convection should be studied with numerical simulations, as it may have important effects anytime overmetallic layers are above matter with a normal chemical composition.

References

Gargett, A., Ruddick, B. 2003, ed., Progress in Oceanography, vol 56, issues 3-4, pages 381-570, special issues on "Double diffusion in Oceanography",

Gough, D.O., Toomre, J., 1982, J. Fluid Mech, 125, 75

Huppert, H.E., Manins, P.C. 1973, Deep-Sea Research, 20, 315

Kato, S., 1966, PASJ, 18, 374

Kunze, E. 2003, Progress in Oceanography, 56, 399

Kippenhahn, R., Ruschenplatt, G., Thomas, H.C. 1980, A&A, 91, 175

Santos, N.C., Israelian, G., Mayor, M., Rebolo, R., Udry, S. 2003, A&A, 398, 363

Spiegel, E. 1969, Comments on Ap and Space Phys, 1, 57

Stern, M.E., 1960, Tellus, 12, 2

Turner, J.S., 1973, Buoyancy effects in fluids, Cambridge University Press

Turner, J.S., Veronis, G. 2000, Journal Fluid Dynamics, 405, 269

Ulrich, R.K., 1972, ApJ, 172, 165

Vauclair, S., 1975, A&A, 45, 233

Vauclair, S., 2004, ApJ, 605, 874

Vauclair, S., Dolez,N., Gough, D.O. 1991, A&A, 252, 618

Veronis, G.J., 1965, Marine Res., 21, 1

Discussion

SHIBAHASHI: Thermohaline convection in a star that you discussed seems to be the nonradial thermal instability. If so, I think the instability condition ($\nabla_\mu < 0$ and $\nabla_{ad} > 0$) shoud be the necessary condition but not the sufficient one, and whether or not the instability occurs indeed is not determined locally but by the gloval analysis (as in the case of overstability due to $\nabla - \nabla_{ad} > 0$ and $\nabla_\mu > 0$). Is your conclusion based on the local stability analysis?

VAUCLAIR: The instability criterion is determined locally but the extent to which the fingers develop and the efficiency of the resulting mixing depends on the situation in subjacent layers: this is a nonlocal treatment.

NOELS: I agree with you that with a gradient of helium the thermohaline convection can be very efficient. However, in the case of an accumlation of iron for example, don't you think that the change in μ is too small that it cannot trigger such a mixing?

VAUCLAIR: Of course metals contribute much less than helium to the mean molecular weight. But in some cases their accumulation can be large enough for this process to occur. It is the case for accretion, for example. In the case of element diffusion, generally helium falls down while metals go up. We may expect that the overall μ–gradient remains stabilizing. However, if metals accumulate in a region where the helium profile is flat, this process may occur and limit the metal abundance.

PRESTON: In your salt finger example what causes the convection to continue after the heavy overlying material has sunk?

VAUCLAIR: It continue if the process which forces the heavy matter to go up continues. The difficulty is to compute correctly the competitive process. For example in helium rich stars, helium accumulates due to the stellar wind but falls down due to fingers. The observations show that a factor of two overabundance remains but it is difficult to compute this result theoretically.

The A-Star Puzzle
Proceedings IAU Symposium No. 224, 2004   ⓒ 2004 International Astronomical Union
J. Zverko, J. Žižňovský, S.J. Adelman, & W.W. Weiss, eds.   DOI: 10.1017/S1743921304004521

Panel discussion section C

Chair: **John D. Landstreet**

Section organizer & key-note speaker: Friedrich Kupka
Invited speakers: B. Smalley, B. Freytag
Contribution speakers: S. Vauclair, M.K. Browning, R. Trampedach

Discussion

Kubát: What are the differences between both codes for convection simulations without magnetic fields which we have just seen?

Freytag: Both codes solve the same set of basic equations and rely on very similar fundamental assumptions. However, the algorithms used to solve the hydrodynamics or the radiation transport equations differ. The codes have no routines in common. There is an ongoing project to perform a simulation of solar granulation with both codes, relying on the very same settings (grid, model extension, equation of state, opacities, ray system, etc.). The remaining differences are tiny, for example, much smaller than the difference between the 2D and the 3D models. Both codes have some extensions (for instance dust or magnetic field) not (yet) found in the other.

Trampedach: I can only agree with Bernd Freytag. I am actually not taking part in the detailed comparison between the results of the two codes, but everything looks very similar. I think most of it has to do with concerns about stability. I believe they have higher inherent stability in their code than we have for example. That makes it easier to run. But the results are quite similar.

Piskunov: How deep is the temperature inversion below the photosphere?

Trampedach: It is not very deep, about a megameter or something below the optical depth one. The temperature is about 8000 K to 10000 K.

Cunha: To Smalley: Have you any idea how the prescription for the efficiency of convection as a function of effective temperature changes for the magnetic stars? If not what is preventing one from determining this using similar techniques?

Smalley: The prescription has not been tested for magnetic stars, but in principle one could be developed. However, the effects of a magnetic field on the atmospheric structure and the surface inhomogeneities may make this difficult. Nevertheless, given sufficient fundamental parameters for these stars one should be able to investigate this.

Vauclair: Could you comment a little more about the possible constraint on convection we may have from asteroseismology?

Kupka: Showing two figures skipped during his talk: For the Sun, helioseismology has already taught us a lot about the convection zone of the Sun. Inversion techniques tell us about the lower boundary of the solar convection zone and its location. We obtained information about the transport of angular momentum and differential rotation in its interior. The numerical simulations by Miesch *et al.* (2000) already reproduce a lot of the inferred properties at equatorial to mid latitudes. It would be marvellous if asteroseismology could help us to gain such kind of insight into the stellar interior of other stars.

VAUCLAIR: As asteroseismology can only detect low order modes, I do not think the resolution will be sufficient to get such a detailed picture, but it may tell us about the deep interior.

KUPKA: It may be also useful to study the interaction between convection and pulsation, as it may be expected for δ Scuti stars.

COWLEY: Can anyone comment on the possible relevance of solar umbral granules? It can be seen that we can have convection or something causing these granules even in the presence of kilogauss fields.

LANDSTREET: In the Sun the filling factor for most of the surface is very small. The gas motions easily dominate the average field and the result is that the gas motions force the field in some way into small flux tubes. The situation in the magnetic A stars is very different in that the filling factor of the field as far as we can tell from observations is 1.0. The field at least has a comparable energy density to the gas that in many cases dominates in the photosphere. So the situation is very different from the Sun.

TRAMPEDACH: If you just look at pictures of sunspots, close-ups, you can see how the granules are distorted and heavily affected by the field. No question about that. I do not know whether there have been hydrodynamic simulations of that yet. Because as you increase the field strength you really shorten the time scale and the problem becomes very hard to handle and simulate. Just with the solar field strength, for the normal fairly quiet Sun, you get a smaller granular effect. So, the field definitely changes the dynamics. We have also tried to increase the overall field strength and actually saw convection pretty much shut off. So magnetic fields have a profound effect on convection and have to be included.

MICHAUD: To Freytag: In your simulations you showed that the zone between the He II and H I convection zones is efficiently mixed in similar stars. Can you say something on what to expect for the mixing above the Fe convection zone at $T \approx 200000$ K up to the hydrogen convection zone. If we try to extrapolate in it a little further, to slightly higher temperature stars, where there is no H convection zone, is it then compatible that the difference between the Am/Fm stars and the HgMn stars will be that about an Fe convection zone? There will be complete mixing when there is also a hydrogen convective zone, but not when the H convective zone disappears completely, at 10000 K.

FREYTAG: The layers between the He II and H convetion zones are easily mixed by overshoot from below and above. The Fe and H I convection zones are much further apart. The layers in between are much harder to mix completely. Overshoot from the H I convection zone would reach down some pressure scale heights. The rest of the mixing has to come from the Fe I convection zone below. That may perhaps be possible if this convection zone has large velocities and spatial scales AND the relative layers above are a "soft" boundary. A rough estimate could be derived from an envelope constructed relying on the classical mixing length theory.

KUPKA: As far as I have seen from the paper by Richard *et al.* (2005), these Fe convection zones in the early A stars are just border-line to being unstable. In the paper it was said that if there is an Fe overabundance by a factor of 3 the zone is not yet convective, only for a factor of 4 it is. So I would expect that this is sort of not very efficient convection and when we estimate the velocities to see what happens, we can be very far off using a simple model such as classical mixing length theory. Also, if you include both the H zone and the Fe zone in a big simulation box, then the stratification is nearly radiative throughout most of the box and to obtain results not affected by the initial conditions, you may have to run simulations for that case for long timescales, which is probably unaffordable at the moment. Non-local models of the type I have discussed in my talk might be useful in that case: at least for a set-up containing the H and He II convection zones we found a qualitative agreement and a quantitative one within a factor of 2 when comparing to simulations and they always underestimated the extent of convective mixing, so their results might be considered a lower limit.

LANDSTREET: Changing a topic somewhat. When you look at the middle A stars, you try to measure the photospheric convection using microturbulence, you find normal A stars having microturbulence of the order of 2 km s^{-1} and Am stars with microturbulence of 4 - 5 km s^{-1}. Has anybody an idea what could produce this kind of rather important velocity difference between these two not that different atmospheres?

KUPKA: When we computed the non-local models, one of the runs was for 3 times the solar metallicity and indeed velocities did change in the right direction [increasing], but not to the extent that we had expected. It would be interesting to repeat these experiments using more realistic opacities [i.e. suitable for Am stars].

SMALLEY: Just a comment on rotation. Am stars have slower rotation compared to "normal" stars. Interestingly, the microturbulence of those [Am] stars is higher. Maybe microturbulence decreases with rotation?

RYABCHIKOVA: I always read in the literature about 4 to 6 km s^{-1} microturbulences in Am stars. However, fitting the profiles of strong lines in extremely sharp-lined Am-stars is impossible with such high microturbulence, it needs 0 to 1 km s^{-1} and no more.

LANDSTREET: I can comment on it. It is not that you can fit with some combination of microturbulence and rotational velocity. You just do not fit with those very strong line profiles very correctly. So when you determine the microturbulence as we do it simply by fitting the equivalent widths, we do not correctly reproduce the actual shapes with any combination of microturbulence or rotational velocity for the very strongest lines in very sharp lined stars.

RYABCHIKOVA: But this means that the real microturbulence is not 4 km s^{-1}.

TRAMPEDACH: Yes, if we are not looking at the line profile, and just fitting the equivalent width. There is not necessarily a relation between the actual velocity fields and the derived equivalent widths and derived microturbulence. I find that a little dangerous, actually. So, please, look at the line profiles when you fit. Fit the spectrum instead of the equivalent widths. We will get a lot more physics out of that.

LANDSTREET: There is hardly any A star which has sharp enough lines to actually see these details in the line profile. So even very slowly rotating ones for 5 - 10 km s^{-1} wash out the shape of the line profiles with the rotation profile.

DWORETSKY: Microturbulence is a fitting parameter, not a direct measure of any actual microturbulence or streaming motion. Many years ago we astronomers found that, for example, including more realistic opacity yielded $T - \tau$ relationships that reduced considerably the microturbulence fitting parameters for curves-of-growth. So it may or may not have anything to do with actual gas motions.

COWLEY: Numerical models explain the need for a microturbulence in a 1D model which does not have velocity fields (by assumption). But if the calculated profile does not fit the observed ones with an assumed $v_{turb} \sim 5$ or 6 km s^{-1} or whatever, then there is clearly something wrong with its determination. Years ago, Myron Smith got $v_{turb} \sim 7$ or 8 km s^{-1} in Am stars and in that case it is clear the problem was with the f-values. The log gf's were OK for strong lines but too small for weak ones. I suggested an error in the way v_{turb} was determined. This does not mean one does not need a v_{turb} in a classical 1D calculation.

LANDSTREET: There are also large scale velocity flows and what is missing for the microturbulence velocity is basically macroturbulence. That produces a kind of asymmetry that we observe in a few of very sharp lines. So it is that what is completely missing from even the model with microturbulence.

COWLEY: Yes, the classical macroturbulence does not affect the curve-of-growth, it only affects profiles. Whereas the classical microturbulence affects the look of the curve-of-growth and the line profile. So if you do not get a correct line profile, you have something wrong. You have not accounted for the velocity correctly.

LANDSTREET: The classical model does not have macroturbulence, the one that I use for the fit is incorrect. It does not have macroturbulence in it. When I put in an empirical macroturbulence, a completely parametrized macroturbulence plus a variation of microturbulence with height, I come much closer to fitting the observed profiles.

COWLEY: I am curious about these sharp lined stars. Are there stars with lines sharper than those of say, 63 Tau and 32 Aqr? Because when I look at those stars I always see a lot of lines with a little square. Just only DAO spectra. Now there are much better spectra than what I had. It is a big impression, but it was a pretty definite one relative to sharp lined HgMn stars. I never saw such a squarish profile. The profiles are wider, although at least strong line profiles are certainly wider than in the HgMn stars, but it is not a rotational velocity in the very sharpest line stars.

LANDSTREET: The numerical models do not yet reproduce the line profiles in the A stars.

TRAMPEDACH: I have not calculated lines yet, so I do not know. We might be missing something essential in the simulations.

PISKUNOV: I was wondering, if we can put some efforts together and bring the magnetic fields generated in the core to the surface, with perhaps the help of David Moss. I will be more than happy, if the residual spatial frequencies of these magnetic fields on the surface will have a maximum in spherical harmonics between 10 and 20.

MOSS: Certainly, the expectation is that for a field to rise from the core to the surface within a Main Sequence lifetime requires a strong field in thin tubes. Then somehow this "spaghetti-like" field has to organize itself into the observed kG+ strength global fields that we observe. This looks like a real difficulty. Remember that some Bp stars are observed to be strongly magnetic at ages $\sim< 10^7$ yrs! Here the problem is accentuated. Other problems are the lack of universality of A star fields, given the core dynamo mechanism should be universal, the lack of correlation of field strength with rotation period (in general, one might expect $|B|$ to increase with Omega for a dynamo), etc. Core dynamos almost certainly operate, but their connection with the observed fields seems unlikely.

PRESTON: One word about damping constants: We are talking about the fitting of strong lines. Were the damping constants included in the calculations?

LANDSTREET: We have got three different independent codes that agree on line profiles to very close tolerance. I think we all included the damping parameters correctly.

References

Richard O., Talon, S. & Michaud, G. 2005, *These Proceedings*, 215

Miesch, M.S., Elliott, J.R., Toomre, J., Clune, T.L., Glatzmeier, G.A. & Gilman, P.A. 2000, *ApJ* 532, 593

The A-Star Puzzle
Proceedings IAU Symposium No. 224, 2004
J. Zverko, J. Žižňovský, S.J. Adelman, & W.W. Weiss, eds.
© 2004 International Astronomical Union
DOI: 10.1017/S1743921304004533

Atomic diffusion in stellar surfaces and interiors

G. Michaud

Département de physique, Université de Montréal, Montréal, PQ, Canada, H3C 3J7

Abstract. Atomic diffusion may play a significant role for the Sun and Population I Main Sequence stars up to some 25000 K, Population II turnoff stars and cluster age determinations, horizontal branch stars (including sdOs and sdBs), white dwarfs and neutron stars. In all these cases, radiative accelerations play a significant role. A stars are, however, arguably those that show most prominently the effects of atomic diffusion. In so far as the effects of accretion, mass loss, turbulence and meridional circulation may be neglected in the evolutionary models of A stars, the effects of atomic diffusion in them have now been calculated from first principles and are presented using complete evolutionary models of 1.7 and 2.5 $M_\odot$ stars. Their abundance anomalies are not only superficial, but extend over a significant fraction of the stellar radius. Iron convection zones appear at a temperature of about 200000 K. Abundance anomalies similar to those observed in Am stars are produced. However the comparison with the observations requires linking atmospheres to interior evolution. Models that have been proposed to take into account atomic diffusion in atmospheric regions to explain observations are critically reviewed. They depend on a number of parameters. Unfortunately the atmospheric regions are imperfectly modeled, the magnetic field is not taken into account, and important hydrodynamic processes currently require arbitrary parameters for their description.

Keywords. Convection, diffusion, turbulence, stars: abundances, stars: atmospheres, stars: chemically peculiar, stars: evolution, stars: horizontal-branch, stars: interiors, stars: magnetic fields, stars: mass loss

---

1. Atomic diffusion in stellar evolution

As the accuracy of abundance determinations improves, atomic diffusion potentially has observational implications for more stars. In the best known case, helioseismology has confirmed the importance of gravitational settling in modifying the He concentration in the Sun's external regions (Guzik & Cox 1992, Christensen-Dalsgaard *et al.* 1993, Proffitt 1994, Guenther *et al.* 1996, Richard *et al.* 1996, Brun *et al.* 1999). In a more speculative paper, Bildsten *et al.* (2003) suggested that radiative accelerations, g_{rad} could be responsible for the strength of Fe lines seen during thermonuclear flashes on neutron stars.

In Population II stars atomic diffusion is responsible, according to VandenBerg *et al.* (2002), for M 92 to have an age 2 Gyr less than determined by Grundahl *et al.* (2000) in the absence of diffusion processes. It also potentially plays a role in the surface composition of turnoff stars (Richard *et al.* 2002) where anomalies might have been observed in M 92 (King *et al.* 1998) though those observations remain to be confirmed. The Li abundance in the stars of the Spite & Spite (1982) plateau and the age of M 92 are compatible (see Richard *et al.* , submitted) with the cosmological age and the Li abundance determination of WMAP (Cyburt *et al.* 2002, 2003).

One of the most striking examples of the effects of atomic diffusion driven by g_{rad} has recently been confirmed in horizontal branch stars. It was originally noticed by Sargent & Searle (1967, 1968) that the field halo stars with the same T_{eff} and the $\log g$ as the

Main Sequence HgMn stars, also appear to have very similar abundance anomalies. This occurs where the horizontal branch crosses the Main Sequence. It was then argued that the small He abundance in those stars could not be used to suggest that some stars had a He abundance smaller than the *cosmological* abundance. The observation of a relative overabundance of ^{3}He by Hartoog (1979) confirmed the link with the Population I Main Sequence star 3 Cen A which had been strengthened by the comparison of the observations of 14 chemical species in both stars by Baschek & Sargent (1976).

Michaud *et al.* 1983) considered atomic diffusion processes in the presence of g_{rad} in horizontal branch stars. They showed that large overabundances of the metals were to be expected in the hotter HB stars, where He underabundances are observed. A limiting equatorial rotation velocity is also expected for the HB stars with anomalies when one considers the 100 km s^{-1} upper limit to the equatorial rotation velocity of the HgMn stars and the link to meridional circulation (Michaud *et al.* 2004a).

Observationally, Glaspey *et al.* (1989) measured an underabundance of He and an overabundance of Fe by a factor of 50 in a T_{eff} =16000 K horizontal branch star of NGC 6752 but not in the cooler HB stars of the same cluster. This observation has now been strikingly confirmed by Behr *et al.* (1999, 2000a, 2000b) and Moehler *et al.* (2000) who observed many horizontal branch stars of NGC 6752, M15 and M13 and found that, whereas those cooler than about 11000 K have the same composition as giants, those hotter than 11000 K usually have larger abundances of some metals by large factors. They observed in particular Fe to be overabundant by a factor of 50 (see Fig. 1 of Behr *et al.* 1999). Since such anomalies cannot have been produced inside these stars and all HB stars must have had very similar original compositions. This is a striking confirmation of the importance of transport processes and of the role of g_{rad} in that region of the HR diagram. The link to transport processes is further strenghtened by that the higher T_{eff} HB stars rotate more slowly than the cooler ones which show no abundance anomalies (Behr *et al.* 2000a, 2000b, Recio-Blanco *et al.* 2002).

In the white dwarfs, ever since the original suggestion of Schatzman (1945), atomic diffusion has been recognized as the main process causing surface abundances. In the hotter white dwarfs ($T_{eff} > 30000$ K), g_{rad} was later suggested to play a role (Vauclair *et al.* 1979, Fontaine & Michaud 1979, Chayer *et al.* 1995).

Am and Ap stars appear as the most evident manifestation of phenomena (Michaud 1970) that are very widespread.

2. Interior of A stars

The availability of large atomic data bases has made it possible to calculate stellar evolution models from first principles. Evolutionary models taking into account g_{rad} , thermal diffusion, and gravitational settling for 28 elements, including all those contributing to the OPAL stellar opacities, have been calculated for a number of Population I stars: the Sun (Turcotte *et al.* 1998), F stars (Turcotte *et al.* 1998), AmFm stars (Richer *et al.* 2000) and solar metallicity stars of 0.5 to 1.4 $M_\odot$ (Michaud *et al.* 2004b). Stellar models of 1.7 and 2.5 $M_\odot$ are used here to describe the interiors of A stars. From Figure 1 of Richard *et al.* (2001), the 2.5 $M_\odot$ star starts its Main–Sequence evolution with T_{eff} = 10500 K, has $T_{eff} > 10000$ K for the first quarter of its Main–Sequence life and ends it at 8500 K, while the 1.7 $M_\odot$ one starts at 8000 K and ends its Main–Sequence life at 6500 K. These stars cross and bracket the T_{eff} range of interest.

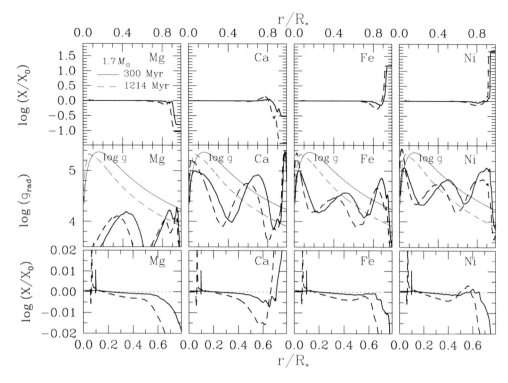

Figure 1. Abundances and g_{rad} as a function of radius, at 300 and 1214 Myr, in a 1.7 $M_\odot$ model for four of the 28 calculated species. For the two upper rows, the abscissa is above and covers the whole radius. The larger anomalies, by factors of more than 10, involve the outer 10% of the star. They are linked to g_{rad} shown on the second row. Note the vertical and horizontal scale changes for the bottom row. As may be seen on the lower row, at the 0.02 dex, or the 5%, level, anomalies cover the outer 40% of the radius. Metals are also overabundant, just outside the central convective zone, by a factor of 1.05. This is caused by interaction of inward settling metals with He diffusing outward from the central convective core (see §3.1.3 of Richard *et al.* (2001) for a more detailed discussion).

2.1. *Calculations*

The models were calculated as described by Turcotte *et al.* (1998) and Richard *et al.* (2001). They were assumed to be chemically homogeneous on the pre-Main Sequence with relative concentrations as defined in Table 1 of Turcotte *et al.* (1998) and most had a solar metallicity. The g_{rad} are from Richer *et al.* (1998) with a correction for redistribution from Gonzalez *et al.* (1995) and LeBlanc *et al.* (2000). The atomic diffusion coefficients were taken from Paquette *et al.* (1986) (see also Michaud & Proffitt 1993).

The g_{rad} calculations use the same OPAL data as was found to best reproduce the solar structure. Some questions have been raised recently after the proposed reduction of the O abundance in the Sun (Asplund *et al.* 2004). Basu & Antia (2004) concluded that such a reduction led to structural changes in clear disagreement with heliosismic measurements unless the opacity tables were revised upwards by 3.5% or that the metal abundance was very arbitrarily assumed to be larger *in* the solar convection zone than in the solar photosphere.

2.2. *Interior structure*

The concentrations of the chemical species are modified by particle transport in the interior of A stars (see Fig. 1). One notes that: 1) Overabundances of metals occur locally and are linked to the variations of g_{rad} and so to electronic shells. 2) These local abundance variations are superimposed on a generalized overabundance of metals at the level of up to 0.3% in the inner 40% by mass (inner 15% in radius) and a generalized underabundance in the outer region. 3) In the hotter A stars, some g_{rad} are larger than gravity over the outer 10% of the mass. 4) Iron convection zones occur at $T \simeq 200000$ K. The He convection zone disappears (Vauclair *et al.* 1974). The formation of Fe convection zones is described in detail in §3.1.2 of Richard *et al.* (2001). It appears in all solar metallicity stars with $M_* \geqslant 1.4 \, M_\odot$ (see Fig. 2). 5) As may be seen in Fig. 1, the abundances are modified by diffusion processes in the outer 40% of the radius at the 5% level, for most species. For a few species, up to the outer 40% by radius is affected by overabundances caused by g_{rad} . This is limited partly by the region where $g_{rad} \geqslant g$ but more importantly by the region where there is enough time for anomalies to develop during the stellar lifetime. The effect of atomic diffusion is not a static but a dynamic process. 6) Atomic diffusion leads to increases in the size of central convection cores. This is related to the appearance of semi-convection zones and is more important for F than A stars (see Richard *et al.* 2001, Michaud *et al.* 2004b). A peak at the 5% level also appears just outside the central convective core for some stellar masses.

The limitations of these models come from the exterior regions. The transport is treated from first principles from the regions where the Fe convection zone forms, down to the center. However, in most calculations, the star is assumed mixed above the Fe convection zone all the way to the surface. In Figure 3 of this paper and Figures 4 and 16 of Richard *et al.* (2001) are shown results where this constraint is relaxed.

The Fe mass fraction, $X(Fe)$, at which the Fe convection zone appears, depends on the Fe contribution to the opacity. Is the Fe contribution to opacity correctly calculated by OPAL? Given the remark made in §2.1, we have evaluated the uncertainty by arbitrarily decreasing the Fe contribution to the opacity by 30% for a 2.0 $M_\odot$ model and found that it increased by approximately 5% the $X(Fe)$ at which the convection zone appears. Increasing the Fe contribution to opacity by 20% decreases by about 2% the $X(Fe)$ at which the convection zone appears. One may find a detailed discussion of the dependence of convection on abundance variations in §3.3 of Michaud *et al.* (2004).

2.3. *Competing processes*

Competing processes include turbulence, meridional circulation, accretion and mass loss. Each of those processes, except some meridional circulation models, requires some adjustable parameter (and more generally parameters) for its description.

Turbulence has been studied in detail in conjunction with evolutionary models of A stars and is mentioned below in §3.1.2. Mass loss is mentioned below in conjunction with Am, Ap and HgMn star models. Meridional circulation has been suggested by Michaud (1982) to lead to a triggering mechanism for the Am and HgMn phenomena. No arbitrary parameter is involved and the limiting equatorial velocity of ~ 100 km s^{-1} is obtained, but the meridional circulation model assumes a boundary layer whose details have been questioned. One would like to see this model confirmed by numerical simulations from first principles (see Talon *et al.* 2003, Théado & Vauclair 2003) but no simulation has yet been done for A stars. For the results to be reliable, the simulations need to be large enough to cover many scale heights with sufficient resolution.

Accretion was first suggested by Havnes & Conti (1971) and Havnes & Goertz (1984) but, except for the λ Booti stars (Turcotte & Charbonneau 1993, Turcotte (2002)), it is

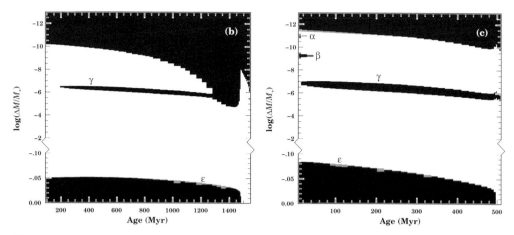

Figure 2. Convection zones in 1.7 and 2.5 $M_\odot$ models. Convection zones are in black. The ones labeled γ are the Fe convection zones, those labeled α and β are respectively He I and He II convection zones. In gray, labeled ϵ, are semi-convection zones.

unlikely to play the dominant role on a large fraction of the peculiar stars because of the relatively large fraction of M_* that needs to be accreted given the dynamical instability that heavy matter causes when accreted on lighter matter (Proffitt & Michaud 1989, see also Vauclair 2004).

3. Modelling atmospheres

To what extent can we currently explain the abundance anomalies observed on Am, Ap and HgMn stars? I will treat these in turn.

The simplest model assumes that the difference between the Ap stars and the others, comes from the presence of magnetic fields on the surface of Ap stars but not on the surface of the others. Where the magnetic field is horizontal, even ionized species whose g_{rad} are larger than g in the outer atmosphere are bound to the stars, since they cannot cross field lines. The magnetic field then guides diffusive transport causing patches or rings or complicated structures on Ap stars.

3.1. *Am stars*

There are currently two quite different models involving atomic diffusion that have been proposed to explain AmFm stars. One involves separation close to the surface so that only the outer 10^{-10} of M_* needs be affected. Mass loss then has to be assumed. The other occurs deeper in and involves the outer $10^{-6} - 10^{-5}$ of M_* . In this case turbulence has been suggested to play a role.

3.1.1. *Separation below the H convection zone*

Watson (1971) (see also Smith 1973) noticed that, immediately below the H-convection zone, Ca is in the Ca II (argon like) state, in which it has a small g_{rad} . He suggested that the separation occurs there, in which case Ca would be underabundant while iron peak elements would be overabundant, in accordance with observations. This requires the disappearance of the He convection zones which occurs after He has settled gravitationally. It has been shown to explain the absence of δ Scuti pulsators (Baglin 1972) among the AmFm stars. It is also compatible with the upper limit to the rotation velocity of AmFm stars (Michaud 1982, Charbonneau & Michaud 1988, 1991).

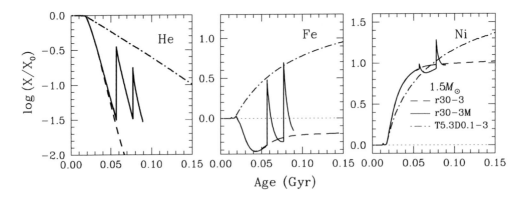

Figure 3. Effect of incomplete mixing above the Fe convection zone in 1.5 $M_\odot$ models. The continuous line is for a model where complete mixing is imposed above a Fe convection zone as soon as it appears. The Fe concentration is homogenized with the whole region above the Fe convection zone. It then disappears and later reappears leading to spikes in the various concentrations. The dashed line assumes no mixing above the Fe convection zone. The dot-dashed line assumes that the whole zone above $log\ T = 5.3$ is mixed throughout evolution, even when the Fe convection zone has not yet appeared. Observations of many species could constrain the mixing processes.

It has also been shown that when g_{rad} are included, atomic diffusion can explain the envelope of the observed abundance anomalies (Michaud *et al.* 1976). However the expected anomalies are generally much larger than those observed in individual stars, suggesting that at least one competing hydrodynamical process is strong enough to reduce the effects of atomic diffusion. Currently the most favored competing process is mass loss (Michaud *et al.* (1983). These calculations were done in static envelope models, for a limited number of species. Time dependent calculations only for Ca and Sc have been made in detail using g_{rad} calculated with Topbase data (Alecian 1996). It is difficult to tune such a model to reproduce the relative uniformity of anomalies observed on Am stars when the separation involves such a small mass and the time scales are as short as in this model. No calculation based on the preceding model has ever successfully reproduced the detailed abundance anomalies of individual stars, including the small anomalies of some chemical species such as Mg. This does not however prove it is impossible.

Turbulence has been investigated (Vauclair *et al.* 1978) but it does not appear to reduce sufficiently the overabundances of iron peak elements and rare earths if helium and calcium are to be underabundant.

3.1.2. *Separation at 200000 K*

As an alternative to the suggestion that separation occurs immediately below the H-convection zone, Richer *et al.* (2000) proposed that the separation occured much deeper. Detailed evolutionary models were calculated as described in §2 including the effect of atomic and turbulent diffusion for stars of 1.45 to 3.0 $M_\odot$. Since Richer *et al.* (2000) used turbulent diffusion coefficients that decrease rapidly as ρ increases within the star, they found their result to depend only on one parameter, the mass mixed by turbulence. The physical cause of the turbulence is not discussed. The zone mixed by turbulence is deeper than the iron convection zone, reducing the abundance anomalies to values which are too small for iron peak convection zones to develop in some models.

These authors compared the calculated surface abundances to observations of a number of AmFm stars. For Sirius A, 16 calculated abundances were compared to observations (including 4 upper limits). Of these, 12 are well reproduced by the model, while 3 are not so well reproduced and one is a very uncertain observation (see Fig. 18 of Richer *et al.* 2000).

In cluster AmFm stars, the age and initial abundances are known. There is then less arbitrariness in the calculations but fewer chemical species have been observed than in Sirius. The available observations (Hyades, Pleiades and Praesepe stars were compared) agree reasonably well with the calculated models for the five stars which they compared (see Fig. 19 to 23 of Richer *et al.* 2000). For most species, calculated abundance anomalies are within one error bar of the observed anomalies with only one fitted parameter, the mass of the mixed zone. All other quantities are from first principles. The pulsational properties of these models appear to be compatible with the observed pulsations of δ Scuti and AmFm stars (see Turcotte *et al.* 2000).

There is considerable scatter in the observations between different observers so that it is premature to conclude that hydrodynamical processes other than turbulence are needed to explain the observations. We are not ruling out that this be the case but the observations do not appear to us good enough to establish it; but neither are they good enough to establish this model. Would observations be compatible with a turbulent transport coefficient decreasing less rapidly than used by Turcotte *et al.* (2000) (see Richard *et al.* 2005)?

Even if turbulence can explain the observations, could mass loss do the same? Or at what level can mass loss be present without modifying the agreement with observations? Would such mass loss flux be chemically differentiated? To what extent are the surface abundances modified by additional separation above the Fe convection zone (see Fig. 3)?

3.2. *Ap stars*

The magnetic Ap stars are the most difficult to model. In the simplest model, Babel & Michaud (1991a) followed the diffusion of half a dozen chemical species, taking the magnetic structure as observed. They assumed mass loss and turbulence to be negligible. There is then no arbitrary parameter but it is still required to make an assumption about the state of convection in the presence of magnetic fields. Is all convection suppressed by the magnetic field or is convection suppressed only where magnetic field lines are vertical?

Using relatively detailed but LTE calculations of g_{rad} (Ca), g_{rad} (Sc), g_{rad} (Ti), g_{rad} (Mn), g_{rad} (Cr) and g_{rad} (Sr), Babel & Michaud (1991a) concluded that the simple model explained the *average* abundance anomalies of those species on 53 Cam, assuming that convection was partially suppressed. However the geographic distribution of the anomalies could not be explained in the measured magnetic field configuration (Landstreet 1988). This result was not changed when the effects of polarization are introduced as described in Babel & Michaud (1991b). Mass loss guided by the magnetic field needs to be introduced Babel (1994, 1995). These calculations should be redone with more accurate g_{rad} (Alecian & Stift 2004) and better magnetic field determinations and abundance maps (Kochukhov *et al.* 2004). It appears likely that mass loss and chemical separation in the wind will also need to be included though that remains to be shown.

3.3. *HgMn stars*

An attempt was made to determine what abundance anomalies would be expected from a parameter free model for HgMn stars (Michaud 1981). It involved detailed calculations of g_{rad} (He) (Michaud *et al.* 1979), g_{rad} (Be), g_{rad} (Mg), g_{rad} (Ba) (Borsenberger *et al.* 1981, 1984), g_{rad} (B) (Borsenberger *et al.* 1979), g_{rad} (Ca), g_{rad} (Sr) ((Borsenberger *et al.* 1981) and g_{rad} (Mn) (Alecian & Michaud 1981) in NLTE model atmospheres.

Once one can calculate g_{rad} , one determines the concentration of each species that can be supported by g_{rad} in the atmosphere. One checks that g_{rad} decreases as the species leaves the atmosphere, so that it is trapped there and finally one checks that the g_{rad} is larger than gravity deep enough into the star for the required amount of the species to be pushed into the atmosphere. All this is determined from first principles. The envelope of the abundance anomalies of He, B, Mg, Ca, Mn, Sr and Ba are explained by this model but the model fails for Be. The triggering of the HgMn phenomenom may be done by meridional circulation as mentioned in §2.3. This may be viewed as reasonnable success given the absence of any arbitrary parameter. However this does not lead to a detailed understanding of the spectrum of individual stars. No star has been found to have all those anomalies at the level calculated and no attempt was made to have a calculated spectrum completely consistent with observations.

Proffitt et $al.$ (1999) carried out the more demanding task of calculating g_{rad} (Hg) and the abundances of Hg that could be supported by g_{rad} in the atmospheres of two HgMn stars. They determined atomic data (transition rates and collision rates) of Hg for that study (Brage et $al.$ 1999) since their NLTE calculation required a large amount of atomic data that was poorly known but needed for abundance determinations, g_{rad} and isotope shift calculations. They used atmospheric models that reproduced the atmosphere as possible although these models did not include stratified concentrations. They found that the g_{rad} (Hg) were too weak to lead to Hg lines as strong as observed and suggested that a stellar wind of some 10^{-14} $M_{\odot}$ y^{-1} was needed. They did not study in detail the possibility that Hg be pushed out of the atmosphere by g_{rad} (Hg).

One problem with such parameter free calculations is that they produce the same abundance anomalies for all stars of a given T_{eff} and $\log g$. According to, for instance, Woolf & Lambert (1999) there is considerable variation from star to star even at a given T_{eff} and $\log g$.

Mass loss is bound to play a role for HgMn stars. It is observed in slightly hotter stars than HgMn. It has been suggested by Michaud et $al.$ (1974) to play a role in the creation of isotope anomalies. These authors suggested that, in so far as g_{rad} (Hg) is close to equilibrium with gravity, the lighter isotopes may be pushed out of the star while the heavier ones would remain bound. This would be further emphasized by the shading of the lines of the heavier isotopes as they tried to acquire an outward velocity to leave the star in a wind. Only the lighter isotopes would then leave. One would expect isotope anomalies mainly for atomic species who become noble gases after only few ionizations[†], such as Hg, Pt and now Ca (Castelli & Hubrig 2004). This process depends on the details of the wind structure and it has never been calculated in detail. To go beyond the suggestion that was made, one needs a detailed wind model calculated from first principles and taking differentiation in the wind into account along with the detailed NLTE g_{rad} (Hg) calculations. This was never attempted.

Light induced drift (Atukov & Shalagin 1988) could play a role. It is likely to be small but a better evaluation of the scattering crosssections of the excited states of ionized species appears needed to include it in calculations (LeBlanc & Michaud 1993).

Seaton (1999) attempted time dependent calculations of the anomalies of Fe group species on HgMn stars in presence of mass loss using g_{rad} calculated with OP project data (Seaton 1995, 1997). He does envelope calculations similar to those of Richer et $al.$ (2000) except that he extends them to $\tau \simeq 1$. He does not include the effect of other species on the opacities except for the one he is considering nor does he include evolutionary effects.

---

† So that their g_{rad} decrease below gravity before they leave the star.

4. Prospects

In §3.1.2, it was seen possible, using current computing power, to calculate stellar evolution models that include the abundance variations of 28 species. Our poor understanding of stellar hydrodynamics is the main source of uncertainty of these models. On the other hand, comparison of the results of the models to observations becomes a source of information on stellar hydrodynamics. Am star observations of up to 16 of those species have been made and it is certainly possible to observe more by observing the whole visible and far UV spectrum. That would also allow improving the accuracy of observations. On the modeling side, it would be very important to have $g_{\rm rad}$ for more species, for instance Sr, some Rare Earth Elements, and Hg. These would add important observational constraints on the models.

It would be in principle possible to do similar calculations for HgMn and Ap stars. It would mainly involve the atmospheric regions. The computing power is there to do the calculations, but a large amount of code development is needed.

Acknowledgements

This work has been supported by an Operating Grant from the Natural Sciences and Engineering Research Council of Canada. The Réseau Québécois de Calcul de Haute Performance (RQCHP) provided the computational resources.

References

Alecian, G. 1996, A&A, 310, 872
Alecian, G. & Michaud, G. 1981, ApJ, 245, 226
Alecian, G. & Stift, M. J. 2004, A&A, 416, 703
Asplund, M., Grevesse, N., Sauval, A. J., Allende Prieto, C., & Kiselman, D. 2004, A&A, 417, 751
Atukov, S. N. & Shalagin, A. M. 1988, SvA, 14, L284
Babel, J. 1994, A&A, 283, 189
—. 1995, A&A, 301, 823
Babel, J. & Michaud, G. 1991a, ApJ, 366, 560
—. 1991b, A&A, 241, 493
Baglin, A. 1972, A&A, 19, 45
Baschek, B. & Sargent, A. I. 1976, A&A, 53, 47
Basu, S. & Antia, H. M. 2004, ApJ, 606, L85
Behr, B. B., Cohen, J. G., & McCarthy, J. K. 2000aa, ApJ, 531, L37
Behr, B. B., Cohen, J. G., McCarthy, J. K., & Djorgovski, S. G. 1999, ApJ, 517, L135
Behr, B. B., Djorgovski, S. G., Cohen, J. G., McCarthy, J. K., Côté, P., Piotto, G., & Zoccali, M. 2000bb, ApJ, 528, 849
Bildsten, L., Chang, P., & Paerels, F. 2003, ApJ, 591, L29
Borsenberger, J., Michaud, G., & Praderie, F. 1979, A&A, 76, 287
—. 1981a, ApJ, 243, 533
—. 1984, A&A, 139, 147
Borsenberger, J., Radiman, I., Praderie, F., & Michaud, G. 1981b, in Chemically Peculiar Stars of the Upper Main Sequence (Liège: Université de Liège), 389–394
Brage, T., Proffitt, C. R., & Leckrone, D. S. 1999, ApJ, 513, 524
Brun, A. S., Turck-Chieze, S., & Zahn, J. P. 1999, ApJ, 525, 1032
Castelli, F. & Hubrig, S. 2004, A&A, 421, L1
Charbonneau, P. & Michaud, G. 1988, ApJ, 327, 809
—. 1991, ApJ, 370, 693
Chayer, P., Fontaine, G., & Wesemael, F. 1995, ApJS, 99, 189
Christensen-Dalsgaard, J., Proffitt, C. R., & Thompson, M. J. 1993, ApJ, 403, 75

Cyburt, R. H., Fields, B. D., & Olive, K. A. 2002, Astroparticle Physics, 17, 87

Cyburt, R. H., Fields, B. D., & Olive, K. A.. 2003, Physics Letters B, 567, 227

Fontaine, G. & Michaud, G. 1979, ApJ, 231, 826

Glaspey, J. W., Michaud, G., Moffat, A. F. J., & Demers, S. 1989, ApJ, 339, 926

Gonzalez, J.-F., LeBlanc, F., Artru, M.-C., & Michaud, G. 1995, A&A, 297, 223

Grundahl, F., VandenBerg, D. A., Bell, R. A., Andersen, M. I., & Stetson, P. B. 2000, AJ, 120, 1884

Guenther, D. B., Kim, Y.-C., & Demarque, P. 1996, ApJ, 463, 382

Guzik, J. A. & Cox, A. N. 1992, ApJ, 386, 729

Hartoog, M. R. 1979, ApJ, 231, 161

Havnes, O. & Conti, P. S. 1971, A&A, 14, 1

Havnes, O. & Goertz, C. K. 1984, A&A, 138, 421

King, J. R., Stephens, A., Boesgaard, A. M., & Deliyannis, C. F. 1998, AJ, 115, 666

Kochukhov, O., Bagnulo, S., Wade, G. A., Sangalli, L., Piskunov, N., Landstreet, J. D., Petit, P., & Sigut, T. A. A. 2004, A&A, 414, 613

Landstreet, J. D. 1988, ApJ, 326, 967

LeBlanc, F. & Michaud, G. 1993, ApJ, 408, 251

LeBlanc, F., Michaud, G., & Richer, J. 2000, ApJ, 538, 876

Michaud, G. 1970, ApJ, 160, 641

Michaud, G. 1981, in Chemically Peculiar Stars of the Upper Main Sequence (Liège: Université de Liège), 355–363

—. 1982, ApJ, 258, 349

—. 1991, Ann. Phys. (Paris), 16, 481

Michaud, G. & Charland, Y. 1986, ApJ, 311, 326

Michaud, G., Charland, Y., Vauclair, S., & Vauclair, G. 1976, ApJ, 210, 447

Michaud, G., Montmerle, T., Cox, A. N., Magee, N. H., Hodson, S. W., & Martel, A. 1979, ApJ, 234, 206

Michaud, G. & Proffitt, C. R. 1993, in Inside the Stars, IAU COLLOQUIUM 137, Vienna, April 1992, ASP Conference Series, 40, ed. W. W. Weiss & A. Baglin (San Francisco: ASP), 246

Michaud, G., Reeves, H., & Charland, Y. 1974, A&A, 37, 313

Michaud, G., Richard, O., Richer, J., & VandenBerg, D. A. 2004, ApJ, 606, 452

Michaud, G., Richer, J., & Richard, O. 2004, in IAU Symposium, Vol. 215, xxx

Michaud, G., Tarasick, D., Charland, Y., & Pelletier, C. 1983a, ApJ, 269, 239

Michaud, G., Vauclair, G., & Vauclair, S. 1983b, ApJ, 267, 256

Moehler, S., Sweigart, A. V., Landsman, W. B., & Heber, U. 2000, A&A, 360, 120

Paquette, C., Pelletier, C., Fontaine, G., & Michaud, G. 1986, ApJS, 61, 177

Proffitt, C. R. 1994, ApJ, 425, 849

Proffitt, C. R., Brage, T., Leckrone, D. S., Wahlgren, G. M., Brandt, J. C., Sansonetti, C. J., Reader, J., & Johansson, S. G. 1999, ApJ, 512, 942

Proffitt, C. R. & Michaud, G. 1989, ApJ, 345, 998

Recio-Blanco, A., Piotto, G., Aparicio, A., & Renzini, A. 2002, ApJ, 572, L71

Richard, O., Michaud, G., & Richer, J. 2001, ApJ, 558, 377

Richard, O., Michaud, G., Richer, J., Turcotte, S., Turck-Chieze, S., & VandenBerg, D. A. 2002, ApJ, 568, 979

Richard, O., Vauclair, S., Charbonnel, C., & Dziembowski, W. A. 1996, A&A, 312, 1000

Richard, O., Talon, S., Michaud, G. 2005, *These Proceedings*, 215

[5 Richer, J., Michaud, G., Rogers, F., Iglesias, C., Turcotte, S., & LeBlanc, F. 1998, ApJ, 492, 833

Richer, J., Michaud, G., & Turcotte, S. 2000, ApJ, 529, 338

Sargent, W. L. W. & Searle, L. 1967, ApJ, 150, L33

—. 1968, ApJ, 152, 443

Schatzman, E. 1945, Annales d'Astrophysique, 8, 143

Seaton, M. J. 1995, J. Phys. B, 28, 3185

—. 1997, MNRAS, 289, 700

—. 1999, MNRAS, 307, 1008

Smith, M. A. 1973, ApJS, 25, 277

Spite, F. & Spite, M. 1982, A&A, 115, 357

Talon, S., Vincent, A., Michaud, G., & Richer, J. 2003, Journal of Computational Physics, 184, 244

Théado, S. & Vauclair, S. 2003, ApJ, 587, 784

Turcotte, S. 2002, ApJ, 573, L129

Turcotte, S. & Charbonneau, P. 1993, ApJ, 413, 376

Turcotte, S., Richer, J., & Michaud, G. 1998a, ApJ, 504, 559

Turcotte, S., Richer, J., Michaud, G., & Christensen-Dalsgaard, J. 2000, A&A, 360, 603

Turcotte, S., Richer, J., Michaud, G., Iglesias, C., & Rogers, F. 1998b, ApJ, 504, 539

VandenBerg, D. A., Richard, O., Michaud, G., & Richer, J. 2002, ApJ, 571, 487

Vauclair, G., Vauclair, S., & Greenstein, J. L. 1979, A&A, 80, 79

Vauclair, G., Vauclair, S., & Michaud, G. 1978, ApJ, 223, 920

Vauclair, G., Vauclair, S., & Pamjatnikh, A. 1974, A&A, 31, 63

Vauclair, S. 2004, ApJ, 605, 874

Watson, W. D. 1971, A&A, 13, 263

Woolf, V. M. & Lambert, D. L. 1999, ApJ, 521, 414

Discussion

CORBALLY: In Population II horizontal branch stars, do you find relative underabundances (besides the overabundances you mentioned)?

MICHAUD: The most conspicuous underabundance is that of He. When it occurs in hot HB stars, there is also an overabundance of Fe. CNO are generally overabundant according to Behr *et al.* (1999). This is also expected from the calculations (Michaud *et al.* 1983).

NOELS: Do you think that the Fe accumulation zone that you find in A stars could also be present in more massive stars, let us say about 9 or 10 $M_\odot$ stars? I ask you this because when analyzing the stability of ν Eri, a β Cep star, it seems that Fe should be overabundant in the Fe opacity bump to excite the observed modes of pulsation.

MICHAUD: If there were no mass loss in 9 or 10 $M_\odot$ stars, a similar accumulation of Fe would occur in them as in A stars. Because of the large mass loss rate expected, however, no large Fe overabundance is expected in such stars. We are planning to introduce mass loss in our evolutionary code and determine quantitatively how abundance anomalies vary with mass loss rate. This has not yet been done.

LANDSTREET: The comparison you have shown between your computation for Sirius and observed abundances highlights a major problem in abundance analyses. Individual investigators can determine abundances with precisions of better than 0.1 dex, but different investigators often differ by far more than the uncertainties suggest. This problem needs much more attention from stellar spectroscopists.

The A-Star Puzzle
Proceedings IAU Symposium No. 224, 2004   © 2004 International Astronomical Union
J. Zverko, J. Žižňovský, S.J. Adelman, & W.W. Weiss, eds.   DOI: 10.1017/S1743921304004545

Diffusion in magnetic fields

G. Alecian

LUTH, Observatoire de Paris-Meudon, F-92195 Meudon Cedex, France
email: georges.alecian@obspm.fr

Abstract. Diffusion of elements in a stellar plasma is strongly modified by the presence of magnetic fields for two primary reasons. The first is that the average motions of ions in outer atmospheres are, because of their charge, substantially constrained by the magnetic field. Both its intensity and orientation play a role. The second is the Zeeman desaturation of absorption lines that often produces amplifications of the radiative accelerations. These effects are important and must lead to the building of complex surface abundance structures. I will present how these two effects are generally modeled and what results have, up to now, been obtained. Future developments will also be considered.

Keywords. Diffusion, stars: abundances, stars: atmospheres, stars: magnetic fields

1. Introduction

When one speaks about the *diffusion of elements* in stars, one designates the macroscopic consequences of a microscopic process which involves a large number of erratic motions of charged particles. Therefore, the presence of a magnetic field has obviously to be taken into account in the description of that process. An approximate theory has been given by Chapman & Cowling (1970). A typical effect of the magnetic fields on the diffusion velocity of metals in the atmospheres of magnetic Ap stars was first discussed by Vauclair, Hardorp & Peterson (1979) for the case of Si accumulation. I will present some aspects of diffusion across magnetic lines in Sec. 2.

There is another effect which was often neglected: the Zeeman desaturation of absorption lines. Among the important terms in the equation of the diffusion velocity, there is the radiative acceleration which has a particular status: it depends strongly on the atomic properties (transitions) of diffusing species and also on the ionic concentrations. One reason why the concentration of a given ion is involved is that the saturation of absorption lines determines the efficiency of the momentum transfered from radiation field to the considered species. One then understands how the Zeeman effect can influence the radiative accelerations and in turn the diffusion process itself. This effect was often neglected because in the few studies which were done, Zeeman desaturation was found to modify radiative accelerations by a smaller amount than their expected accuracies. The numerical cost to compute it appeared too high with regard to the benefits, and large atomic databanks were not available at that time. These considerations are no longer valid. After the pioneering work of Babel & Michaud (1991b), the importance of the Zeeman desaturation has been confirmed by the detailed study and the extensive numerical computations of Alecian & Stift (2004). I will discuss Zeeman amplification in Sec. 3.

Of course, magnetic fields can affect element stratifications in stars in many other ways, for instance through interactions with macroscopic motions of the plasma (see also Leblanc, Michaud & Babel (1994)) who have considered the effect of Lorentz forces on model atmospheres). We consider in this talk only the effects of the magnetic fields on the microscopic diffusion velocities and the radiative accelerations. Some new results on diffusion in magnetic fields, with both effects (particle motions and Zeeman desaturation) will be shown in Sec. 4.

2. Diffusion velocity

To model diffusion processes in Ap stellar atmospheres (Michaud 1970), one needs to compute the diffusion flux for each element. This flux is obtained through a weighted sum of the diffusion velocities corresponding to the various ionisation states (including neutral) for each species considered. We do not detail here how to obtain these velocities, rather we limit ourselves to the effects of the magnetic field.

2.1. Diffusion across horizontal magnetic lines

The diffusion velocity of a given type of particles in a partially ionised plasma is the average quantity which remains after the summation of a large number of microscopic movements of particles. These movements are unaffected by the magnetic field when particles are neutral. But charged particles have a circular motion in the direction transverse to the magnetic field. The resultant motion during the time between two collisions is a spiral along magnetic lines, but the diffusion velocity remains rectilinear. An approximate theory of diffusion in magnetic fields can be found in Chapman & Cowling (1970). These authors have shown that the diffusion velocity of charged particles, when it is orthogonal to magnetic lines, is reduced by the factor:

$$f_{slow,i} = \left(1 + \omega_i^2 t_i^2\right)^{-1} \tag{2.1}$$

where t_i is the collision time (the time needed to a particle i with mass m_i and charge Ze, to deviate through collisions by $\frac{\pi}{2}$ from its initial motion), $\omega_i/2\pi$ is the cyclotron frequency in a field with intensity H:

$$\omega_i = \frac{ZeH}{m_i c} \tag{2.2}$$

The average diffusion velocity of an element can be approximated by the following expression (the sums are over the ionisation states i):

$$V_D \approx \frac{\sum\limits_i N_i\, f_{slow,i}\, V_{Di}}{\sum\limits_i N_i} \tag{2.3}$$

This formula was used by Vauclair, Hardorp & Peterson (1979) to study the Si accumulation in magnetic Ap stars. Silicon is found to be strongly overabundant in magnetic Ap stars and more or less normal in nonmagnetic peculiar stars. Vauclair, Hardorp & Peterson (1979) have shown that the radiative acceleration on ionized Si is not very strong and cannot lead to high overabundances, even if the acceleration of the neutral state is large. This explains the quasi-normal abundance of Si in the nonmagnetic stars. But in the magnetic Ap stars, the downward diffusion flux due to the settling of Si ions is slowed down by the magnetic field (small $f_{slow,i}$ in Eq.(2.1)), while for neutral state, $f_{slow,0} = 1$ ensures a positive velocity V_D and leads to Si overabundances. These authors concluded that their diffusion model is compatible with the Si abundances observed in peculiar stars. This study has been extended to the oblique rotator model by Michaud, Megessier & Charland (1981) assuming a dipole plus quadrupole magnetic structure. They have shown how elements could accumulate according to the orientation of the magnetic field (at the magnetic poles and/or the magnetic equator).

2.2. *Oblique magnetic lines*

In the previous section, we have essentially considered the average diffusion velocity across horizontal magnetic lines. At the surface of magnetic stars the direction of the magnetic field lines varies from 0 to $\frac{\pi}{2}$, between magnetic poles(s) and equator(s). This angular dependence contributes to explaining the inhomogeneous abundance distribution over the stellar surface: the sensitivity of the diffusion velocity to the inclination of the magnetic lines determines the size of abundance structures [see Michaud, Megessier & Charland (1981), and Alecian & Vauclair (1981)]. In oblique magnetic fields lines, the diffusion velocity vector is no longer vertical, rather its inclination depends on the angle θ of the magnetic lines to the vertical according to the expressions given by Alecian & Vauclair (1981). The vertical (Z axis) component is:

$$V_{H,Zi} \approx V_i \left(f_{\text{slow},i} + \frac{1}{f_{\text{slow},i}} \cos^2 \theta \right), \tag{2.4}$$

and the horizontal one (X axis) is:

$$V_{H,Xi} \approx \frac{f_{\text{slow},i} V_i}{2} \sin 2\theta. \tag{2.5}$$

V_i is the diffusion velocity for zero magnetic field.

The horizontal component given by expression (2.5), represents an horizontal diffusion of ions only. The diffusion velocity of the element will be affected through the averaged velocity (2.3). Does this means that horizontal diffusion could play a role for surface abundance structures observed in magnetic Ap stars (see for instance Kochukhov *et al.* (2004))? The answer is no! In the case of silicon, Megessier (1984) has estimated the time scale for Si migration from the pole to the equator. This time scale appears to be of the order of 10^7 years! This is smaller than the lifetime of Ap stars on the main sequence. However, it implies that the lifetime of the magnetic structures should be also of the same order of magnitude if one would hope to have an observable effect due to horizontal diffusion. On the other hand, and this is more constricting, the time scales of vertical diffusion are much smaller (4 or 5 orders of magnitude smaller), because vertical height scales are much smaller too. Vertical particle diffusion flux dominates. One can then conclude that its seems rather unlikely that horizontal diffusion may have a significant effect. This means that the abundance inhomogeneities on magnetic Ap stars (patches or rings) are mainly due to the angular dependence of the vertical component of the velocity vector as shown in Eq.(2.4), combined with the local mass-loss velocity.

3. Radiative acceleration

A pioneering work, based both on analytical approximations to the polarised radiative transfer problem and on numerical solutions, was carried out by Babel & Michaud (1991a). More recently, Alecian & Stift (2004) have investigated the same problem in detail by numerical means. Their exhaustive computations were carried out for 329 ions, using the VALD atomic database (Piskunov *et al.* 1995), solving numerically the polarised radiative transfer equation by means of a new formulation of the Zeeman Feautrier method (Auer, Heasley, & House 1977, Rees, Murphy, & Durrant 1989) includes magneto-optical effects, and provides for the correct treatment of line blending.

The radiative acceleration is a vector which usually is not more vertical when a magnetic field is present. It is obtained through the following integral over the solid angle Ω

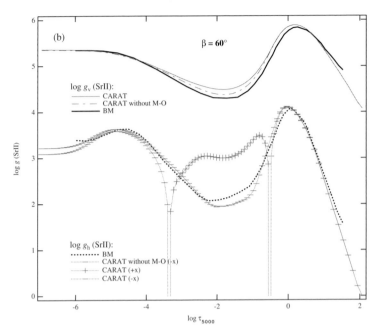

Figure 1. Radiative accelerations due to the Sr II $\lambda\,4077$ line in the non-magnetic case and in a 1.5 T field, based on a 8500 K, $\log g = 4.0$ ATLAS9 (Kurucz 1993) model atmosphere (from Alecian & Stift 2004). This plot shows a comparison between the accelerations obtained by Babel & Michaud (1991a)(BM) and results of CARAT code, both with and without magneto-optical effects. The field is vertical. The vertical acceleration is shown by the upper curves and the horizontal one by the lower curves. The field is inclined by $60°$ with respect to the vertical. Note that the horizontal acceleration changes sign twice over optical depth when m.-o. effects are taken into account.

and frequency ν (from Alecian & Stift 2004):

$$\mathbf{g}_i^{\mathrm{rad}} = \sum_{k,\,m>k} \frac{n_{i,k}}{n_i A m_p c} \iint (\mathbf{e} \cdot \mathbf{I}) \mathbf{\Omega} d\Omega d\nu \tag{3.1}$$

- with n_i the total number density (cm^{-3}) of ions A^{+i}, n_{ik} the number density of ions in initial lower level k,
- and the summation extending over all transitions from initial level k to higher levels m.
- $A m_p$ is the mass (g) of the ion,
- $(\mathbf{e}\cdot\mathbf{I})$ denotes the inner product of the vector $\mathbf{e} = \kappa_o\,\{\phi_I, \phi_Q, \phi_U, \phi_V\}$ (line absorption matrix elements, see Alecian & Stift (2004) for details), with the Stokes vector $\mathbf{I}$.

3.1. *Zeeman amplification*

The results presented here have been obtained by Alecian & Stift (2004) with their new LTE diffusion code CARAT (Code pour les Accélérations Radiatives dans les ATmosphères) that solves the polarized radiation transfer equations for Eq.(3.1) on massively parallel multiprocessor machines.

A comparison of a result obtained by CARAT with the curves of Babel & Michaud (1991a) is shown in Fig. 1. The role of magneto-optical effect is highlighted and also the particular behavior of the horizontal component of the radiative acceleration. This later appears negligible compared to the vertical component, at least for the cases Alecian & Stift (2004) have considered (see their paper for more details).

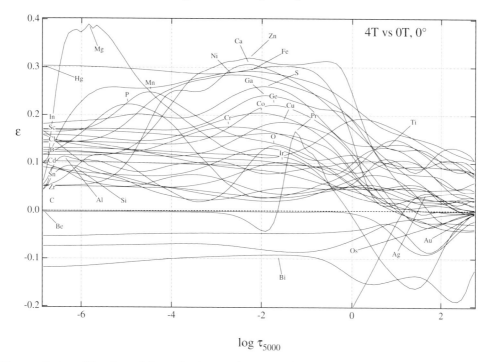

$\log \tau_{5000}$

Figure 2. Amplification of the radiative accelerations of the chemical elements due to Zeeman splitting as a function of optical depth and of the magnetic field strength (from Alecian & Stift 2004). Logarithmic amplifications are displayed for a field strength of 4 T for the vertical magnetic lines. Curves are plotted only when $|\varepsilon| > 0.0414$ dex (10% amplification) at any depth point.

Some amplifications are shown in Fig. 2. They have been computed for a $T_{\rm eff} = 12000$ K and $\log g = 4.0$ ATLAS9 (Kurucz 1993) model atmosphere . Amplifications are based on the simplified expression for the total radiative accelerations $g^{\rm rad}$ given by a sum of ion accererations (3.1), weighted by their relative population. The magnetic field is vertical. The amplifications ($\varepsilon = \log g^{\rm rad}$(magnetic) $- \log g^{\rm rad}$(nonmagnetic)) are at most about 0.40 dex for 4 T. This corresponds to a maximum increase of radiative accelerations by a factor 2.5. This increase is by no means negligible.

4. Preliminary results on diffusion in magnetic Ap stars

As emphasized above, the study of microscopic diffusion of elements in magnetic fields needs to compute the vertical components (Eq. 2.4) of the diffusion velocities of ions, using the radiative acceleration given by Eq. (3.1). Actually, one of the main goals of Ap stars modelling is to know what abundance stratification can be obtained. Unfortunately, this is beyond present abilities because the stratification process is non-linear and time-dependent, and needs much more computing power than presently available as well as theoretical and software developments. However, a first guess of what stratifications could be expected can be obtained by considering the fluxes of the diffusing species at the beginning of the process (when the elements have still their normal solar abundances and when no competing processes are considered). This kind of study was first carried out by Alecian & Vauclair (1981) who neglected the Zeeman effect. A preliminary detailed

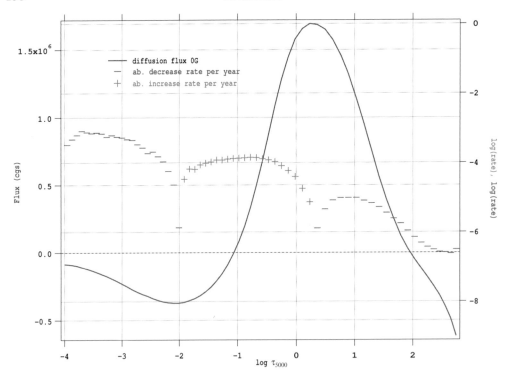

Figure 3. Diffusion flux of Al (left axis) for zero magnetic field (for solar abundance). The logarithm of the abundance change rate per year is also shown (right axis). The + curve indicates abundance increase, and the − one the abundance decrease.

calculations of aluminum fluxes (Alecian & Stift 2005) are shown in Fig. 3 and Fig. 4 in the case of the $T_{\rm eff} = 12000$ K model atmosphere noted previously.

For zero magnetic field (Fig. 3), aluminum is pushed upwards in the atmosphere in layers with $-1 < log\,\tau_{5000} < +2$. This allows us to predict (curves with symbols + and −) that the abundance of Al will start to increase in layers with $-2 < log\,\tau_{5000} < +0.2$ by −4 dex per year (the abundance of Al will be doubled in about 10^4 years if one neglects the non-linearity of the process). This increase rate is rather small compared to that of many iron peak elements. The effect of a magnetic field (1 Tesla) with various inclinations are shown in Fig. 4. The difference between the 0 T and 1 T (vertical) fluxes is due to the modest Zeeman amplification of Al radiative acceleration (these two curves would be superimposed if Zeeman effect were neglected). When the magnetic field is horizontal, the flux is significately reduced as one goes up in the atmosphere. This is due to the increase of the collision time t_i which allows charged particles to spiral more lengthly around magnetic field lines.

5. Conclusions

We have presented here a short overview of the main effects of magnetic fields on microscopic diffusion. One can reasonably expect that in the short term, one should be able to compute the trends of abundance stratifications in magnetic atmospheres, assuming LTE and assuming that equilibrium stratifications can exit (this is still unknown). Similar modelling is already in progress for the case without a magnetic field (see LeBlanc & Monin 2005). In the magnetic case, the next steps should be to address self-consistent

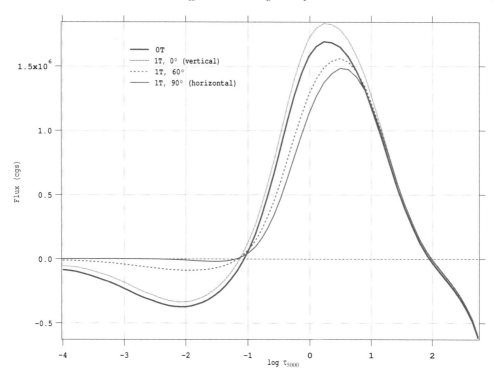

Figure 4. Effect of a magnetic field (1 Tesla) on the diffusion flux of Al is shown for various inclinations of the magnetic lines. The "0T" heavy solid curve is the same as the one of Fig. 3

modelling (feedback of stratifications on the model atmospheres), NLTE, 2D oblique rotator modelling, and take into account additional physical processes such as ambipolar diffusion, hydrodynamics, etc.

Thus many investigations remain to be done before reaching the 'mythical' goal of a realistic time-dependent numerical simulation of the abundance stratification process.

Acknowledgements

I thank Martin J. Stift who provided me in time with a new version of the CARAT code (Alecian & Stift 2005) which has contributed to the computation of the diffusion fluxes shown in Sec. 4. His work has the support of the *Austrian Science Fund (FWF)*, project P16003-N05 "Radiation driven diffusion in magnetic stellar atmospheres".

References

Alecian G., Artru M.-C., 1987, A&A, 186, 223
Alecian G., Stift M.J., 2004, A&A, 416, 703
Alecian G., Stift M.J., 2005, in preparation
Alecian G., Vauclair S., 1981, A&A, 101, 16
Auer, L.H., Heasley, J.N., & House, L.L. 1977, ApJ, 216, 531
Babel J., Michaud G., 1991a, A&A, 241, 493
Babel J., Michaud G., 1991b, A&A, 248, 155
Chapman S., Cowling T.G., 1970, The Mathematical Theory of Non-uniform Gases, Cambridge University Press, 3rd ed.
Hui Bon Hoa A., Alecian G., Artru M.C., 1996, A&A, 313, 624

Kochukhov, O., Bagnulo, S., Wade, G.A., Sangalli, L., Piskunov, N., Landstreet, J.D., Petit, P.
    & Sigut, T.A.A., 2004, A&A, 414, 613
Kurucz, R. 1993, CDROM Model Distribution, Smithsonian Astrophys. Obs.
Kuschnig, R., Ryabchikova, T.A., Piskunov, N.E., Weiss, W.W., Gelbmann, M.J. 1999, A&A,
    348, 924
Leblanc F., Michaud G., Babel J., 1994, ApJ, 431, 388
LeBlanc F., Monin D. 2005 *These Proceedings*, 193
Megessier C., 1984, A&A, 138, 267
Michaud G., 1970, ApJ, 160, 641
Michaud G., Megessier C., Charland Y., 1981, A&A, 103, 244
Piskunov, N.E., Kupka, F., Ryabchikova, T.A.,Weiss, W.W., & Jeffery, C.S. 1995, A&AS, 112,
    525
Rees, D.E., Murphy, G.A., & Durrant C.J., 1989, ApJ, 339, 1093
Vauclair S., Hardorp J., Peterson D.M., 1979, *ApJ*, **227**, 526

Discussion

PISKUNOV: Rapidly oscillating Ap stars show element stratifications at very high atmospheric layers ($\tau < 10^{-8}$). Can your procedure handle such conditions and what should be done to compute magnetic diffusion in these conditions?

ALECIAN: As I mentioned, the CARAT code uses LTE physics. Therefore its validity domain is limited to layers with $\tau > 10^{-5}$ or 10^{-4}. On the other hand, the magnetic field is a very strong barrier in high atmospheric layers. Then the computation of diffusion velocities at such high layers ($\tau < 10^{-8}$) in strong magnetic fields using the same approximations as in the lower atmosphere is probably not justified.

KUBÁT: What is the role of elastic collisions in your model? How are they included into your calculations?

ALECIAN: Elastic collisions are one of the hypotheses imposed to establish the usual equations for diffusion velocities in the framework of the kinetic theory of gases (see Chapman & Cowling 1970). The diffusion coefficients we use are determined assuming elastic collisions.

DWORETSKY: In HgMn stars the observed range of iron abundances is nearly two orders of magnitude. Are these more likely to be equilibrium abundances or "snapshots" in which Fe abundances evolve with time in a complicated way?

ALECIAN: The building of abundance stratifications by diffusion is a non-linear process and some instabilities might occur. In my opinion, clouds of metals in optically thin layers could disappear and form again. Therefore, the abundance scatter that you mention should not correspond to any equilibrium abundances.

The A-Star Puzzle
Proceedings IAU Symposium No. 224, 2004   © 2004 International Astronomical Union
J. Zverko, J. Žižňovský, S.J. Adelman, & W.W. Weiss, eds.   DOI: 10.1017/S1743921304004557

Modelling of stratified atmospheres of CP-stars

F. LeBlanc and D. Monin

Département de physique et d'astronomie, Université de Moncton, Moncton, NB, E1A 3E9,
Canada
e-mail: leblanfn@umoncton.ca

Abstract. Several observational anomalies seem to confirm the presence of abundance gradients as a function of depth in different types of chemically peculiar stars. Results emanating from the construction of model atmospheres that take into account the abundance gradients caused by radiative diffusion will be presented. The atmospheric structure, which is calculated self-consistently along with the abundance gradients, will be compared to models with homogeneous abundances. Recent improvements brought to these models will be discussed, along with the intricacies of these calculations and the remaining uncertainties. Several possible applications of such models will also be presented.

Keywords. Diffusion, stars: abundances, stars: atmospheres, stars: chemically peculiar

---

1. Introduction

Mounting observational evidence seem to show that element stratification is present in the atmospheres of several types of stars. For atomic diffusion (Michaud 1970) to be able to create abundance stratifications, an hydrodynamically stable atmosphere is required. The accumulation (or depreciation) of the elements as a function of depth will modify the atmospheric structure of such stars. It is then imperative, not only to obtain the stratification profiles of the various elements, but also to include the back effect of these abundance gradients on the structural calculation of the atmosphere. Up until now, the vast majority of photometric and spectroscopic studies have employed homogeneous models.

Recently, model atmospheres including stratification due to diffusion have been constructed for white dwarf stars (Dreizler & Wolff 1999) and horizontal-branch stars (Hui-Bon-Hoa *et al.* 2000). These models were successful in explaining certain observational anomalies. For example, for white dwarf stars, it was shown that the models with diffusion could explain a flux depression in the ultraviolet (see Fig. 6 of Dreizler & Wolff 1999). Meanwhile, the observed photometric jumps (e.g., Grundhal *et al.* 1999) and gaps (e.g., Caloi 1999) that are observed in hot horizontal-branch stars are qualitatively reproduced by stratified models (see Fig. 1 and 2 of Hui-Bon-Hoa *et al.* 2000), assuming that diffusion becomes dominant in stars hotter than $T_{\rm eff} \simeq 11{,}500$ K. For more details concerning the observed anomalies of horizontal branch stars see Moehler (2005).

Since stratification seems to be present in Ap stars (e.g., Wade *et al.* 2001, Babel 1994) and might also be present in HgMn stars (e.g., Savanov & Hubrig 2003), self-consistent model atmospheres including diffusion are urgently needed to properly study these stars (for a review of observational evidence of stratification in stellar atmospheres see Ryabchikova *et al.* 2003). In this paper we will present results of recent modelling of the atmospheres of an A type star with such self-consistent models. These models are calculated with an improved version of the Hui-Bon-Hoa *et al.* (2000) models, which are based on the PHOENIX (e.g., Hauschildt *et al.* 1999) code. We will briefly review the method and ingredients used in our models as well as describing the improvements recently included. Examples of such models will be shown and future applications, actual limitations and possible improvements will be discussed.

2. Theory

The diffusion velocity (V_i) of an ion i of a trace element A can be approximated by (Burgers 1960, Vauclair & Vauclair 1982, Alecian & Vauclair 1983, Landstreet *et al.* 1998):

$$V_i \approx D_i \left[-\nabla \ln C_i + \frac{A_i m p g_{\mathrm{rad}}}{kT} - \{(A_i - 1) + (A_i - Z_i) f_{\mathrm{p}}\} \nabla \ln P + \kappa_T \nabla \ln T \right]. \quad (2.1)$$

Here D_i is the diffusion coefficient, C_i is the concentration of the ion, A_i its atomic mass, g_{rad} its radiative acceleration, and Z_i its ionic charger. The local temperature, pressure and ionization fraction of H II are designated by T, P and f_{p}. The factor f_{p} appears to give the proper asymptotic values of the electric field in the medium (e.g., Landstreet *et al.* 1998). The term including $\nabla \ln T$ is the thermal diffusion term, which was neglected in our calculations since it is negligible in the atmosphere. The abundance gradient term was also neglected in the models presented here.

The most important ingredient in the diffusion equation is g_{rad}. Contrarily to diffusion calculations in stellar interiors (e.g., Turcotte *et al.* 1998) where the radiative flux used is the one given by the so-called diffusion approximation (Milne 1927) which is only valid at large optical depths, in the atmosphere the flux must be explicitly calculated by resolving the radiative transfer equation at a large number of frequency points. The g_{rad} values used here are thus calculated using the opacity sampling method (e.g., LeBlanc *et al.* 2000) which is the preferred calculation method to properly evaluate line blending and saturation effects.

A major source of uncertainty in g_{rad} is related to the possible redistribution of momentum among the ions (e.g., Gonzalez *et al.* 1995). Since the various ions have different mobility (or diffusion coefficients), if an ion that acquires momentum following a bound-bound transition, for example, ionizes or recombines before losing this momentum, the acceleration can thus be modified. Since the appropriate evaluation of this redistribution effect is extremely difficult, this causes an uncertainty in g_{rad}. This is particularly important in the upper atmosphere since there is a large difference in the diffusion coefficients of the neutral state as compared to the once ionized species. In the calculations presented here, we will approximate the redistribution effect with the method used in Hui-Bon-Hoa *et al.* (1996) which is based on the formalism described by Montmerle & Michaud (1976).

Magnetic fields can also affect atomic diffusion. First, the diffusion of charged ions is modified when they cross magnetic field lines. Also, the Zeeman effect can also change the value of the radiative accelerations (Alecian & Stift 2004). Neither of these magnetic effects are included in the models presented here.

3. Modelling and results

Our model atmospheres try to simultaneously calculate the abundance stratifications of various elements and the atmospheric structure. We thus modify the abundance of each element at each depth in the atmosphere to converge the diffusion velocity to zero (i.e., equilibrium abundances). The magnitude of the change brought to the abundance is a function of the diffusion velocity, and is greater where the velocity is higher. The diffusion velocities obtained in a previous iteration are used to calculate correction factors. The abundances are then changed by these factors.

An iteration scheme that alternates between abundance and temperature corrections is used. The atmospheric physical structure has to be adjusted each time the abundances are changed. It is necessary to iteratively converge the diffusion velocity towards zero first, and then the physical structure to insure that these structures are self-consistent. A series of six abundance corrections is followed by a series of temperature corrections. We repeat this several times or until we see successful convergence. The robustness of the scheme depends on the number of abundance iterations in a series. Changes in the physical structure must be slow enough to ensure proper convergence. About 10 to 15 temperature iterations are typically required to reach convergence to better than a few K. We find that this iteration approach provides excellent convergence speed and stability.

The radiative transfer equation is solved before each abundance correction or temperature correction iteration. The computing time to obtain a converged model is considerable since to obtain precise g_{rad} the radiative transfer equation is solved at almost a quarter of a million frequency points.

The elements included in the PHOENIX code are H–Ga, Kr–Nb, Ba and La. The line atomic data used are from Kurucz (1994). The models presented here are plane-paralel atmospheres that are calculated in LTE. Both bound-bound and bound-free transitions are included in the g_{rad} calculations.

Several improvements to the models presented in Hui-Bon-Hoa *et al.* (2000) have been added to the atmospheric code. Most are related to that they supposed a completely ionized hydrogen medium. Since we now use this code for cooler stars, such as A-type stars, it was necessary to more precisely evaluate diffusion in a partially ionized buffer gas. Changes were thus made to the diffusion coefficients and to the equation of the electric field that appears in the diffusion equation. These forced us to use a different convergence scheme to modify the abundances to obtain null diffusion velocities. The code was also modified to evaluate the effect of mass loss on the abundance stratification of the elements.

3.1. *Stratification profiles*

The abundance of a given element that can be supported by radiative pressure varies as a function of depth since different ions have different opacities and thus different g_{rad} values . Stratification profiles can then be very different from one element to another.

Wade *et al.* (2001) found that they could better fit the observed lines of several elements with a two zone empirical stratification model for the Ap star β CrB. They found that for Ca, Fe and Cr, that their lines are better fitted with a strong underabundance in the outer atmosphere and an overabundance in deeper layers, than with a vertically homogeneous abundance. It should be noted that the underlying atmospheric model used was a homogeneous model and the two zone stratification profile was only used for their line synthesis. They found that the transition zone for the elements considered (Ca, Fe and Cr) for the Ap star β CrB is found at $\log(\tau_{5000}) \simeq -0.7$.

Figure 1 shows the one step abundance profiles for Fe and Cr found by Wade *et al.* (2001) as compared to the ones found in various self-consistent models for β CrB. We chose $T_{eff} = 7700\,\mathrm{K}$ for β CrB while realizing that uncertainties exist in determining T_{eff} for Ap stars. The abundance increases in our model assuming no convective mixing are deeper than those found by Wade *et al.* (2001). In the models shown, the abundance gradient term in the diffusion equation was neglected. However, preliminary results show that this term cannot account for the deeper increase of the abundances in the model

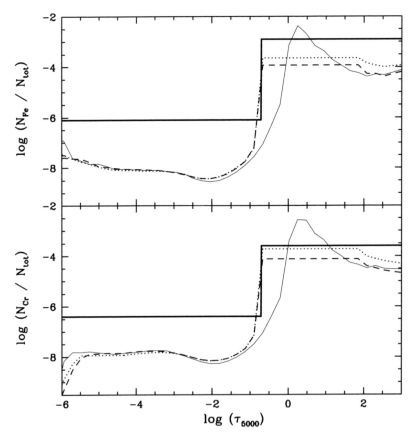

Figure 1. Abundance profiles of Fe and Cr in a $T_{\mathrm{eff}} = 7700\,\mathrm{K}$ model as a function of optical depth at 5000 Å. The heavy solid line is the empirical profile of Wade *et al.* (2001) found for $\beta\,$CrB, the thin solid line is the profile in a self-consistent model with no convective mixing, the long dashed line is the profile in a self-consistent model with convective mixing, while the dotted line is the profile in a model with convective mixing and a mass loss rate of $3 \times 10^{-15} M_{\odot}\,\mathrm{yr}^{-1}$.

with no convective mixing from those found in observations. It should be noted that in the theoretical models, this depth depends on the T_{eff} of the model.

Since Wade *et al.* (2001) found that the abundance jumps abruptly at $log(\tau_{5000}) \simeq -0.7$ for the three elements studied and this is where one would expect for convection to begin, according to the Schwarzschild criterion, therefore we calculated models in which convective mixing was present. In these models we supposed that the presence of a strong magnetic field suppresses overshooting since this would erase elemental stratification. Indeed, $\beta\,$CrB possesses a strong magnetic field of 5 kG (Mathys *et al.* 1997). Even though it is widely believed that the presence of magnetic fields at least partially suppresses convection, it could be possible for weak convective mixing to persist and dominate atomic diffusion. A convection velocity of the order of $1\,\mathrm{cm\,s}^{-1}$, which is well below the detection treshold, is sufficient to dominate diffusion where a convection zone could possibly persist. The supported abundance in the convection zone was taken as the abundance sustained by radiative pressure at its bottom. The abundance profiles are then similar to the two plateau model used by Wade *et al.* (2001). Figure 1 shows the stratification in models with a convective mixing zone. It should be noted that the abundance supported in the convective zone is very sensitive to the position of the bottom of this zone since

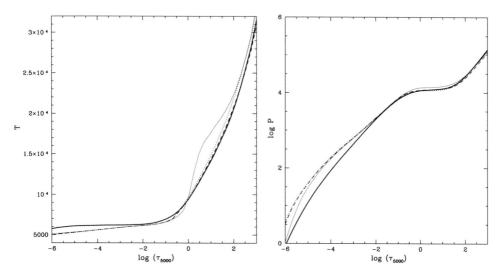

Figure 2. Temperature and pressure in a $T_{\text{eff}} = 7700\,\text{K}$ model as a function of optical depth at $5000\,\text{Å}$. The heavy solid line is for an homogeneous model, the thin solid line is for a self-consistent model with no convective mixing, the long dashed line is for a self-consistent model with convective mixing, while the dotted line is for a self-consistent model with convective mixing and a mass loss rate of $3 \times 10^{-15}\,M_{\odot}\,\text{yr}^{-1}$.

radiative accelerations can vary quickly as a function of depth there. These results seem to show that in Ap stars, some slow convective mixing might persist. As for the abundance supported at shallow layers, our results are quite different from those found by Wade *et al.* (2001).

Babel (1992) calculated the stratification of Ca in an Ap star, but without calculating the atmospheric structure. His model included a convective mixing zone. He showed that the value of the upper plateau strongly depends on mass loss. We therefore calculated a model with a mass loss rate of $3 \times 10^{-15}\,M_{\odot}\,\text{yr}^{-1}$. We can see in Figure 1 that the mass loss increases the abundance that can be supported in the convective mixing zone for the two elements considered and gives values closer to the empirical stratifications found by Wade *et al.* (2001).

3.2. *Atmospheric structure of stratified atmospheres*

Figure 2 shows the temperature and pressure as a function of depth for self-consistent models including diffusion with and without convective mixing as compared to a homogeneous stellar atmosphere of a $T_{\text{eff}} = 7700\,\text{K}$ star. The abundances used for the homogeneous models are the average abundances observed in $\beta\,\text{CrB}$ (Ryabchikova, private communication).

We can see that in the diffusion models the temperature is lower in the outer atmosphere, as compared to the homogeneous model, while it is larger deeper in the atmosphere, where large accumulations of the elements occur. This temperature increase is related to the increase in opacity due to accumulation of several elements in these deep layers. The maximum temperature increase occurs at depths larger than $\tau_{5000} = 1$. This maximum increase is of the order of 30% in the model without convective mixing, while it is only approximately 6% in the model with convective mixing and mass loss.

Meanwhile the pressure in the outer atmosphere is up to two to three times larger in the models with diffusion as compared to the homogeneous model. This is due to that many elements are very underabundant there making the real physical depth larger for the same optical depth in the models with stratification. In deep layers, the pressure difference between the homogeneous model and the stratified model with convective mixing and mass loss is less than 7%.

4. Conclusion

The results from modelling which include the stratification of the elements due to diffusion have shown that the stratification profiles found are similar to those observed in the Ap star β CrB, but convection mixing is needed to get better fitting. It should be noted, that in the models with convective mixing, the value of the upper abundance plateau is very sensitive to the depth of the convection zone and also depends on the mass loss rate. Stratification affects the physical structure of the atmosphere moderately in the models with convective mixing, and stongly in the models assuming no convective mixing. However, the modelling of magnetic stars is quite difficult since abundances patches are observed on their surface. Therefore, one dimensional models like those presented here can only have limited success in reproducing observational anomalies related to the accumulation (or depreciation) of the various elements in the atmospheres of magnetic stars.

The effect of the magnetic field on diffusing atoms as well as that of the Lorentz force on the structure of the atmosphere and the interaction of the diffusing atoms with the magnetic field could have an important effect on the atmosphere (Valyavin et al. 2004, LeBlanc et al. 1994). Missing opacity sources, like those of rare-earth elements which are very overabundant in Ap stars, should also be included in future modeling. The interaction between the atmosphere and the interior, during the diffusion process, as well as the possible importance of the initial conditions could also play an important role in the final state of the atmosphere. Other physical phenomena such as light-induced drift (Atutov & Shalagin 1988, LeBlanc & Michaud 1993) should also be included in HgMn models to attempt to explain certain observed isotopic anomalies (Dworetsky & Vaughan 1973).

The results shown here in which the elemental stratification is calculated self-consistently with the atmospheric structure in CP-stars are encouraging. However, many other physical phenomena that were neglected here and mentioned above can come into play. Much more work needs to be done in modelling stratified atmospheres to elucidate some of the many observational anomalies observed in CP-stars. Also, abundance stratifications derived for a greater number of Ap stars such as those presented by Ryabchikova et al. (2005) will also enable a better constraint of the models.

References

Alecian, G., & Vauclair, S. 1983, Fund. of Cosmic Physics, 8, 369

Alecian, G., & Stift, M. J. 2004, A&A, 416, 703

Babel, J. 1992, A&A, 258, 449

Babel, J. 1994, A&A, 283, 189

Burgers, J.M. 1960, in Plasma Dynamics, ed. F.H. Clauser (Reading: Addison-Wesley), 119

Caloi, V. 1999, A&A, 343, 904

Dreizler, S., & Wolff, B., 1999, A&A, 348, 189

Dworetsky, M. M., & Vaughan, A. H., 1973, ApJ, 181, 811

Gonzalez, J.-F., LeBlanc, F., Artru, M.-C., & Michaud, G. 1995, A&A, 297, 223

Grundahl, F., Catelan, M., Landsman, W. B., Stetson, P. B., & Andersen, M. I. 1999, ApJ, 524, 242

Hauschildt, P. H., Allard, F., & Baron, E. 1999, ApJ, 512, 377

Hui-Bon-Hoa, A., Alecian, G., & Artru, M.-C., 1996, A&A, 313, 624

Hui-Bon-Hoa, A., LeBlanc, F., & Hauschildt, P.H., 2000, ApJ, 535, L43

Kurucz, R. L., 1994, Atomic Data for Opacity Calculations (Kurucz CD-ROM No.1)

Landstreet, J. D., Dolez, N., & Vauclair, S. 1998, A&A, 333, 977

LeBlanc, F., & Michaud, G., 1993, ApJ, 408, 251

LeBlanc, F., Michaud, G., & Babel, J. 1994, ApJ, 431, 388

LeBlanc, F., Michaud, G., & Richer, J. 2000, ApJ, 538, 876

Mathys, G., Hubrig, S., Landstreet, J.D., Lanz, T., & Manfroid, J., 1997, A&ASS, 123, 353

Montmerle, T., & Michaud, G., 1976, ApJS, 31, 489

Michaud, G., 1970, ApJ, 160, 641

Milne, E.A., 1927, M. N., 87, 687

Moehler, S., 2005, *These Proceedings*, 395

Ryabchikova, T.A., Leone, F., Kochukhov, O., & Bagnulo, S., 2005, *These Proceedings*, DP1

Ryabchikova, T.A., Wade, G.A., & LeBlanc, F., 2003, in Modelling of Stellar Atmospheres, ed. N. Piskunov, W.W. Weiss & D. F. Gray, IAU Symp. 210, 301

Savanov, I., & Hubrig, S., 2003, A&A, 410, 299

Turcotte, S., Richer, J., & Michaud, G., 1998, ApJ, 504, 559

Valyavin, G., Kochukhov, O., & Piskunov, N., 2004, A&A, 420, 993

Vauclair, S., & Vauclair, G. 1982, ARA&A, 20, 37

Wade, G. A., Ryabchikova, T. A., Bagnulo, S., & Piskunov, N. 2001, in Magnetic Fields across the Hertzsprung-Russell Diagram, ed. G. Mathys, S. K. Solanki & D. T. Wickramasinghe, ASP Conf. Series 248, 373

Discussion

TALON: How unique is the solution you calculate?

LEBLANC: Diffusion phenomena is dependant on the initial conditions since for example, blending effects, which are included in our calculations, modify the radiative accelerations of the elements. We suppose that initially, the atmosphere is homogeneous. Tests show that the final solution depends weakly on the initial atmosphere used provided that the starting model is 'reasonable'.

VAUCLAIR: Large accumulation of heavy elements in the outer layers of stars create an inverse μ-gradient which is unstable and leads to a double-diffusive or thermohaline convectio). The models should include such physical process in the future.

LEBLANC: I agree that the large overabundances obtained, for example in the models assuming no convective mixing, could lead to unstabilities and will have to be investigated.

MATHYS: It is an excellent thing that, on the theoretical side, you treat element stratification with self-consistent models. We should be careful that on the observational side, much less consistency is achieved. We know (e.g., from the core-wing anomaly in Balmer lines of hydrogen) that standard model atmospheres do not adequately represent the atmosphere of Ap stars). Yet we use such a model to find different abundances from different lines of a given ion and conclude from there to the existence of vertical stratification of the ion. Without necessarily questioning the latter, this calls for caution against overinterpreting the quantitative details of the abundance gradient that are currently derived.

LeBlanc: The observed abundance stratification shown here as compared to those obtained in the self-consistent model atmospheres are only indicative. As mentioned in several talks, the abundances and even the value of the effective temperatures of Ap stars are very difficult to evaluate precisely.

Wade: The situation is indeed quite bad (notice there were no error bars on the empirical stratification distribution). Not only is the temperature a 'free parameter', but we can study stratification best in those stars for which the least information about other complications (e.g., surface abundance + magnetic field structures). Investigations are underway to try to understand the influence of these factors.

The A-Star Puzzle
Proceedings IAU Symposium No. 224, 2004
J. Zverko, J. Žižňovský, S.J. Adelman, & W.W. Weiss, eds.
© 2004 International Astronomical Union
DOI: 10.1017/S1743921304004569

Multicomponent stellar winds and chemical peculiarity in A stars

Jiří Krtička[1] and Jiří Kubát[2]

[1]Ústav teoretické fyziky a astrofyziky PřF, Masarykova univerzita, CZ-611 37 Brno, Czech Republic, email: krticka@physics.muni.cz

[2]Astronomický ústav, Akademie věd České republiky, CZ-251 65 Ondřejov, Czech Republic, email: kubat@sunstel.asu.cas.cz

Abstract. We calculate multicomponent radiatively driven stellar wind models suitable for A stars. We discuss the possible decoupling of individual elements from the stellar wind and its influence on the chemical peculiarity of these stars. We obtain a range of stellar parameters for different types of multicomponent flow.

Keywords. Hydrodynamics, stars: mass loss, stars: early-type, stars: winds

1. Introduction

Stellar winds of hot stars are accelerated mainly due to the absorption of radiation in the resonance lines of heavier elements like carbon, nitrogen, oxygen or iron, and due to light scattering by free electrons. Since the stellar wind is composed mostly of hydrogen and helium, the process of wind acceleration consists basically of two steps. In the first step momentum is transferred from the radiation field to heavier elements, which have enough absorbing lines. Passive components (hydrogen and helium) are accelerated in the following step due to Coulomb collisions with heavier elements (Castor *et al.* 1976). Clearly, stellar winds of hot stars have a multicomponent nature. For a relatively high density stellar wind, the multicomponent wind nature does not influence the wind structure, thus high-density stellar winds (e.g.,those of galactic O stars) may be adequately modelled as one-component ones. However, this is not the case for low-density stellar winds (Krtička & Kubát 2001, hereafter KKII) for which the velocity difference between the wind components becomes comparable with the thermal speed. Many new physical effects occur in this case, which may influence the wind structure, ranging from frictional heating, wind decoupling, and hydrogen fallback to a pure metallic wind. Here we study multicomponent effects in the A star winds and discuss their consequences on the chemical peculiarity of these stars.

2. Wind models

2.1. Model equations

Published multicomponent wind models have been described in detail by KKII and Krtička (2003). Here we only summarize their basic properties. We assume a stationary and spherically symmetric stellar wind, which is composed of four components, namely metallic ions, hydrogen, helium, and free electrons. For the calculation of wind models we solve the continuity equation, the momentum equation, and the energy equation for each component of the flow. The continuity equation has the form

$$\frac{\mathrm{d}}{\mathrm{d}r}\left(r^2 \rho_a v_{ra}\right) = 0,$$

(2.1)

where ρ_a is the density, r is the radius, and v_{ra} is the velocity of a component a. The momentum equation is

$$v_{ra}\frac{dv_{ra}}{dr} = g_a^{\text{rad}} - g - \frac{1}{\rho_a}\frac{d}{dr}\left(a_a^2\rho_a\right) + \frac{q_a}{m_a}E + \sum_{b\neq a}g_{ab}^{\text{fric}}, \quad (2.2)$$

where g_a^{rad} is the radiative acceleration, g is the gravity acceleration, E is the electric polarisation field, and g_{ab}^{fric} is the frictional force (Burgers 1969)

$$g_{ab}^{\text{fric}} = \frac{1}{\rho_a}K_{ab}G(x_{ab})\frac{v_{rb} - v_{ra}}{|v_{rb} - v_{ra}|}, \quad (2.3)$$

where $G(x_{ab})$ is the Chandrasekhar function. The frictional coefficient is

$$K_{ab} = n_a n_b\frac{4\pi q_a^2 q_b^2}{kT_{ab}}\ln\Lambda, \quad (2.4)$$

where the mean temperature $T_{ab} = (m_b T_a + m_a T_b)/(m_b + m_a)$ is calculated using temperatures T_a and T_b of individual wind components with atomic masses m_a and m_b. The radiative force in the CAK approximation (Castor et al. 1975) due to line-absorption acts on the metals and the radiative force due to Thomson scattering acts on free electrons.

The energy equation for each component of the flow is

$$\frac{3}{2}v_{ra}\rho_a\frac{da_a^2}{dr} + \frac{a_a^2\rho_a}{r^2}\frac{d}{dr}\left(r^2 v_{ra}\right) = Q_a^{\text{rad}} + \sum_{b\neq a}(Q_{ab}^{\text{ex}} + Q_{ab}^{\text{fric}}), \quad (2.5)$$

where Q_{ab}^{ex} is the heat exchange, Q_{ab}^{fric} is the frictional heating and Q_a^{rad} is the radiative heating calculated using the thermal balance of electrons method (Kubát et al. 1999).

The system of hydrodynamic equations is closed using the equation for the electric polarisation field (see KKII) and by the equations of ionization equilibrium (we assume a nebular approximation, see Mihalas (1978), Eq. 5.46).

2.2. The possibility of wind decoupling

The frictional force Eq. (2.3) between two components depends on the velocity difference between these two components via the so-called Chandrasekhar function,

$$G(x_{ab}) = \frac{1}{2x_{ab}^2}\left[\Phi(x_{ab}) - x_{ab}\frac{d\Phi(x_{ab})}{dx_{ab}}\right], \quad (2.6)$$

where the dimensionless velocity difference x_{ab} is given by

$$x_{ab} = \frac{|v_{rb} - v_{ra}|}{\alpha_{ab}} = \frac{\Delta v_{ab}}{\alpha_{ab}}, \quad (2.7)$$

where α_{ab} is the mean thermal speed, $\alpha_{ab}^2 = 2k(m_a T_b + m_b T_a)/(m_a m_b)$. A plot of the Chandrasekhar function is given in Figure 1. For relatively low velocity differences, $\Delta v_{ab} \lesssim \alpha_{ab}$, the Chandrasekhar function is an increasing function of the velocity difference Δv_{ab} and components a and b are well coupled. However, for higher velocity differences, $\Delta v_{ab} \gtrsim \alpha_{ab}$, the Chandrasekhar function is a decreasing function of the velocity difference and the decoupling of the components may occur.

Important simplifications. Our models used some important simplifications which may affect the validity of results obtained, especially in the case of A stars:

• the radiative force is calculated in the CAK approximation with force multipliers after Abbott (1982), thus wind parameters may not be adequate,

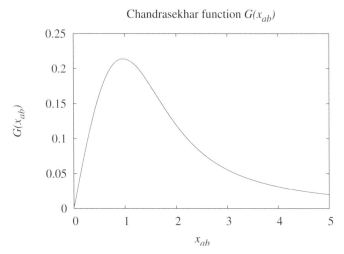

Figure 1. The run of the Chandrasekhar function. Note that Eq. (2.7) yields $x_{ab} \sim \Delta v_{ab}$, where Δv_{ab} is the velocity difference between the wind components. If the flow is well coupled, $x_{ab} \lesssim 0.97$, $G(\Delta v_{ab}) \sim \Delta v_{ab}$. If the drift velocity is large, $x_{ab} \gtrsim 0.97$, $G(\Delta v_{ab}) \sim \Delta v_{ab}^{-2}$, and the wind may decouple. Note that the point $x_{ab} \approx 0.97$, corresponds to the maximum of G.

- the ionization equilibrium is approximated using a "nebular approximation" after Mihalas (1978), this may influence the frictional force,
- we neglect wind instabilities (Owocki & Puls 1999),
- we neglect magnetic fields (ud-Doula & Owocki 2002),
- only Coulomb collisions are accounted for the calculation of the frictional force.

3. Multicomponent wind models

3.1. *Types of hot star winds*

There are several types of hot-star winds with respect to multicomponent effects:
- winds with negligible multicomponent effects (these winds may be adequately described by one-component models, e.g., winds of galactic O supergiants),
- winds with the temperature influenced by frictional heating (e.g., winds of Main-Sequence B stars),
- winds where the decoupling of the components occurs (which may result in helium decoupling or in both hydrogen and helium decoupling),
- the decoupling of wind components in the atmosphere (i.e., at the onset of the wind); for helium decoupling, a helium-free wind may exist, this may cause a helium overabundance in the atmosphere; for the case of both hydrogen and helium decoupling, a pure-metallic stellar wind may be present.

3.2. *Helium decoupling*

Helium decoupling was proposed by Hunger & Groote (1999) as the explanation of chemical peculiarity of Bp stars. Helium decoupling in the atmosphere occurs when the helium frictional acceleration is lower than the absolute value of gravity acceleration acting on helium,

$$g_{\alpha p}^{\text{fric}} < g, \tag{3.1}$$

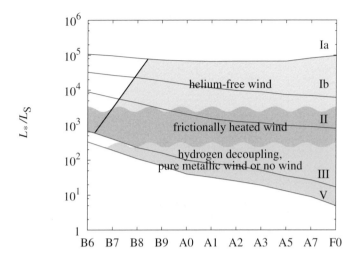

Figure 2. The domains in HR diagram where stars with different types of stellar wind exist. The fundamental stellar parameters of individual spectral types are taken from Straižys & Kuriliene (1981).

thus, for solar metallicity stars helium decouples when the mass-loss rate is lower than

$$\dot{M} \lesssim 2 \cdot 10^{-16} \, M_\odot \, \text{year}^{-1} \left(\frac{M}{M_\odot} \right) \left(\frac{T_{\text{eff}}}{10^4 \, \text{K}} \right)^{3/2} z_\alpha^{-2}. \tag{3.2}$$

which can easily be found from the Eq. (3.1) using Eqs. (2.3) and (2.4).

In A stars helium may be neutral in the stellar atmosphere, thus it is not sufficiently accelerated by friction, hence a helium-free wind may be present in this case (see Figure 2). Consequently, some A stars may have an enhanced abundance of helium in their atmospheres. However, this effects occurs for late B stars outside the domain of the He rich B stars, consequently our models do not support the model of helium enrichment in the atmospheres of He rich B stars proposed by Hunger & Groote (1999).

3.3. *Frictional heating*

Multicomponent effects are found to be important if the velocity difference is comparable with the thermal speed (see Krtička *et al.* 2003),

$$\frac{v_{\text{ri}} - v_{\text{rp}}}{\sqrt{\frac{2kT}{m_{\text{p}}}}} \gtrsim 0.1. \tag{3.3}$$

For solar metallicity stars multicomponent effects become important for mass-loss rates lower than

$$\dot{M} \lesssim 10^{-10} \, M_\odot \, \text{year}^{-1} \left(\frac{v_\infty}{10^8 \, \text{cm s}^{-1}} \right)^3 \left(\frac{R_*}{R_\odot} \right) \left(\frac{T_{\text{eff}}}{10^4 \, \text{K}} \right) \frac{1}{z_{\text{H}}^2 z_{\text{i}}^2}. \tag{3.4}$$

In such a case the stellar wind may be heated by friction. An example of such frictionally heated wind of an A5 II star is given in Figure 3. The domain in HR diagram where stars with frictionally heated wind may exist is displayed in Figure 2.

Note that in a low-density stellar wind another type of heating, the so-called Gayley-Owocki heating (Gayley & Owocki 1994, KKII), may be important for the wind temerature balance.

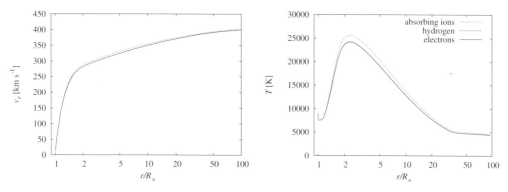

Figure 3. An example of the frictionally heated wind of an A5 II star ($T_{\text{eff}} = 8\,300\,\text{K}$, $M_* = 5.5\,M_\odot$, $R_* = 15.1\,R_\odot$). *Left panel:* The velocities of the individual components of the stellar wind. Note that velocities of wind components are nearly equal. *Right panel:* The temperatures of the individual components. The stellar wind is frictionally heated around $r \approx 2\,R_*$.

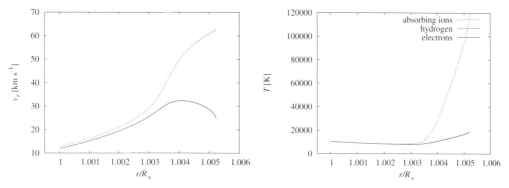

Figure 4. An example of hydrogen decoupling in the wind of an A0 III star ($T_{\text{eff}} = 9\,600\,\text{K}$, $M_* = 2.7\,M_\odot$, $R_* = 3.63\,R_\odot$). *Left panel:* The velocities of the individual components of the stellar wind. Stellar wind decouples around $r \approx 1.004\,R_*$ in this case the hydrogen component deccelerates. *Right panel:* The temperatures of the individual components. Note the strong frictional heating in the outer parts of the wind.

3.4. *Hydrogen decoupling*

Hydrogen decoupling occurs if the velocity difference is equal to the thermal speed,

$$\frac{v_{ri} - v_{rp}}{\sqrt{\dfrac{2kT}{m_p}}} \approx 1.$$

After decoupling there are several possibilities for further hydrogen movement. Namely
- hydrogen leaves the star if the hydrogen velocity is greater than the escape velocity,
- hydrogen falls back onto the stellar surface or forms clouds above the surface (Porter & Skouza 1999) if the hydrogen velocity is lower than the escape velocity,
- hydrogen decouples in the atmosphere, only a pure metallic wind exist (Babel 1995, 1966) if the frictional force is small already in the stellar atmosphere.

An example of hydrogen decoupling in the wind of an A0 III star is given in Figure 4. The domains in HR diagram where stars with hydrogen decoupling, a pure metallic wind or without any wind may exist are displayed in Figure 2.

4. Conclusions and discussion

We have shown that multicomponent effects may be very important for those A stars which have a stellar wind. We have indicated the domains of different types of multicomponent stellar wind in the HR diagram (see Figure 2). First, it is possible that for A stars helium decouples from the stellar wind already in the stellar atmosphere and, consequently, that A stars have helium-free winds. Secondly, for stars with lower luminosities than those of supergiants, frictional heating may influence the wind temperature. Stars with even lower luminosities have a fractionated stellar wind due to hydrogen decoupling. Finally, A stars with the smallest luminosities have a pure metallic wind or do not have a stellar wind at all.

However, the detailed location of stars with different types of stellar wind in HR diagram depends on the model assumptions, especially on ionization equilibrium and wind parameters. Consequently, more advanced (NLTE, see Krtička & Kubát, 2005) models are necessary to study these effects in detail.

Acknowledgements

This work was supported by grants GA ČR 205/02/0445, 205/03/D020, 205/04/1267, MVTS SR-CR 128/04. JKr is grateful to IAU for travel grant. The Astronomical Institute Ondřejov is supported by projects K2043105 and Z1003909.

References

Abbott, D. C. 1982 *ApJ* 259, 282–301
Babel, J. 1995 *A&A* 301, 823–839
Babel, J. 1996 *A&A* 309, 867–878
Burgers, J. M. 1969 *Flow equations for composite gases*, Academic Press, New York
Castor, J. I., Abbott, D. C., & Klein, R. I. 1975 *ApJ* 195, 157–174
Castor, J. I., Abbott D. C., & Klein R. I., 1976 In *Physique des mouvements dans les atmosphères stellaires* (eds. R.Cayrel & M.Sternberg) CNRS Paris, p.363
Gayley, K. G. & Owocki, S. P. 1994 *ApJ* 434, 684–694
Hunger, K., & Groote, D. 1999 *A&A*, 351, 554–558
Krtička, J. 2003 In *Stellar Atmosphere Modelling* (eds. I. Hubeny, D. Mihalas & K. Werner) ASP Conf. Ser., Vol. 288, 259–262
Krtička, J. & Kubát, J. 2001 *A&A* 377 175–191 (KKII)
Krtička, J. & Kubát, J. 2005, *These Proceedings*, 23
Krtička, J., Owocki, S. P., Kubát, J., Galloway, R. K., & Brown, J. C. 2003 *A&A*, 402, 713–718
Kubát, J., Puls, J. & Pauldrach, A. W. A. 1999 *A&A* 341, 587–594
Mihalas, D. 1978 *Stellar Atmospheres*, 2nd ed., W. H. Freeman & Comp., San Francisco
Owocki, S. P., & Puls, J. 1999 *ApJ* 510, 355–368
Porter, J. M., & Skouza, B. A. 1999 *A&A* 344, 205–210
Straižys, V., & Kuriliene, G. 1981 *Ap&SS* 80, 353–368
ud-Doula, A., & Owocki, S. P. 2002 *ApJ* 576, 413–428

Discussion

LANDSTREET: Your slide shows metallic or no winds for late B and early A Main Sequence stars. Do you have models yet to predict whether winds occur in such stars, and how the mass loss rate would vary with spectral type?

KRTIČKA: Unfortunatelly, we do not have such models yet. However, this is an interesting problem and we plan to calculate these models in the future.

VAUCLAIR: In case of a decoupled wind in A stars, with helium remaining behind, I think that helium would accumulate below the atmosphere, at the place of the helium ionization (Cunha *et al.* 2005, *These Proceedings*, 359) while it would occur in the atmospheres of B stars, around $T_{\rm eff} = 20000$ K, where helium rich stars are actually observed.

KRTIČKA: Yes, I agree. We would like to mention that our models do not support the explanation of peculiar helium abundances in some B stars proposed by Hunger & Groote (1999) and probably some diffusion mechanism (as you discuss above) shall be invoked.

PRESTON: Does your formalism permit the calculation of the removal of metallic dust grains in the cooler (RV Tauri) stars?

KRTIČKA: Yes, it would be possible to include a dust grain component into our models. As we have calculated wind models of hot stars only, we have not included it.

The A-star puzzle
Proceedings IAU Symposium No. 224, 2004   © 2004 International Astronomical Union
J. Zverko, J. Žižňovský, S.J. Adelman, & W.W. Weiss, eds.   DOI: 10.1017/S1743921304004570

Abundances of A/F and Am/Fm stars in open clusters as constraints to self-consistent models including transport processes

Richard Monier and Olivier Richard

GRAAL, Universite Montpellier II, Place E. Bataillon, 34095 Montpellier, France
email: Richard.Monier@graal.univ-montp2.fr, Olivier.Richard@graal.univ-montp2.fr

Abstract. We describe the current status of a programme we started a few years ago to observe a large number of A/F and Am/Fm stars in open clusters of various ages. Spectra were obtained with the AURELIE and ELODIE spectrographs at a resolving power of about 40000 and S/N ratios from 100 up to 500. Abundances of 11 chemical elements have been derived by using Takeda's (1995) procedure. A short review on previous abundance determinations of A and F dwarfs in open clusters and a progress report on the current status of this project are presented. New abundance determinations for 24 A and F dwarfs in the Coma Berenices cluster are presented. These abundance determinations serve to set constraints to self-consistent evolutionary models of A and F stars including transport processes.

Keywords. Diffusion, stars: abundances, stars: chemically peculiar, open clusters and associations: individual: (Coma Ber)

1. Introduction

A few recent studies have addressed the chemical composition of the A and F dwarfs in the brightest open clusters. High quality high resolution spectroscopy is now feasible with 2 meter-class telescopes for the nearest clusters. Spectrum synthesis is the most appropriate method to derive abundances for A and F stars, some of which are fast rotators. Ultimately these abundance determinations can help constrain models of the internal structure of these stars. The Montreal group has recently developed models predicting the evolution of 28 chemical elements for A and F stars consistently with the internal structure (Turcotte *et al.* 1998, Richer *et al.* 2000, Richard *et al.* 2001). These new models include the effects of radiative diffusion and gravitational settling for all species having OPAL opacities. Provided overshooting homogenizes the surface regions, the predicted abundances are larger by a factor of 3 than the ones derived from the observations. Turbulent transport has been added to improve the agreement. For stars in clusters, the comparison of the abundance pattern (i.e., [X/H] versus Z) predicted by these models with those derived from observed spectra helps to set constraints on the processes included into the models.

The main thrust of this paper is to review the current knowledge on the abundances of A and F dwarfs in various open clusters, present a progress report on our programme to observe a large number of these stars well distributed in masses in selected open clusters. New abundance determinations for 13 A and 11 F dwarfs in the Coma Berenices cluster are presented. We also briefly discuss comparisons of predicted abundances with those derived from the spectra.

2. Previous abundance determinations of A and F stars in open clusters

A and F main sequence stars in open clusters are useful objects to constrain transport processes in time-dependent models. In contrast with the field stars, stars in clusters all have the same age and the same initial chemical composition (which can be derived from the G dwarfs in the cluster). Very few studies have so far addressed the chemical composition of A stars in open clusters and they usually have focused on a very limited numbers of stars not necessarily well distributed in mass along the Main Sequence. For the Hyades cluster, Burkhart & Coupry (1989) found the abundances of Li, Al, Si and Fe for 5 Am and 1 A star. Takeda & Sadakane (1997) derived the abundances of O and Fe for 8 Am and 10 normal stars and Hui-Bon-Hoa & Alecian (1998) those of Mg, Ca, Sc, Cr, Fe and Ni for 4 Am and 2 normal A dwarfs. Varenne & Monier (1999) determined the abundances of 11 chemical elements ($C I$, $O I$, $Na I$, $Mg I$, $Si I$, $Ca I$, $Sc II$, $Fe I$, $Ni I$, Y II, Ba II) for a much larger sample of stars: 19 A/Am dwarfs and 29 F dwarfs using AU-RELIE spectra centered at three wavelengths ($\lambda 6160$, $\lambda 5080$ and $\lambda 5530$). These stars are regularly distributed in spectral type along the Main Sequence to sample the expected masses uniformly. All these stars were analysed in a uniform manner using spectrum synthesis as a few of them are fast rotators. The effective temperatures and surface gravities were derived using Napiwotzki *et al.*'s (1993) revision of the UVBYBETA code. The line list was constructed from lists retrieved from VALD and from Kurucz's web site. We carefully checked the quality of the atomic parameters using the NIST database and also by synthesizing the spectrum of the Moon and that of Procyon. Given an appropriate model atmosphere, Takeda's (1995) iterative procedure yields the abundances, the microturbulent velocity and $v_e \sin i$. For the Hyades, Varenne & Monier (1999) found large star-to-star variations for the normal A stars·in particular for O, Na, Ni, Y and Ba. The Am stars are almost all deficient in Sc and Ca and overabundant in Fe, Ni, Y and Ba and also show star-to-star variations. In contrast, the F stars show very little scatter in their abundances. The A stars show a much larger scatter than the F stars on graphs displaying the abundances of individual elements [X/H] versus the effective temperature (T_{eff}). No convincing anticorrelation between the abundances and rotationnal velocity ($v \sin i$) was found. Note that significant abundance differences have been found among the few field A stars analysed so far (Hill 1995, Hill & Landstreet 1993, Holweger *et al.* 1986, Lambert *et al.* 1986, Lemke 1989,1990, Rentzsch-Holm 1997, Varenne 1999).

For convenience, we have gathered previous abundance determinations for the A and F dwarfs of the Coma Berenices cluster in Table 1. A similar list collecting all abundance determinations in the Hyades can be found in Table 1 of Varenne & Monier (1999).

3. Status of the project and data reduction

In our observational programme, we ultimately intend to acquire high resolution ($R = 40000 - 60000$) high S/N spectra of all A and F stars in several open clusters: IC 2391, α Persei, the Pleiades, Coma Berenices, Praesepe and the Hyades. We have excluded Ap stars since the models we wish to constrain do not include the effects of a magnetic field. In these clusters, the A and F dwarfs which range in magnitude from $V = 5$ to $V = 11$ are observed with 2-m telescopes. They sample the expected mass range from A0 V to F9 V along the main sequence. Spectra with S/N ratios ranging from 100 to 500 were obtained first with the AURELIE monorder spectrograph at the Observatoire de Haute Provence (OHP, France) for α Per, Coma Berenices and the Hyades. Three spectral regions centered on $\lambda 6160$ (region 1), $\lambda 5080$ (region 2) and $\lambda 5530$ (region 3)

Table 1. Previous abundance determinations for the Coma Ber A, F and G dwarfs.

Reference	Stars studied	Chemical Elements
Savanov (1996)	13 A-Am and F-Fm	C, O, Si, Ca, Fe, Ba
Hui-Bon-Hoa *et al.* (1997)	2 A-Am	Mg, Ca, Sc, Cr, Fe, Ni
Hui-Bon-Hoa & Alecian (1998)	4 A-Am	Mg, Ca, Sc, Cr, Fe, Ni
Burkhart & Coupry (2000)	7 A-Am	Li, Al, Si, S, Fe, Ni, Eu
Boesgaard (1987)	22 A and F	Li
Friel & Boesgaard (1992)	14 F	Fe, C
Jeffries (1999)	15 F, G, K	Li
Soderblom *et al.* (1990)	28 G	Li

have been selected as they include several lines having accurate oscillator strengths for the chemical elements we study. These regions were those observed by Edvardsson *et al.* (1993) in their spectroscopic survey of F dwarfs in the galactic disk. We are currently reobserving these three clusters and the Pleiades and Praesepe with the ELODIE echelle spectrograph (also at OHP) in the range λ3920 to λ6800 to have access to a much larger number of spectral lines. The selected open clusters of our programme and the current status of their observations and analysis are collected in Table 2. Note that we have also started analysing the UVES spectra for IC 2391 which we retrieved from the POP archive (Bagnulo 2005).

Table 2. Selected open clusters

Cluster ID	Age (Myrs)	Mag	$N_\star$	Spectrograph	Observed	Analysed
IC 2391	45	7.0-9.0	20	UVES	POP	yes
α Per	71	7.6-11.0	52	AURELIE	19	no
Pleiades	135	6.5-9.0	49	ELODIE	19	no
Coma Ber	447	5.0-8.6	68	AURELIE	24	yes
				ELODIE		no
Praesepe	730	6.3-10.0	80	ELODIE	10	no
Hyades	787	4.2-6.9	48	AURELIE	48	yes

4. New results for the Coma Berenices cluster

The results presented here were obtained by synthesizing AURELIE spectra in the 3 spectral regions 1, 2 and 3 mentioned above. The MIDAS software was used to reduce all the AURELIE spectra following the standard procedure (offset removal, division by the flat field, wavelength calibration). The spectra of the narrow lined stars were used to select continuum windows through which cubic splines were interpolated. The spectra were then normalised to this continuum. Continuum windows are very narrow for the fast rotators (but there are very few in Coma Berenices) and the normalisation is less secure for these stars. To derive abundances, we have adjusted synthetic spectra to the observed spectra using Takeda's (1995) iterative procedure as described in Varenne & Monier (1999).

We have first checked the spectral synthesis on Procyon whose abundances are fairly well known (Steffen 1985).The abundances we derived have been compared to those determined by Steffen (1985) and Edvardsson *et al.* (1993). The differences

$$\Delta[\text{X/H}] = [\text{X/H}]_{\text{this study}} - [\text{X/H}]_{\text{other}} \tag{4.1}$$

are less than 0.1 dex for 8 out of 11 elements and less than 0.15 dex for the 3 remaining (C , Y and Ba) (see Varenne & Monier 1999).

The abundances for the 24 Coma Berenices stars relative to the Sun

$$([\text{X/H}] = \log[\text{X/H}]_\star - \log[\text{X/H}]_\odot) \tag{4.2}$$

are collected in Figure 1.

An accurate assessment of the uncertainties of the abundances is mandatory if one wishes to constrain usefully the predictions of the models. The main sources of uncertainties for the abundances stem a priori from those of the effective temperatures, surface gravities, microturbulent velocities, oscillator strengths and apparent rotational velocities which are assumed to be independent. A change of $\Delta \log g = +0.2$ dex induces a very small change in the abundances ($\simeq 0.04$ dex) whereas an increase of 200 K of T_{eff} induces a larger change (~ 0.10 dex). Adding the inaccuracy on the oscillator strengths and apparent rotational velocities and continuum placement for fast rotators typically enhances the uncertainties of the relative abundance up to at least $0.30 - 0.40$ dex. The reader should be aware that the uncertainties in Figure 1 are lower limits (see Varenne & Monier (1999) for a complete discussion of the errors).

Table 3. Abundances for A and F dwarfs in Coma Berenices

HD	SpT	C I	O I	Na I	Mg I	Si I	Ca I	Sc II	Fe I	Ni I	Y II	Ba II
106103	F5V		-0.40	-0.21		-0.14	-0.19		-0.32	-0.24		-0.08
106293	F5V		-0.11	-0.14		-0.22	-0.21		-0.31	-0.34		0.00
106691	F5IV	(-0.1)	-0.46	-0.12	-0.11	-0.10	-0.16	-0.33	-0.23	-0.31	-0.29	-0.23
106946	F2V		0.00	-0.04		-0.20	-0.27		-0.38	-0.45		-0.16
107611	F6V		-0.44	-0.29		-0.15	-0.18		-0.18	-0.22		0.04
107877	F5V	-0.15	-0.35	-0.13		-0.11	-0.11		-0.14	-0.20	-0.15	0.13
108154	F8V		-0.13	-0.26	0.04	-0.13	-0.16	-0.05	-0.18	-0.19		0.21
108226	F5	-0.10	-0.12	-0.19		-0.09	0.06		-0.14	-0.13	-0.17	0.43
108976	F6V		-0.60	-0.14	0.08	-0.05	0.05	0.03	-0.06	-0.13		0.37
109069	F0V	-0.18	-0.32	-0.56	-0.26	-0.21	-0.41	0.04	-0.41	-0.29	-0.26	-0.39
109530	F2V	-0.20	-0.06	-0.08	-0.10	-0.19	-0.32	-0.07	-0.36	-0.30	-0.18	-0.20
106887	A4m					-0.22	(-0.42)	-0.73	0.06			
106999	Am	(0.00)	-0.15	-0.02	0.04	0.02	0.01	0.07	-0.02	0.27	0.07	0.03
107131	A6IV-V/A4IV-V		-0.14	-0.92		-0.13	-0.18		-0.28	-1.00		-0.36
107168	A8m/kA5hA5mF0	-0.61	-0.46	0.07	0.18	0.22	-0.02	-0.02	0.20	0.69	0.78	0.96
107276	Am/kA5mA7	0.10	-0.21	-0.25	-0.25	0.07	-0.23	-0.07	-0.19	-0.62	0.00	-0.12
107513	Am/kA7hF0mF0	-0.10	-0.16	-0.55	-0.45	-0.39	-0.44	-0.24	-0.53	-0.58	-0.16	-0.45
107655	A0V	-0.12	-0.55		-0.12	-0.15	-0.26	-0.69	0.11	0.68	0.63	0.80
107966	A3V/A3IV	-0.05	-0.04	0.53	0.01	-0.19	0.02	-0.02	-0.13	0.01	0.05	-0.06
108382	A4V/A3IV	-0.05	-0.12	0.07		-0.20	-0.13		-0.23	-0.46	0.01	-0.16
108486	Am/kA3hA5mA7	-0.40			0.07		-0.16	-0.83	0.28	0.48		0.94
108642	A2m/kA2hA7mA7	-0.61	-1.00	0.24		-0.07	-0.56		0.03	0.52	0.50	0.60
108651	Am	-0.60	-0.57	0.24		0.10	-0.61		0.34	0.59	0.64	1.50
109307	A4m/A3IV-V	-0.29	-0.30	-0.27	-0.25	-0.11	-0.14	0.10	-0.10	0.17	0.44	0.22

As for the Hyades, we find very little scatter for the abundances of the F dwarfs in Coma Berenices. The mean iron abundance for the F dwarfs in Coma Ber is slightly below solar in very good agreement with Boesgaard (1989). The A dwarfs exhibit large star-to-star variations in [Na/H] and [Ni/H], but less in [Si/H]. The Am stars all are deficient in oxygen. Most have overabundances in Ni, Y and Ba. No convincing trend of the abundances [X/H] versus $T_{\rm eff}$ nor [X/H] versus $v_e \sin i$ are found. The A stars exhibit a much wider scatter than the F stars in [X/H] versus $T_{\rm eff}$.

5. Constraints on transport processes

Several types of processes are expected to affect the surface abundances of A and F dwarfs: radiative diffusion, gravitationnal settling, meridional circulation, wind, accretion (Michaud 2005). The most recent models (Richer *et al.* 2000, Richard *et al.* 2001, Michaud 2005) take into account the atomic diffusion of metals and the radiative accelerations for all species having OPAL opacities. In these models, iron-peak convection zones appear at temperatures close to 200000 K. Assuming that overshooting homogenizes the surface regions, the abundances predicted by the models are usually 3 times larger than the abundances derived from the spectroscopy. Turbulent transport has been added to improve the agreement. The anomalies appear then to depend on the depth of the zone mixed by turbulence. Detailed comparison of the model predictions with derived abundances for individual stars are presented in Richer *et al.* (2000) (see for instance their figure 19 for the Hyades star 68 Tau). While the models fail to predict the abundances for each species, they do reproduce the shape of the abundance pattern. For a given star, there usually is considerable scatter between the abundances derived by different authors. The sources for this discrepancy are many: use of different methods to derive the abundances (curve-of-growth versus spectral synthesis), use of different codes and/or assumptions to model the atmosphere and the spectrum, use of different effective temperatures, surface gravities, atomic parameters. Processes not yet included in the models such as differential rotation or differential mass loss (Michaud *et al.* 1983) might also account for the disagreements.

6. Conclusions

To constrain the current models, abundances must be determined in a uniform manner for a large number of A and F stars well distributed in mass in clusters of various ages. Abundances are needed for as many chemical species as possible for $Z \leqslant 30$ with errors properly assessed. In Coma Berenices, we find evidence for large star-to-star variations in the normal A stars and in the Am stars as in the Hyades. There is very little scatter in the F dwarfs in both clusters. We believe that the differences in abundances among the A stars born from the same original interstellar matter may be the signature of the occurence of transport processes in their interiors. Observations with UVES of A and F dwarfs in more distant clusters sampling the age sequence a few 10 Myrs up to 800 Myrs are highly desirable.

References

Bagnulo S., 2005, *These proceedings*, 473
Boesgaard A.M., 1987, ApJ 321, 967
Boesgaard A.M., 1989, ApJ 336, 798
Burkhart C., Coupry M.F., 1989, A&A 220, 197

Burkhart C., Coupry M.F., 2000, A&A 354, 216

Edvardsson B., Andersen J., Gustafsson B., Lambert D.L., Nissen E., Tomkin J., 1993, A&A
       275, 101

Friel E.D., Boesgaard A.M., 1992, ApJ 387, 170

Hill G.M., 1995, A&A 294, 536

Hill G.M., Landstreet J.D., 1993, A&A 276, 142

Holweger H., Gigas D., Steffen M., 1986, A&A 155, 58

Hui-Bon-Hoa A., Burkhart C., Alecian G., 1997, A&A 323, 901

Hui-Bon-Hoa A., Alecian G., 1998, A&A 332, 224

Jeffries R.D., 1999, MNRAS 304, 821

Lambert D.L., McKinley L.K., Roby S.W., 1986, PASP 98, 927

Lemke M., 1989, A&A 225, 125

Lemke M., 1990, A&A 240, 331

Michaud G., 2005, *These Proceedings*, 173

Michaud G., Tarasick D., Charland Y., Pelletier C., 1983, ApJ 269, 239

Napiwotzki R., Schönberner D., Wenske V., 1993, A&A 268, 653

Rentzsch-Holm I., 1997, A&A 317, 178

Richard O., Michaud, G., Richer, J., 2001, ApJ 558, 377

Richer J., Michaud G., Turcotte S., 2000, ApJ 529, 338

Savanov I.S., 1996, Astronomy Reports 40, 196

Steffen M., 1985, A&AS 59, 403

Takeda Y., 1995, PASJ 47, 287

Takeda Y., Sadakane K., 1997, PASJ 49, 367

Turcotte S., Richer J., Michaud G., 1998, ApJ 504, 559

Soderblom D.R., Oey M.S., Johnson D.R., Stone R.P.S., 1990, AJ 99, 595

Varenne O., 1999, A&A 341, 233

Varenne O., Monier R., 1999, A&A 351, 247

Discussion

NOELS: Have you checked and analysed the abundances in the lowest mass stars of
your clusters? This would be interesting because it would give better knowledge of the
composition of the original cloud and by comparing them with the A and F stars of your
samples, it would give a quantitative estimation of the effect of diffusion.

MONIER: The late type F stars are the most difficult to observe on 2-m class telescopes
because they are faintest. I have indeed observed them for the closest, i.e., brightest
clusters.

The A-Star Puzzle
Proceedings IAU Symposium No. 224, 2004
J. Zverko, J. Žižňovský, S.J. Adelman, & W.W. Weiss, eds.

© 2004 International Astronomical Union
DOI: 10.1017/S1743921304004582

Surface abundances of Am stars as a constraint on rotational mixing

Olivier Richard[1], Suzanne Talon[2] and Georges Michaud[2]

[1]GRAAL UMR5024, Université Montpellier II, CC072, Place E. Bataillon, 34095 Montpellier, France
email: richard@graal.univ-montp2.fr

[2]Département de Physique, Université de Montréal, Montréal, PQ, H3C 3J7, Canada
email: talon@astro.umontreal.ca, michaudg@astro.umontreal.ca

Abstract. Abundance determinations obtained from spectroscopic observations of Am stars provide information concerning the transport processes present in these stars. In this paper we have used models of Am stars which include gravitational settling, thermal diffusion, and radiative accelerations for 24 elements. We used a specific model of rotation induced mixing which has reproduced anomalies in other types of stars. For this preliminary study, models of 1.7 $M_\odot$ and 1.9 $M_\odot$ have been computed. A comparison of the predicted abundances to the observed ones for the Praesepe star HD 73045 sets constraints on rotational mixing.

Keywords. Diffusion, hydrodynamics, stars: abundances, stars: atmospheres, stars: chemically peculiar, stars: individual (HD 73045), stars: rotation

1. Introduction

Am stars were first recognized by Titus & Morgan (1940) as chemicaly peculiar stars. Since then all observed Am stars have been found to be slowly rotating stars ($V_{eq} \sim$ 100 km s^{-1}). In these stars C, N, O, and Ca are generally underabundant compared to the Sun while iron-peak elements are overabundant. Different physical processes have been investigated to account for these abundance anomalies such as planet absorption, mass loss, turbulence, ... (see Michaud 2005).

Models of Am stars including gravitational settling, thermal diffusion, and radiative accelerations for 24 elements have been computed by the Montréal group (Riche *et al.* 2000). Using a simple parameterization of turbulent transport, they have shown that these stars need to be mixed from the surface down to at least $10^{-5}M_*$ to reproduce the observed surface abundances. It is interesting to use Am star anomalies to test the effects of a specific model of rotation induced mixing which has reproduced anomalies in other types of stars. In this paper, we will test a parameterization of rotation induced mixing for various rotational velocities in 1.7 $M_\odot$ and 1.9 $M_\odot$ models.

2. Models

2.1. *Physics*

The models were computed using the CEFF equation of state (Eggleton *et al.* 1973; Christensen-Dalsgaard & Däppen 1992) and the Bahcall & Pinsonneault (1992) nuclear energy generation routine. We used the OPAL monochromatic opacities for 24 elements to compute the Rosseland opacity at each time step, for each mesh point in the model for the current local chemical composition (Turcotte *et al.* 1998). All the models are self consistent models taking into account gravitational settling, thermal diffusion and radiative accelerations, which come from first principles. The diffusion velocity of each species is computed with the equations developed in Burgers (1969) which take into account the interactions among all these diffusing species. The radiative accelerations are from Richer *et al.* (1998) with correction for redistribution from Gonzalez *et al.* (1995) and LeBlanc *et al.* (2000). Convection, semi-convection, and the mixing processes are modeled as diffusion processes as described in Richer *et al.* (2000) and Richard et al. (2001).

2.2. *Mixing model*

In this preliminary study, we will use the simplest version of the turbulent diffusion obtained from a self-consistent treatment of meridional circulation following Zahn (1992). The reference rotation model is for a 1.7 $M_\odot$ star with a surface velocity of $\sim 50 \ \mathrm{km\,s^{-1}}$, and we consider the turbulent diffusion associated with the internal rotation profile at 500 Myr. Turbulence is assumed to be dominated by the shear instability, and the stabilizing effect of mean molecular weight gradients is ignored, both in the development of turbulence and of meridional circulation. Furthermore, no cutoff has been imposed based on a Reynolds number criterion. Details on the model physics are described in Talon (2005).

The complete expression for the turbulent diffusion taking into account thermal diffusivity and horizontal diffusion is

$$\nu_v = \frac{8Ri_c}{5} \frac{(r \, d\Omega/dr)^2}{N_T^2/(K+D_h) + N_\mu^2/D_h} \tag{2.1}$$

(Talon & Zahn 1997) and the rotation profile $d\Omega/dr$ is the instantaneous one and evolves towards the equilibrium profile.

2.3. *Computation*

All models were assumed homogeneous on the pre-main sequence with the abundance mix defined in Table 1 of Turcotte *et al.* (1998). We have computed models of $1.7M_\odot$ and $1.9M_\odot$ from the PMS to the subgiant branch, in around 1000 time steps. Each model has about 1500 mesh points. The effect of the mixing is included in our evolutionary models with a parametric mixing diffusion coefficient: $D_T = \omega \rho^n$. In V50 models, ω and n are chosen to give the best fit with the mixing diffusion coefficient obtained with the mixing model for the same mass and surface velocity of $\sim 50 \ \mathrm{km\,s^{-1}}$ (see Figure 1).

For the V15 models and V5 models the mixing coefficents are obtained from the $\Omega(r)$ equilibrium profiles of the surface velocity of, respectively, $\sim 15 \ \mathrm{km\,s^{-1}}$ and $\sim 5 \ \mathrm{km\,s^{-1}}$. All V50, V15, and V5 models are assumed fully homogenized between the surface and the depth where $\log(T) = 5.3$, which is the temperature where the convective zone due to iron accumulation occurs (Richard *et al.* 2001). The $1.9M_\odot$ V5N model have the same mixing coefficient as the V5 one but we do not assume full mixing between the surface and $\log(T) = 5.3$. In all models $n = -0.7$ and in the $1.9M_\odot$ V50, V15, and V5 models ω equals respectivly 180, 18, and 1.8.

3. Results

Figure 2 show the effective temperature and surface gravity range covered by our 1.7 $M_\odot$ and 1.9 $M_\odot$ models. At the age of the Pleiades ($\sim 100 \, \mathrm{Myr}$) and of Praesepe ($\sim 800 \, \mathrm{Myr}$) the 1.7 $M_\odot$ models have respectively $T_{\mathrm{eff}} = 8000 \, \mathrm{K}$ and $T_{\mathrm{eff}} = 7400 \, \mathrm{K}$, and the 1.9 $M_\odot$ models have $T_{\mathrm{eff}} = 8600 \, \mathrm{K}$ and $T_{\mathrm{eff}} = 7600 \, \mathrm{K}$.

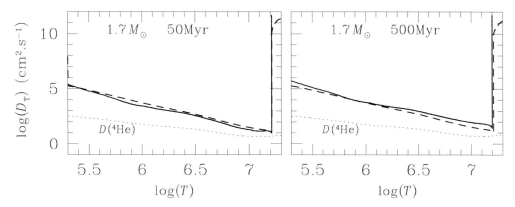

Figure 1. Profile of the mixing diffusion coefficient as a function of temperature in the $1.7\,M_\odot$ models with a surface velocity of $\sim 50\ \mathrm{km\,s^{-1}}$ at two ages. The full line represent the coefficient computed in evolutionary models including a self-consistent treatment of the meridional circulation (see section 2.2), the parametric mixing coefficient used in our models of $1.7\,M_\odot$ is also shown (dashed line). The dotted line represent the ^{4}He microscopic diffusion coefficent.

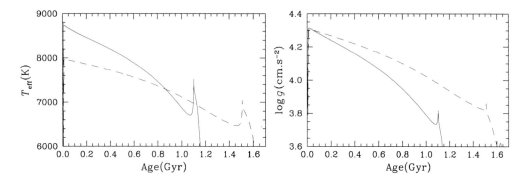

Figure 2. Evolution of the effective temperature and surface gravity in the $1.7\,M_\odot$ (dashed line) and $1.9\,M_\odot$ (full line) models.

Figure 3 presents the mixing coefficent profile as a fonction of depth in our $1.9\,M_\odot$ models and for two models of Richer *et al.* (2000) which reproduce quit well the surface abundances of the Praesepe star HD 73045. Our V15 and V5 models and the Richer *et al.* (2000) models have similar mixing coefficients around $\log(\Delta M/M_*) = -5.0$. The increase of D_T around $\log(\Delta M/M_*) = -6.0$, in the V5N model is due to the appearance of the convective zone due to the iron accumulation (Richard *et al.* 2001).

The radiative acceleration profile (upper panels) and abundance profile (lower panels) for Ca and Fe are shown in Figure 4 in our different models at 800 Myr. The right part and the left part of Figure 4 correspond, respectively, to $1.9\,M_\odot$ and $1.7\,M_\odot$ models. One can see the effect of saturation on the Fe radiative acceleration for $\log(\Delta M/M_*) \lesssim -4.5$, reducing the mixing coefficent from V50 to V5 reduces the mixed mass to the depth where Fe is supported by radiative acceleration so its abundance increases, which reduces the radiative acceleration due to saturation. The same effect of saturation is present for Ca in the $1.9\,M_\odot$ V5N model for $\log(\Delta M/M_*) \lesssim -6.6$.

From Figure 4 we also see that to obtain a Ca underabundance, one needs to have a mixed mass of $\sim 10^{-10} M_*$ to $\sim 6\times10^{-9} M_*$, which is even less mixing between the surface convective zone and the iron convective zone than in our V5N model, or $\sim 3\times10^{-7} M_*$ to

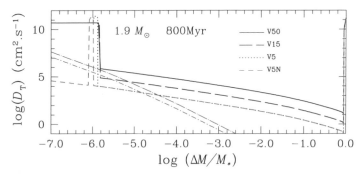

Figure 3. Mixing coefficient vs depth, $\log(\Delta M/M_*) = \log((M_* - m_r)/M_*)$, in the 1.9 $M_\odot$ models at the age of Praesepe cluster ($\sim$ 800 Myr). Richer et a.l (2000) 1.9R300-2 (dot-dashed line) and 1.9R1K-2 (dot-long dashed line) models are also plotted for comparison.

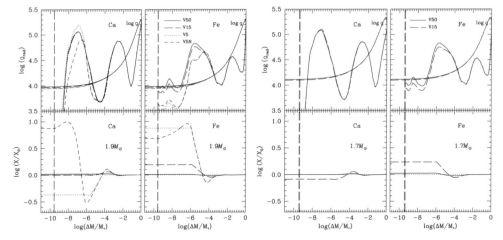

Figure 4. Profiles of radiative acceleration (upper panels) and abundance (lower panels) of Ca and Fe at 800 Myr in our different models (left part: 1.9 $M_\odot$; right part: 1.7 $M_\odot$). The vertical long dashed lines show the bottom of the surface convective zone.

$\sim 10^{-4} M_*$. The V5N model show a Ca overabundance due to the partial mixing between the surface convective zone and the iron convective zone which is not compatible with AmFm stars.

Figure 5 shows the comparison between the predicted abundances of our models at 800 Myr and the abundances determined from spectroscopic observation of the Praesepe star HD 73045 (Hui-Bon-Hoa *et al.* 1997, Burkhart & Coupry 1998). The 1.7 $M_\odot$ and 1.9 $M_\odot$ models have compatible effective temperature with HD 73045 ($T_{\rm eff} \simeq 7500$ K) at the age of Praesepe. The left panel shows the predicted abundances in our 1.7 $M_\odot$ and 1.9 $M_\odot$ V15 models which have similar abundance anomalies. The central panel shows the predicted abundances in our 1.9 $M_\odot$ models with different mixing parametrization and in Richer *et al.* (2000) R1K-2 model. The V15 and V5 models bracket the abundances of HD 73045. The right panel shows the effect of the different prescription on the mixing between the surface convective zone and the Iron convective zone, elements between Cl and Mn are more supported by radiative acceleration in the V5N model than in the V5 model while elements between He and S settle faster in the V5N model than in the V5 model.

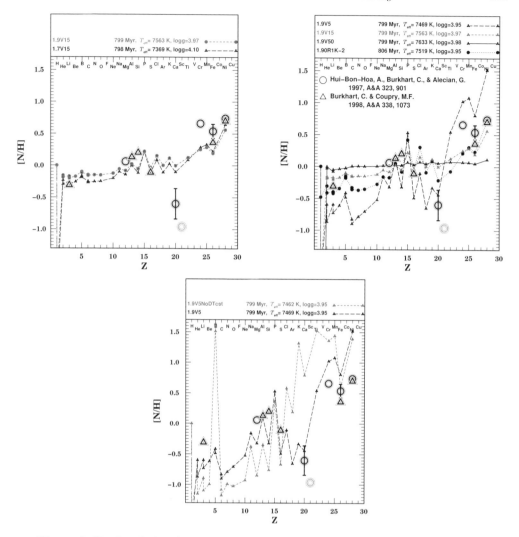

Figure 5. Predicted abundances around 800 Myr in different models compared to the abundance determined from spectroscopic observation of the Praesepe star HD 73045. (HIP 42247)

4. Discussion

The results presented here show that, in the model, turbulence seems to be too strong to allow the Am phenomenon to occur at a reasonable velocity (i.e., 50 km s^{-1}). This could be related to several features that have been neglected here:

• The leading coefficient of the turbulent diffusion coefficient $8Ri_c/5$ (cf. Eq. 2.1) is not determined from first principles and could vary by a factor of a few;

• The effect of the build up of mean molecular weight gradients has been overlooked here;

• Strong horizontal diffusion could reduce the efficiency of mixing (cf. Vincent *et al.* 1996).

The first factor is most probably not sufficient to explain the discrepancy. However, the stabilizing effect of mean molecular weight gradients on turbulence as well as the role of strong horizontal diffusion should be examined in more details.

Acknowledgements

This research was partially supported at the Université de Montréal by NSERC. We thank the Réseau Québécois de Calcul Haute Performance (RQCHP) for providing us with the computational resources required for this work.

References

Bahcall, J.N. & Pinsonneault, M.H., 1992 *Rev. Mod. Phys.* **64**, 885

Burgers, J. M., 1969, *Flow equations for composite gases* (New York: Academic Press)

Burkhart, C. & Coupry, M.F. , 1998, *A&A* **338**, 1073

Christensen-Dalsgaard, J. & Däppen, W., 1992 *ARA&A* **4**, 267

Eggleton, P.P., Faulkner, J. & Flannery, B.P., 1973 *A&A* **23**, 325

Gonzalez, J.-F., LeBlanc, F., Artru, M.-C. & Michaud, G. , 1995, *A&A* **297**, 223

Hui-Bon-Hoa, A., Burkhart, C. & Alecian, G. , 1997, *A&A* **323**, 901

LeBlanc, F., Michaud, G. & Richer, J., 2000, *ApJ* **538**, 876

Michaud G. 2005, *These Proceedings*, 173

Richard, O., Michaud, G. & Richer, J., 2001, *ApJ* **558**, 377

Richer, J., Michaud, G., Rogers, F., Iglesias, C., Turcotte, S. & LeBlanc, F., 1998, *ApJ* **492**, 833

Richer, J., Michaud, G. & Turcotte, S., 2000, *ApJ* **529**, 338

Talon, S. & Zahn, J.-P., 1998 *A&A* **329**, 315

Talon S. 2005, *These Proceedings*, 59

Titus, J. & Morgan, W.W., 1940 *ApJ* **92**, 256

Turcotte, S., Richer, J., Michaud, G., Iglesias, C. A. & Rogers, F. J., 1998 *ApJ* **504**, 539

Vincent, A., Michaud, G. & Meneguzzi, M., 1996 *Phys. Fluids* **8 (5)**, 1312

Zahn, J.-P., 1992 *A&A* **265**, 115

The A-Star Puzzle
Proceedings IAU Symposium No. 224, 2004
J. Zverko, J. Žižňovský, S.J. Adelman, & W.W. Weiss, eds.
ⓒ 2004 International Astronomical Union
DOI: 10.1017/S1743921304004594

Panel discussion section D

CHAIR: **Charles R. Cowley**

SECTION ORGANIZER & KEY-NOTE SPEAKER: G. Michaud
INVITED SPEAKERS: G. Alecian, F. LeBlanc
CONTRIBUTION SPEAKERS: J. Krtička, R. Monier, O. Richard

Discussion

NOELS: What are the uncertainties in the diffusion velocities? What can we do if we would like to multiply or divide it by something? How can we play with that? What would be your best guess?

MICHAUD: The atomic diffusion coefficients have been confirmed by different methods one of which is by numerical simulations. They should be relatively accurate. I would think a factor of 1.3 uncertainty is reasonable. The largest uncertainties are probably from the thermal diffusion coefficients, which are not, however dominant in these stars in general. They play a larger role when elements are highly ionized and so in central regions of stars.

Thermal diffusion is caused by an integration of the interaction of say Fe with the up going and down going protons. In their random motions the protons moving upwards have slightly larger energy than those moving downwards because the former come from higher T regions. The net flux of up going and down going protons is of course 0.0 at equilibrium. It is the energy variation of the scattering cross section that causes thermal diffusion. This crosssection is dominated by small far away collisions. They are affected by Debye shielding which is energy dependent. In the classical Chapman and Cowling expression for thermal diffusion that dependence was not included. They are largely overestimated and should not be used. The Paquette *et al.* (1986, ApJS, 61, 177) expressions for thermal diffusion are more accurate but they remain more uncertain than the atomic diffusion coefficients. One could have very large diffusion in the central regions of stars due to thermal diffusion if one used the low density limits of the thermal diffusion coefficients as used by Aller & Chapman (1960, ApJ, 132, 461), for instance, but these are clearly wrong and should not be used. The thermal diffusion coefficients, are very difficult to simulate precisely from first principles and in so far as I know, no statistical physicist is trying to improve the thermal diffusion coefficients we currently have although I have often mentioned the problem. For more details see Michaud (1991, Ann. Phys. (Paris), 16, 481).

ALECIAN: There are some other sources of uncertainties. For instance, the turbulent diffusion coefficients are not well known. This makes it difficult to correctly describe the transition between convection zones and radiative zones. On the other hand, there are still some uncertainties about radiative accelerations.

PISKUNOV: At which point in your calculations do you actually recalculate the ionization equilibrium? Is that during the abundance adjustment or during the temperature correction step?

LeBlanc: During the temperature correction step.

Piskunov: How much effort would it take to convert the results by Michaud into some kind of tables and then incorporate them in your models?

LeBlanc: I think it would be probably quite complicated. Michaud is working on the same stars as I, but he needs our results as outer boundary conditions. For PHOENIX, of course, we could use pretabulated tables, but in the atmosphere, with blends, it could also be dangerous.

Alecian: The Vienna group is presently preparing a new code which is a paralel-lized version of the Kurucz Atlas12 model (see Bischof 2005, *These Proceedings*, AP1). Presently the model atmosphere is an input of our code for the accelerations. A next step will consist in joining both the codes (the one for the accelerations and the other one for the model atmosphere).

Landstreet: Both Monier and Hill find that, even among rapidly rotating ($v_{eq} \geqslant 100$ km s^{-1}) A stars, abundance anomalies exist. If turbulent mixing increases with rotation velocity, as some models predict, how can these anomalies be understood? (Alternatively, one should perhaps try to understand both A and Am stars at the same time.)

Michaud: In the evolutionary model calculations including turbulent diffusion which I duscussed we are not considering a potential dependence of turbulent transport coef-ficients on rotation velocity. We are using abundance anomalies to constrain turbulent transport coefficients. In a second step the results may be used to constrain meridional cir-culation, or other sources of instability. In earlier work, Charbonneau & Michaud (1991, ApJ, 370, 693) had concluded that some overabundances persisted above 100 km s^{-1} when atomic diffusion competed with meridional circulation. They were smaller than below that limit. That was in static models but I would expect the same result to hold in evolutionary models.

Now, in evolutionary models described by Talon and Richard where turbulence is effectively coupled with rotation, the results are not very encouraging in this case with the coefficients used. Many of the coefficients of those models, and presumably the turbulent transport coefficients, are increasing as the square of the rotational velocity. That would mean a relatively rapid decrease of the anomalies as you go beyond about 100 km s^{-1} but not necessarily their disappearance. On the other hand one has to be careful, mass loss could also be involved at the same time.

Talon: Rotational mixing according to the simplest model of meridional circulation gives too strong mixing. It is even worse when you consider that the Eddington-Sweet timescale for stars with rotation velocities below 50 km s^{-1} is as long as the Main Se-quence lifetime. However this long timescale implies that the slow rotators have stronger differential rotation, and leads to turbulent diffusion coefficients that do not scale in Ω^2 as expected for solid body rotation but are rather much more similar for various velocities (below 100 km s^{-1}). There remains the problem that turbulence is at least a factor 100 too strong. This could be due to several neglected features namely the stabilizing effect of the mean molecular weight gradient, the erosion of vertical turbulence by horizontal turbulence and/or a cutoff effect due to radiative viscosity. If we find a self consistent model for A and Am stars, we should go back and look at the Geneva group's calculations of abundance anomalies in O stars to see if the modifications introduced in the model still permit one to explain the observations.

VAUCLAIR: With Sylvie Théado (2003, ApJ, 587, 777) we have computed the coupling effect of helium diffusion and rotation-induced mixing. The induced μ-gradient has a feed-back effect on the mixing which is not trivial. It depends strongly on the ratio of the horizontal to the vertical diffusion coefficient. In any case there is no single scaling between the effective diffusion coefficient and the rotation velocity.

NOELS: What was the age range of the galactic clusters you analyzed? It would be very interesting to sort out a dependence of the peculiarities in A stars as a function of time by comparing them from young clusters to older ones.

MONIER: The clusters range in ages from 45 Myrs to 700 Myrs. The search for a putative dependence of abundances versus age is indeed one of the ultimate goals of this project.

GRIFFIN: The third magnitude binary o Leo consists of 2 Am-like stars, both having *classical* Am-type abundances. The secondary is indeed an A star, but the primary is evolving. It is crossing the Hertzsprung gap and has $T \approx 6100$ K. Does the persistence of the Am phenomenom in such a (relatively speaking) cool giant influence your interpretation of the effects of diffusion? The range of anomalies is just the same.

MICHAUD: There are still effects of diffusion in the cooler stars but they are much smaller. Diffusion could not explain the abundance anomalies by the same factor in the Sun and in the 7000 K stars. It is completely excluded. However, when you look at a star cooling on the subgiant branch, the situation is a little different. In Fig. 14 of Richard *et al.* (2001, ApJ, 558, 377) you may see that abundance anomalies disappear only at $\log g \sim 3.2$ as a 2.5 $M_\odot$ star cools. We also have figures for the evolution as a function of $T_{\rm eff}$ (unpublished) and in the same star the disappearance of the anomalies is at 6000 K.

YILDIZ: It is generally adopted that the elements are following the magnetic field lines during the diffusion process. But, I remember a paper from 1980s in which it is found that the equatorial region of young stars, the elements are moving upward crossing the field lines. Can you conclude that how this can be?

ALECIAN: As I explained in my talk, horizontal diffusion is not the dominant process in element stratifications because of large time scales involved. It is the angular dependence of the vertical component of the diffusion velocity which determines the abundance patches at the surface of magnetic Ap stars. Magnetic fields reduces the diffusion velocity of ions, but not completely. On another hand, elements can often diffuse across magnetic lines because of the contribution of neutral states.

WADE: What are the impediments to developing a coupled, time-dependent, self-consistent model of the interior and atmosphere including diffusion?

LEBLANC: The first thing are timescales. In the atmospheres the timescales are very short as compared to the interior. So we would need a longer calculation time at least a factor 1000. Also in the atmosphere we need to calculate the radiative field differently than in the interior. In the interior we use the diffusion approximation. While in the atmosphere we absolutely need the detailed radiative field for a large number of points.

MICHAUD: To properly predict anomalies, one would like to couple atmosphere and mass loss models to evolutionary models. It is probably possible to make the assumption that the atmospheric part has come to some equilbrium with what is coming from below. Simply modify that atmospheric equilbrium abundance progressively.

At the moment what we do for the atmospheres in the evolutionary model is very approximate. But if you wish to evaluate the possibility of selective mass loss in particular, we have to be much more sophisticated than we currently are in the calculations of the atmospheric region.

WADE: But what is the solution? Is it a question of bigger and faster computer?

MICHAUD: I think first one must develop the proper selective mass loss models. I do not think that has yet been done to an acceptable degree of accuracy and then we can couple them to the evolutionary models. But those models have to be developed. It is a lot of software development.

BUDAJ: I just want to address Wade's question. A possible way to couple these interior calculations with the radiative transfer envelopes follows. Calculations in the envelope can correlate the boundary condition in the sense that we would know the abundance gradient at the bottom of the atmosphere and this would provide us with the flux of the element at that point and the condition in the atmosphere would not be a zero diffusion velocity but a constant particle flux. This would couple the atmosphere with the interior. So we would just need two numbers. The concentration and gradient at the bottom of the atmosphere.

PAUNZEN: We heard a lot about the stars with overabundances. I am interested in stars with underabundaces, which are the λ Bootis stars. A few years ago there were some papers concluding that diffusion together with mass loss can produce underabundances in 1 Gyr or so. We heard a great deal about new calculations. Is there any news about what underabundances can be generated with diffusion together with mass loss? At earlier stages of the stellar evolution? Is it possible to produce underabundances on timescales smaller than 1 Gyr?

COWLEY: That is actually a question I have been interested in asking as well. What is the current standing from the theoretical point of view of stars with underabundances, the so-called λ Boo stars, which form apparently a very inhomogeneous group. Can we understand now, young objects with underabundances?

MICHAUD: The diffusion models that have been proposed for λ Boo stars, where there is radiative diffusion and undifferentiated mass loss (Michaud & Charland 1986, ApJ, 311, 326) required very large timescales, longer than some of the clusters where some λ Boo stars had been observed. For that reason atomic diffusion does not appear as the most likely explantion for that. A differentiated mass loss as has been discussed today might succeed, in combination with atomic diffusion, but that requires a credible model of the differentiated mass loss. Separation would then occur partly in the wind. So that is something that should be investigated, but the explanation for these stars could not be only atomic diffusion in the stellar interior. That requires too long timescales. As an alternative, accretion has been investigated with some success by Turcotte & Charbonneau (1993, ApJ, 413, 376).

The A-Star Puzzle
Proceedings IAU Symposium No. 224, 2004   © 2004 International Astronomical Union
J. Zverko, J. Žižňovský, S.J. Adelman, & W.W. Weiss, eds.   DOI: 10.1017/S1743921304004600

Magnetic fields of A-type stars

G. Mathys

European Southern Observatory, Casilla 19001, Santiago 19, Chile

Abstract. The current status of our knowledge of magnetic fields in A-type stars is reviewed with special emphasis on the progress achieved for "classical" Ap stars over the last four years. For those stars, the distribution of the strength of the field, its three-dimensional structure, and its relation with rotation and evolution are discussed.

Keywords. Stars: magnetic fields, stars: chemically peculiar, stars: rotation, stars: evolution

1. Introduction

A-type stars occupy a special place in the history of the study of stellar magnetism. Indeed, it is in such a star, 78 Vir, that the first detection of a magnetic field in a star other than the Sun was achieved (Babcock 1947). That Babcock had selected a peculiar star (78 Vir was classified A2p) as his target of choice did not reflect any *a priori* assumption about the existence of a connection between peculiarity and magnetic field. Instead it relied on the hypothesis that fast rotating stars could generate strong magnetic fields and it was driven by the consideration that the small effects of magnetic fields could be better detected in sharp spectral lines than in broad ones. Babcock originally believed that 78 Vir was a fast rotating star whose rotation axis lies (almost) parallel to the line of sight. To this date (fairly) sharp-lined stars remain privileged targets for magnetic field studies. Among the stars in which magnetic fields have been found, Ap stars still represent the most numerous group. However, the overall picture of stellar magnetism that has progressively emerged since Babcock's ground-breaking discovery corresponds to a physical reality very remote from the reasoning underpinning his detection attempt.†
An overview of our current knowledge and understanding of A star magnetic fields, based on the wealth of observational results that have been obtained over the years, is given in this presentation. Important open questions, whose answers require additional observational constraints, are emphasised.

For a long time, Ap and Bp stars were the only non-degenerate stars besides the Sun in which magnetic fields had been definitely detected and measured. Occasionally, detections of magnetic fields in stars of other types in the same temperature range were reported, but in general they could not be confirmed in subsequent, independent studies. The continued development of instrumentation that brought about the breakthrough discovery of solar-like magnetic fields in the late-type dwarfs ξ Boo A and 70 Oph A (Robinson *et al.* 1980) triggered renewed interest in attempts to detect magnetic fields in non-peculiar A stars, taking advantage of the ever increasing accuracy achievable. Naturally, considering the apparent relation of their chemical peculiarities to those of Ap stars and their predominantly slow rotation, Am stars and HgMn stars were primary targets for such investigations. The first "modern" study of this kind was conducted by Mathys & Lanz (1990), who concluded to the probable presence of a magnetic field of the

---

† It is an amazing, and quite fortunate coincidence that the order of magnitude of the Ap star magnetic fields is similar to the one Babcock had come to expect on the basis of an incorrect reasoning.

order of 2 kG in the hot Am star *o* Peg, from consideration of the differential broadening of spectral lines with different magnetic sensitivities and of what appeared to be differential magnetic intensification between the two Fe II lines $\lambda\lambda$ 6147.7 and 6149.2. Analysis of the same line pair led Lanz & Mathys (1993) to suggest that magnetic fields are also present in two other hot Am stars, HD 29173 and HD 195479A. More recently, the same method was applied by Hubrig *et al.* (1999) and by Hubrig & Castelli (2001) to look for magnetic fields in a sample of HgMn stars and of superficially normal late B stars. The analysis relies on modern spectrum synthesis techniques and on recent determinations of the *gf* values to show how critically the conclusions depend on the accuracy of the latter. Yet, it suggests that detectable magnetic fields may be present in some of the studied stars. If so, they must have a geometric structure sufficiently complex so as to remain undetected with the unprecedented sensitivity achieved in the recent circular spectropolarimetric observations of Shorlin *et al.* (2002). The median 1σ uncertainty of the mean longitudinal magnetic field‡ determinations achieved by these authors are 18 G for Am stars, 39 G for HgMn stars, and 13 G for normal A and B stars. However, Moninet *et al.* (2002) report the probable detection of a longitudinal field of -54 ± 12 G in the normal F7V star χ Dra. On the other hand, the recent discovery (Wahlgren *et al.* 2001, Adelman *et al.* 2002) of periodic profile changes in the line Hg II λ 3984 of the HgMn star α And, if interpreted as due to inhomogeneities in the distribution of Hg on the surface of this star (as seems most plausible), requires the existence of a mechanism of horizontal differentiation of the atmospheric composition, which suggests that a magnetic field with sufficient large-scale organisation may be present in this star.

In summary, despite the increase in the achievable sensitivity and accuracy for magnetic field detection and determination that has taken place over the last 15 years, the question of the presence of magnetic fields in non-peculiar A stars, in Am stars, and in HgMn stars remains open. Accordingly in the rest of this review, I shall focus my attention on what is still the only class of A stars where the existence and the properties of magnetic fields are well established and documented: the Ap stars. More specifically, I shall consider explicitly only the "classical" Ap stars, or CP2 stars, in Preston's (1974) naming convention, to the exclusion of the hotter He weak and He strong stars, which as B stars fall outside the scope of this symposium. Yet one should keep in mind the close relation they bear to the Ap stars proper. Most of what refers to the latter also apply to the former. On the other hand, the sequence of the classical Ap stars actually stretches from late B to early F spectral types, and it would make no sense to exclude its coolest and hottest members from the discussion even though they are not strictly speaking A-type stars.

2. Ap star magnetic fields: overview

A review of the status of knowledge and understanding of Ap star magnetic fields as of end of 2000, with some emphasis on the way in which the presented conclusions were reached, can be found in Mathys (2001). I shall only summarise here the main results reported in this paper, referring the reader to it for more details. In this presentation, I emphasise the progress that has been made and the questions that have been debated since then.

Magnetic fields of Ap stars cover the whole stellar surface. They have a significant degree of large-scale organisation, such that the range of field strengths over the stellar surface is relatively narrow. In first approximation, their structure is roughly symmetric

---

‡ The mean longitudinal magnetic field is the line-intensity weighted average over the visible stellar hemisphere of the component of the magnetic field along the line of sight.

about an axis that is inclined to the stellar rotation axis, Ap stars are *oblique rotators*. In most cases this structure bears some resemblance to a simple dipole encompassing the whole star. Intrinsic variations of Ap star magnetic fields have not been definitely observed so far. The only variations that are observed are periodic and result from the changing aspect of the visible stellar hemisphere as the star rotates.

The magnetic field moment for which the largest number of measurements has been obtained is the mean longitudinal magnetic field. Its measured values range (in absolute value, since it can be either negative or positive) from less than 100 G up to 20.5 kG in Babcock's star HD 215441 (Borra & Landstreet 1978). Bychkov *et al.* (2003) show that the observed distribution of the rms longitudinal field (Borra *et al.* 1983) of Ap stars is consistent with a decreasing exponential. It is quite plausible, but cannot be fully ascertained at present, that the breakdown of this distribution below 100 G reflects the currently achievable limit of sensitivity of longitudinal field determinations. The latter has become progressively lower over the past years (e.g., Wade *et al.* 2000, Shorlin *et al.* 2002), allowing a growing number of weak magnetic fields to be definitely detected and accurately measured. Some of the most accurate longitudinal field measurements obtained to this date suggest that all Ap stars are magnetic and that there may exist a minimum field strength for which Ap-type characteristics are produced (Aurière & Wade 2005). At the other end of the distribution, while Babcock's star, which has been known for many years, still stands by quite some margin as the record holder, a couple of additional stars showing exceptionally large longitudinal fields have recently been discovered: HD 178892 (B9p SrCrEu), with measured values of up to 8.5 kG (El'kin *et al.* 2003), and NGC 2244-334 (a Bp He weak star), for which a single measurement yielded a longitudinal field of -9.0 kG (Bagnulo *et al.* 2004). The longitudinal fields of these two Bp stars are stronger than any other known so far, save Babcock's star. El'kin *et al.* (2002, 2003) found three more stars with longitudinal fields greater than 3 kG, and four others where the largest (in absolute value) measurement of the longitudinal field is between 1.5 and 3.0 kG. Their contribution considerably increases the number of Ap stars known to have strong longitudinal fields. Another recently found member of this group is HD 66318 ($=$ NGC 2516-24) for which Bagnulo *et al.* (2003) report a longitudinal field value of 4.5 kG. Strongly magnetic Ap stars are interesting because they provide opportunities to study the effect of the magnetic field with the greatest detail and accuracy.

Due to its strong dependence on the geometry of the observation, and its resulting often large variability over a rotation period, the mean longitudinal magnetic field is poorly suited to characterise the actual stellar field. To this effect, a much more appropriate field moment is the mean magnetic field modulus, that is, the line-intensity weighted average over the visible stellar hemisphere of the modulus of the magnetic field. It is determined from measurement of the separation of the σ (or σ and π) components of lines that show resolved magnetic splitting. Such lines can be observed only in stars that have a sufficiently strong field and that rotate slowly enough so that the magnetic splitting is not smeared out by rotational Doppler effect. This, of course, restricts the number of stars in which the mean field modulus can be determined. At the time of writing, to the best of this author's knowledge, this number is 47. It includes the 42 stars discussed by Mathys *et al.* (1997), HD 18610 (Stütz *et al.* 2003), HD 66318 (Bagnulo *et al.* 2003), HD 51684 and HD 213637 (Mathys *et al.* , in preparation†), and HD 94427 (unpublished;

† Resolved magnetically split lines were found in HD 213637 by these authors on a spectrum taken on November 11, 1997. This finding was announced by the present author in his review presented at the conference on "Magnetic fields across the Hertzsprung-Russell diagram" held in Santiago in January 2001. Mathys (2003) reports that a total of 7 high-resolution spectra of

the line Fe II λ6149.2 was observed to be resolved in a spectrum taken on February 26, 2000). Mathys (2003) pointed out that 7 of the stars where magnetically resolved lines have been observed belong to the group of the rapidly oscillating Ap (roAp) stars. The most recent addition, HD 116114 (Kurtz et al. 2005), also shows lines resolved by the magnetic field. The roAp stars with resolved magnetically split lines represent more than 20% of all known roAp stars. This indicates an unusually high frequency of occurrence of strong magnetic fields and of slow rotation among roAp stars, compared to Ap stars in general.

A remarkable result of the extensive study of stars with magnetically resolved lines conducted by Mathys et al. (1997) is the apparent existence of a low-field cutoff around 2.8 kG in the distribution of the mean field modulus, averaged over the rotation period (for more details see the original reference, and Mathys 2001). Four of the five additional stars with resolved magnetically split lines have mean field moduli exceeding 5 kG. The only exception is HD 94427, where the field modulus measured in February 2000 is 2.2 kG. This single measurement does not by itself question the existence of a low-field cutoff in the field modulus distribution. Indeed it may just correspond to a rotation phase close to the minimum of the field modulus, such that the average of the latter over the rotation period is significantly higher and above the cutoff. This is, as matter of fact, observed in other stars (e.g., HD 59435: see Wade et al. 1999). Mathys et al. (1997) established that the 2.8 kG lower limit of the distribution of the mean field modulus (averaged over the rotation period) is well above the limit of resolution of the diagnostic line Fe II λ6149.2. But they could not definitely decide if this lower limit represents a cutoff or a discontinuity in the distribution. They mention the observation of a number of stars in which no magnetic line splitting could be resolved observationally. At least 14 of the latter have $v \sin i$ small enough so that smearing out of magnetically resolved line components by rotational Doppler effect can be ruled out. Of these 14 stars, only two have been observed in circular polarisation to measure their mean longitudinal magnetic field: HD 8441 and HD 176232. Mathys & Hubrig (1997) did not definitely measure a non-zero longitudinal field in the latter, and the significance of Babcock's (1958) measurements of this field moment depends on the correct estimation of their uncertainties. Suspicion that the latter are underestimated for HD 176232 is strengthened by the consideration that this is almost certainly the case for Babcock's (1958) determinations of the longitudinal field of HD 8441, as they would otherwise be irreconcilable with its 69.5 d rotation period, which is securely established from analysis of photometric variations (Wolff & Morrison 1973). On the other hand, Preston (1970) refers to 9 high quality measurements of the longitudinal field of HD 176232 obtained on 10 consecutive nights in 1970, which indicate a constant value of 500 ± 100 G; the lack of further details does not allow a critical evaluation of their significance. In other words, in a couple of cases, circular spectropolarimetry gives some hint of, but does not definitely establish, the presence of weak detectable magnetic fields in sharp-lined Ap stars where no magnetic line splitting or broadening has been observed in high-resolution spectra recorded in unpolarised light. On the other hand, line synthesis suggests the presence of a magnetic field of the order of 1 kG in HD 176232 (Ryabchikova et al. 2000), which is fully consistent with the upper limit resulting from the study of Mathys et al. (1997). A somewhat larger value, 1.5 kG, was derived through a similar analysis by Kochukhov et al. (2002). But the most definite proof so far of the presence of a fairly weak magnetic field in HD 176232 is due to the high-resolution infrared observations of Leone et al. (2003) from which a mean

this star had been obtained by end of 2002. An independent report of the discovery of resolved lines, based on a spectrum taken on April 17, 2002, was published by Kochukhov (2003).

field modulus of 1.4 kG is diagnosed. As a matter of fact, high-resolution spectroscopy in the infrared is the royal way for future studies of the low end of the magnetic field strength distribution. Indeed the quadratic wavelength dependence of the Zeeman effect (opposed to the linear dependence of the Doppler effect) implies that for a given $v \sin i$, resolved magnetically split lines can be observed for lower fields than in the visible, and conversely, that magnetic line resolution is observable at a given field strength for faster rotators in the infrared. With high-resolution infrared spectrographs now coming of age, such as CRIRES at ESO's VLT, a new era opens for the study of stellar magnetism.

3. Magnetic field structure

Modelling of the geometric structure of the magnetic fields of Ap stars has long been limited by the crudeness of the observational constraints that could be obtained. Until the advent of modern CCD detectors, magnetic field diagnosis was for most stars restricted to the determination of their mean longitudinal magnetic field. At the achievable accuracy, its variation in a stellar rotation period did not significantly depart from a sinusoid. With the latter as the only constraint, no model more complex than a single dipole at the stellar centre, with its axis inclined to the stellar rotation axis, can be uniquely derived. In the few fortunate cases where magnetic splitting of spectral lines could be resolved and followed throughout a stellar rotation cycle on spectra recorded on photographic plates, the deficiencies of the centred dipole model were revealed. Additional free parameters were required, leading to the introduction of "perturbations" of the centred dipole, such as the superposition of a dipole and a collinear quadrupole at the centre of the star, with their common axis inclined to the stellar rotation axis, or a dipole offset from the centre of the star by a (relatively small) fraction of the stellar radius in the direction of its axis (not aligned with the stellar rotation axis). Again, models of this type are the most complex ones that could be unambiguously constrained with the observational data available at the time. Systematic exploitation of the information contents of line profiles recorded in circular polarisation began in the second half of the 1980s with this author's observations of a sample of magnetic Ap stars with ESO's Cassegrain Echelle Spectrograph (CASPEC) attached to ESO's 3.6 m telescope (Mathys 1991). Curves of variations along the rotation cycle of moments of the stellar magnetic fields, as derived from measurements of low-order moments of the observed Stokes I and V line profiles (Mathys 1995a, 1995b, Mathys & Hubrig 1997, Mathys *et al.* in preparation), possibly combined with curves of variation of the mean magnetic field modulus, for those stars with magnetically resolved lines (Mathys *et al.* 1997), or with broad band linear polarisation measurements (Leroy 1995), proved amenable to the derivation of models of the field structure of greater complexity, such as the superposition of collinear dipole, quadrupole and octupole components at the star centre (Landstreet & Mathys 2000) or that of a dipole and a (nonlinear) quadrupole at the star centre (Bagnulo *et al.* 2000, 2002b). However, the shortcomings of these models soon became evident when the linear polarisation line profiles that they predicted were confronted with newly obtained high resolution, high signal-to-noise, Stokes Q and U observations (Bagnulo *et al.* 2001). This work provides the most compelling *concrete* demonstration that observations in all four Stokes parameters are necessary to fully constraint the magnetic field geometry. This had been realised, in principle, for a long time, but the extent to which simple models based only on Stokes I and V spectra fail to represent the actual structure of the magnetic field as revealed by the addition of Stokes Q and U observations had not been previously proven. This underlines the need for observations in all four Stokes parameters. While circular polarisation is in first approximation proportional to the (line-of-sight component) of the magnetic field, linear polarisation is a second-order effect in the field strength (see e.g., Mathys 2002). Furthermore, in disk-integrated observations, Stokes Q and U signals from various parts of the visible stellar hemisphere cancel out more effectively than their Stokes V counterparts. Accordingly, considerably higher signal-to-noise is required for significant detection and characterisation of line profiles in linearly polarised light than in circularly polarised light. Only recently have the required observations become feasible. The publication by Wade *et al.* (2000) of the first systematic spectropolarimetric measurements in all four Stokes parameters of a sizable sample of Ap stars with, for a fraction of them, adequate phase coverage, represents a breakthrough in the study of the magnetism in this type of stars. This takes place at a time when the Magnetic Doppler Imaging (MDI) techniques have reached the stage of maturity (Piskunov & Kochukhov 2002) and developments in computer technology allow one for the first time to fully exploit them to synthesise simultaneously in all four Stokes parameters high-resolution (e.g., $\lambda/\Delta\lambda \sim 30000$) profiles of a representative number of spectral lines recorded at enough phases to sample adequately their variations along the stellar rotation period.

The analysis of 53 Cam by Kochukhov *et al.* (2004a) represents the state of the art in this area. Besides the simultaneous consideration of all four Stokes parameters, the essential difference between MDI and earlier attempts at deriving models of Ap star magnetic fields is that the former does not make *a priori* assumption about the field structure (e.g., some superposition of low-order multipoles). Yet, a regularisation function is required to ensure consistency between the number of free parameters and the information contents of the available observational data, and to maintain numerical stability. The choice of this function is based on an overall knowledge of the type of field structure under consideration (large-scale organisation for Ap stars, high fractionation for late-type stars). But the derived model seems independent of the details of this choice, so that the representation of the stellar magnetic field that it yields is *unambiguous*.

Kochukhov *et al.* (2004a) find that the structure of the magnetic field of 53 Cam is much more complex than any low-order multipole expansion. They regard this result as questioning the validity of such expansions as used so far. However, the existence of some rough axial symmetry in the vast majority of Ap stars seems inescapable unless one adopts the view that statistical correlations established through use of predominantly dipolar models, such as between the inclination of the rotation axis with respect to the rotation axis and the stellar rotation period (see Sect. 5), are purely coincidental, which is not a very sustainable scientific attitude.

4. The third dimension

Modelling attempts as described in the previous section are mostly restricted to mapping the *horizontal* distribution of the strength and orientation of the magnetic vector over the stellar surface. While the possible existence of observable radial gradients of the magnetic field in Ap star photospheres has been debated since more than a quarter of a century (see, e.g., Wolff 1978), this question has in the past few years taken increased relevance, both because of the progress achieved in the knowledge of the vertical variation of other physical parameters and thanks to the improved observational prospects opened by the developments that have taken place in both the instrumental and theoretical areas. Interpretation of the impossibility to account for the strengths and shapes of observed spectral line profiles by adopting a unique value for the abundance of a given element as the manifestation of vertical abundance stratification (e.g., Bagnulo *et al.* 2001, Ryabchikova et al. 2002) must be regarded with some caution as long as the (vertical) atmospheric structure is not securely established. Indeed, observations of Core-Wing Anomalies in Balmer lines of cool Ap stars indicate that the latter may depart considerably from standard model atmospheres (Cowley *et al.* 2001).

Recently, Nesvacil *et al.* (2004) tried to determine the mean magnetic field modulus on both sides of the Balmer jump for a small sample of Ap stars with resolved magnetically split lines. This is a similar strategy to that of Wolf (1978), except that the field modulus, rather than the longitudinal field, is considered. The significance of the results is limited by the number of suitable diagnostic lines on the short wavelength side of the Balmer jump, but there is a clear hint of stronger fields (with a difference of few hundred Gauss) below the Balmer discontinuity, suggesting that the field increases from lower to higher atmospheric layers. By contrast, Romanyuk (2005) using Wolf's (1978) original approach (albeit with a CCD rather than photographic plates), finds that the magnetic field of α^2 CVn decreases with height. On the other hand, the short vertical wavelength of the pulsation modes of roAp stars (Kurtz *et al.* 2005 and references therein) allows one to use the pulsation amplitudes and phases derived for individual spectral lines as atmospheric depth proxies. This opens the prospect to investigate the depth dependence of the magnetic field by relating values determined from such lines with their formation depth. While this approach has not been applied yet, controversy has arisen in the last year about the results obtained through a different use of pulsation observations to diagnose the possible existence of vertical magnetic field gradients. Namely, as a result of the pulsation, the line-forming region moves up and down, hence samples different depths in the atmosphere. If the magnetic field strength vertical variation is sufficient over the amplitude of this motion, field values derived from consideration of a given line should vary with the pulsation period. Attempts to detect such variations based on longitudinal field determinations yielded inconsistent results. Hubrig *et al.* (2004), applying the method introduced by Bagnulo *et al.* (2002a) to diagnose the field from low-resolution spectropolarimetric observations with FORS1 at the VLT, did not find any variation at a level of the order of 100 G over the pulsation cycle of six roAp stars. This sample includes γ Equ, for which Leone & Kurtz (2003) report the discovery in Nd III lines of longitudinal field variations with amplitudes of 110-240 G and a period of 12.1 min. This discovery was subsequently questioned by Kochukhov *et al.* (2004), who obtained an upper limit of 40-60 G for variations with the pulsation period in 13 Nd III lines. Bychkov *et al.* (2005) using Balmer line photopolarimetry, also fail to detect any variation of the longitudinal field of γ Equ with its pulsation period. On the other hand, attempts to detect variations of the field modulus of the same star over its pulsation cycle from Fe II lines observations set upper limits of 5-10 G for such variations (Kochukhov *et al.* 2004b, Savanov *et al.* 2005). This is hardly constraining, though, since a rough theoretical estimate suggests that the field modulus variation with the pulsation period might have an amplitude of the order of 11 G.

5. Rotation, binarity, and evolution

An extensive review of the rotation of A and B type chemically peculiar stars has recently been presented elsewhere (Mathys 2004). It also includes some considerations about binarity. Here I shall only summarise the results that appear most relevant in relation to the magnetic fields of Ap stars. Magnetic Ap stars rotate in average significantly slower than their normal counterparts of similar temperature. Most of them have rotation periods between 1 and 10 days. But there also exists a non-negligible population (possibly of the order of 10% of all Ap stars) of stars with rotation periods longer than 100 days. The longest periods may exceed one century. These extraordinary slow rotators are interesting because they represent the most extreme manifestation of the braking process that distinguishes Ap stars from normal A stars. One can hope to gain insight into this process by identifying other aspects in which they differ from shorter period stars. Some results of this kind have started to emerge. The conclusion reached by Mathys *et al.* (1997) that the mean magnetic field modulus exceeds 7.5 kG in more than half of the Ap stars with resolved magnetically split lines that have a period shorter than 150 days, but is always smaller than 7.5 kG in longer period stars, remains valid even though a considerable number of additional measurements of this field moment have been accumulated since its publication, and new rotation periods have been determined. Also, the (approximate) symmetry axis of the magnetic field appears to be nearly aligned with the rotation axis in the slowest rotators, while in shorter period Ap stars, the angle between the two axes is usually large (Landstreet & Mathys 2000, Bagnulo *et al.* 2002b). A possibly related, intriguing result has been presented at this meeting by Bychkov *et al.* (2005) who found a bimodal distribution of the angle between the magnetic and rotation axes for a large sample of Ap stars, based on magnetic field measurements from the literature.

On the other hand, while the deficiency of short-period binaries among magnetic Ap stars (compared to normal A stars) is well known (Gerbaldi *et al.* 1985; Leone & Catanzaro 1999, Carrier *et al.* 2002), it has more recently been found (Mathys *et al.* 1997) that this deficiency is even more pronounced among very slowly rotating Ap stars. Of the 12 Ap stars with rotation period longer than 30 days known to be binaries, only one may have an orbital period (not yet unambiguously determined) shorter than 100 days. In other words, the effectiveness of the achievable rotational braking of Ap stars in binaries appears to be somehow related to the orbital period.

The understanding of the origin and evolution of the magnetic fields of Ap stars, as well as of their slow rotation, requires their evolutionary status to be established. This remains a controversial issue, which cannot be fully discussed within the scope of this review. I shall restrict myself to pointing out some recent results. The conclusion drawn by Hubrig *et al.* (2000), that magnetic Ap stars with masses below $3\,M_\odot$ have completed at least 30% of their main sequence lifetime, was based on the consideration of a sample comprising stars with magnetically resolved lines and stars for which quadratic magnetic field determinations were available. This sample accordingly consists of stars that have magnetic fields above average, and rotation velocities below average. This bias is overcome in the recent work of Hubrig *et al.* (2004), who fully confirm the original result of Hubrig *et al.* (2000), on the basis of two different samples of magnetic Ap stars: stars with longitudinal field measurements in the literature and stars whose longitudinal fields have been recently determined in a dedicated project from spectropolarimetric observations with FORS1 at the VLT. The discovery by Bagnulo *et al.* (2003) of a strong magnetic field in HD 66318, a $2.1\,M_\odot$ member of the open cluster NGC 2516 that has completed only 16% of its main sequence lifetime does not fundamentally challenge the general result of Hubrig and collaborators, since the latter is statistical and does not rule out the existence of some outliers. Furthermore, one can note in the figures of the papers of Hubrig and collaborators that the isochrone $\log t = 8.3$ defines in the Hertzsprung-Russell diagram the early evolution envelope of the region where magnetic Ap stars with masses below $3\,M_\odot$ yr are found, and that the age limit that it places on such stars is not significantly different from the age of NGC 2516, $\log t = 8.2 \pm 0.1$ yr.

In a search for progenitors of magnetic Ap stars, an attempt by Druin *et al.* (2005) to detect magnetic fields in Herbig Ae/Be stars has so far remained inconclusive. At the other end of stellar evolution, observations of the magnetic white dwarfs from the Sloan Digital Sky Survey are consistent with the view that the highest field white dwarfs evolved from main-sequence Ap and Bp stars (Schmidt *et al.* 2003). However it has not been possible so far to trace the evolution of magnetic Ap stars from the end of their main sequence lifetime to the white dwarf sage. On the other hand, growing evidence for the existence of a large population of white dwarfs with low and moderate magnetic fields (e.g., Aznar Cuadrado *et al.* 2004) strongly suggests that the bulk of the magnetic white dwarfs cannot be accounted for as descendants of magnetic Ap stars, hence that they must have another origin.

References

Adelman, S.J., Gulliver, A.F., Kochukhov, O.P., Ryabchikova, T.A. 2002, *ApJ* 575, 449

Auriere, M., Wade, G. 2005. *These Proceedings*, EP12

Aznar Cuadrado, R., Jordan, S., Napiwotzki, R., Schmid, H.M., Solanki, S.K., Mathys, G. 2004, *A&A* 423, 1081

Babcock, H.W. 1947, *ApJ* 105, 105

Babcock, H.W. 1958, *ApJS* 3, 141

Bagnulo, S., Landolfi, M., Mathys, G., Landi Degl'Innocenti, M. 2000, *A&A* 358, 929

Bagnulo, S., Szeifert, T., Wade, G.A., Landstreet, J.D., Mathys, G. 2002a, *A&A* 389, 191

Bagnulo, S., Landi Degl'Innocenti, M., Landolfi, M., Mathys, G. 2002b, *A&A* 394, 1023

Bagnulo, S., Wade, G.A., Donati, J.-F., Landstreet, J.D., Leone, F., Monin, D.N., Stift, M.J. 2001, *A&A* 369, 889

Bagnulo, S., Landstreet, J.D., Lo Curto, G., Szeifert, T., Wade, G.A. 2003, *A&A* 403, 645

Bagnulo, S., Hensberge, H., Landstreet, J.D., Szeifert, T., Wade, G.A. 2004, *A&A* 416, 1149

Borra, E.F., Landstreet, J.D. 1978, *ApJ* 222, 226

Borra, E.F., Landstreet, J.D., Thompson, I. 1983, *ApJS* 53, 151

Bychkov, V.D., Bychkova, L.V., Madej, J. 2003, *A&A* 407, 631

Bychkov, V.D., Bychkova, L.V., Madej, J. 2005, *These Proceedings*, EP13

Carrier, F., North, P., Udry, S., Babel, J. 2002, *A&A* 394, 151

Cowley, C.R., Hubrig, S., Ryabchikova, T.A., Mathys, G., Piskunov, N., Mittermayer, P. 2001, *A&A* 367, 939

Drouin, D., Bagnulo, S., Landstreet, J. W. *et al.* 2005, *These Proceedings*, EP1

El'kin, V.G., Kudryavtsev, D.O., Romanyuk, I.I. 2002, *Astron. Lett.* 28, 169

El'kin, V.G., Kudryavtsev, D.O., Romanyuk, I.I. 2002, *Astron. Lett.* 29, 400

Gerbaldi, M., Floquet, M., Hauck, B. 1985, *A&A* 146, 341

Hubrig, S., Castelli, F. 2001, *A&A* 375, 963

Hubrig, S., Castelli, F., Wahlgren, G.M. 1999, *A&A* 346, 139

Hubrig, S., North, P., Mathys, G. 2000, *ApJ* 539, 352

Hubrig, S., North, P., Szeifert, T. 2004, in: *Astronomical Polarimetry – Current Status and Future Directions*, ASP Conf. Ser. (in press)

Hubrig, S., Kurtz, D.W., Bagnulo, S., Szeifert, T., Schöller, M., Mathys, G., Dziembowski, W.A. 2004, *A&A* 415, 661

Kochukhov, O. 2003, *A&A* 404, 669

Kochukhov, O., Ryabchikova, T., Piskunov, N. 2004, *A&A* 415, L13

Kochukhov, O., Landstreet, J.D., Ryabchikova, T., Weiss, W.W., Kupka, F. 2002, *MNRAS* 337, L1

Kochukhov, O., Bagnulo, S., Wade, G.A., Sangalli, L., Piskunov, N., Landstreet, J.D., Petit, P., Sigut, T.A.A. 2004a, *A&A* 414, 613

Kochukhov, O., Ryabchikova, T., Landstreet, J.D., Weiss, W.W. 2004b, *MNRAS* 351, L34

Kurtz, D.W., Elkin, V.G., Mathys, G., Riley, J., Cunha, M.S., Shibahashi, H., Kambe, E. 2005, *These Proceedings*, 343

Landstreet, J.D., Mathys, G. 2000, *A&A* 359, 213

Lanz, T., Mathys, G. 1993, *A&A* 280, 486

Leone, F., Catanzaro, G. 1999, *A&A* 343, 273

Leone, F., Kurtz, D.W. 2003, *A&A* 407, L67

Leone, F., Vacca, W.D., Stift, M.J. 2003, *A&A* 409, 1055

Leroy, J.L. 1995, *A&AS* 114, 79

Mathys, G. 1991, *A&AS* 89, 121

Mathys, G. 1995, *A&A* 293, 733

Mathys, G. 1995, *A&A* 293, 746

Mathys, G. 2001, in: G. Mathys, S.K. Solanki & D.T. Wickramasinghe (eds.), *Magnetic fields across the Hertzsprung-Russell diagram*, ASP Conf. Ser. vol. 248, p. 267

Mathys, G. 2002, in: J. Trujillo-Bueno, F. Moreno-Insertis & F. Sánchez (eds.), *Astrophysical spectropolarimetry* (Cambridge: University Press), p. 101

Mathys, G. 2003, in: L.A. Balona, H.F. Henrichs & R. Medupe (eds.), *Magnetic fields in O, B and A stars: origin and connection to pulsation, rotation and mass loss*, ASP Conf. Ser. vol. 305, p. 65

Mathys, G. 2004, in: A. Maeder & P. Eenens (eds.), *Stellar rotation*, IAU Symp. 215, ASP Conf. Ser. (in press)

Mathys, G., Hubrig, S. 1997, *A&AS* 124, 475

Mathys, G., Lanz, T. 1990, *A&A* 230, L21

Mathys, G., Hubrig, S., Landstreet, J.D., Lanz, T., Manfroid, J. 1997, *A&AS* 123, 353

Monin, D.N., Fabrika, S.N., Valyavin, G.G. 2002, *A&A* 396, 131

Nesvacil, N., Hubrig, S., Jehin, E. 2004, *A&A* 422, L51

Piskunov, N., Kochukhov, O. 2002, *A&A* 381, 736

Preston, G.W. 1970, *PASP* 82, 878

Preston, G.W. 1974, *ARA&A* 12, 257

Robinson, R.D., Worden, S.P., Harvey, J.W. 1980, *ApJ* 236, L155

Romanyuk, I. I. 2005, *These Proceedings*, EP4

Ryabchikova, T.A., Savanov, I.S., Hatzes, A.P., Weiss, W.W., Handler, G. 2000, *A&A* 357, 981

Ryabchikova, T.A., Piskunov, N., Kochukhov, O., Tsymbal, V., Mittermayer, P., Weiss, W.W. 2002, *A&A* 384, 545

Savanov, I., Hubrig, S., Mathys, G. *et al.* 2005, *This meeting*, http://www.ta3.sk/IAUS224/PDF/GP10.pdf

Schmidt, G.D., Harris, H.C., Liebert, J., *et al.* 2003, *ApJ* 595, 1101

Shorlin, S.L.S., Wade, G.A., Donati, J.-F., Petit, P., Sigut, T.A.A., Strasser, S. 2002, *A&A* 392, 637

Stütz, Ch., Ryabchikova, T., Weiss, W.W. 2003, *A&A* 402, 729

Wade, G.A., Mathys, G., North, P. 1999, *A&A* 347, 164

Wade, G.A., Donati, J.-F., Landstreet, J.D., Shorlin, S.L.S. 2000, *MNRAS* 313, 851

Wahlgren, G.M., Ilyin, I., Kochukhov, O. 2001, AAS Mtg. 199, #135.04

Wolff, S.C. 1978, *PASP* 90, 412

Wolff, S.C., Morrison, N.D. 1973, *PASP* 85, 141

Discussion

KOCHUKHOV: The roAp star HD 166473 has a rotation period of the order of 10 years and the maximum of its mean magnetic field modulus is about 9 kG. Does this star fall in the empty region of the field modulus vs. period diagram?

MATHYS: No, because the result that no fields stronger than 7.5 kG are found for stars with rotation periods longer than 150 d refers to the average of the field modulus over the rotation period. In the case of HD 166473, this average is of the order of 7 kG.

PRESTON: In constructing maps of surface magnetic fields, can observational errors introduce spurious fine structure in the maps?

PISKUNOV: Doppler Imaging (and Magnetic Doppler Imaging) is based on a model of rotational modulation which implies certain changes of the line profiles as a function of phase. For example, solid body rotation would predict that a surface feature at a given longitude and latitude should produce certain distortions to the line profiles. The amplitude and the Doppler shift of these distortions are well defined functions of rotational phase. Spurious details in the real observations are presumably random and thus ignored by the inversion. This situation is similar to fitting a sine curve to a long but noisy data sequence when the period is known a priori.

The A-Star Puzzle
Proceedings IAU Symposium No. 224, 2004
J. Zverko, J. Žižňovský, S.J. Adelman, & W.W. Weiss, eds.

© 2004 International Astronomical Union
DOI: 10.1017/S1743921304004612

Stellar magnetic fields: The view from the ground and from space

Gregg A. Wade

Department of Physics, Royal Military College of Canada, P.O. Box 17000 Station Forces,
Kingston, ON, Canada K7K 7B4
email: wade-g@rmc.ca

Abstract. This article reviews the methods of measurement which allow us to infer the presence of magnetic fields in (A) stars. Beginning with the basic observational consequences of the Zeeman effect, we describe various modern spectroscopic and polarimetric techniques which allow us to directly detect and characterise magnetic fields in stellar photospheres. Sometimes, nature conspires to make such detections difficult, forcing us to rely on indirect (proxy) indicators of magnetism. This talk will also briefly discuss a number of these indirect field indicators, some of which demand space-based observations.

Keywords. Methods: data analysis, techniques: polarimetric, techniques: spectroscopic, stars: magnetic fields

1. Introduction

We are gathered here in Poprad to discuss "The A-Star Puzzle", the scientific issues associated with stars of spectral type A, with effective temperatures between about 7500 and 10000 K. If we draw vertical lines on the HR diagram according to these temperature extremes, we see that "A stars" include a tremendous variety of objects, on various evolutionary paths. At the earliest evolutionary stages, we have pre-Main Sequence A stars, both those which remain enshrouded in their nascent cocoons (pre-birthline) and those which have recently shed their formative gas and dust clouds (post-birthline) and are observable in visible light. These stars are significantly reddened, are typically rapidly-rotating, and exhibit emission lines due to remaining circumstellar matter. On the Main Sequence, the A stars include chemically peculiar stars, both magnetic and nonmagnetic, as well as pulsating variables and "normal" stars. These stars exhibit a tremendous variety of physical and spectral characteristics: line profile shape and polarisation variations, peculiar energy distributions, vertically and laterally non-uniform abundance distributions, and strong magnetic fields. Evolved A stars include higher-mass supergiants, whose spectra are dominated by the effects of mass-loss and wind variability, and intermediate-mass giants which include the radially, and possibly non-radially, pulsating RR Lyrae stars.

The huge variety of physical properties of A-type stars makes the potential presence of magnetic fields in their atmospheres and immediate circumstellar environments extremely interesting. At the same time (and as we shall see in the following sections), the large range of spectroscopic characteristics demands a sensitive and flexible suite of techniques for detecting and characterising such fields.

2. Overview of the Zeeman effect in spectral lines

Stellar magnetic fields are detected and characterised using the Zeeman effect. When an atom is immersed in an external magnetic field B, individual atomic levels (with energy

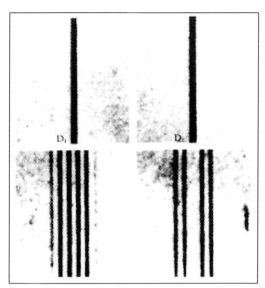

Figure 1. *Left:* Dr. Pieter Zeeman, discoverer of the effect to which his name is attached. *Right:* Photograph of the splitting of the Na D doublet under the influence of an external magnetic field, as reported by Zeeman (1897).

E_0, total angular momentum J, and Landé factor g) are split into $(2J + 1)$ substates, characterised by their magnetic quantum number M $(-J \leqslant M \leqslant +J)$. The energies of these atomic states are given by:

$$E(M) = E_0 g M \hbar \omega_L \tag{2.1}$$

where $\omega_L = eB/(2m_e c)$ is the Larmor frequency (e.g., Mathys 1989).

As a result of this splitting, each transition that generates a single line in the stellar spectrum when no field is present leads, in the presence of a field, to a group of closely-spaced spectral lines, or *Zeeman components*. For transitions in LS coupling, these lines may be grouped into two different types with different properties. Those lines resulting from transitions in which M does not change (i.e. $\Delta M = 0$) are spread symmetrically about the zero-field wavelength $\lambda_0 = hc/E_0$ of the line. These are called the π components. Those resulting from transitions in which $\Delta M = \pm 1$ have wavelengths to the red (+) and blue (−) of the zero-field wavelength. These are called the σ components. The wavelength separation between the centroids of the π and σ component groups can be calculated from Eq. (2.1), and is:

$$\Delta\lambda = \frac{eN\bar{g}}{4\pi m_e c^2}\lambda_0^2 \equiv \Delta\lambda_Z B\bar{g} \tag{2.2}$$

where $\bar{g}$ is the effective Landé factor of the transition and $\Delta\lambda_Z$ is called the *Lorentz unit*. Note that the splitting is directly proportional to the strength of the applied field B. For illustration, a 5 kG magnetic field produces a splitting of ~ 0.1 Å at 5000 Å.

Pieter Zeeman was the first to report these splitting properties in his seminal paper of 1897, *The Effect of Magnetisation on the Nature of Light Emitted by a Substance*:

> Sodium was strongly heated in a tube of biscuit porcelain, such as Pringsheim used in his interesting investigations upon the radiation of gases. The tube was closed at both ends by plane parallel glass plates.... The tube was placed horizontally between the poles, at right angles to the lines of force. The light of an arc lamp was sent through. The absorption spectrum showed both D lines. The tube was

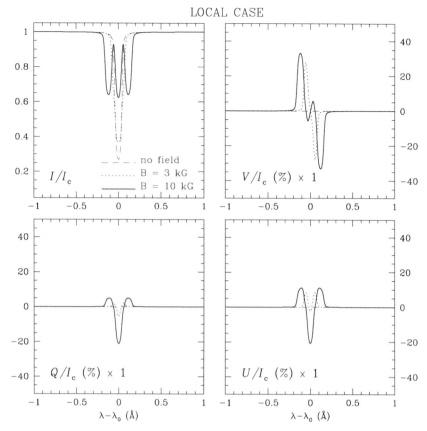

Figure 2. Influence of the intensity of the applied magnetic field on a spectral line, in natural light (Stokes I), circular polarisation (Stokes V) and linear polarisation (Stokes Q and U). Each set of these "local" profiles corresponds to a single intensity and direction of the magnetic field, as might be observed in a laboratory or on the sun. Profiles are shown for 3 different field intensities.

continuously rotated round its axis to avoid temperature variations. Excitation of the magnet caused immediate widening of the lines. It thus appears very probable that the period of sodium light is altered in the magnetic field.

(Zeeman 1897)

Zeeman and his observations are illustrated in Fig. 1. At the suggestion of Hendrik Lorentz, Zeeman furthermore investigated the polarisation properties of the π and σ components. Zeeman found that in a magnetic field aligned parallel to the observer's line-of-sight (a *longitudinal* field), the π components vanish, whereas the σ components have opposite circular polarisations. On the contrary, when the magnetic field is aligned perpendicular to the observer's line-of-sight (a *transverse* field), the π components are linearly polarised parallel to the field direction, while the σ components are linearly polarised perpendicular to the field direction:

I have since found by means of a quarter-wave plate and an analyser, that the edges of the magnetically-widened lines are really circularly polarised when the line of sight coincides in direction with the lines of force... On the contrary, if one looks at the flame in a direction at right angles to the lines of force, then the edges of the broadened sodium lines appear plane polarised, in accordance with theory.

(Zeeman 1897)

In summary, Zeeman observed that application of a magnetic field leads to the splitting

STELLAR CASE (DIPOLAR FIELD)

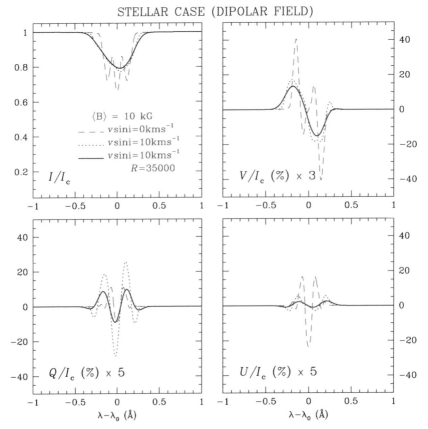

Figure 3. Disc-integrated line profiles (solid and dotted lines) as might be encountered in the spectrum of an unresolved star with a distribution of field strengths over the observed hemisphere (in this case, a dipole corresponding to a mean integrated field modulus of 10 kG). Profiles are shown for $v \sin i = 0$ km s^{-1}, $v \sin i = 10$ km s^{-1}, and $v \sin i = 10$ km s^{-1} with additional instrumental broadening corresponding to a spectral resolution of $R = 35000$. Note the significant decrease in the diagnostic content of the Stokes I profile, and the decrease in the amplitude of the polarisation profiles, as a consequence of disc integration and rotation.

of spectral lines into multiple Zeeman components, and that the components are polarised in particular ways corresponding to the orientation of the field. Thus, both the *intensity* and the *geometry* of the magnetic field in the line-forming region are encoded in the line profile as a consequence of Zeeman effect.

In the "weak-field limit", when the ratio of the Zeeman splitting to the intrinsic width of the line $\Delta\lambda_Z/\Delta\lambda_I \ll 1$, under the assumption of a linear Milne-Eddington source function, one obtains the first-order solution for the Stokes profiles emergent from a stellar photosphere:

$$I(\tau = 0, \lambda) = B_0 + B_0\beta_0\mu[1 + \eta(\lambda)]^{-1} \tag{2.3}$$

$$Q(\tau = 0, \lambda) = 0 \tag{2.4}$$

$$U(\tau = 0, \lambda) = 0 \tag{2.5}$$

$$V(\tau = 0, \lambda) = -\bar{g}\Delta\lambda_Z B_Z \frac{dI(\lambda)}{d\lambda} \tag{2.6}$$

where B_0 and β_0 are the Milne-Eddington parameters, μ is the limb angle, $\eta(\lambda)$ is the line opacity variation in the absence of a magnetic field, and B_Z is the longitudinal component of the magnetic field.

Eqn. (2.3) will be immediately recognised as the Milne-Eddington solution for the emergent Stokes I profile in the *absence of a magnetic field*, i.e., to first order, the Stokes I profile is unaffected by the presence of a weak magnetic field. Furthermore, Eqs. (2.4) and (2.5) show, to first order, that the line profile is not linearly polarised for weak fields. In fact, *the only first-order influence of the magnetic field in the weak-field regime is on the amplitude of the Stokes V profile.* For any spectral line, the shape of Stokes V is determined entirely by the shape of the corresponding Stokes I profile, via the derivative $dI/d\lambda$. However, the amplitude of Stokes V is proportional to the intensity of the longitudinal component of the magnetic field. This implies that for most stars, the most easily accessible Zeeman diagnostic will be line circular polarisation, and for such stars only the longitudinal field component can be practically measured. The linear polarisation Stokes parameters, which constrain the transverse components of the field, will be in general much weaker (appearing only at second order), and consequently much more difficult to detect. Synthetic line profiles, illustrating the influence of field strength on the splitting and polarisation profile amplitudes, are shown in Fig. 2.

In real stars, our ability to diagnose magnetic fields depends on various (nonmagnetic) physical and spectroscopic attributes (e.g., visual magnitude, S/N, spectral line depth and density, rotational velocity, spectral resolving power of the spectrograph), as well as the magnetic field. In many situations, the observable signal may be near the practical limits of available instrumentation. This contrast is illustrated in Fig. 3, in which we examine disc-integrated line profiles as they might actually appear in the spectrum of an unresolved, rotating, magnetic star.

3. Application

And so, as we have seen in section 2, the Zeeman effect leads to various observational effects in both natural and polarised light. Depending on the particular spectroscopic characteristics of a star, as well as the intensity and the geometry of the magnetic field, some of these effects may provide a sensitive diagnosis of the magnetic field, and some may not.

3.1. *Line broadening, desaturation and splitting in natural light*

In the absence of a polarimetric diagnosis (i.e., in the natural light or Stokes I spectrum), the presence of a magnetic field in the stellar photosphere can be detectable in certain special situations. For magnetic fields sufficiently intense to induce an appreciable separation of the Zeeman components of spectral lines (splitting $\Delta\lambda_Z$ of the same order of magnitude as the *observed* line width $\Delta\lambda_{\rm obs}$ accounting for all broadening effects, including rotation), broadening or even resolved splitting of spectral lines is possible. Roughly, the field intensity required to produce a resolved splitting is of order 1 kG per km s^{-1} of line broadening. Hence, fields in excess of about 3 kG are required to produce a resolved magnetic splitting in Fe lines of a non-rotating A star; a field in excess of 13 kG is necessary to split the lines of an A star rotating at 10 km s^{-1}. Therefore broadening and splitting tend only to be observable in the unpolarised spectra of the sharpest-lined A stars, and are essentially only unambiguously detectable for relatively large fields. These effects have been recently exploited in studies by e.g., Leone *et al.* (2003), Stütz *et al.* (2003) and Landstreet & Mathys (2000). Notably, Mathys and collaborators employed these effects in large scale studies of the magnetic properties of Ap stars during the 1990s (Mathys 1995, Mathys *et al.* 1997).

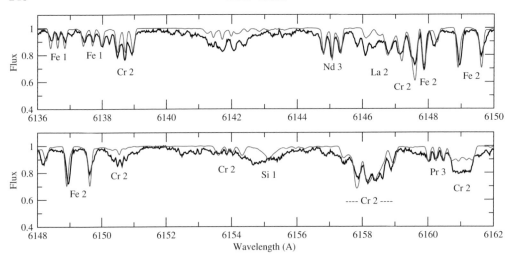

Figure 4. Observed and calculated spectra of HD 66318, reported by Bagnulo *et al.* (2003). Note the resolved splitting of many lines into Zeeman doublet and triplet patterns.

Fig. 4 illustrates the influence of magnetic splitting on the spectrum of a sharp-lined star. It shows observed (ESO Coudé Echelle Spectrograph) and calculated (ZEEMAN, Landstreet 1988, Wade *et al.* 2001) spectra of the young cluster Ap star HD 66318 reported by Bagnulo *et al.* (2003). Note the splitting of most lines into double and triplet Zeeman patterns.

3.2. *Circular polarisation in metal and hydrogen lines*

As discussed in Sect. 2, the most easily accessible polarised diagnostic of stellar magnetic fields is circular polarisation (resulting from the longitudinal Zeeman effect and indicative of the presence of a line-of-sight component of the magnetic field in the stellar photosphere (a longitudinal magnetic field)). Typically, Zeeman circular polarimetry (and linear polarimetry, to be discussed in the section following) is obtained *differentially*, by recording the two orthogonal polarisation states (left and right circularly polarised light, separated using a $\lambda/4$ retarder and a polarising beamsplitter) simultaneously on a CCD†. The sum of the recorded fluxes gives the natural light spectrum (Stokes I), whereas their difference gives the net circularly polarised flux (Stokes V). This procedure is described in some detail by, e.g., Bagnulo *et al.* (2002).

Zeeman circular polarimetry has the advantage of being sensitive even under conditions of moderate or rapid stellar rotation. Using metal lines, practical sensitivity up to about 60 km s^{-1} can be achieved for Ap stars with strong fields (e.g., Wade *et al.* 2000a)‡. When metallic-line spectropolarimetry is coupled with the powerful Least-Squares Deconvolution multi-line technique (LSD, Donati *et al.* 1997), remarkably high precision can be attained, particularly for the cooler A stars. Using intrinsically broad lines such as Balmer lines, sensitivity can furthermore be extended to several hundreds of km s^{-1} (e.g., Bagnulo *et al.* 2002, see also, e.g., Bohlender *et al.* 1987). These techniques have been recently employed in studies by, e.g., Shorlin *et al.* (2002), Leone & Catanzaro

---

† In fact, most modern methods employ a rotation of the retarder to permit switching of the positions of the two beams on the CCD, allowing the removal of systematic detector effects with multiple exposures; see, e.g., Semel *et al.* (1993).

‡ Although limited in sensitivity to moderate $v \sin i$ objects, metallic line spectropolarimetry can exploit stellar rotation to allow the detection of magnetic fields *even when the mean longitudinal magnetic field is null* (see, e.g., Shorlin *et al.* 2002).

(2004), Chadid *et al.* (2004), and Bagnulo *et al.* (2004), achieving in some cases precision better than 10 G. Notably, circular polarimetry of metallic lines by Babcock, Preston and collaborators, and in Balmer lines by Landstreet, Borra and collaborators, have led the way to the fundamental understanding of magnetism in intermediate mass stars that we enjoy today.

Fig. 5 illustrates the remarkable circular polarisation signatures detected in the lower Balmer series of the very young Rosette cluster member NGC 2244-334 (Bagnulo *et al.* 2004, observations using FORS1 at the ESO VLT).

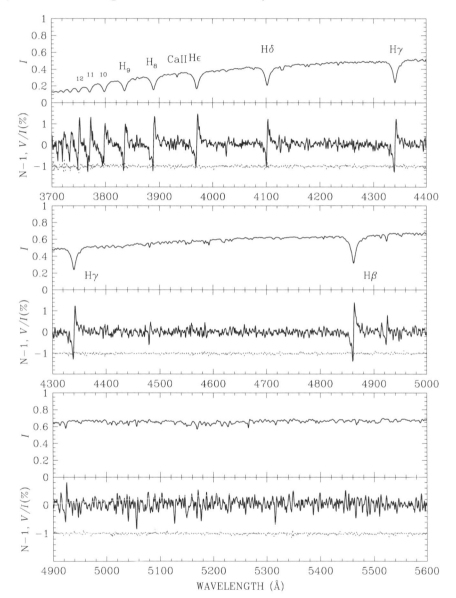

Figure 5. Stokes *I* and *V* spectra of the lower Balmer series of NGC 2244-334, obtained by Bagnulo *et al.* (2004) using the FORS1 spectropolarimeter at the ESO VLT. The longitudinal field diagnosed using these data is -9 kG, the second-largest longitudinal field ever measured in a non-degenerate star.

3.3. *Linear polarisation in the continuum and in metal lines*

By observing the transverse Zeeman effect (Zeeman linear polarisation) in combination with the circular polarisation longitudinal Zeeman effect, qualitatively new constraints on the magnetic field geometry of A stars can be obtained. The first systematic observations of the transverse Zeeman effect in A stars were obtained by Leroy and collaborators (e.g., Leroy 1995) during the early 1990s. These authors measured time variations of low levels of linear polarisation in the broadband light of Ap stars, produced as a result of the saturation of Zeeman-broadened metallic absorption lines. Upon comparison with the predictions of theoretical models, they found that the variations were not reproducible using symmetric magnetic field models. More recently, observations of transverse Zeeman effect *within* spectral absorption lines has become possible (Wade *et al.* 2000b obtained observations of several stars using the MuSiCoS spectropolarimeter at Pic du Midi observatory). Modeling of the phase variations of both circularly and linearly polarised line profiles for the Ap star 53 Cam has been performed using the Magnetic Doppler Imaging method (Kochukhov *et al.* y 2004) resulting in the first assumption-free high-resolution vector map of an Ap star magnetic field constructed from spectroscopic data in all four Stokes parameters. The maps of 53 Cam support the conclusions of Leroy and collaborators, revealing a remakably complex magnetic field: underlying dipolar characteristics, but with significant contributions by multipolar components of higher order.

4. Indirect detection of magnetic fields

Despite the flexibility and power of modern magnetic field diagnostic techniques, there remain stars in which magnetic fields cannot be detected directly, due to the limitations of the techniques themselves. Many of these limitations apply primarily to the hotter B and O type cousins of A stars; they are, however, important because of known or potential relationships between these objects and A stars.

Indirect methods for inferring fields in hard-to-study stars exploit phenomena typically associated with magnetic fields, such as surface structures, non-thermal radio, UV and X-ray emission, and azimuthal structuring of winds and other circumstellar matter. For example, the presence of periodic line profile variability in Main Sequence A stars, particularly when accompanied by particular patterns of photospheric abundance peculiarities (overabundances of rare eaths, Fe peak elements (especially Cr) or Si), are strongly indicative of the presence of magnetic fields (see, e.g,. Auriere & Wade 2005). In A supergiants and hotter stars, variability of line profiles formed in stellar winds, along with non-thermal radio emission, UV line emission and X-ray emission, may indicate the influence of magnetic fields in imposing structure on stellar winds and other circumstellar matter and energy (e.g., Wade 2001). Finally, observations of Hanlé effect (e.g., Ignace 2003) may provide a new window to magnetic fields in wind-obscured stars. Many of these effects are only, or most effectively, observed from space.

5. Conclusion

Modern Zeeman magnetic field diagnostics provide a powerful and flexible suite of tools for studying magnetism in stellar atmospheres. The introduction of sensitive new spectropolarimeters on ground-based telescopes, both large (such as FORS1 at ESO/VLT, ESPaDOnS at CFHT) and small (SARG at TNG, NARVAL at TBL/Pic du Midi), as well as space-based platforms (the Far Ultraviolet Spectro-Polarimeter, FUSP) will provide a much more detailed picture of the characteristics of magnetic fields over a much larger fraction of all A stars.

References

Auriere, M., Wade, G.A. 2005, *These Proceedings*, EP12

Bagnulo S., Szeifert T., Wade G.A., Landstreet J.D., Mathys G., 2002, A&A 389, 191

Bagnulo S., Landstreet J.D., Lo Curto G., Szeifert T., Wade G.A., 2003, A&A 403, 635

Bagnulo S., Hensberge H., Landstreet J.D., Szeifert T., Wade G.A., 2004, A&A 416, 1149

Bohlender D.A., Landstreet J.D., Brown D.N., Thompson I.B., 1987, ApJ 323, 325

Chadid M., Wade G.A., Shorlin S.L.S., Landstreet J.D., 2003, A&A 413, 1087

Donati J.-F., Semel M., Carter B.D., Rees D.E., Cameron A.C., 1997, MNRAS 291, 658

Ignace R., 2003, in the proceedings of *Magnetic fields in O, B and A stars*, ASP conference series #305, L. Balona, H.F. Henrichs & R. Medupe (eds.), p.28

Kochukhov O., Bagnulo S., Wade G.A., Sangalli L., Piskunov N., Landstreet J. D., Petit P., Sigut T. A. A., 2004, A&A 414,613

Landstreet J.D., 1988, ApJ 326, 967

Leone F. & Catanzaro G., 2004, A&A 425, 271

Leroy J.-L., 1995, A&AS 114, 79

Mathys G. 1989, Fund. Cosmic Phys. 13, 143

Mathys G. 1995, A&A 293, 746

Mathys G., Hubrig S., Landstreet J.D., Lanz T., Manfroid J., 1997, A&AS 123, 353

Semel M., Donati J.-F., Rees D. E., 1993, A&A 278, 231

Shorlin S.L.S., Wade G.A., Donati J.-F., Landstreet J.D., Petit P., Sigut T.A.A., Strasser S., 2002, A&A 392, 637

Wade G.A., Donati J.-F., Landstreet J.D., Shorlin S.L.S., 2000a, MNRAS 313, 851

Wade G.A., Donati J.-F., Landstreet J.D., Shorlin S.L.S., 2000b, MNRAS 313, 823

Wade G.A., 2001, in the proceedings of *Magnetic fields across the HR diagram*, ASP conference series # 248, G. Mathys & S. Solanki (eds.), p. 403

Wade G.A., Bagnulo S., Kochukov O., Landstreet J.D., Piskunov N., Stift M.J., 2001, A&A 374, 265

Zeeman P., 1897, Nature 55, 347

The A-star puzzle
Proceedings IAU Symposium No. 224, 2004   © 2004 International Astronomical Union
J. Zverko, J. Žižňovský, S.J. Adelman, & W.W. Weiss, eds.   DOI: 10.1017/S1743921304004624

Magnetic fields in A stars

David Moss

Department of Mathematics, University of Manchester, Manchester M13 9PL, UK
email: moss@ma.man.ac.uk

Abstract. Some theoretical problems associated with the presence of large-scale magnetic fields in A stars are reviewed. Possible implications of some recent theoretical and observational results are discussed, with particular attention to the survival of fields from the ISM and their long-term stability. A more coherent picture of the origin of the strong large-scale fields seen in the magnetic CP stars may be beginning to emerge.

Keywords. Stars: magnetic fields, MHD

1. Introduction

After several decades of effort by astrophysicists, the strong large-scale magnetic fields seen in a minority of near-Main Sequence A-stars continue to present a number of problems. Although this paper (as the meeting) explicitly addresses A-type stars, most of what I write will apply also to the fields of the magnetic Bp stars, i.e., to all the magnetic CP stars. In this short paper I will not attempt to give yet another comprehensive review of the current situation (but see, e.g., Mestel 2003a,b, Moss 2001, 2003a for recent accounts), but rather I will focus on a few issues where I feel that some progress has been made in the last year or two.

A rather personal viewpoint leads to the following:
'Fundamental problems'
- *Origin of large-scale fields.
- The evolutionary status of the magnetic A-stars.
- *The long-term stability and survival of large-scale magnetic fields in stars.
- The '10%' problem, why are only about 10% of Main Sequence A-stars observably magnetic?
'Second tier problems'
- *The striking difference in distributions of the obliquity angle between the rotational and magnetic axes between slower and faster rotators (Landstreet & Mathys 2000).
- Why do apparently similar stars sometimes have quite different magnetic properties (including, e.g., close binary systems where both members are A stars)? Perhaps this is really part of the '10% problem' mentioned above.
'Third tier'
- The relatively slow rotation (as a class) of the magnetic A stars, Stepien (2000) suggests a plausible scenario involving the interaction of the field with a protostellar disc.
- The compositional anomalies and their relationship to the magnetic fields. An associated question, sharpened now by Aurière *et al.* (2004), is that all Ap and Bp stars appear to have surface fields in excess of about 100 G, and that no normal A stars are seen with fields $\lesssim 100$ G (down to the observational limit of O(10) G).
- The roAp phenomenon.

This is not intended to suggest any hierarchy of interest or difficulty, rather some sort of structuring, filtered by my personal interests. Below, I will concentrate mostly on the items above that are preceded by an asterisk.

2. Origin of the fields

The longstanding debate between fossil and dynamo origins of the fields is still undetermined, although the possibility of a 'hybrid' field, essentially a relic from a pre-Main Sequence dynamo, and more recently, a field resulting from a magnetorotational instability in the radiative envelope (Rüdiger *et al.* 2001, Arlt *et al.* 2005) have also been discussed. (It is of some interest to note that both of the latter would also appear as fossil relics from an earlier evolutionary phase.) In brief, the convective cores of Main Sequence A stars appear to be suitable sites for dynamo action. The dynamo theory for the magnetic A stars suggests that this dynamo generates the field that is observed at their surfaces. It is certainly to be expected that dynamos do operate in the convective cores of A stars (e.g., Browning *et al.* 2005). However, given the overlying deep radiative envelopes, any core-generated dynamo field has to rise passively from the core to the surface, the rise being mediated by buoyancy, compositional gradients, etc. Whilst arguments have been made lately that such a mechanism may be relevant in bringing some field to the surface, especially in stars rather more massive than being considered here, quite general arguments suggest that fields that are steady in the rotating frame and have the strength and spatial coherence of the observed fields are unlikely to be produced in such a manner (see Moss 1989, 2001, also McDonald & Mullan 2004). Also why all, or nearly all, A stars are not observably magnetic, and the lack of correlation between magnetic and other properties, in particular angular velocity, remain difficult to explain with either core-dynamo or envelope magneto-rotational instability theories.

In contrast, the fossil theory proposes that the field seen on these stars on the Main Sequence is a relic of the field present in the ISM at the time of star formation, and subsequently amplified during contraction. This allows the 'initial conditions' of the ISM as an additional degree of freedom. There seems little doubt that there is 'enough' magnetic flux present in the pre-stellar material (see below). Nevertheless, this picture does encounter various problems, some of which will now be discussed.

Since the work of Hayashi and collaborators in the early 1960s, it has largely been believed that lower and middle Main Sequence stars initially contract down almost vertical paths in the Hertzsprung Russell diagram, where they are wholly or largely convective, see the broken part of the path in the schematic Fig. 1. Kinematically, such turbulence will distort the field, reducing its length-scale down to something like a diffusion scale, with consequent field expulsion or decay on a timescale very much shorter than a contraction time. Dynamically, general arguments and simulations suggest that the turbulence can concentrate the field into ropes, of strength about or rather larger than the equipartition strength, and these ropes may then be able to resist further shredding and decay. It can then be argued reasonably plausibly that significant signed flux can survive until a radiative core begins to form, at which time the flux ropes will become anchored in a stable region. As this radiative core grows, the field will diffuse to a more uniform geometry.

Quantification of such a scenario is difficult: it cannot be treated analytically, and it is still beyond any realistic simulation (the situation may change in the next few years). Moss (2003a,b) constructed an order of magnitude model to describe the contraction of a magnetic star down a Hayashi track, concluding that if the survival in strong ropes idea is valid, more massive stars may be able to retain stronger fields to the Main Sequence than stars of about solar mass, but with a smooth dependence on mass that is unlike the turning on of the magnetic CP star phenomenon at $M \gtrsim 2M_\odot$. The lack of evidence for a field in the solar radiative core of more than O(1) G argues against the presence of a significant convective envelope on the Main Sequence being a decisive factor.

2.1. *A more recent view of pre-Main Sequence stellar evolution*

Palla & Stahler (1993, see also Marconi & Palla 2005) published a model of pre-Main Sequence evolution in which more massive stars do not necessarily experience a largely convective phase. Their initial condition, rather than being an isolated gas sphere which first found equilibrium at the top of the Hayashi track, was a sphere embedded in an accreting envelope, such as might be expected from the late stages of the transition from a more-or-less freely falling gas cloud to a 'star' in dynamical equilibrium. Effectively, this envelope changes the outer boundary condition in such a way that stars of mass more than about $2M_\odot$ (depending somewhat on choice of parameters) start their contraction as radiative objects along the almost horizontal part of the evolutionary track (solid in Fig. 1). Thus they do not experience an episode when they are largely or wholly convective. Interestingly, this is essentially the 'Henyey track', which pre-dated Hayashi's work (see, e.g., Schwarzschild 1958).

If this picture is valid, it provides a clear cut distinction between stars with $M \lesssim 2M_\odot$, which experience large-scale turbulence and subsequent field destruction, and so do not appear as strongly magnetic on the Main Sequence, and those with $M \gtrsim 2M_\odot$, which do not suffer this field destruction and so can retain their primordial fields. Nevertheless, this still leaves outstanding problems about the survival of flux from the ISM, especially the dynamical stability of large-scale field structures, why observed fields are all nonaxisymmetric (i.e., the field axis and rotation vector are non-parallel), and the 10% problem. These are all discussed, briefly at least, in the following sections.

3. Nonaxisymmetric fields

In general, the gravitational contraction of an initially rotating gas sphere can be expected to result in a state of differential rotation. Then, from the winding up of nonaxisymmetric field lines and the bringing together of field lines of opposite sign, enhanced decay can be expected to result (Fig. 2). This does not happen to axisymmetric fields, and so nonaxisymmetric fields are discriminated against (e.g., Rädler 1986). This is a purely kinematic argument. When dynamical effects are taken into account, the picture may change. The azimuthal stretching of field lines results in an azimuthal component of Lorentz force which acts to reduce the differential rotation (e.g., Mestel 2003a). Thus a strong enough nonaxisymmetric field can survive and dominate differential rotation. Rädler (1986) showed that, ignoring dynamical effects, field winding will be limited by ohmic dissipation after n_c windings, where $n_c \sim (\Delta\Omega L^2/2\pi\eta)^{1/3}$, i.e. after a time $t_w \sim 2\pi n_c/\Delta\Omega$. If a typical dynamical timescale is $t_D \sim L/v_A$, with v_A the Alfvèn velocity, where $\Delta\Omega$ is an estimate of the variation in angular velocity Ω, L a length-scale for the field and η is magnetic diffusivity in units of cm^2 s^{-1}, then the condition $t_w \approx t_D$ defines a field $B_c \sim \rho^{1/2}(L\eta(\Delta\Omega)^2)^{1/3}$G (Moss 1992). This can be rewritten as

$$B_c \sim 4 \times 10^4 \rho^{\frac{1}{2}} \left(\frac{\Omega}{10\Omega_\odot}\right)^{\frac{2}{3}} \left(\frac{R}{3R_\odot}\right)^{\frac{1}{3}} \left(\frac{\eta}{10^{12}}\right)^{\frac{1}{3}} \left(\frac{\Delta\Omega}{\Omega}\right)^{\frac{2}{3}} \left(\frac{L}{R}\right)^{\frac{1}{3}} \text{G}, \qquad (3.1)$$

and fields in excess of B_c can be expected to reduce rotational shear. ρ is a typical density (in cm^{-3}) and $\Delta\Omega/\Omega$ is an estimate of the fractional differential rotation. Putting $\rho = 10^{-1}$gm cm^{-3}, $\Delta\Omega/\Omega = 1$, $L \sim R$, $\eta = 10^9$ cm^{2}s^{-1}, $\Omega = 10\Omega_\odot$ gives $B_c \sim 250$ G, suggesting that nonaxisymmetric fields of moderate strength might survive and even control differential rotation. However, strong differential rotation and nonaxisymmetric fields cannot coexist.

We can make an experiment by 'rewinding' a kilogauss field in a Main Sequence star of about $3M_\odot$ back through the pre-Main Sequence contraction. This value of the mean

Table 1. Mean field strengths $< B_1 >$ and $< B_2 >$ (defined in text) for a $3M_\odot$ star along its pre-Main Sequence evolutionary path, assuming flux freezing with $B \propto R^{-2}$. B_{c1} is given by Eq. 3.1, with $\eta = 10^9 \mathrm{cm^2 s^{-1}}$, whereas for B_{c2}, $\eta = 10^{12} \mathrm{cm^2 s^{-1}}$ when the star is largely convective (C), and $\eta = 10^6 \mathrm{cm^2 s^{-1}}$ where it is radiative (R). $\Omega = 10\Omega_\odot$, $L = R$, $\Delta\Omega/\Omega = 0.1$. All magnetic fields are in G, and the mean density $< \rho >$ is in $\mathrm{cm^{-3}}$

$R/R_\odot$	$< B_1 >$	$< B_2 >$	$< \rho >$	B_{c1}	B_{c2}
2 (R)	10^3	2×10^4	0.5	500	50
5 (R)	160	3.2×10^3	0.03	200	20
10 (C)	40	800	0.004	80	800
20 (C)	10	200	0.0005	40	400

field strength, referred to as $< B_1 >$, is proportional to R^{-2} assuming flux conservation, see Table 1. An alternative viewpoint is to take an estimate of field strength at the top of the Hayashi track, as suggested by star formation models with approximate flux freezing after reionization (Desch & Mouschovias 2001, Nakano *et al.* 2002). Moss (2003b, Table 1) estimated this to give a field of about 200 G there, although the figure is rather uncertain (and some flux loss can certainly be expected to occur during the subsequent contraction process). This field, also proportional to R^{-2} assuming flux conservation, is denoted by $< B_2 >$. Table 1 also lists values B_{c1} and B_{c2}, calculated from Eq. (3.1): B_{c1} is given using $\eta = 10^9 \mathrm{cm^2 s^{-1}}$ throughout, whereas for B_{c2}, $\eta = 10^{12} \mathrm{cm^2 s^{-1}}$ on the convective part of the track ($R \gtrsim 5R_\odot$) and $10^6 \mathrm{cm^2 s^{-1}}$ on the radiative part ($R \lesssim 5R_\odot$). $< \rho >$ is a mean density. In this Table, $< B_2 >$ exceeds $B_{c1,2}$ everywhere, whereas $< B_1 >$ is of the order of, or somewhat smaller, than, B_{c1}, and much less than B_{c2} when $R > 5R_\odot$.

Thus it appears that with 'turbulent' values of $\eta (\gtrsim 10^{12} \mathrm{cm^2 s^{-1}}?)$, the field $< B_2 >$ is about of the order of magnitude necessary to control differential rotation, so preventing excessive winding and enhanced decay, but that the field $< B_1 >$ is too small. Note that $< B_1 >$ is likely to be something of an underestimate, since some flux loss is probably unavoidable. But with values of the diffusivity appropriate to a radiative region ($\eta \lesssim 10^6$), both these fields are large enough to limit differential rotation. Thus stars whose pre-Main Sequence evolution is completely along a radiative track might be expected to retain their large-scale nonaxisymmetric fields, with near-uniform rotation.

In passing, note that if large-scale fields of the strength discussed here were to be present thoughout the radiative part of the pre-Main Sequence star, then the magneto-rotational instability discussed by Rüdiger *et al.* (2001) and Arlt *et al.* (2005) would be inhibited.

4. Dynamical stability of large-scale fields

Since the work of Tayler (1973), Markey & Tayler (1973) and Wright (1973) it has been known that purely poloidal and purely toroidal fields in stars will be dynamically unstable. Essentially, poloidal fields near the neutral line, and toroidal fields near their axis have the topology of the z-pinch, which is subject to the kink instability (see e.g., Fig. 1 of Tayler 1973). This instability will reduce field length scales, presumably resulting in a greatly enhanced dissipation rate. It was shown that that a field topology with linked poloidal and toroidal field lines has improved stability properties, analogous to the stabilization of the z-pinch by a longitudinal field. The basic arguments are mostly simply presented in terms of axisymmetric fields, but the instability depends only on general properties of the field and its geometry (and is independent of field strength), and so is not restricted to axisymmetric fields. Rotation, especially when combined with nonaxisymmetric fields, and finite resistivity are further complicating features (see, e.g.,

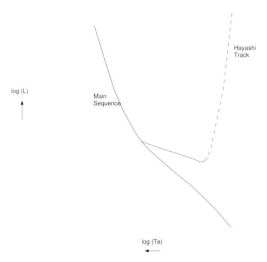

Figure 1. A schematic Hertzsprung Russell diagram, showing the Main Sequence and a pre-Main Sequence evolutionary track for a star of several solar masses. The largely convective Hayashi track is shown as a broken curve. According to Palla & Stahler's work, such a star would only trace the radiative part of the track, shown as solid.

Pitts & Tayler 1985). Given the difficulties of making further analytic progress it has been implicitly assumed that stable linked poloidal-toroidal configurations exist and are commonly found.

Very recently Braithwaite & Spruit (2005) have simulated the dynamical evolution of the initial fields embedded in gas spheres. They find that dynamically stable magnetic fields with linked poloidal and toroidal fluxes naturally emerge, providing a very welcome confirmation for the picture described above, which had been widely accepted as plausible, but had lacked detailed confirmation. Moreover, depending on initial conditions, the fields may in some cases take about a classical diffusion time of $O(10^9)$ yr to reach the stellar surface (i.e., of order of or longer than the Main Sequence lifetime of A stars). This last figure is rather uncertain, as computationally some rescaling is necessary, and there is also a dependence on unknown initial conditions. Further, this timescale might be shortened by the effects of unconsidered (and uncertain) microinstabilities, which would lead to an effective enhanced diffusivity. The initial field strength remains a free parameter. Importantly, for these estimates to be relevant in the context of the magnetic A star problem, it seems that the star must be largely radiative (see Sect. 2.1 above).

5. The angle between the rotation and magnetic axes

Rather unexpectedly, Landstreet & Mathys (2000) found that in the more slowly rotating magnetic CP stars ($P_{\rm rot} \gtrsim 25$ d) the angle β between magnetic and rotation axes is small, whereas when $P_{\rm rot} \lesssim 25$ d it is large, with a significant fraction having β near $90°$. The origin of this distinctly nonrandom distribution is quite unclear.

Three mechanisms that could operate to modify an initial β-distribution have been identified. In an oblique rotator, magnetically controlled stellar winds exert both braking and precessional torques. Only rather idealized cases have been worked out in detail. These suggest that $\beta \to 0°$ or $90°$, depending on the surface field distribution (Mestel & Selley 1970). This mechanism would have to operate before the arrival of the star on the Main Sequence.

Mestel *et al.* (1981) studied the motions resulting from the Eulerian nutation that occurs when $\beta \neq 0, 90°$, and deduced that the resulting dissipation of energy would

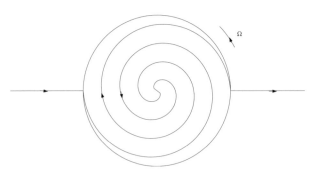

Figure 2. Cartoon, showing how an initial field line of small curvature lying perpendicular to the rotation axis will be wound up by a differential rotation, so that fields of opposite sense are brought close together, facilitating reconnection and decay. (Here, $d\Omega/dr < 0$.)

again result in asymptotic states of $\beta = 0°$ or $90°$, now depending on the position of the axis of maximum moment of inertia.

The timescales for both of these processes to operate depend on several ill-known factors, but it is conceivable that they could be relevant. The third mechanism is the advection of field by the rotationally driven (and magnetically modified) meridional circulation, which was shown by Moss (1977, 1984) to affect the observed angle β. This mechanism has the appealing property that for sufficiently rapid rotators, β becomes smaller when the initial value of $\beta < 55°$, whereas for larger initial β, $\beta \to 90°$. Rather disappointingly perhaps, it seems that the mechanism will only be effective for rotation periods of less than about 5 d, and maybe substantially shorter.

Another, very speculative, possibility perhaps now emerges. Braithwaite & Spruit's simulations do not take into account the effects of rotation. This can be argued to be unimportant if the Alfvèn travel time is shorter than the rotational period. For a $3M_\odot, 5R_\odot$ protostar, an internal field of 10^5 G has an Alfvèn time of order 25 d. 10^5 G is rather a large field in the context of Table 1, but $< B_1 >, < B_2 >$ are mean poloidal fields, and toroidal fields that are comparable with or locally stronger than the poloidal fields will be present in stable configurations. Also it is plausibly an Alfvèn time for an interior region, where fields are larger than their mean values, that is relevant. Could the emergence of a stable field configuration result in different values of β for faster and slower rotators? Again, it seems probable that only large-scale simulations will be able to answer this question.

6. Conclusions

It now seems possible to give a plausible broad-brush account of the history of the anomalously strong fields of the magnetic CP stars.

- If some (at least) stars follow the Palla & Stahler (1993) pre-Main Sequence evolutionary history, then the presence of strong global-scale fields in a fraction of Main Sequence stars with $M \gtrsim 2M_\odot$ can be explained, and also the absence of any such fields in less massive Main Sequence stars.

- Nonaxisymmetric fields of the strengths characteristic of magnetic CP stars are strong enough to resist winding by differential rotation, and indeed can bring about a state of near uniform rotation.

- Fields naturally evolve to dynamically stable configurations, which can survive for times as long as the Main Sequence lifetime of A stars. For most of this time these fields are likely to be predominantly dipolar at the surface.

• It is possible that in some cases fields may emerge after a star reaches the Main Sequence – this can depend on a number of rather uncertain factors.

• It still seems to be necessary to appeal to variations in the initial conditions to resolve the '10%' problem. There is now the additional feature that not only the total flux present in the protostellar medium, but also its spatial distribution, may play a role.

Crucially, this picture requires that the pre-Main Sequence evolution of the magnetic CP stars is substantially radiative, i.e., that global turbulent convection does not occur.

Of course, there are still a number of outstanding problems. Readers will have their own lists of priorities and concerns but I would like to emphasize again the question of the distribution of angles β between magnetic and rotational axes (Sect. 5):

• Why do relatively rapid rotators ($P_{\rm rot} \lesssim 25$ d) have large values of β (i.e. axes close to perpendicularity), whereas slower rotators have small β and so axes that are more nearly aligned (Landstreet & Mathys 2000)?

Acknowledgements

I am grateful to Leon Mestel for comments on a draft of this paper.

References

Aurière, M., Silvester, J., Wade, G.A. and 10 others 2004, preprint
Arlt, R. 2005, *These Proceedings*, 103
Braithwaite, H., Spruit, H. 2005, *This meeting*, http://www.ta3.sk/IAUS224/PDF/ET1.pdf
Browning, M. K., Brun, A. S., Toomre, J. 2005, *These Proceedings*, 149
Desch, S.J. & Mouschovias, T. Ch. 2001, *ApJ*, 550, 314
Landstreet, J.D. & Mathys, G. 2000, *A&A*, 359, 213
McDonald, J. & Mullan, D.J. 2004, *MNRAS*, 348, 702
Marconi, M., Palla, F. 2005, *These Proceedings*, 69
Mestel, L., 2003a, *Stellar Magnetism* (paperback edition, Clarendon Press; Oxford)
Mestel, L., 2003b, in: L.A. Balona, H. Henrichs & T. Medupe (eds.), *Magnetic fields in O, B and A stars: origin and connection to pulsation, rotation and mass loss*, ASP Conf. Ser., 305, p. 3
Mestel L. & Selley, C.S. 1970, *MNRAS*, 149, 197
Mestel, L., Nittman, J., Wood, W.P., & Wright, G.A.E., *MNRAS*, 195, 979
Moss, D. 1977, *MNRAS*, 178, 61
Moss, D. 1984, *MNRAS*, 209, 607
Moss, D. 1989, *MNRAS*, 236, 629
Moss, D. 1992, *MNRAS*, 257, 593
Moss, D. 2001, in: G. Mathys, S.K. Solanki, D.T. Wickramasinghe (eds.), *Magnetic Fields across the HR Diagram*, ASP Conf. Ser., 248, 305
Moss, D. 2003a, in: J. Arnaud & N. Meunier (eds.), *Magnetic activity of the Sun and stars*, EAS Publ. Ser., Vol. 9, p. 21
Moss, D. 2003b, *A&A*, 403, 693
Nakano, N., Nishi R. & Umbeyashi T., 2002, *ApJ*, 573, 1999
Palla, F. & Stahla, S.W., 1993, *ApJ*, 418, 414
Rädler, K.-H. 1986, in: *Plasma Astrophysics*, ESA, SP-251, p. 569
Rüdiger, G., Arlt, R. & Hollerbach, R. 2001, in: G. Mathys, S.K. Solanki & D.T. Wickramasinghe (eds.), *Magnetic Fields across the HR Diagram*, ASP Conf. Ser., 248, 315
Stępień, K. 2001, *A&A*, 353, 227
Tayler, R.J. 1973, *MNRAS*, 161, 365
Markey, P. & Tayler, R.J. 1973, *MNRAS*, 163, 77
Pitts, E. & Tayler, R.J. 1985, *MNRAS*, 216, 139
Schwarschild, M. 1958, The Structure and Evolution of the Stars (Princeton University Press)
Wright, G.A.E., 1973, *MNRAS*, 162, 339

Discussion

WADE: I want to stress the point that the 10% problem is not confined to the statistical incidence of fields in random, unrelated field stars. Rather, a more fundamental problem is that A stars which appear to be explicitly related (e.g., in open clusters and SB2 systems), and that have undergone similar/identical evolutionary scenarios, can have qualitatively different surface magnetic field characteristics (i.e., one is magnetic, one is not). This is a real puzzle!

MOSS: I agree! I certainly did not intend to play down this aspect.

PISKUNOV: The two mechanisms of magnetic field generation, fossil and dynamo, must coexist in stellar interiors, at least during some stages of stellar evolution. Can we expect some significant interaction between the two fields? Can a fossil field affect the dynamo, or a dynamo field perturb a simple dipolar structure of the fossil field?

MOSS: Turbulent convection in the core would by itself be expected to exclude the large-scale field from the central regions, without any very large global effects. I cannot see any reason why a dynamo that generated relatively small-scale fields, as in the simulations of Browning *et al.* (2005), should change this significantly: it is hard to see how such a field in the core would reconnect to the global field. In reverse, the situation is less clear. I recently (2004, A&A, 414, 1065) suggested that a field external to a dynamo with significant differential rotation could significantly affect the operation of the dynamo. Of course, someone should do the required simulations!

COWLEY: Regarding the question of whether a collapsing star is radiative or convective: is the primordial field itself strong enough to influence this question? To put this another way: if there is a field in the interstellar medium, is the star more likely to follow a radiative track than if there were no such field?

MOSS: This problem was more-or-less the topic of my first published paper, in MNRAS in 1968! The result was that for lower mass stars the effect of a magnetic field, by slightly inhibiting convection, was to move the Hayashi track a small distance to the red. I would imagine that for more massive stars, and with contemporary ideas about pre-Main Sequence evolution, there would be at most a very marginal difference in the mass at which stars changed from having to not having a largely convective phase (see Sect. 2.1 above).

NOELS: I would like to point out that in the Henyey tracks the fully convective phase was suppressed because it was thought that this phase had no influence on the Main Sequence models. In the Henyey code (in the 1960s) the Hayashi track could be computed, but as computations took so long they were generally ignored and evolution was started as near to the Main Sequence as possible.

MOSS: I was thinking of the paper by Henyey *et al.* (PASP 1955) where the possible effects of envelope convection are explicitly ignored. Hayashi (PASJ 1961) seems to have been the first to realise the dramatic effects of envelope convection on the pre-MS evolutionary tracks of lower mass stars.

The A-Star Puzzle
Proceedings IAU Symposium No. 224, 2004
J. Zverko, J. Žižňovský, S.J. Adelman, & W.W. Weiss, eds.
© 2004 International Astronomical Union
DOI: 10.1017/S1743921304004636

Vertical and horizontal abundance structures of the roAp star HD 24712

T. Lüftinger[1], O. Kochukhov[1], T. Ryabchikova[1,3], I. Ilyin[2] and W.W. Weiss[1]

[1]Department of Astronomy, Vienna University, Tuerkenschanzstrasse 17, A-1180 Vienna, Austria, email: name@astro.univie.ac.at

[2]Astrophysical Institute Potsdam, An der Sternwarte 16, D-14482 Potsdam, Germany email: ilyin@aip.de

[3]Institute for Astronomy, Russian Academy of Sciences, Pyatnitskaya 48, 109017 Moscow, Russia, email: ryabchik@inasan.rssi.ru

Abstract. High-resolution spectroscopic and spectropolarimetric data of the rapidly oscillating Ap star HD 24712 (HR 1217, DO Eri) has been analysed including modelling the vertical elemental abundance structures. We study the interaction and the relation of the vertical (stratification) and the horizontal (spots) abundance characteristics of Fe and the stellar magnetic field. By this synopsis and the relation of our results to the analysis of high resolution and high time resolved observations (Sachkov et al. 2005) we are likely to gain new insights about the atmospheric structure and the geometry, the origin, and the evolution of the magnetic fields of roAp stars.

Keywords. Stars: abundances, stars: stratification, stars: imaging, stars: individual: (HD 24712)

1. Introduction

HD 24712 (HR 1217, DO Eri, $V = 6.00$) is an intensively studied roAp star with light (Wolff & Morrison 1973), spectrum and magnetic variations. Kurtz (1981), after having discovered photometric oscillations of 6.15 min, found that the maximum of the pulsational amplitude of this star corresponds to the maximum of the longitudinal magnetic field (Kurtz 1982). Matthews et al. (1988) discovered radial velocity variations with the main photometric period and an amplitude of 0.4 ±0.05 km s^{-1}. Ryabchikova et al. (2000) detected the greatest line intensity variations for the REE, and found that the region of REE overabundances roughly coincides with the visible magnetic pole. According to their analysis and ours (Lueftinger et al. 2003) Fe varies in antiphase and with a smaller amplitude and seems to be accumulated in a ring-like structure around the magnetic equator. Evidence was found for element stratification in the atmosphere of this star.

2. Observations and fundamental parameters

High resolution ($R = 80000$) spectroscopic and spectropolarimetric observations were obtained with the SOFIN spectrograph attached to the Northern Optical Telescope (NOT) in November 2003. One spectrum obtained with the VLT UVES spectrograph at ESO was obtained near the magnetic minimum of the star. It is from the UVES POP database (Bagnulo et al. 2005). Using the new stellar model atmosphere code LLModels (Shulyak et al. 2004), which implements a direct accounting for line opacities, we obtained the best fit to the Hα line with $T_{\rm eff} = 7350$ K. $\log g = 4.3$ was used as determined by Ryabchikova et al. (1997).

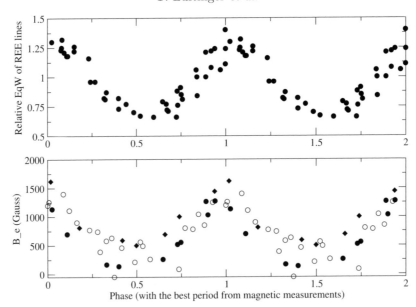

Figure 1. Magnetic field measurements of HD 24712 phased with the best fit period of
12.45965 ± 0.00020 d.

3. Rotational period

The rotational period of HD 24712 is an ongoing quest: Preston (1972) derived a period of $P = 12.448$ d, but Bonsack (1979) needed a slightly longer period 12.46 d to
phase his data with that of Preston. Kurtz & Marang (1987) combined their high-speed
photometry data and that of Wolff & Morrison (1987), which led to an improved period of 12.4572 ±0.0003 d, but the extrema of photometric and magnetic variations were
then separated. Catalano *et al.* (1991) obtaining photometric observations in the JHK
bands and again found a phase shift between the photometric and the magnetic variations. Mathys (1991), combining various measurements of the mean longitudinal magnetic
field, derived a period of $P = 12.4610\pm0.0025$ d, which agrees with that of Bagnulo *et al.*
(1995) $P = 12.4610 \pm 0.0011$ d. Combining measurements of Preston (1972), Mathys and
Landstreet, and MuSiCoS (e.g., Wade *et al.* 2000), we obained a rotational period of
12.45965±0.00020 d. Our search for an improved rotational period has been performed
using the program Period98 (Sperl 1998). The highest peak in the amplitude spectrum
is used to specify the starting frequency and amplitude for a least-squares fit yielding
the period, the amplitude and the phase angle. The corresponding rms errors are the
results of numerical simulations of error propagation as provided by the program EPSim
(Reegen 2004).

4. Stratification analysis

Peculiar A stars, such as HD 24712, usually posess a global magnetic field, which
stabilize their atmospheres and enables the mutual diffusion of the chemical elements
due to different separation processes. Calculations of self consistent model atmospheres
including the effect of diffusion, predict changing abundance profiles for a large number
of elements and the corresponding changes in the atmospheric structure. Thus, if we
want to derive the surface abundances and the magnetic field geometry of Ap and roAp
stars, it is indispensible to account for stratification effects within the atmosphere, which

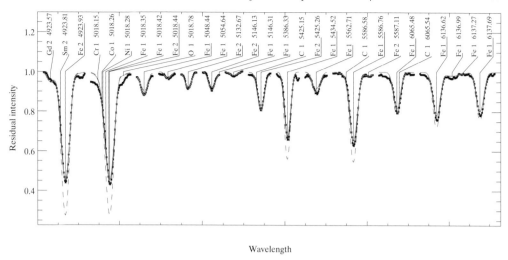

Figure 2. Comparison between observed and calculated line profiles with homogeneous abundances (dashed line) and with stratified abundance distribution (full line) around magnetic minimum.

are also shown in Fig. 5, where we compare model calculations of Intensity- and Stokes profiles of the Fe II line near 5018 Å with and without stratification. Observations that require us to account for the vertical chemical abundance inhomogeneities are

- The impossibility to fit the spectral line wings and the core with the same abundance
- Different abundance values for one element derived from different ionization stages
- Disagreement of the abundances derived from strong and weak lines of the same element
- Unusual behaviour of very high-excitation lines

For HD 24712 we studied the abundance stratification for Fe using DDAFIT, a newly developed procedure for the determination of vertical abundance gradients written in

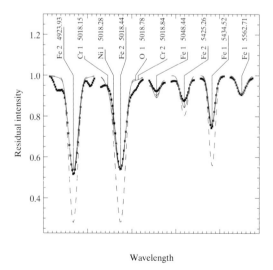

Figure 3. Comparison between observed and calculated line profiles with homogeneous abundances (dashed line) and with stratified abundance distribution (full line) around magnetic maximum.

Fe abundance stratification of HD 24712

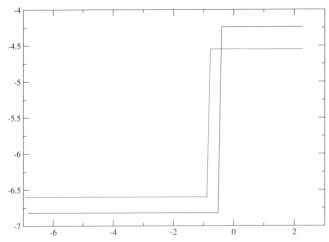

Figure 4. Abundance stratification of Fe in the atmosphere of HD 24712. Abundances in dex are plot versus atmospheric depth in $log\tau_{5000}$.

IDL by Kochukhov, providing an optimization and visualization interface to the spectrum synthesis calculations with SYNTH3, a modification of SYNTH (Piskunov 1992). Vertical abundance distributions were derived using a one step model. The chemical abundance in the upper atmosphere, the abundance in deeper layers, and the vertical position of the abundance step were optimized simultaneously by least squares fitting. With the high resolution spectra obtained around magnetic minimum of the star (UVES) and around magnetic maximum (NOT), we derived stratification profiles for both phases. For the phase around magnetic maximum only six Fe lines could be used and the results can only be regarded as a first approach. In Fig. 2 and Fig. 3 a comparison between observed and calculated profiles of spectral lines selected for the stratification analysis is given. The homogeneous (dashed line) and the stratified (full line) abundances are presented. Our analysis, as presented in Fig. 4, reveals a change in the Fe abundance from -6.6 ± 0.046 dex in the higher atmospheric layers to -4.55 ± 0.022 below $\log \tau_{5000} = -0.7$ around the phase of magnetic minimum (dotted line). Observations around magnetic maximum (full line) suggest the abundance jump occurs in slightly deeper atmospheric layers, around $log\ \tau_{5000} = -0.5$.

5. Conclusions and the future

HD 24712 is the first star, where an analyses of the elemental surface abundances and the magnetic field geometry is being performed, including and accounting for effects of the changing vertical abundance within the stellar atmosphere. It is now possible to simultaneously reconstruct the chemical distribution and the magnetic field geometry on the stellar surface (*INVERS10*, Piskunov *et al.* 2002) including the detailed analysis of the vertical abundance gradients. To reliably determine the elemental surface abundance patterns of various chemical elements on the surface of HD 24712, we will derive the stratification profiles of additional (iron-peak) elements, and further investigate the analysis of the Fe stratification profile around the magnetic maximum of the star, where it seems that the abundance step is moved towards deeper atmospheric layers. Observing a larger sample of Ap and roAp stars in all four Stokes parameters and combining the analysis

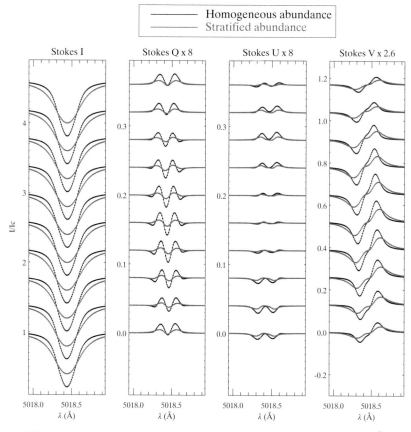

Figure 5. Intensity and Stokes profiles of the Fe II line near 5018 Å.

of the vertical and the horizontal abundance gradients and the magnetic field geometry, we expect to gain new insights into the atmospheric structure of Ap and roAp stars.

Acknowledgements

Financial support for this research was received from the Austrian 'Fonds zur Förderung der wissenschaftlichen Forschung' (projects P14984 and P14546).

References

Bagnulo, S., Landi degl'Innocenti, E., Landolfi, M., Leroy, J. L. 1995, *A&A* **295**, 459

Bagnulo, S., Jehin, E., Melo, C., Ledoux, C., Cabanc, R. 2005, *These Proceedings*, JP6

Bonsack, W.K. 1979, *PASP* 91, 648

Catalano, F.A., Kroll, R., Leone, F. 1991, *A&A* 248, 179

Lueftinger, T., Ryabchikova, T.A., Kochukhov, O., Piskunov, N.E., Kuschnig, R., Wade, G.A., Weiss, W.W. 2003, *International Conference on magnetic fields in O, B, and A stars, ASP Conference Series, Vol. No. 305*, L.A. Balona, H.F. Henrichs, & R. Medupe, eds. 305, 92

Kochukhov, O., Piskunov, N. 2002, *A&A* 388, 868

Kurtz, D.W. 1981, *Inform. Bull. Var. Stars* 1915, 1

Kurtz, D.W. 1982, *MNRAS* 200, 807

Mathys, G., 2000, *A&AS* 89, 121

Martinez, P., Girish, V., Joshi, S., Kurtz, D. W., Ashoka, B. N.,Chaubey, U. S., Gupta, S. K., Sagar, R. 2000, *IBVS* 4853

Martinez, P., Kurtz, D.W., Ashoka, B.N., *et al.* 2001, *A&A* 371, 1048

Matthews, J.M., Wehlau, W.H., Walker, G.A.H., Yang, S. 1988, *ApJ* 324, 1099

Piskunov, N.E., 1992, *in Glagolevskij Yu. V., Romanyuk I.I., eds., Stellar Magnetism. Nauka, St. Petersburg* 92

Piskunov, N.E., Kochukhov, O. 2002, *A&A* 381, 736

Preston G.W. 1972, *ApJ* 175, 465

Reegen, P. 2004, *In: Second Eddington Workshop: Stellar structure and habitable planet finding, 9 - 11 April 2003, Palermo, Italy. Edited by F. Favata, S. Aigrain and A. Wilson. ESA SP-538, Noordwijk: ESA Publications Division, ISBN 92-9092-848-4* 389

Ryabchikova, T.A., Landstreet, J.D., Gelbmann, M.J., Bolgova, G.T.,Tsymbal, V.V., Weiss, W.W. 1997, *A&A* 327, 1137

Sachkov, M., Ryabchikova, T., Ilyin, I., Kochukhov, O., Lueftinger, T. 2005, *These Proceedings*, GP2

Shulyak, D., *et al.* 2004, *A&A* accepted

Sperl, M. 1998, *CoAst.* 111, 1

Wade, G.A., Donati, J.-F., Landstreet, J.D., Shorlin, S.L.S. 2000, *MNRAS* 313, 851,

Wade, G.A., Bagnulo, S., Donati, J.-F., Lueftinger, T., Petit, P., Sigut, T.A.A. 2001, *APN* 35

Wolff, S.C., Morrison N.D. 1973, *PASP* **85**, 141

Discussion

MKRTICHIAN: Fe lines in HR 1217 show small intensity variations over the rotation period. Can you show the results of stratification analyses for lines that show a strong intensity variation over the rotation period?

LUEFTINGER: So far, we derived stratification profiles for Fe, but indeed, stratification analysis for further elements, of course and especially including lines with strong intensity variations, is one of the very next steps in our investigation.

WEISS: You showed two stratification profiles for minimum and maximum magnetic field. Can you please comment comment on the error bar?

LUEFTINGER: The error bar of abundances are below ± 0.05 dex. The abundance jump was determined with an accuracy of ± 0.009 in rhox

MONIER: How accurate is the location of the optical depth ($log\ \tau_{5000}$) at which the abundance discontinuity occurs in your analysis of HD 24712?

KOCHUKHOV: The accuracy of the derivation of the position of vertical abundance jump is about 0.1 dex in the $log\ \tau_{5000}$ scale.

BAGNULO: You have adopted a step function to approximate the stratification of the chemical elements. Is that real? Have you tried to use for instance a second order polynomial?

LUEFTINGER: It is possible to apply, e.g., second order polynomials, and also vary the width of the transition region, but we found that in most cases a one step model is a very good and evidently realistic approximation.

RYABCHIKOVA, REMARK TO BAGNULO'S QUESTION: Many experiments with stratification calculations found that steep abundance jumps (about zero width of abundance transition zone) frequently occur in Ap atmospheres. However, wider transition zones seems to be needed in the case of strongly magnetic stars, not excluding a polynomial shape of abundance transition zone.

The A-Star Puzzle
Proceedings IAU Symposium No. 224, 2004 © 2004 International Astronomical Union
J. Zverko, J. Žižňovský, S.J. Adelman, & W.W. Weiss, eds. DOI: 10.1017/S1743921304004648

Panel discussion section E

CHAIR: **S. Bagnulo**

SECTION ORGANIZER & KEY-NOTE SPEAKER: Gautier Mathys
INVITED SPEAKERS: G.A. Wade, D. Moss
CONTRIBUTION SPEAKERS: J. Brathwaite, Th. Lüftinger

Discussion

BAGNULO: I would like to ask Mathys and Moss to comment about the possible detection of the radial gradients of the magnetic fields in Ap stars. What would be the consequence of such a feature on the stellar atmosphere? I refer in particular to that the length scale for horizontal variability of the field is not less than 10^5 km, whereas the depth of the photosphere may be 100 times less.

MOSS: As I understand it, the suggestion is that field strengths vary quite dramatically in the vertical direction on much shorter length scales than any horizontal field variation. This would imply the presence of strong currents and dissipation, and would appear to be hard to reconcile with long-term stability of the magnetic fields.

BAGNULO: Is it possible that the apparent detection of a magnetic field gradient actually reflects an imperfect knowledge of the Landé factors?

MATHYS: I do not think so. Landé factors are fairly secure in general, at least at the kind of level of accuracy that we are talking about. But for longitudinal field determinations on both sides of the Balmer discontinuity, one should keep in mind that possible systematic effects may be introduced by the quarter-wave plate, which may not have exactly the same retardance throughout the whole wavelength range of interest. Romanyuk (2005, *These Proceedings*, EP4) pointed out to me that his measurements blueward of the Balmer jump are hampered by the low sensitivity of the spectrograph in that spectral region. On the other hand, for the field modulus measurements of Nesvacil *et al.* (2005, *These Proceedings*, EP9), the main limitation arises from the very small number of diagnostic lines available on the blue side of the Balmer jump.

WADE: There are two things to remember concerning vertical gradients. First, the length scale over which the gradient is presumed to occur is very short, hundreds or thousands of kilometres. Second, the magnetic fields are measured with spectral lines. We know very well that surfaces of Ap stars are nonuniform. Therefore, lines of different elements routinely produce different field strengths, whether these lines are formed in the same region of the spectrum, or in different regions. Similarly, lines of different strengths can produce different field strengths, whether or not they are formed in different regions of the spectrum. Thus one has to be extremely careful in interpreting measurements particularly from small samples of lines.

RYABCHIKOVA: Before drawing any conclusions about magnetic field gradients in Ap star atmospheres, one has to estimate the line formation depth, taking into account abundance stratification effects. One cannot take for granted that lines much before ($\sim \lambda 3300$) and much after ($\sim \lambda 5000\text{-}6000$) the Balmer jump are formed at different optical depths.

NOELS: I have a question concerning the evolutionary status of Ap stars. As I understand it, the mechanism proposed by the theory for the generation of the magnetic fields tries to explain what happens at the ZAMS, and some observations suggest that magnetic Ap stars are somewhat evolved away from the ZAMS. Can you comment on this?

BRAITHWAITE: The field strength increases with time for stable configurations. So it may well be that a certain time is required for the field on the surface to become strong enough to be visible. Or maybe not. The timescale on which the field strength increases is a little uncertain. The timescale found from my work is 2×10^9 yr, which is somewhat longer than the Main-Sequence lifetime of an Ap star. To explain Hubrig *et al.* 's result that Ap stars become observably magnetic only once they have completed 30% of their Main-Sequence lifetime, we would need the surface field strength to increase on a shorter timescale.

MOSS: What I would like to emphasise again is that I do not understand why the behaviour of the more massive Bp stars appears to be so different. In these stars we see strong surface fields at ages of the order of 10^7 yr, much less than the claimed age at which surface fields manifest themselves in the lower mass Ap stars. If we think that the Bp stars form a continuum with the Ap stars, then something very strange would appear to be happening.

I would also like to point out the idea that the surface field changes on timescales of 10^7 to 10^8 yr is not new, as it is present in models of the faster rotators that included meridional circulation, published some 20 years ago. However, we now have a further theoretical mechanism to drive such changes.

WADE: I just want to underscore that the situation is not that there are no known magnetic Ap stars that have ages consistent with a small fraction of their Main Sequence evolution, but rather that only a very small number of the low-mass Ap stars are relatively young. There are certainly a larger number of well established Bp stars with masses between 3 and 5 $M_\odot$ that have completed less than a few percent of their Main Sequence evolution. If you believe that the higher mass magnetic stars and the lower mass magnetic stars are all part of the same sequence that is connected by the same physics, then there is still a need for the presence of fields at the ZAMS and, I am sure, before the ZAMS too.

PISKUNOV: Can the changes and the evolution of the near-surface magnetic fields affect the anticipated chemical stratification? If this is the case, we should see characteristic anomaly patterns for stars of similar evolutionary status.

MICHAUD: In the outer atmosphere, vertical diffusion timescales are very short, of the order of 10^4 yr, so that diffusion adapts rapidly to the changing magnetic configuration, as long as the stability of the atmosphere is not too affected. On the other hand, if the structural changes of the magnetic field required horizontal diffusion, this could take much longer.

VAUCLAIR: Emerging magnetic fields can have some influence on the abundances if they sweep up regions where some elements are mostly neutral while others are mostly ionised. The magnetic lines bring the ionised atoms upward, but not the neutrals. Of course, this depends on the timescales and on the exact structure of the field, since the ions can also fall back along the magnetic lines if they are inclined.

CUNHA: We are now starting to be able to use the global oscillations of Ap stars to probe the magnetic structure of the interior, both in terms of the average magnetic field intensity and of its geometry (although for the geometry, we are limited to those layers that are magnetically dominated, that is, right below the photosphere, because these are the layers that influence the oscillations). In relation with this, it was mentioned that the magnetic fied structure could be different in the stellar interior from what is observed at the star's surface. What do you mean by stellar interior? Is it very deep in the star?

MATHYS: It could be just below the surface.

WADE: Observationally, we do not know anything about the field below the line-forming region. We have to leave that to people like Braithwaite and Moss. Something that may potentially be useful to look into is the "10% problem". That is, let us tentatively interpret the fact that we observe magnetic fields in only 7% of the Main Sequence A stars not as meaning that only 7% of these stars are magnetic, but that only 7% show their field on their surface. This may not make things simpler for the theoreticians, though.

ADELMAN: I have a comment and a question. My comment: if you build an H-R diagram and look at the distribution of rotation periods, there is a tendency for the rotation to become slower as one proceeds from the ZAMS to the giants and supergiants, upon which a stochastic distribution of periods is superimposed. A proper statistical study of this effect remains to be performed. As to my question, how well known are the Landé g factors?

MATHYS: I have been limiting myself to iron peak elements, for which I was using Landé factors from Kurucz's CD-ROM. When there is a measured Landé factor, it is used in this list. Otherwise Kurucz derives the Landé factors from the same approximate model atoms that he uses to compute the other pieces of information included in his table. I have shown (1990, A&A 236, 527) that the Landé factors that he obtains in this way are considerably better than those computed by application of the LS-coupling formula. I do not know what VALD uses as a source for the Landé factors. Perhaps one of the people involved in this project who are present here could comment on this.

LANDSTREET: I would like to dispel the impression that there are only a few Landé factors in the literature. Most laboratory spectroscopists measured Landé factors to classify energy levels, and Charlotte Moore obtained these values and included them in the NBS *Atomic Energy Levels* volumes. There are thousands of values in these volumes.

MATHYS: I concur, but there are still some cases of great practical interest, such as the high-excitation Fe II lines (which are useful diagnostic lines for observations of Ap stars in the red spectral region), for which no laboratory values are available.

RYABCHIKOVA: Our studies of the Zeeman structure of different spectral lines show that the surface fields obtained from nearby lines (such as the Fe II $\lambda 6149$ doublet and the pure triplet Fe I $\lambda 6173$) may differ by 100–200 G. We find this for instance in γ Equ, which has a surface field of 4 kG. I also obtained surface magnetic field estimates using other lines. All other pure triplets yield field values closer to that obtained through the analysis of the $\lambda 6173$ triplet. My feeling is that the famous Fe II $\lambda 6149$ doublet tends to provide field values slightly smaller than the other lines. I checked the Landé factors in the NIST publications, and they seem to be correct. This situation should be taken into acccount in all studies of the fine behaviour of magnetic fields, such as magnetic gradients.

MATHYS: I think that you are right. The Fe II $\lambda 6149$ line is a doublet only in first approximation, because it is formed in a regime of the partial Paschen-Back effect. Therefore I would not be surprised if the field values derived from measurements of this doublet showed some systematic effect at the level of a couple hundred Gauss; the magnitude of the effect may actually depend on the strength of the field in the considered star. The main interest in using this diagnostic line is that it is visible in virtually all Ap stars and it has a really large splitting. If you use it consistently in a given star, you can trace the variation of the magnetic field with phase in a meaningful way. But you should exert great caution, I agree, in using it for the study of effects like magnetic gradients.

KOCHUKHOV: I have a comment concerning the relation between the magnetic field and the surface inhomogeneities. According to our most recent abundance Doppler imaging studies, there does not seem to be any clear systematic correlation between the surface chemical inhomogeneities and the magnetic field structure. Therefore there is no reason to believe that only the magnetic field can induce nonuniformities in the distribution of chemical elements over the stellar surface. Thus in my opinion, the use of nonuniformities (e.g., of Hg in α And) as a proxy for the existence of a magnetic field is questionable, because it seems that other processes besides magnetic effects can modify the chemical diffusion in Ap stars. This is easy to understand: if for a certain ion, the radiative and gravitational forces are balanced, then any external perturbation such as, e.g., rotation or the effect of a binary companion, can potentially induce changes in the diffusion over the stellar surface.

KURTZ: There is another sequence of peculiar stars, the Am stars, which have similar rotation properties, similar structures, and similar ages. But they do not show any horizontal abundance differentiation. The only known difference is that the Ap stars have magnetic fields, which implies that the latter are the major cause of the horizontal abundance inhomogeneities in Ap stars.

VAUCLAIR: I just want to point out that the effect of the magnetic field on diffusion is not trivial. It may be much more complicated than simply having a spot at the magnetic poles or a ring at the magnetic equator. The observed difference between abundance and magnetic patterns may be due to this nontrivial coupling (which may be different for different elements).

ADELMAN: There are a few papers suggesting that young Am stars have abundance distributions different from older Am stars.

WADE: As the only example of well-established HgMn stars with clear, undisputed line profile variations, α And, is extremely important for the understanding of the role of the magnetic field and of other processes in producing abundance structures. On the other hand, in older abundance Doppler images (e.g., Hatzes), you could see very coherent geometrical structures (e.g., rings, caps) that were apparently associated with the underlying magnetic geometry in a clear and "natural" way (e.g., rings at the magnetic equator, spots at the poles). Are *any* of these kinds of relationships evident in the latest generation of Doppler images?

KOCHUKHOV: Only few chemical elements show obvious correlations with the magnetic field geometry. For the majority of the species, we find fairly complex chemical patterns, without rings or spots symmetric relative to the magnetic field. This observation is based on the mapping of 17 chemical elements in the roAp star HR 3831 (Kochukhov *et al.* 2004, A&A 424, 935).

BAGNULO: North (1998, A&A 336, 1072) has shown convincing evidence that magnetic Ap stars do not loose angular momentum during their main sequence life. I should like to ask Saul Adelman if he finds a different result.

ADELMAN: My paper (2002, Baltic Astron. 11, 475) suggests that most, but not all, of the most rapidly rotating magnetic CP stars are close to the ZAMS, and some of the least rapidly rotating magnetic CP stars are the furthest from the ZAMS. The problem when you do this kind of study is that you really want to make sure that you know the periods quite well. For a fair fraction of the Ap stars for which periods have been reported, these periods are not correct. So this study is based on those stars with periods determined in my own studies and in those with my associates, so that I was absolutely certain that the periods were correct. I also made sure that I had reliable distance estimates, so that I have used Hipparcos parallaxes. If you mix this data with that from other sources, you would have to worry about different systematic errors.

The A-Star Puzzle
Proceedings IAU Symposium No. 224, 2004   © 2004 International Astronomical Union
J. Zverko, J. Žižňovský, S.J. Adelman, & W.W. Weiss, eds.   DOI: 10.1017/S174392130400465X

The CP Stars, an overview: Then and now

Charles R. Cowley[1] and Donald J. Bord[2]

[1]Department of Astronomy, University of Michigan, Ann Arbor, MI 48109-1090, USA
email: cowley@umich.edu

[2]Department of Natural Sciences, University of Michigan-Dearborn,
4901 Evergreen Rd., Dearborn, MI 48128-1491, USA
email: dbord@umd.umich.edu

Abstract. We briefly review the traditional classifications of CP stars. The current availability of large numbers of abundances now make it possible to use multivariate techniques, both to supplement traditional classification methods and probe the abundance patterns. We discuss cluster analysis and correlation matrices for sample material. We review the historical resistance to the notion that CP stars were indeed chemically peculiar. Modern work shows that while these objects do indeed have atmospheric anomalies, they are nevertheless chemically peculiar.

Abundance patterns are an important clue to the origin of the abundance peculiarities. We contrast patterns due to nuclear and chemical differentiation processes. The roAp and related stars show vertical as well as horizontal abundance variations, and abnormal line profiles. Photospheric abundances in these stars are surely abnormal (nonsolar), but as long as the models are uncertain the derived abundances will be very crude.

For more than two decades, observations of CP stars in the X-ray and radio regimes have been made with increasing sensitivity and pointing accuracy. We discuss the current evidence linking magnetic and nonmagnetic CP stars to sources of galactic X-rays and radio radiation. There seems no doubt that high energy phenomena are associated with, if not produced by, some CP stars. This circumstance admits the possibility that the release of high energy particles (p's, n's and α's) during such events may initiate nuclear reactions on the surfaces of the CP stars. We briefly reconsider the viability of such processes for producing exotic species like Pm by proton bombardment using recent data for solar and stellar flares.

Keywords. Stars: abundances, stars: chemically peculiar, stars: individual: (Vega, Merak, α And, 32 Aqr, σ Boo, κ Cnc, θ Cyg, π Dra, 46 Dra, γ Equ, 112 Her, ϕ Her, υ Her, θ Leo, μ Lep, 33 Lib, χ Lup, 21 Lyn, σ Ori E, ι Psc, o Peg, 53 Tau, 68 Tau, 15 Vul, HR 178, HR 465, HR 1728, HR 1890, HR 4072, HR 7143, HR 7245, HR 7361, HR 7664, HR 7775, HR 8349, HD 965, HD 29647, HD 101065, HD 166473, HD 213637, HD 215441, HD 217522)

---

1. The classes of CP Stars–visual spectroscopy

Spectroscopic anomalies among B- and A-stars were known before astronomers made the distinction between upper and lower Main Sequences. The early history of this work has been amply reviewed (Bidelman 1967, Wolff 1983). Preston (1974) introduced the designation CP for chemically peculiar stars on the upper Main Sequence. His CP1 stars were previously called Am or Fm. The magnetic CP2 stars had anomalous lines of Sr, Cr, Eu, and Si, which could be seen on classification (125 Å mm^{-1}) spectra. CP3 stars were typically late B objects. Their detection was difficult at classification resolution. The extent of their peculiarities became increasingly apparent only on higher dispersion spectra. They have anomalously strong manganese lines and usually also λ3984 of Hg II, and are commonly called HgMn stars.

Jaschek & Jaschek (1974) describe the various classes of CP stars and introduce the notion of "visual spectroscopy." This can be done at both high and low dispersion. Workers who used high dispersion, like the Jascheks themselves, Bidelman (1966), or Cowley and coworkers (cf. Cowley 1976) made qualitative inferences about the abundances from the characteristics of the spectra. While this procedure lacked the rigor of a fine analysis, it was closely connected to the observations, and therefore not subject to the uncertainties of atomic parameters or poor determinations of effective temperature. Visual spectroscopy was often more valuable than ostensibly more rigorous analyses. We eschew an invidious listing of specific examples, but there are some interesting cases for which this is true.

Modern abundance work, especially supplemented by the atomic data bases made available by R. L. Kurucz (cf. Kurucz 2004), VALD (Kupka *et al.* 1999), DREAM (cf. Biemont 2004), etc., is less likely to go astray. Modern workers have now produced a rather large body of abundances of CP stars. We propose they be classified and analyzed by multivariate techniques, which should become standard tools. They will provide powerful supplements to the traditional classification methods.

2. Cluster analysis

This method uses the individual abundances as points in an n-dimensional space, where n is the number of abundances. Initially, if N stars are considered, there are N clusters. Various algorithms are used to then merge the clusters into a smaller number of objects, which may be used as a basis of classification. We have employed a simple method that finds the closest pair of points, and merges them into a single cluster with a position that is the average of the two points. The next closest points are then found, and similarly merged. We have used Euclidian distances, though other choices are possible. The progress of the merging is illustrated on a tree or *dendrogram*, where the vertical

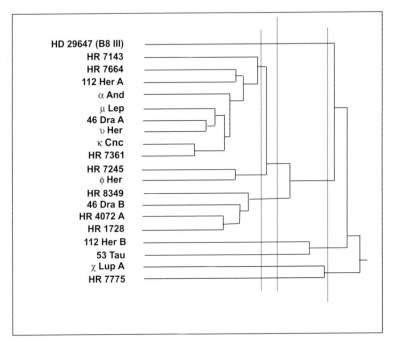

Figure 1. Dendrogram for CP stars. Abundances from Adelman, Castelli, Ryabchikova, Wahlgren, and coworkers. See text.

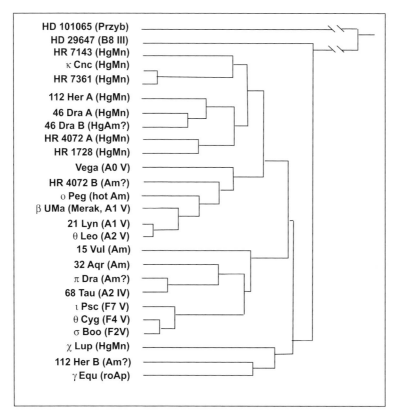

Figure 2. Dendrogram for CP stars. Abundances from Adelman, Castelli, Ryabchikova, Wahlgren, and coworkers. See text.

dimension shows the clusters. The horizontal dimension is a measure of the distance of the clusters from one another.

In most multivariate studies, some decision must be made to allow for missing data. There is no unique choice of how to deal with this. In the provisional studies below, we have filled in missing abundances with "solar" values. One must keep in mind that this can bias the proximity of stars in their multidimensional space.

2.1. HgMn stars

Fig. 1 shows a dendrogram for selected HgMn stars. The vertical cuts show possible means of separating the objects into classes. Cuts further to the right result in fewer classes. References to the data for individual stars in this and the following figure that are *not* cited elsewhere in the text are: Adelman (1994), Adelman *et al.* (1998, 2001), Ryabchikova *et al.* (1997, 1999), Leckrone *et al.* (1999), and Wahlgren *et al.* (2000).

2.2. CP1, CP2, and CP3 objects

In Fig. 2, we apply the cluster method to a more heterogeneous group of stars. Those familiar with the individual spectra will not be surprised by the groupings, in most cases. The pairing of 112 Her B with γ Equ may seem odd. The brighter star of the SB2, 112 Her A, is a well-known HgMn star, while γ Equ is a cool roAp star (see Kurtz 1990, 2003 for a full description of the properties of the roAp stars). However, Ryabchikova *et al.* (1996) have already noted the similarity of 112 Her B to Am and roAp stars.

Table 1. Correlation matrix for selected elements based on Erspamer-North abundances.

	Ca	Sc	Ti	V
Sc	0.54	1.00		
Ti	0.65	0.26	1.00	
V	0.09	-0.08	0.41	1.00
Cr	0.47	0.03	0.82	0.48
Mn	0.49	0.23	0.62	0.37
Fe	0.57	0.05	0.85	0.47
Co	0.05	0.01	0.09	0.12
Ni	0.34	-0.07	0.67	0.51
Sr	0.17	-0.25	0.51	0.44
Y	0.20	0.07	0.54	0.49
Zr	0.06	-0.26	0.27	0.38
Ba	0.06	-0.25	0.48	0.51
Ce	-0.17	-0.39	0.06	0.26

2.3. *Automated abundances of A and F stars*

We performed cluster analysis and made correlation matrices using the results of Erspamer & North's (2003) automated abundance calculations for 140 stars. The full dendrogram is too large to display here. It is available on the web at

    http://www.astro.lsa.umich.edu/users/cowley/tree.htm

Likewise, the full correlation matrix may be seen at

    http://www.astro.lsa.umich.edu/users/cowley/table6.htm

A portion is shown in Table 1.

The formal significance of a correlation coefficient depends on the number of points used in its calculation. However, to actually judge the reality of a correlation, one should exclude the default, solar values. In Figs. 3 and 4, stars with such default abundances are not plotted.

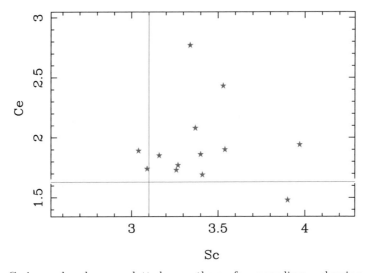

Figure 3. Cerium abundances plotted vs. those for scandium, showing a marginal anti-correlation at best. Vertical and horizontal lines show the solar abundance of Sc and Ce, respectively. Data from Erspamer & North (2003).

Erspamer–North AA, 398

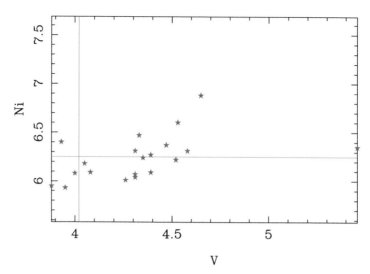

Figure 4. Ni abundances plotted vs. those for vanadium, showing a good correlation. Data from Erspamer & North (2003).

3. AA or AA: anomalous abundances or anomalous atmospheres

IAU Symposium 26 was on *Abundance Determinations in Stellar Spectra*. In it, Anne Underhill (1966) gave a critique of analytical methods, and ended with the controversial conclusion that many of the traditional abundance anomalies might instead be the result of anomalous atmospheric structure. At that time, most abundances were determined using highly schematic model stellar photospheres. The slab, or Schuster-Schwarzschild model assumed single, characteristic values of the temperature and pressure. Milne-Eddington models allowed some depth dependence, but assumed constant ratios of the line to continuous absorption coefficient. Differential methods were widely used, where a single set of parameters, temperature, pressure, and surface gravity, was assumed to apply to each of two stars. The methods are described in classical textbooks.

Depth-dependent numerical models were just coming into their own, and it seemed possible that some variations from the classical temperature-pressure structure might give rise to effects that could be interpreted as abundance anomalies.

Abundance determinations at that time rested on the assumption of local thermody-namic equilibrium (LTE). This assumption was generally consistent with observations of stellar spectra. However, it was in principle subject to confirmation by a more fun-damental non-LTE approach, which was practically beyond the techniques of the 1960s. Many astronomers wondered about the extent to which a rigorous non-LTE treatment would remove abundance anomalies in stars. Underhill put the question trenchantly: Do we have AA (anomalous abundances) or AA (anomalous atmospheres). In this regard, it should be noted that in the first half of the 20th century, it was widely believed that almost all stars had the same abundance distribution as the Sun. A. Unsöld (1955), the doyen of analytical stellar spectroscopy, wrote in his classical text that "it therefore seems justified to speak with caution of a *cosmic abundance distribution of the elements.*" [tr. CRC]. A few obvious discordant cases, such as the R, N, and S stars were apparently set aside.

Detailed LTE-modelling and realistic non-LTE calculations have mostly confirmed the abundance peculiarities of a wide variety of stars, including the CP stars (cf. Van't Veer-Menerret 1963, Conti 1965, and various papers in Piskunov *et al.* 2003). Yet it is now clear that the spectra of many of these stars cannot be described by classical, plane-parallel, chemically homogeneous model atmospheres.

Within a decade of IAU Symposium 26, it was generally accepted that CP2 stars had horizontal abundance variations, chemically distinct continents. Stellar rotation, when combined with the surface abundance variations could account for the spectral, color, and light variations of these stars.

Nevertheless, it was thought the atmospheres were reasonably well described by classical models, at least within an abundance patch. Spectra were still analyzed ignoring complications due to horizontal abundance variations, which would essentially render the spectra composite. These analyses were reasonably consistent, insofar as the effective temperatures of the models were in general agreement with the degrees of excitation and ionization determined from various elemental species. This remained true for the observational and analytical tools available for much of the latter half of the 20th century.

Recently, it has become clear that classical models do not describe the *vertical* structure of at least CP2 stars, even within a putative abundance patch. The rapidly-oscillating or roAp stars are good examples. The cores of their low-Balmer members show anomalous structure (Cowley *et al.* 2001), and anomalous ionization is found for several of the lanthanides (Ryabchikova *et al.* 2004). The catchword, used through much of this symposium, is "stratification."

There is nothing new about a chemically stratified atmosphere. It has been known for years that the Earth's upper atmosphere, above about 100 km, is chemically stratified, according to the molecular weights of atoms, ions, and molecules. Numerous studies (cf. Feldman & Widing 2003) have shown that the solar atmosphere is vertically stratified. Elements with first ionization potentials below about 10 eV are some 2 to 4 times more abundant in the chromosphere and corona than in the photosphere.

What is new is that calculations are now being made of CP star spectra, based on models with explicit elemental stratification. These models have achieved a modicum of success. For example, Kochukhov *et al.* (2003) have shown how the core-wing anomaly of cool CP2 stars could be accounted for in terms of a localized hot region in the high photospheres. Likewise, ionization and excitation anomalies may be accounted for in terms of chemical stratification (cf. Ryabchikova *et al.* 2004, and contributions in this Symposium).

4. Abundance patterns

The history of matter is written into its abundance patterns. We measure abundances today in stars, nebulae, and planets, and try to discern the processes that yielded them. Abundance patterns arise from nuclear and non-nuclear (e.g., chemical) processes. To some extent, each leave characteristic signatures (cf. Guthrie 1971). Nuclear processes are readily recognized. Well before the nature of atoms and nuclei were understood, the remarkable William Harkins (1917) wrote of "...a relation between the abundance of the elements and the structure of the nuclei of atoms." His work presaged the later and more influential compilation of Suess & Urey (1956) of elemental abundances. These, and subsequent refinements were called, as we have noted above, *cosmic abundances*. Audouze & Tinsley (1976) suggested the more useful and accurate term *standard abundance distribution*, or SAD. It is shown in Fig. 5.

Logarithmic SAD Abundances: Log(H) = 12.0

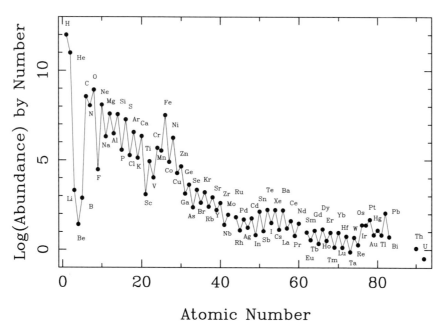

Figure 5. The standard sbundance sistribution (SAD), formerly known as the "cosmical abundance distribution;" data are from Grevesse & Sauval (1998).

Logarithmic Crustal Abundances: Log(Si) = 6.0

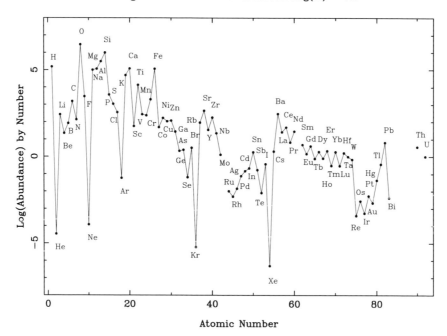

Figure 6. Abundances in the Earth's crust. The result of differentiation from the SAD. Data are from Mason (1966).

Sun and HR 7143

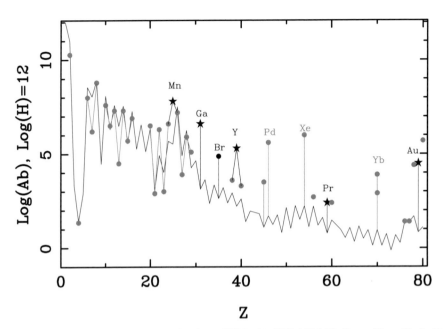

Figure 7. SAD abundances along with those HR 7143 (HD 175640) from Castelli & Hubrig (2004). Vertical lines are drawn to help indicate whether the points belong to an odd- or even-Z element. Here, and in the two following figures, filled circles and stars indicate the stellar abundances; stars are used to indicate variations from a nuclear pattern. The two points for Yb come from Yb II and Yb III lines.

Standard astronomical practice is to concentrate on *departures* from the SAD. We note that plots of [El/H] vs. Z *obscure* one of the most notable and basic characteristics of nuclear patterns, namely the odd-even alternation. There are certainly times when it is desirable to suppress this pattern. In the following figures, we plot logarithms of the stellar abundances directly, and not their ratio to the SAD.

Fig. 6 shows a decidedly non-nuclear pattern, that of the Earth's crust. Note the huge even-Z anomalies that occur at the noble gasses. These and other characteristics of the crustal abundance pattern are well understood in terms of the thermodynamics of Earth materials.

5. CP star abundance patterns

5.1. *HR 7143 = HD 175640*

The CP stars that show the most obvious departures from a nuclear abundance pattern are the CP3, or Mercury-Manganese stars. Given that the SAD abundance of manganese is roughly 2 dex less than that of iron, a star can show a substantial manganese excess without violating the expectations of nuclear processing in the region of the iron peak. For example, Cowley & Aikman (1975) showed that quasi-equilibrium calculations could lead to a Mn (Z = 25) abundance greater than that of Cr (Z = 24), with Fe (Z = 26) only a few tenths of a dex greater than that of Mn. However, in 53 Tau, Mn is surely more abundant than either Cr or Fe, severely violating an expected nuclear pattern (Smith & Dworetsky 1993). Oddly, the star 53 Tau shows no λ3984 Hg II line.

Sun and 32 Aqr

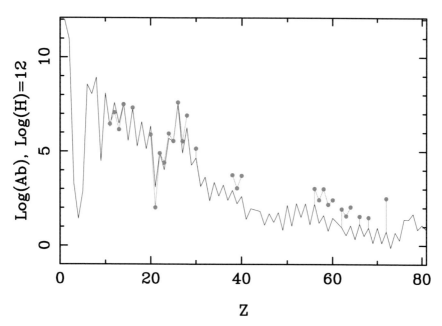

Figure 8. SAD abundances along with those of a quintessential Am star, 32 Aqr (HR 8410, HD 209625). Abundances from Adelman *et al.* (1997) show no obvious odd- or even-Z anomalies. See caption to Fig. 7

Castelli & Hubrig (2004) found that the HgMn star HR 7143 (HD 175640) also has more Mn than either Cr or Fe. Fig. 7 is based on their paper. The odd-Z anomalies at Mn, Ga, Br, Y, Pr, and Au are indicated. The abundance patterns in the HgMn stars provide the most obvious signs of non-nuclear processes.

5.2. *32 Aqr*

We have considered 32 Aqr (HR 8410) to be a quintessential metallic-lined (Am) star. HR 178 and 63 Tau, which have been studied extensively by Van't Veer-Menneret and her colleagues (cf. Van't Veer-Menneret *et al.* 1988, Van't Veer-Menneret 1963), are similar Am stars with relatively sharp spectral lines. The Am (or CP1) stars are not identical, but their compositions are more uniform than those of the magnetic (CP2) sequence.

Michaud (2004) and his coworkers (cf. Richer *et al.* 2000) have made detailed diffusion calculations for stars in the mass range appropriate to the Am and cool CP2 stars. These calculations provide a basis for understanding the lower abundances of the CNO elements, which is generally characteristic of CP stars. Like previous diffusion calculations, they tend to predict abundance enhancements of elements with naturally low abundances, and thus a kind of inverse odd-even effect. While such patterns are realized among the HgMn stars, there is little or no indication of them in the Am stars. This is illustrated in Fig. 8.

5.3. *HD 213637*

Kochukhov (2003) determined abundances for 36 elements in the remarkably cool (T_{eff} = 6400 K) roAp star HD 213637. They are displayed graphically in Fig 9. Abundances for several of the lanthanides have two values, with the higher value coming from the

Sun and HD 213637

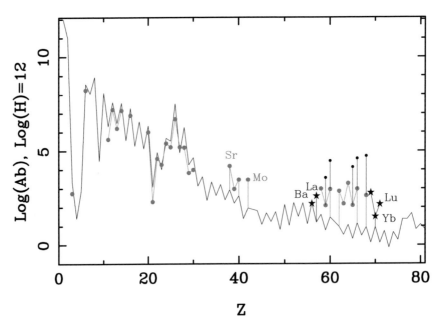

Figure 9. SAD abundances along with those of the cool roAp star HD 213637. Abundances from Kochukhov (2003). See caption to Fig. 7

third spectra. A possible even-Z anomaly is seen at $Z = 70$, where Yb is less abundant than its odd-Z neighbors. This must be confirmed with better oscillator strengths and a definitive model. Kochukhov finds different abundances from weak and strong lines of numerous species, as well as from high- and low-excitation lines. This is attributed to stratification of the elemental abundances.

The point for Ba $(Z = 56)$ could also be an even-Z anomaly, though no abundance is available for Cs $(Z = 55)$. The Ba abundance varies considerably among the CP2 stars, unlike that of Sr $(Z = 38)$, which is typically enhanced.

6. High-energy phenomena associated with CP Stars

6.1. Radio observations

Drake (1998) reviewed the subject of radio continuum observations of magnetic CP stars. We will briefly summarize the principal conclusions from his review, and then touch on developments that have occurred since its publication. We will also consider the status of radio observations of the non-magnetic CP stars.

Examination of magnetic CP stars in the radio regime began in the 1970s at centimeter wavelengths, but failed to produce any evidence for strong radio emission from this class of objects. Out of some two dozen stars surveyed, only Babcock's star (HD 215441) showed a weak 50 mJy (2σ) radio excess (Kodaira & Fomalont 1970). By the end of the 1980s, with the advent of the VLA and similar arrays, five magnetic CP stars (out of 50 investigated) were identified as radio sources, including three He-strong stars and two Si stars (cf. Drake *et al.* 1985, 1987). No later-type SrCrEu magnetic CP stars were detected in these early surveys. Phillips & Lestrade (1988) demonstrated that the radio emission from the high luminosity He-strong stars was nonthermal in origin.

Leone (1991) and Leone & Umana (1993) discovered that the radio emission from σ Ori E (HD 37479) and HR 1890 (HD 37017), two of the first He-strong radio sources identified, varied with the phase of the magnetic field as the stars rotated. The observed modulation suggested that the radio emission arose from a stable co-rotating magnetosphere. Linsky *et al.* (1992) reported detections of three out of nine He-strong stars and 13 out of 38 He-weak/Si-type stars in the Orion OB1 and Sco OB2 associations, but no detections of cooler Ap stars out of 14 sampled. These authors also developed a model for producing the radio radiation involving gyrosynchrotron emission by mildly relativistic, nonthermal electrons in optically thick tori near the magnetic equators of these stars.

Further work by Leone *et al.* (1994) and Drake and his collaborators (cf. Drake 1998) raised the total of radio emitting He- and Si-type CP stars to nearly three dozen. Correlations between the radio luminosity at the source, and fundamental stellar parameters for members of this group led Drake (1998) to the following empirical relation, accurate to ± 0.6 dex:

$$\log(L_R) \approx 15.1 + (6.85 \cdot \log(T_{\mathrm{eff}}/10^4)) + (1.20 \cdot \log(H_{S/\mathrm{kG}})) - (0.60 \cdot \log(P_{\mathrm{rot}}/\mathrm{days})).$$

Here, L_R in in ergs^{-1}Hz^{-1} for observations at 6 cm.

Recent work pertaining to the magnetic CP stars as radio sources has focused on confirming and expanding previous models to explain the radio emission. For example, Trigilio *et al.* (2004) have developed a 3D model in which they substantiate quantitatively the qualitative models proposed by Linsky *et al.* (1992) and others by reproducing the observed modulation in the 2 cm light curves for σ Ori E and HR 1890. These authors emphasize the need for multi-frequency radio light curves to further test and/or constrain their model. Such improvements not withstanding, the number of *bona fide* radio emitting magnetic CP stars is substantial, the mechanism that produces such radiation appears to be largely understood, and there exists a means of predicting, to within reasonable limits, the radio luminosity of magnetic CP stars on the basis of their fundamental properties.

Searches for radio emission from non-magnetic CP stars (Am and HgMn stars) using the VLA by Drake *et al.* (1994) have proved fruitless. Twenty-three stars (11 Am's and 12 HgMn's) were surveyed at 3.6 cm for nonthermal radio flux and none were found to emit with an upper limit of $\leqslant 0.20$ mJy. This result corroborated earlier efforts to detect radio emission from Am and HgMn stars conducted by Linsky *et al.* (1992) and White *et al.* (1993). The non-detection of these stars at radio wavelengths may not be too surprising if they follow an empirical relation like that quoted above, given their relatively cool temperatures and weak magnetic fields.

6.2. *X-ray observations*

Among the chemically peculiar stars, the magnetic CP stars (CP2s) have provided obvious targets for observation in the X-ray regime going back to the era of the *Einstein* and EXOSAT missions. Drake et al. (1987) discovered two X-ray sources, both early type He-strong stars, with emission rates comparable to those of hot, non-magnetic OB stars and concluded that the emitting regions in these sources are related to stellar winds and not to the presence of magnetospheres. The launch of ROSAT in 1990 provided detection sensitivities 10 to 100 times greater than previous X-ray observatories, and studies of the ROSAT All-Sky Survey by Drake *et al.* (1994) yielded ten detections out of 100 magnetic CP star positions examined. Five of the sources possessed properties indicating that the X-rays originated from a stellar wind-type mechanism and were unrelated to the magnetic peculiarities of the host star. Of the remaining five sources, Drake and coworkers concluded that two He-weak and two Si stars *may* represent *intrinsic* X-ray sources whose emission is of magnetospheric origin. Their final source, the SrCrEu star 52 Her, is a triple star system wherein the the X-ray emission can plausibly be associated with one of the other cooler components. A similarly low detection rate for magnetic CP stars as X-ray emitters was found by Leone (1994) in his re-examination of *Einstein* images: in a sample of 90 sources associated with B-type stars, only four were known magnetic CP stars, and of those, three were identified as binary systems.

Drake (1998) has extended the program of Drake *et al.* (1994) to include archival, pointed ROSAT data and has culled out an additional 18 magnetic CP star candidates with intrinsic X-ray emission. Included in this group is Babcock's star, as well as another 13 He-weak and Si stars, plus four SrCREu-type objects. Despite the increase in potential X-ray emitting CP stars, Drake finds no correlation between the X-ray emission characteristics of these stars and any other of their known physical properties. This is in striking contrast to the radio emission from magnetic CP stars which is predictable from their magnetic field strength, temperature and rotation. Drake concludes that the case for X-ray emission from CP2 stars via magnetospheric activity (as opposed to stellar wind-induced effects) has yet to be persuasively made.

Observations of galactic clusters and associations have also revealed some CP2 stars as potential X-ray emitters. ROSAT investigations of the young open cluster NGC 2516 were made by Dachs & Hummel (1996) and by Jeffries *et al.* (1997) that identified 4 to 6 magnetic CP cluster stars as X-ray sources. Jeffries and coworkers concluded that the emission was most probably intrinsic to the CP stars, although contributions from an unseen companion in each case could not be ruled out. *Chandra* X-ray observations of NGC 2516 by Damiani et al. (2003) have yielded a detection rate for dwarf A stars as X-ray emitters of only about 21% (12 out of 58) as a whole, but fully one-half of those discovered are chemically peculiar. Within the class of CP stars alone, the detection fraction is very high: six or seven out of eight. The observed X-ray luminosity function for the A star detections is compatible with those found for later-type stars, however, indicating that the X-ray emission could be associated with lower-mass companions.

Padovani (2005), using the extensive data handling and analysis capabilities of the Virtual Observatory, has found that CP A-type stars are 3 to 4 times more likely to be associated with X-ray sources than normal A-type stars, but that the detection rate for the magnetic CP A-types is comparable to that of all CP A-type stars. He concludes that the presence of a magnetic field does not play a significant role in triggering X-ray emission in this class of star. This result, together with characteristic X-ray luminosity function for the A-star detections, tends to corroborate Drake's (1998) assessment of the situation regarding the CP2 stars as intrinsic X-ray emitters.

The roAp stars may hold special promise for elucidating the potential for intrinsic X-ray emission within the CP2 class. In addition to the clear presence of a magnetic field (see, e.g., Mathys *et al.* 1997 and Hubrig *et al.* 2004), these stars seem to possess shallow convection zones that are instrumental in exciting their pulsation. Moreover, none of the 32 or so known roAp stars has been proven to be a member of a binary system (Hubrig *et al.* 2000), so that any detected X-ray emission is likely to be associated with the magnetic star itself. Stelzer (2004, private communication) and colleagues have proposed to make *Chandra* observations of five roAp stars in an effort to address this and related questions.

Among the non-magnetic HgMn (or CP3) stars, eleven were detected as soft X-ray emitters in the ROSAT All-Sky Survey by Berghöfer *et al.* (1996). Hubrig *et al.* (1998) carefully examined each of these systems using all available data and concluded that in every case, the X-ray emission had a natural explanation in terms of a companion star: for seven stars, the X-ray emitter was likely a late-type main sequence star, while in four instances, a pre-main sequence object was required. Thus, like their cooler CP2 congeners, the HgMn stars do not appear to be intrinsic X-ray emitters.

6.3. *Surface nuclear reactions in Ap stars*

Recently, Cowley *et al.* (2004) have reported the identification of promethium in the spectra of HD 101065 and HD 965. Like an earlier report of Pm in the spectrum of HR 465 (Aller & Cowley 1970), this one again raises questions about the possible source(s) of this short-lived ($T_{1/2} \leqslant 17.7$ yrs) radioactive element. Conventional theories of nucleosynthesis fail to provide plausible mechanisms for the existence of observable amounts of Pm in the atmospheres of unevolved stars. This has led to the consideration of nuclear reactions occuring on the surfaces of Ap stars as a means of producing measurable quantities of this species. Although scenarios involving such reactions are generally believed to require larger energy inputs than are sustainable by typical stars to work (cf. Tjin A Djie *et al.* 1973), Cowley (1973) demonstrated that some models, for example, ones involving Nd(p,n)Pm reactions, could be made viable, assuming Pm/Nd abundance ratios of 10^{-2} and 10 MeV protons with energy flux densities of only $\approx 5 \cdot 10^9$ erg cm^{-2} s^{-1} over times of the order of 10 years. Padovani's (2005) discovery that Ap stars are 3 to 4 times more likely to be associated with X-ray sources than normal A stars suggests that high energy events commonly occur in the vicinity of (if not on) many Ap stars. Such events may provide the necessary flux of high energy particles (p's, n's, and α's) to initiate nuclear reactions in their atmospheres. (In studies of 19 solar flares, Ramaty *et al.* (1995) found that the energy associated with expelled particles equalled or exceeded that of the nonrelativistic electrons that produce the hard X-rays in these events.) Similar arguments have been made for the production of Li in low mass secondaries by CNO spallation reactions caused by high energy protons generated in transient X-ray events from the compact object in these systems (Martín *et al.* 1994a,b).

In the light of this possibility, we review some recent results from solar and stellar flare studies. To provide order-of-magnitude estimates, we will assume the Ap star to be part of an X-ray emitting binary with a 10-day period. The value adopted for the period is a compromise choice: periods for CP3 binaries are generally less than about 5 days, while most binaries containing CP2 stars have periods greater than 10 days (Preston 1974). Although we will also focus our attention on proton-induced reactions only, it must be said that other production reactions exist.

Neutron addition to stable Nd isotopes to produce Nd-147 and/or Nd-149 that then beta decay to Pm is a possible production mechanism for this exotic species. However, the particle densities in the atmospheres of low mass X-ray stars are sufficiently high to effectively thermalize the neutrons emitted in the high energy events. In this case, the neutrons are consumed primarily in reactions producing deuterium in H(n, γ)D reactions or tritium according to ^{3}He(n, p)T (Hua & Lingenfelter 1987), resulting in escaped neutron fluxes that are several orders of magnitude smaller than those of the protons. As Audouze (1970) remarks: "The action of thermal neutrons is negligible in the alteration of abundances of elements heavier than helium."

Promethium production by α-addition on Pr-141 (the only stable isotope of the element) to yield Pm-143 is also a possibility. However, the α/p ratio in typical Sun-like flares is only 0.1-0.5 for energies greater than about 1 MeV per nucleon (MacKinnon & Toner 2003). For particle energies in the range 10-50 MeV per nucleon, the smaller flux of α's, coupled with the larger Coulomb barrier to α reactions, makes this scenario of less importance than one involving protons.

Leya *et al.* (2003) have modeled the production of short-lived nuclides by solar energetic particles, adopting a proton fluence of 10^{23} cm^{-2}. Assuming a typical flare area of about 10^{17} cm^2 (Paesold *et al.* 2003), the mean proton fluence over the solar surface is $1.6 \cdot 10^{17}$ cm^{-2}. Ten large flares per year then yield an average proton flux of about $1.6 \cdot 10^{18}$ cm^{-2} yr^{-1}. For a CP binary with a 10-day period, the proton flux in the neighborhood of the Ap star will be $\approx 3 \cdot 10^{15}$ cm^{-2} yr^{-1}. Over ten years, the proton fluence will be $\approx 3 \cdot 10^{16}$ cm^{-2}. It could be much larger if the protons were guided along magnetic field lines.

Following Audouze's (1970) methods for estimating the required proton fluence to produce heavy and rare earth elements by surface reactions, we adopt a mean inverse formation cross-section for production by 50 MeV protons to be $10^{35} - 10^{36}$ cm^{-2} (cf. Andouze's Table 3.). Taking the Nd-to-H ratio to be $3 \cdot 10^{-11}$ in the Sun (Grevesse & Sauval 1998), we find a mean proton fluence for Nd formation in the Sun of $3 \cdot 10^{24-25}$ cm^{-2}. Assuming a Pm-to-Nd ratio of 10^{-2} and noting that Nd overabundances in Ap stars can reach 10^3 or more relative to the Sun, we predict formation fluences for observable amounts of Pm to be in the range $3 \cdot 10^{23-24}$ cm^{-2}. This is 10^{7-8} times larger than the 10-year proton fluence calculated above.

Proton fluences from flare activity on young stellar objects (YSOs) can be as much as 10^5 times greater than on the Sun (Feigelson *et al.* 2002), so that if the companion to the Ap star were a YSO, then the available proton fluence might be as high as about $3 \cdot 10^{21}$ cm^{-2}. Alternatively, if the magnetic fields associated with the flare regions were 1000 times greater than those on the Sun, then the expected fluence might be as much as a factor of 10^6 larger or about $3 \cdot 10^{22}$ cm^{-2}. In either of these cases, if the Nd abundance were to be a factor of 10 greater than assumed above (as it is in the case of HD 101065 [Cowley *et al.* 2000]), then it may just be marginally possible to produce measurable quantities of Pm in this manner.

From an evolutionary standpoint, it is difficult to reconcile the presence of a YSO in the company of a Main Sequence Ap star, although it has been shown that T Tauri stars possess multipolar magnetic fields like the Sun only with greatly enhanced strengths and surface coverage (Valenti & Johns-Krull 2001) as needed to make heavy element production possible in the manner described above. Thus, the validity of this mechanism, even for stars as unusual as HD 101065, remains in question, and further work on this problem will be required to satisfactorily account for the characteristics of stars like HD 101065.

Acknowledgements

We would like to acknowledge the useful comments from many colleagues. S. J. Adelman, W. P. Bidelman, and T. A. Ryabchikova have provided us with help and insights into the CP star puzzle for many years.

References

Adelman, S. J. 1994, *MNRAS* 266, 97.
Adelman, S. J., Ryabchikova, T. A. & Davydova, E. S. 1998, *MNRAS* 297, 1.
Adelman, S. J., Caliskan, H., Kocer, D. & Bolcal, C. 1997, *MNRAS* 288, 470.
Adelman, S. J., Snow, T. P., Wood, E. L., *et al.* 2001, *MNRAS* 328, 1144.
Aller, M. F. & Cowley, C. R. 1970, *ApJ* 162, L145.
Audouze, J. 1970, *A&A* 8, 436.
Audouze, J. & Tinsley, B. M. 1976, *ARA&A* 14, 43.
Bearman, P.W. & Graham, J.M.R. 1980, *J. Fluid Mech.* 99, 225.
Berghöfer, T.W., Schmitt, J.H.M.M. & Cassinelli, J.P. 1996, *A&AS* 118, 481.
Bidelman, W. P. 1966, in *Abundance Determinations in Stellar Spectra, IAU Symp. 26* ed. H. Hubenet (New York: Academic Press) p. 118.

Bidelman, W. P. 1967, in *The Magnetic and Related Stars* 1967, R. C. Cameron, ed. (Baltimore: Mono Book Corp.) p. 29.

Biémont, E. 2004, in *8th International Colloquium on Atomic Spectra and Oscillator Strengths for Astrophysical and Laboratory Plasmas*, to appear in *Phys. Scr.*, T Ser.

Castelli, F. & Hubrig, S. 2004, *A&A* in press.

Conti, P. S. 1965, *ApJS* 11, 47.

Cowley, C.R. 1973, *The Observatory* 93, 195.

Cowley, C. R. 1976, *ApJS* 32, 631.

Cowley, C. R. & Aikman, G. C. L. 1975, *ApJ* 196, 521.

Cowley, C.R., Bidelman, W.P., Hubrig, S., *et al.* 2004, *A&A* 419, 1087.

Cowley, C. R., Hubrig, S., Ryabchikova, T. A., *et al.* 2001, *A&A* 367, 939,

Dachs, J. & Hummel, W. 1996, *A&A* 312, 818.

Damiani, F., Flaccomio, E., Micela, G., *et al.* , 2003 *ApJ* 588, 1009.

Drake, S.A. 1998, *Contrib. Astron. Obs. Skalnaté Pleso* 27, 382.

Drake, S.A., Abbott, D.C., Linsky, J.L., Bieging, J.H. & Churchwell, E. 1985, in *Radio Stars*, eds. R.M. Hjellming & D.M. Gibson (D. Reidel: Dordrecht), p. 247.

Drake, S.A., Abbott, D.A., Bastian, T.S., *et al.* , 1987, *ApJ* 322, 902.

Drake, S.A., Linsky, J.L., Schmitt, J.H.M.M. & Rosso, C. 1994, *ApJ* 420, 387.

Drake, S.A., Linsky, J.L. & Bookbinder, J.A. 1994, *AJ* 108, 2203.

Erspamer, D. & North, P. 2003, *A&A* 398, 1121.

Feigelson, E.D., Garmire, G.P. & Pravdo, S.H. 2002, *ApJ* 572, 335.

Feldman, U. & Widing, K. G. 2003, *Sp. Sci. Rev.* 107, 665.

Grevesse, N. & Sauval, A. J. 1998, *Sp. Sci. Rev.* 85, 161.

Guthrie, B. N. G. 1971, *Ap&SS* 13, 168.

Harkins, W. D. 1917, *J. Am. Chem. Soc.* 39, 856.

Hubrig, S., Berghöfer, T.W. & Mathys, G. 1998, *Contrib. Astron. Obs. Skalnaté Pleso* 27, 464.

Hubrig, S., Kharchenko, N., Mathys, G. & North, P. 2000, *A&A* 355 1031.

Hubrig, S., Kurtz, D.W., Bagnulo, S., *et al.* 2004, *A&A* 415, 661.

Hua, X.-M. & Lingenfelter, R.E. 1987, *ApJ* 319, 555.

Jaschek, M., and Jaschek, C. 1974, *Vistas in Astron.* 16, 131.

Jeffries, R.D., Thurston, M.R. & Pye, J.P. 1997, *MNRAS* 287, 350.

Kodaira, K. & Fomalout, E.B. 1970, *ApJ* 161, 1169.

Kochukhov, O. 2003, *A&A* 404, 669.

Kochukhov, O., Bagnulo, S. & Barklem, P. S. *Modelling of Stellar Atmospheres, IAU Symp. 210*, N. Piskunov, W. W. Weiss & D. F. Gray, eds. (Provo Utah: ASP-CS & IAU), D17.

Kupka, F., Piskunov, N., Ryabchikova, T. A., Stempels, H. C. & Weiss, W. W. 1999, *A&AS* 138, 119.

Kurtz, D.W. 1990, *ARA&A* 28, 607.

Kurtz, D.W. 2003, *Ap&SS* 284, 29.

Kurucz, R. L. 2004, http://kurucz.harvard.edu.

Leckrone, D. S., Proffitt, C. R., Wahlgren, G. M. *et al.* 1999, *AJ* 117, 1454.

Leone, F. 1991, *A&A* 252, 198.

Leone, F. 1994, *A&A* 286, 486.

Leone, F. & Umana, G. 1993, *A&A* 268, 667.

Leone, F., Trigilio, C. & Umana, G. 1994, *A&A* 283, 908.

Leya, I., Halliday, A.N. & Wieler, R. 2003, *ApJ* 594, 605.

Linsky, J.L., Drake, S.A. & Bastian, S.A. 1992, *ApJ* 393, 341.

MacKinnon, A.L. & Toner, M.P. 2003, *A&A* 409, 745.

Martín, E.L., Spruit, H.C. & van Paradijs, J. 1994a, *A&A* 291, L43.

Martín, E.L., Rebolo, R., Casares, J. & Charles, P.A. 1994b, *ApJ* 435, 791.

Mason, B. 1966, *Principles of Geochemistry*, 3rd. ed. (New York: John Wiley), Table 3.3.

Mathys, G., Hubrig, S., Landstreet, J.D., *et al.* 1997, *A&AS* 123, 353.

Michaud, G. 2005, *These Proccedings*, 173

Padovani, P. 2005, *These Proccedings*, 485

Paesold, G., Kallenbach, R. & Benz, A.O. 2003, *ApJ* 582, 495.

Piskunov, N., Weiss, W. W. & Gray, D. F., eds. 2003, *Modelling of Stellar Atmospheres, IAU Symp. 210* (Provo Utah: ASP-CS & IAU).

Phillips, R.B. & Lestrade, J.-F. 1988, *Nature* 334, 3239.

Preston, G.W. 1974, *ARA&A* 12, 257.

Ramaty, R., Mandzhavidze, N., Kozlovsky, B. & Murphy, R.J. 1995, *ApJ* 455, L193.

Richer, J., Michaud, G. & Turcotte, S. 2000, *Ap. J.* 529, 338.

Ryabchikova, T. A., Malanushenko, V. P. & Adelman, S. J. 1999, *A&A* 351, 963.

Ryabchikova, T. A., Zakharova, L. A. & Adelman, S. J. 1996, *MNRAS* 283, 1115.

Ryabchikova, T. A., Adelman, S. J., Weiss, W. W. & Kuschnig, R. 1997, *A&A* 322, 234.

Ryabchikova, T. A., Nesvacil, N., Weiss, W. W., Kochukhov, O. & Stütz, Ch. 2004, *A&A* 423, 705.

Smith, K. C. & Dworetsky, M. M. 1993, *A&A* 274, 335.

Suess, H. E. & Urey, H. C., 1956, *Rev. Mod. Phys.* 28, 53

Tjin A Djie, H.R.E., Takens, R.J. & van den Heuvel, E.P.J. 1973, *Astrophys. Letters* 13, 215.

Trigilio, C., Leto, P., Umana, G., Leone, F. & Buemi, C.S. 2004, *A&A* 418, 593.

Underhill, A. B. 1966, in *Abundance Determinations in Stellar Spectra, IAU Symp. 26* H. Hubenet, ed. (New York: Academic Press) p. 118.

Unsöld, A. 1955, *Physik der Sternatmosphären*, 2nd ed. (Berlin: Springer-Verlag), see p. 433.

Valenti, J. & Johns-Krull, C. 2001, in *Magnetic Fields Across the Hertzsprung-Russell Diagram*, eds. G. Mathys, S.K. Solanki, & D.T. Wichramasinghe, ASP Conf. Ser. 248 (San Francisco: ASP), p. 179.

Van't Veer-Menneret, C. 1963, *Ann. d'Ap.* 26, 289.

Van't Veer-Menneret, C., Burkhart, C. & Coupry, M. F. *A&A* 203, 123.

Wahlgren, G. M., Dolk, L., Kalus, G., Johansson, S. & Litzén, U. 2000, *ApJ* 539, 908.

White, S.M., Jackson, P.D. & Kundu, M.R. 1993, *AJ* 105, 563.

Wolff, S. C. 1983, *The A-Stars: Problems and Prospectives* (Washington DC: NASA).

Discussion

MACERONI: Cluster Analysis Algorithms are usually applied in a new parameter space derived from Principal Component Analysis on the dataset (a space whose axes contain a decreasing amount of the sample variance going from the first to the last). The physical meaning of the clustering is sometimes trivial, sometimes rather difficult to extract. What is in your case the main physical parameter(s) producing the main branches of your dendrogram?

COWLEY: This is a very interesting idea. We must try it! In the current work, which was never intended to be definitive, we simply considered each abundance as variable. Therefore, clusters that are close just mean the abundances are similar. We filled in missing data with "solar" values. There are other ways of dealing with data. Also, there are a variety of clustering algorithms that could be explored. Especially for the HgMn stars, it would be interesting to try to identify the different clusters or classes with stages in the development of anomalies.

KURTZ: Have you tried applying your cluster analysis to say Georges' theoretical abundances to see if the clusters look the same as the real stars, are there any examples you (can describe?) How (about) doing the theoretical data?

COWLEY: What I could say immediately from looking at the theoretical calculations is that you can see for example that when you look at some of those abundance patterns done at a fixed time with a given set of mixing parameters you get the same family that you would get if you consider one mixing parameter for different groups of ages. In other words, you can simulate the effect of a star aging by just having a different mixing parameter. So in that sense the cluster analysis would show you that the two methods are giving you the same or similar results. In many cases, when you look at these things, most of what you see are things that you already know. And that should you give some confidence that you are not doing something wrong. But you see a few new things. Now it could be that some other person would say "that's not anything new, I knew that all along." But some results may still be new to me, and therefore valuable for me to recognize. What I just described is an example of a case when we would see something we already knew.

VAUCLAIR: I was surprised to see so many gaps in the abundance determinations in well known stars like HR 7143, ξ Lup or 32 Aqr.

COWLEY: Yes, the gaps are unfortunate. Usually it means that there are no usable lines in the regions studied. I think that abundance workers should make a spcial effort to try to fill some of these gaps.

MATHYS: In the dendogram with various types of CP stars, where Przybylskyi's star stands out, is HD 965 included?

COWLEY: No. But some stars are certainly more closely related to HD 101065 than in those shown.

MATHYS: Could you comment about the possible presence of unstable isotopes in the atmospheres of CP stars?

COWLEY: I ran out of time. We have found evidence for some very surprising elements. Dr. Bidelman has a poster that you may peruse . For Pm II, the sepctroscopic evidence is quite strong. However, if you wish to say it is impossible, a reasonable possibility is that the atomic line lists, we used several, are contaminated with lines from other elements. The laboratory spectroscopists tell me that is unlikely. Still, it seems the easiest way for me wiggle out of an otherwise very interesting result.

ALECIAN: This is rather a comment. The building of abundance anomalies by diffusion is a non-liear process. We are not sure than a time-dependent diffusion will lead to a sationary solution (this is at least true for HgMn stars). Several stars with the same properties could then exhibit various abundance patterns. Therefore, methods of classification only based on abundance patterns are perhaps inadequate.

COWLEY: This is a comment rather than a question.

The A-Star Puzzle
Proceedings IAU Symposium No. 224, 2004
J. Zverko, J. Žižňovský, S.J. Adelman, & W.W. Weiss, eds.

© 2004 International Astronomical Union
DOI: 10.1017/S1743921304004661

Observations of magnetic CP stars

T. Ryabchikova[1,2]

[1]Institute of Astronomy, Russian Academy of Science, 48 Pyatnitskaya str., 119017 Moscow, Russia
email:ryabchik@inasan.rssi.ru

[2]Institute for Astronomy, Vienna University, Türkenschanzstraße 17, A-1180 Vienna, Austria

Abstract. Important results of magnetic CP star research in the last decade from photometric and spectroscopic observations are discussed. Hipparcos parallaxes confirm CP stars are Main Sequence stars. Photometric monitoring of the rapidly rotating stars provides evidences for the rotational braking on the Main Sequence. High signal-to-noise spectra with high resolution and time-resolution give strong support for chemical separation processes operating in stellar atmospheres (abundance stratification). There are also observational evidences for the departure of the temperature structure of cool CP star atmospheres from that for normal stars.

Keywords. Stars: photometry, stars: spectroscopy, stars: chemically peculiar.

1. Introduction

Due to the growing power of modern telescopes on Earth and in space astronomers now have an opportunity to study the phenomenon of magnetic (and nonmagnetic) chemically peculiar (CP) stars in greater detail. We penetrate deeper into the Ap star puzzle. The development of observational techniques and research methods permit us to solve old problems and contradictions, providing us at the same time with new ones. In the next sections I will try to review briefly the latest important results obtained from the observations of magnetic CP (CP2) stars.

2. Evolutionary status of magnetic CP stars from observations

The analysis of Hipparcos parallaxes for CP stars (Gomez *et al.* 1998) showed that these stars are uniformly distributed across the whole width of the Main Sequence (MS), and their distribution does not differ from normal stars. Based also on Hipparcos data, but exploiting a specifically chosen group of CP2 stars (stars with $M \leqslant 3 M_\odot$, with resolved Zeeman Splitting in non-polarized spectra, and with small rotational velocities), Hubrig *et al.* (2000a) concluded that a star posesses a substantial magnetic field and chemical peculiarites only after it has completed at least 30% of its MS lifetime. An obvious contradiction between the two studies was successfully overcome by the recent investigations of the CP2 stars in open clusters. Pöhnl *et al.* (2003) on the basis of Δa photometry classified a few CP stars in young open cluster NGC 2516 (log $t \approx 8.1$), and Bagnulo *et al.* (2003) have measured a strong magnetic field $B_s = 14.5$ kG in one of these stars, HD 66318, which has completed less than 16% of MS lifetime. The survey of magnetic fields in stars in open clusters with the FORS1 multi-object spectrograph revealed another supermagnetic ($B_\ell = -9$ kG) CP star NGC 2244-334 in the Rosette Nebula cluster, which has completed less than 3% of its MS lifetime (Bagnulo *et al.* 2004). These direct observations show that a star reaches ZAMS with slow rotation and significant magnetic field in agreement with the scenario proposed by Stępień (2000) to explain the slow rotation of CP2 stars, where magnetic fields in Bp-Ap stars are already present in the pre-Main Sequence phases.

3. Period changes

What is going on with a magnetic CP star on the MS ? Does it undego the further magnetic braking? According to scenario by Stępień (2000) only stars with large
magnetic fields become slow rotators, and they loose their angular momentum before
the ZAMS. However, there are stars with typical CP characteristics, slow rotation and
very weak magnetic fields which are already finished their MS lifetime. One example is
HD 204411 (Ryabchikova et al. 2005) in which magnetic field does not exceed by much its
equipotential value. The star might originally be a slow rotator, or seen pole-on, or it lost
somehow its angular momentum during the MS stage. Although statistical investigations
of rotational velocities in A stars give evidence for the conservation of angular momentum
throughout the whole MS lifetime (see, e.g., Hubrig et al. 2000b), at present we have two
CP2 stars, CU Vir=HD 124424 and 56 Ari=HD 19832 with observed retardations of their
rotation. Both are among the fastest rotators of the magnetic CP stars. CU Vir suddenly
changed its rotation period P=0.52 day by only 2 sec in 1984 (Pyper et al. 1998). This
is sufficient to cause a half period phase change ten years later. The second star, 56
Ari, P=0.72 day, demonstrated a constant retardation by 2 sec over 100 years (Adelman
et al. 2001). With such an angular momentum loss, a star will be a typical slow rotating
CP star by the end of its MS lifetime. Both studies were based on careful analysis of
all photometric and spectroscopic observations over 40 years, and were successful due to
photometric monitoring with the FCAPT (Four College Automatic Photometric Telescope). Is the observed loss of the angular momentum typical for other rapidly rotating
CP2 stars? Photometric monitoring of short-period CP stars with automatic telescopes
will provide very important information for the evolutionary studies of CP stars (see also
Pyper & Adelman 2005).

4. Spectroscopy: abundance stratification

Analyses of the high resolution spectroscopic observations of the magnetic CP stars
in wide spectral regions clearly show the presence of abundance stratifications in their
atmospheres. From the early 1970s a competition between radiative acceleration and
gravitational settling in stabilized atmosphere has been considered as the primary mechanism responsible for the observed abundance anomalies in CP stars (Michaud 1970) but
detailed abundance startification studies began with the pioneer work by Babel (1992).
His theoretical diffusion calculations showed that abundance profile may be represented
by a step function with the steep abundance gradient. All empirical abundance stratification studies are based on this simple model (Babel (1992) 53 Cam, Ryabchikova
et al. (2002) γ Equ, Wade et al. (2003) β CrB, Ryabchikova et al. (2005) HD 133792,
HD 188041, and HD 204411). I plotted Cr and Fe empirical stratifications in these five
stars in Fig. 1. The abundance jump occurs in a rather narrow range of optical depths
$-1.5 \leqslant \tau_{5000} \leqslant -0.5$, and the position of this jump moves upwards with temperature.
Abundance limits are also greater for higher temperatures. There is only a small difference in abundance startifications for stars with similar temperatures and very different
magnetic field strength (HD 188041 and HD 133972, see Ryabchikova et al. 2005). However, significant differences exist for stars with comparable magnetic field strengths, but
with different temperatures (γ Equ and HD 188041). The process of element separation
in stellar atmospheres is governed by the temperature rather than by the magnetic field.

Five stars, of course, is insufficient for statistics, but we may look at the whole picture
using averaged abundances which are available for many more stars and which roughly
represent element abundances near the middle of abundance jump. The distribution of
averaged Cr and Fe abundances with effective temperatures for different group of A stars

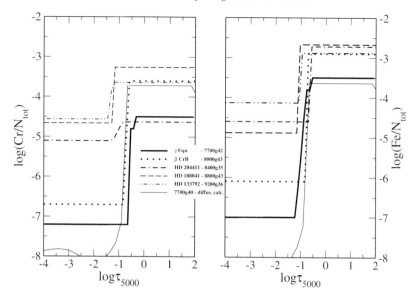

Figure 1. Element distribution in the atmospheres of Ap stars and comparison with self-consistent diffusion model predictions.

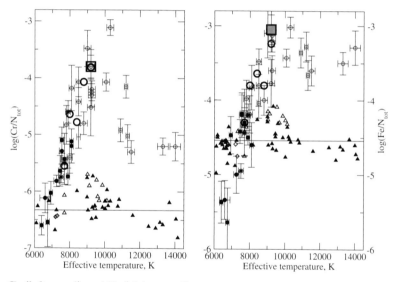

Figure 2. Cr (left panel) and Fe (right panel) average abundances versus effective temperature for 13 roAp stars (filled circles), 23 non-pulsating Ap stars (open circles) and 4 spectroscopic roAp candidates: HD 29578, HD 75445, HD 62140, and HD 115708 (asterisks). Solar abundances are indicated by a horizontal line. Five stars for which stratification analyses are available are shown by large open circles, and HD 66318 by a large open square.

are shown in Fig. 2. I add 3 additional stars to this plot, taken from Ryabchikova *et al.* (2004a - Fig. 5), HD 133792, HD 204411 and HD 66318. For the last star with a very large magnetic field 14.5 kG, abundance data are taken from Bagnulo *et al.* (2003). The five stars with abundance stratification analyses are shown by large open circles, and HD 66318 by a large open square. All six stars follow the overall distribution, where CP stars with very different magnetic fields are plotted. Fig. 2 seems to support the main role of the temperature in the process of element separation in CP star atmospheres.

Stratification analyses of the CP stars with a wider range of effective temperatures and magnetic field strengths is required for a better understanding the separation processes.

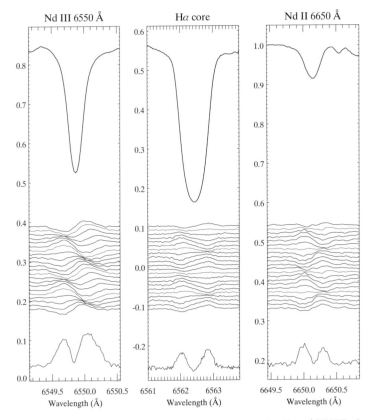

Figure 3. Pulsational line provile variations in spectrum of γ Equ (CFHT observations, 2001). Different behaviour of the residual spectra are easily seen.

Empirical abundance stratifications derived in CP stars are supported by the results of the first self-consitent model calculations, which include radiative diffusion processes for 39 chemical elements (LeBlanc 2003, Ryabchikova *et al.* 2003, LeBlanc & Monin 2005). We simultaneously get the abundance stratifications and the atmospheric model, calculated with stratified chemical composition. Model Cr and Fe distributions discussed by Leblanc & Monin (2005) are plotted in Fig. 1 together with the empirical distributions. The model confirms the realibility of the empirical abundance distributions. Thus empirically derived stratifications are not an artefact, caused by poor knowledge of atmospheric structure in such complex objects as magnetic CP stars. Further we may extend the empirical stratification analysis to the elements not yet calculated. Rare-earth elements (REE) are the most important, because their lines provide extremely important information about the pulsations in the atmospheres of cool CP2 stars (Savanov *et al.* 1999, Kochukhov & Ryabchikova 2001, Ryabchikova *et al.* 2002, Kurtz 2005). In spectra of rapidly oscillating Ap (roAp) stars the REE lines show the largest radial velocity pulsational amplitudes. In some stars only these line are pulsating while the lines of the Fe-peak elements exhibit either constant RV's or pulsate with negligible amplitudes. A joint abundance and pulsational analysis of γ Equ (Ryabchikova *et al.* 2002) permitted us to derive REE abundance distributions with large concentrations of Pr and Nd in the upper atmosperic layers. NLTE Nd calculations presented by Mashonkina *et al.* (2005)

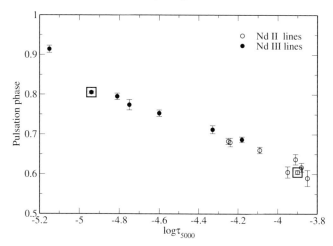

Figure 4. Optical depths of Nd II-Nd III line formation in γ Equ with stratified Nd distribution presented by Mashonkina *et al.* (2005). Nd II $\lambda6650$ and Nd III $\lambda6550$ are marked by large open squares.

demonstrate the importance of NLTE effects for stratification studies of REEs in CP stars.

5. Time-resolved spectroscopy

Time-resolved spectroscopy is an advantage for the study of stellar pulsations, mainly due to higher amplitudes in the RV pulsations, and due to the selective character of the RV pulsations. As mentioned in the previous section, lines of different elements and even different ions of the same element show RV amplitudes which may differ by an order of magnitude. The study of pulsational line profile variations in lines formed at different parts of stellar atmosphere provide a way possibility to understand pulsation wave propagation through the whole atmosphere from the photosphere to the upper layers. Fig. 3 demonstrates significant differences in the pulsation picture observed from the lines of Nd II and Nd III and the similarity between Nd II and the Hα core in γ Equ. According to NLTE Nd calculations in γ Equ (Mashonkina *et al.* 2005), Nd II and Nd III lines are formed in a Nd abundance layer. Therefore any changes in pulsation wave propagation takes place at smaller depths as is illustrated in Fig. 4.

The power of the modern spectrographs allows one to carry out time-resolved Zeeman spectroscopy to search for possible pulsational variations of the magnetic field. The first results for one of the brightest roAp stars, γ Equ, are controversial. Leone & Kurtz (2003) measured 12 min pulsational variations of the longitudinal magnetic field B_ℓ in four Nd III lines with the amplitude $\sim$ 110-240 G, while Kochukhov *et al.* (2004a) put an upper limit 40-60 G for possible B_ℓ amplitude based on the larger set of Nd III lines. The same limit was derived from non-pulsating lines of Fe-peak elements.

Careful measurements of the surface magnetic field (B_s) pulsational variations in the same star show no variations with the amplitude above 5-10 G (Kochukhov *et al.* 2004b). This conclusion was supported by Savanov *et al.* (2005).

6. Lithium problem in cool CP stars

An alternative identification of the famous spectral feature at $\lambda6708$ as Ce II $\lambda6708.099$ instead of to Li I successfully resolved a problem of Li in atmospheres of post-AGB

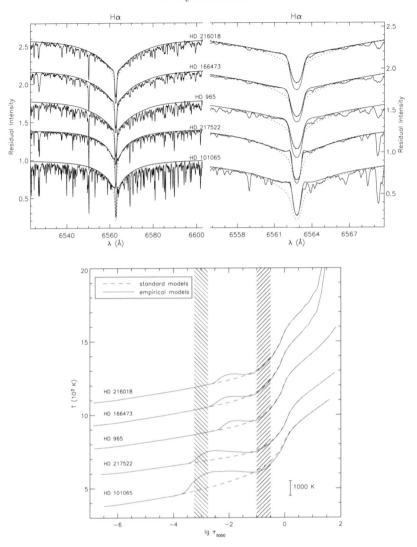

Figure 5. CWA effect in the cool roAp stars (top) and its explanation (bottom), from Kochukhov *et al.* (2003). The dashed line represents the standard model atmosphere while the full line shows the empirically derived temperature distribution. The left shadow region shows the approximate position of the REE jumps in roAp atmospheres obtained in NLTE approximation (see Mashonkina *et al.* 2005) while the right shadow region shows the position of Fe-peak abundance jumps derived empirically.

stars (Reyniers *et al.* 2002). It is tempting to replace Li by Ce in CP stars with their known large overabundances of REE (Nesvacil & Hubrig 2005). Indeed, in some stars, for example, in HD 217522 ($T_{\mathrm{eff}} = 6750$ K) the observed feature is entirely due to Ce II while in many other stars careful spectral synthesis shows that Li II remains the dominant contributer to the observed feature (Shavrina *et al.* 2005). It is also tempting to replace the Li I identification by any other unclassified line of REE because Li tends to be concentrated in spots at magnetic poles as REE do (Polosukhina *et al.* 2005, Kochukhov 2005). But in γ Equ which shows one of the highest RV pulsational amplitudes derived from REE lines, a strong feature at $\lambda6708$ does not show any significant pulsational variations, making doubtful the possible identification of the feature as a REE line.

7. Problems of the modelling of magnetic CP star atmosphere

Until now analyses of CP stars were based on different versions of ATLAS LTE model atmosphere codes by Kurucz with homogeneous chemical compositions. In Section 4 I showed that cool CP star atmospheres are chemically inhomogeneous. Abundance gradients may alter the thermal structure of stellar atmosphere. Hydrogen line profiles may be used for diagnostics of plasma conditions. The study of high-resolution Balmer line profiles in spectra of a few cool CP stars revealed a departure of the observed profiles from those in normal stars and from the best theoretical ones in the transition region between line wing and line core. Instead of a smooth transition between the Stark wing and the Doppler core observed in normal stars an abrupt change in slope takes place in cool CP stars producing an unusually sharp line cores. The inner part of a Balmer line wing in a CP star is shallower than in the normal stars or in theoretical line profiles. This effect called core-wing anomaly (CWA) was first recognized by Cowley *et al.* (2001). Kochukhov *et al.* (2002) calculated empirical models where a temperature increases at intermediate atmospheric layers $-4 \leqslant \log \tau_{5000} \leqslant -1$ to obtain a good fit to the inner and outer wings as well as to the sharp cores of Hα and Hβ lines. Fig. 5 taken from Kochukhov *et al.* (2002) shows the results of their empirical modelling. I add to this plot an approximate range of the optical depths where empirically derived abundance jumps for Fe-peak and rare-earth elements take place. Observed bumps in the empirical temperature distributions fall between these jumps. Thus abundance gradients are responsible for the temperature bumps. The first self-consistent diffusion model atmosphere calculations (see LeBlanc & Monin 2005) did not produce the needed temperature bumps, perhaps because of the lack of REEs in the diffusion calculations. The magnetic field which is not taken into account in these calculations may also alter the atmospheric structure of a CP star. The deviation of the observed Hα line profile from the theoretical one is observed for hotter CP stars with very large magnetic fields (see, e.g., HD 66318, Bagnulo *et al.* 2003, Fig.4). Here the inner part of the line wing is deeper in the observed profile contrary to that observed in cooler CP stars. Evidently, more elaborated atmosphere models are required to describe the most observed features in spectra of magnetic CP stars.

Acknowledgements

This work was supported by the FWF project *P 14984*, by RFBR (grants 03-02-16342, 04-02-16788) and by the Presidium RAS Programme "Nonstationary phenomena in astronomy".

References

Adelman, S.J., Malanushenko, V., Ryabchikova, T.A., Savanov, I. 2001, *A&A* **375**, 982

Babel, J 1992, *A&A* 258, 645

Bagnulo, S., Landstreet, J.D., Lo Curto, G., Szeifert, T., Wade, G.A. 2003, *A&A* 403, 449

Cowley, C.R., Hubrig, S., Ryabchikova T.A., Mathys, G., Piskunov, N., Mittermayer, P. 2001, *A&A* 367, 939

Bagnulo, S., Hensberge, H., Landstreet, J.D., Szeifert, T., Wade, G.A. 2004, *A&A* 416, 1149

Gomez, A. E., Luri, X., Grenier, S., Figueras, F., North, P., Royer, F., Torra, J., Mennessier, M. O. 1998, *A&A* 336, 953

Hubrig, S., North, P., Mathys, G. 2000a, *ApJ* 539, 352

Hubrig, S., North, P., Medici, A. 2000b, *A&A* 359, 306

Kochuknov, O., Bagnulo, S., Barklem, P. 2002, *ApJ* 578, L75

Kochukhov, O., Ryabchikova, T. 2001, *A&A* 374, 615

Kochukhov, O., Ryabchikova, T., Landstreet, J.D., Weiss, W.W 2004b, *MNRAS* 351, L34, 2004

Kochukhov, O., Ryabchikova, T., Piskunov, N. 2004a, *A&A* 415, L13

Kochukhov, O. 2005, *These Proceedings*, 433

Kurtz, D., Elkin, V.G., Mathys, G., Riley, J., Cunha, M.S., Shibahashi, H., Kambe, E. 2005, *These Proceedings*, 343

LeBlanc, F. 2003, in: L.A. Balona, H. Henrichs, & T. Medupe (eds.), *Int. Conf. on Magnetic Fields in O, B and A Stars, ASP Conf. Ser*, v.305, p.206

LeBlanc, D. & Monin D., 2005, *These Proceedings*, 193

Leone, F., Kurtz, D.W. 2003, *A&A* 407, L67

Mashonkina, L., Ryabtsev, A., Ryabchikova, T. 2005, *These Proceedings*, 315

Michaud, G. 1970, *ApJ* 160, 641

Nesvacil, N., Hubrig, S. 2005, *These Proceedings*, FP21

Póhnl, H., Maitzen, H.M., Paunzen, E. 2003, *A&A* 402, 247

Poloshukina, N., Shavrina, A, Drake, N., *et al.* 2005, *These Proceedings*, FP7

Pyper, D. M., Adelman, S. J. 2005, *These Proceedings*, 307

Pyper, D.M., Ryabchikova, T., Malanushenko, V. Kuschnig, R., Plachinda, S., Savanov, I. 1998, *A&A* 339, 822

Reyniers, M., Van Winckel, H., Biémont, E., Quinet, P. *A&A* 395, L35

Ryabchikova, T., Piskunov, N., Kochukhov, O., Tsymbal, V., Mittermayer P., Weiss, W.W. 2002, *A&A* 384, 545

Ryabchikova, T., Wade, G.A., LeBlanc, F. 2003, in: N.E. Piskunov, W.W. Weiss, & D.F. Gray (eds.), *IAU Symp. 210 Modeling of Stellar Atmospheres*, p. 301

Ryabchikova, T., Nesvacil, N., Weiss, W.W., Kochukhov, O., Stütz, Ch. 2004, *A&A* 423, 705

Ryabchikova, T., Leone, F., Kochukhov O., Bagnulo. S. 2005, *These Proceedings*, DP1

Savanov, I., Malanushenko, V.P., Ryabchikova, T.A. 1999, *Astron. Lett.* 25, 802

Savanov, I., Hubrig, S., Mathys, G., Ritter, A, Kurtz, D. W. 2005, *This meeting*, http://www.ta3.sk/IAUS224/PDF/GP10.pdf

Stępień, K. 2000, *A&A* **353**, 227

Wade, G.A., LeBlanc, F., Ryabchikova, T.A., Kudryavtsev, D. 2003, in: N.E. Piskunov, W.W. Weiss, & D.F. Gray (eds.), *IAU Symp. 210 Modeling of Stellar Atmospheres*, CD-D7

Discussion

ALECIAN: In the stratified profiles you have shown, there is a step and then, a plateau. Is it possible to really check the extention of the plateau toward the deep layers because the line profiles are less sensitive to the deeper layers ?

RYABCHIKOVA: An abundance plateau is assumed in our two-step abundance distribution model. However, the wings of strong lines with a large Stark effect are sensitive to element abundances in deeper layers below $\log \tau_{5000} = -1$ and even -0.5. You can see it in our calculations for Si (Dimitrijević *et al.* 2003, A&A 404, 1099, Fig.7).

MATHYS: Note that in Gomeź *et al.* (1998) less than 20% of stars have parallaxes known with relative accuracies better than 20%. As to the way in which HD 66318 fits in the picture of deficiency of Ap stars close to the ZAMS, see my comment in yesterday afternoon's discussion.

BAGNULO: A comment to a question by Mathys: Probably the major problem with Hubrig *et al.* (2000) is that it is <u>not</u> possible to determine the evolutionary state of field stars with sufficient accuracy. At least, one can state weather the star is "young" or "old". By contrast, our work based on open cluster stars, although based on a poorer statistics, allows us to determine the age with much higher precision. In conclusion, I believe that the example of HD 66318 as well as some other ones not published yet, represents a clear counter-example to the Hubrig *et al.* hypothesis.

The A-star puzzle
Proceedings IAU Symposium No. 224, 2004
J. Zverko, J. Žižňovský, S.J. Adelman, & W.W. Weiss, eds.
© 2004 International Astronomical Union
DOI: 10.1017/S1743921304004673

Observations of nonmagnetic CP stars: Crossing boundaries

Glenn M. Wahlgren

Lund Observatory, University of Lund, Box 43, SE-22100 Lund, Sweden
email: glenn.wahlgren@astro.lu.se

Abstract. Observations of chemically peculiar (CP) stars have been conducted for decades and have revealed a variety of spectrum anomalies, most prominent among them are line enhancements of heavy elements. The earlier observations were limited to the optical region and the use of less sensitive detectors, yet are responsible for much of our current characterization of the CP star phenomenon. More recent observations embrace a wider expanse of the electromagnetic spectrum and employ more sensitive detectors that continue to unveil new levels of spectrum peculiarity. The traditional criteria used to distinguish normal from peculiar stars have become blurred, thus in some sense replacing the concept of peculiarity with one of continuity. This presentation will address the observations of the traditional nonmagnetic CP star groups over the past decade, paying particular attention to new avenues of research that have a bearing upon the interpretation of the atmospheres of CP stars and the origins of this phenomenon.

Keywords. Stars: abundances, stars: chemically peculiar, stars: magnetic fields

---

1. Introduction

Traditionally, the CP stars of the upper main sequence extend from spectral type mid-B at the hotter end through the A type stars and into spectral type F. The inclusion of the roAp stars now extends this range to essentially the start of spectral type G. Traditionally, there are two parallel tracks of stars considered, one magnetic and the other nonmagnetic, with the nonmagnetic chemically peculiar (nmCP) stars being comprised of the HgMn, hot-Am, Am and Fm stars. The case of the He-weak stars is particularly troubling in that many of them display magnetic signatures, thereby making them interlopers between the He-strong and Bp stars. Other He-weak stars have spectral signatures that make them hotter analogs of the HgMn stars. As a class, designated mostly by the nonunique characteristic of a helium deficiency in the observable atmosphere, the He-weak stars serve as a buffer zone between the magnetic and nonmagnetic tracks. Traditionally, we have used low and medium spectral resolution data and have been blind to the plethora of peculiarity that pervades this spectral region. As essentially no star in this region is chemically normal (displaying solar-like chemical composition) the boundaries imposed by classification schemes are becoming blurred by discoveries made from high resolution spectra of peculiarities that appear to *cross boundaries*.

Much progress has been made in this field during the past decade. What can be consider to be the last thorough review of B and A star properties occurred in 1992, at IAU Colloquium No. 138. Subsequent conferences related to stars of these spectral types were for the most part dedicated to the subject of magnetic fields, often with solitary contributions to peripheral topics. It was also during the early to mid 1990s that one starts to see benefits from the changes in observation techniques from using photographic emulsions to a more prominent role of the silicon diode arrays and CCDs.

This quantum leap in detector sensitivity allowed for more subtle details of the spectrum to be recorded, and subsequently encouraged new areas of stellar astrophysics.

From these conferences and the published journal literature one decade ago the topics under study can be categorized as follows: basic properties (colors, fluxes, temperatures, classification), binarity (statistics), spectrum analysis (elemental abundances, isotopic composition, the extension into the UV from *HST* observations, and rumblings about stratification), magnetic fields (presence in nmCP stars), variability (both photometric and spectroscopic), and new wavelength regimes for observations (x-ray, radio). Research has matured in each of these areas over the past decade and is essentially the content of this present review, with emphasis placed on spectroscopic analyses. It is not the purpose of this discussion to be complete in its bibliography, but rather to present examples that have advanced our knowledge and display the wealth of new phenomena that need to be considered to provide a complete picture of the atmospheres of nmCP stars.

2. Binarity

As a result of improvements in the quality of spectra, companions to HgMn stars continue to be identified, and we should expect to find additional cases of binarity among currently known Am and HgMn stars (see, e.g., Iliev *et al.* 2005) Most literature still cites older statistics that claim binarity among HgMn stars to be no more prevalent than among the general field stars, which is to say about 50%. However, this conflicts with published statements (Hubrig & Mathys 1995, Mathys & Hubrig 1995) claiming a lower limit perhaps as high as 67% and the general impression that binarity is more prevalent since companions continue to be discovered. A revision in this number is needed to perhaps clear up any misconceptions. What is known is that HgMn stars have a high presence among SB2 systems and the number of known triple systems is also increasing. Am stars have a high presence in binary systems. Further it has been suggested that HgMn star companions with $T_{\mathrm{eff}} < 10000\,\mathrm{K}$ are Am stars (Ryabchikova 1998).

The study of main sequence B and A star companions to Cepheid variables (Evans 1995) has detected CP phenomena in several of these companions. This work has implications for the onset of CP star characteristics through evolutionary dating of the Cepheid variable. An unexpected result of this work has been the detection of a third star as the close companion to the main sequence B star. To date, the systems SU Cyg (HgMn companion, Wahlgren & Evans 1998), T Mon (Bp companion, Evans *et al.* 1999) and AW Per (B8 V companion, Evans *et al.* 2000) have been studied. Although there is still uncertainty in the CP star nature of the companion to AW Per, it is interesting to consider that each proported CP star is accompanied by a close companion.

A final note on the issue of binarity regards the early detection of X-rays in late B type stars by the *EINSTEIN* and the *ROSAT* satellites. Weak X-ray emission from several late B type stars, including HgMn stars, was suggested to be the result of wind processes. Subsequent X-ray observations obtained at higher spatial resolution from the *CHANDRA* satellite observatory and ground based IR imaging have detected red companions to these binary systems (Stelzer *et al.* 2003), in agreement with the idea of an active late-type companion. The idea of wind induced X-rays in late B-type stars seems now to be a less likely explanation for observed emission.

3. Elemental abundances

Much work has been undertaken in the discipline of elemental abundance determination for nmCP stars. However, the nature of the work has not been the same for the HgMn stars as for the Am stars. For the Am stars the majority of studies published during the past decade address members in stellar clusters with the underlying purpose being to determine the onset of peculiarity, its development with cluster age, the frequency of CP stars in clusters and to provide tests for diffusion theory (see, e.g., Burkhart & Coupry 1997, Hui-Bon-Hoa *et al.* 1997, Hui-Bon-Hoa & Alecian 1998, Savanov 1998). This line of inquiry is more natural for the Am than HgMn stars since the latter are found to be a more rare occurrence in open clusters as a result of their faster pace for evolution. The work on cluster Am stars is by no means complete, as can be seen from presentations at this conference (Monier & Richard, 2005).

The Am star elemental abundance work is being conducted using optical region spectra, which means that the basic picture of the abundance pattern remains the same as it has been for the past several decades. Although there exists a range for abundances of individual elements, in general, a trend exists for increasing abundance with atomic number (Smith 1996) with the enhancements reaching up to two orders of magnitude. The sampling of the periodic table is rather poor for elements heavier than the iron group, with the exception of the lanthanides. High quality ultraviolet data are needed to extend these analyses to elements that may best be detected through ultraviolet transitions. Unfortunately, the low resolution of the *IUE* satellite spectra and the extremely limited data obtained from the *HST* for Am stars will not serve to improve this situation. The importance of filling-in the existing abundance pattern lies in following the trends in abundance to both higher and lower temperatures, i.e. to cross boundaries of classification, and in providing theoreticians with additional constraints for modeling processes of atmospheric dynamics.

In contrast to the Am stars, the abundance analyses for HgMn stars tend to focus on stars as individuals or address specific tasks (isotopic composition) from observations of a group of stars. In this way progress has been made on including additional elements to better define the abundance pattern and in identifying new atmospheric phenomena from the spectra. The ultraviolet spectral region has played a key role in defining the elemental abundance distributions of HgMn stars. *IUE* spectra have been collected for a number of HgMn stars and studied with synthetic spectrum techniques for specific elements. A series of papers by K.C.Smith (e.g., 1995, 1997) address abundances for several elements and searches for trends of abundance with temperature. Similar analyses from coadded *IUE* spectra have also been undertaken by Adelman *et al.* (1993), Wahlgren *et al.* (1993) and Adelman *et al.* (2004) for the heaviest elements. Using optical wavelength data Adelman and collaborators have continued to analyse elemental abundances for bright HgMn, Am, and normal main-sequence B and A type stars.

The HgMn star having the highest fidelity for its elemental abundance distribution is χ Lupi. This star has been the focus of *HST* observations that were conducted during the first half of the 1990s, but continue to reveal new results as new atomic data are produced to analyse the spectrum. The χ Lupi Project (Leckrone *et al.* 1999, and subsequent papers on individual elements) has produced a benchmark by which other HgMn stars can be compared. The most recent results for osmium and iridium (Ivarsson *et al.* 2004) have lead to defining a clear abundance peak among the heaviest stable elements (Fig. 1). The peak is independent of ionization stage and its explanation will require input from theoreticians and atomic spectroscopists. It remains to be seen whether this same peak structure is reproduced by other HgMn stars, and if not, then to determine which parameters (temperature, age, etc.) are responsible for the observed peak and its evolution. But we must also look at Fig. 1 with caution because it has been created at a specific time in the evolution of this star, using specific observational tools, specific spectral lines (typically resonance lines or other strong lines), and analysed using specific

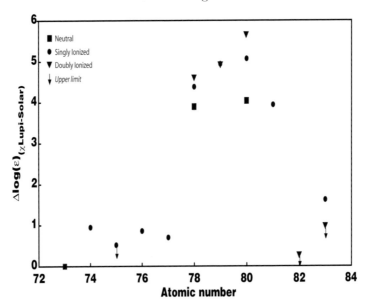

Figure 1. The χ Lupi A abundance peak for the heaviest stable elements. The singly and
doubly ionized points for gold lie on top of each other. Courtesy of S. Ivarsson.

codes (ATLAS9/SYNTHE) and assumptions (LTE). Would this peak and the relationship between abundances of different ions be changed by using different spectral lines or
codes that incorporate different physical assumptions (NLTE, diffusion, magnetic fields)?

Regarding the analysis of spectral line profiles in nmCP stars I would like to point out
three examples of lessons learned, but warn that these lessons may easily be forgotten,
or lost, if they do not become ingrained into our approach to spectrum analysis. 1) The
treatment turbulence is inadequate, misunderstood, and/or incorrect. Velocity fields are
clearly noticeable as asymmetric line profiles for strong lines in the spectra of Am stars
(Landstreet 1998). For the hottest Am stars (for example HD 72660) the asymmetry may
also be present but to only a small degree, while for the late B stars it was not noticed.
The classical LTE approach to spectrum synthesis is not capable of interpreting these
asymmetries. For HgMn stars the turbulent velocity parameter has been shown to influence both the derived abundance and isotopic composition of mercury (Dolk *et al.* 2003).
Therefore, the typical assumption of zero turbulence for late B type stars, or the use of
a constant depth independent value, may lead to a misinterpretation of the spectrum.
2) The reality of ionization anomalies cannot be disputed soley by invoking the 'faulty
atomic data' clause. While certain atomic data do reflect larger uncertainties than desired, it has been shown that the abundance difference for Zr as determined from spectral
lines of Zr II and Zr III for the star χ Lupi A are larger than the uncertainties (Sikström
et al. 1999). While this may seem obvious from other examples, a careful analysis of the
atomic data uncertainties is typically not the concern of most astronomers, leading to
claims of ionization imbalance and elemental stratification in stellar atmospheres that
may not be warranted. Ionization imbalance has yet to be adequately addressed in light
of the assumptions of common (LTE) model atmospheres. 3) Abundance analysis results
are a by-product of the lines chosen to analyse. Lines originating from high versus low
excitation states may lead to very different results. It is a common approach to determine
the abundance of an element for many lines of an ion and then average the results. Since
the spectral lines analysed may be taken from a large wavelength interval the chances

are good that the lines reflect a large range in the excitation potential of their lower energy level. The use of homogeneous, LTE model atmospheres will not treat the different excitation states properly. Also, the quality of atomic data is typically not similar for high versus low excitation lines, with uncertainties increasing for higher excitation. The abundance analysis will therefore have greater uncertainties (one σ fitting errors) when lines of mixed excitation states are used, or may produce different abundances for analyses that use only high or low excitation lines. Obvious examples of problems with the classical modeling of Fe II lines in the He-weak star 3 Cen A are presented by Wahlgren & Hubrig (2004).

4. Isotopic anomalies

The composition of isotopes in stellar atmospheres provides a particularly telling account of dynamic processes at work. Observations of He-weak and HgMn stars made during the 1960s established that the elements helium, platinum, and mercury present line profiles (He) and shifts (Pt, Hg) that can best be explained as variations of the isotopic composition in the region of line formation. As a result of atomic structure there was shown to be a relationship between the shift in the center of gravity of specific spectral lines of Pt II and Hg II with isotope composition. The q-formalism was introduced by White *et al.* (1976), quantifying the shift for the line Hg II λ3984. The q-formalism was useful for many years and is still a useful indicator of the Hg isotopic composition for low to medium resolution spectroscopy. However, it has been shown that the q-formalism is untenable at high spectral resolution (Smith 1997, Wahlgren *et al.* 2000, Dolk *et al.* 2003, Woolf & Lambert 1999). The high resolution spectra show that the heavier mercury isotopes (A = 202, 204) are prominent in most HgMn stars, in contrast to the solar system composition.

Isotopic composition may also be inferred to be different from the measurements of different spectral lines both among and between ions. From optical and *HST* ultraviolet spectra for the stars χ Lupi and HR 7775 the isotopic composition of platinum and mercury is suspected to be different from different ions (Wahlgren *et al.* 2000). The variation of isotopic composition with atmospheric depth and ionization stage is an outcome of diffusion theory (Michaud *et al.* 1974). Additional stars, comprising a range of atmospheric conditions need to be investigated in a manner similar to χ Lupi and HR 7775 to better serve as a constraint for diffusion.

There is also evidence for isotopic anomalies among other elements. From ultraviolet *HST* observations of χ Lupi are measured line shifts and synthetic spectrum fitting of the resonance line Tl II λ1908 to be essentially only from isotope A = 205, with no contribution from A = 203. The same data set tentatively claims an anomaly for lead from a single line of Pb III, with the uncertainty resulting from line blending. For χ Lupi the interesting consistency among the isotopic anomalies is that the heaviest isotope is essentially the only one that is observed.

Most recently, synthetic spectrum fitting of optical lines of Ga II (Nielsen *et al.* 2000, Nielsen 2002) point to the distinct possibility of an anomaly for the star κ Cnc, for which the gallium abundance is highly enhanced. Additionally, wavelength shifts measured for the Ca II infrared triplet have been interpreted as an anomalous isotopic composition in several HgMn stars (Castelli & Hubrig 2004). For both of these elements additional study is warranted to explore isotope anomalies in additional stars. But a certain caution must be interjected into the discussion. Derived isotopic compositions may depend upon the amount of turbulence assumed in the analysis and perhaps the modeling approach. These points have not been adequately documented.

5. Weak emission lines

The detection of very weak emission lines in high resolution, high S/N spectra of mid to late B type main sequence stars is an exciting discovery that imparts a little more reality into the bland thinking of B and A star atmospheres as theoretically simplistic. It is perhaps somewhat surprising that they have escaped detection until now, given that data of a necessary quality to detect them has existed for a couple of decades (post-photographic era). The few published accounts of their existence (Sigut *et al.* 2000, Wahlgren & Hubrig 2000, 2001a, 2001b, 2004, Wahlgren *et al.* 2003) show their equivalent widths to be on the order of a few to tens of mÅ. It is highly probable that all mid to late B main sequence stars create such lines, since those stars without detections are binaries with considerable flux from a secondary star that acts to dilute the visibility of the primary star emission lines. The identified lines originate from the first ionization state of mostly iron-group elements (Cr, Mn, Fe, Ni, Cu) as well as the elements P, Si, Ca, and Hg. The interpretation of these lines has been discussed in terms of NLTE processes, both from a traditional modeling approach (Sigut 2001a,b) and less conventional means (Wahlgren *et al.* 2001a, 2004). These emission lines have not been observed in the few Am stars observed for this purpose.

One interesting aspect of the Mn II emission lines in HgMn star spectra is their relationship with the manganese abundance. For stars of a rather solar-like composition, often referred to as chemically normal stars, emission lines from Mn II multiplet 13 are detected. As the abundance of manganese is increased to the highest levels observed in HgMn stars, these same lines become observed as absorption lines. In cases other than the most Mn-rich or nearly solar-like Mn composition, the relative line strengths (absorption or emission) of lines in the mutiplet are not observed to be the same as the gf-values for the lines. The emission lines are quenched as the abundance increases and since the presence of certain emission lines varies with effective temperature and chemical composition we can be assured that our present atmospheric models do not properly treat the potentially stratified nature of material in the upper atmosphere.

6. Spectrum variability

It is generally presumed that the nmCP stars do not exhibit photometric or spectroscopic variability unless it is related to binarity. Variability is a common phenomenon among the magnetic CP stars, resulting from the inhomogeneous surface distributions of elements caused by the interaction of material with strong, global dipole magnetic fields. Photometric monitoring of CP stars, including nmCP stars, is being undertaken by Adelman and collaborators, but has yet to detect variability for HgMn stars.

Claims of spectroscopic variability are found in the literature but are typically not definitive. The most well known case of reported variability is for the bright, archtype HgMn star α And. Suspicions of spectrum variability date back to early ultraviolet satellite observations. The reality of spectrum variability for this star has only recently been realized through observations of the Hg II λ3984 line (Ryabchikova *et al.* 1999, Ilyin 2000, Wahlgren *et al.* 2001, Adelman *et al.* 2002, Wahlgren *et al.* 2002). This line is the strongest optical transition of Hg II in HgMn stars and is known to display isotopic composition variations among HgMn stars. For α And this line presents undulations on a time scale of 2.38 days, the presumed rotational period. The star is also a binary, with an orbital period of 96.88 days. The application of Doppler imaging techniques to time series observations produces surface maps in the light of Hg II that are similar to those commonly constructed for magnetic CP stars. The Hg II line profile variations can thus be explained as an inhomogeneous surface distribution of mercury with higher concentrations occurring along the rotational equator. But can this superficial comparison between α And and magnetic CP stars imply that a global magnetic field exists in this HgMn star? Similar line profile variability has yet to be verified for other elements in this star, which makes one suspicious of the magnetic origins. Further observations are necessary for α And to extend the search for variable line profiles to other elements and wavelength regions, as well as for other HgMn stars to provide information regarding the parameter space for such phenomena.

7. Magnetic fields

The chemically peculiar stars of the upper main sequence are usually considered in terms of two separate tracks that run parallel in spectral type. They were established by the presence or absence of a magnetic field as detected spectroscopically through Zeeman broadening of magnetically sensitive spectral lines. The attributed differences to these two tracks are stark, despite the fact that the uncertainties of spectral classification for CP stars in the system of Preston (1974), which does not explicitly invoke magnetic field as a classification criterion.

One explanation for the nondetection for magnetic fields among nmCP stars is that the fields are *complex*, meaning that the field is not a global dipole structure, but instead is comprised of a number of magnetic dipoles or complexes that are distributed over the surface of the star. The net effect of these fields would be cancellation, resembling a solar analog. The potential for complex magnetic fields has been suggested through observations of the desaturation of magnetically sensitive line pairs in HgMn and Am stars. Field strengths on the order of 1 to 2 kG were proposed for o Peg (Mathys & Lanz 1990), HD 29173 and HD 195479A (Lanz & Mathys 1993), 74 Aqr A and χ Lupi B (Mathys & Hubrig 1995), and for several stars by Hubrig, Castelli & Wahlgren (1999) and Hubrig & Castelli (2001). (The catalogue of Bychkov *et al.* (2003) is a useful compendium of stars with reported magnetic field measurements.) The analysis of magnetic fields in nmCP stars is particularly difficult due to the small values of the deviation of the equivalent width from the nonmagnetic case. Their study relies on good accuracy for the atomic data, the appropriateness of the present generation of model atmospheres, assumptions of the field structure, and high quality observational data.

The legitimacy of these results is called into question by the nondetection of any magnetic field to the detection limit of the longitudinal Zeeman effect (Shorlin *et al.* 2002), which measures the polarization of starlight. The absence of any such detections implies that the HgMn and Am stars do not possess field structures similar to magnetic CP stars, for which polarization measurements are providing modelers with additional constraints on the magnetic geometry. The polarization measurements are not yet conclusive for nmCP stars because they have not been made for the same targets for which claims of magnetic fields have been made. But even in the eventual absence of any support for magnetic fields from either Zeeman line broadening or polarization we should address the question of whether it is possible for there to be other evidences indicative of a magnetic field. Is it possible that a magnetic field (global dipole or not) be found at sufficient depth below the photosphere so that it is not detected by the aforementioned techniques, but manifests itself in some other manner, such as an enhancement of spectral lines of a particularly sensitive ion? Is attributing line strength enhancements to diffusion processes alone masking the presence of a magnetic field? It is difficult to believe that nmCP stars have no magnetic fields, when magnetic fields are thought to permeate the astrophysics of star and galaxy formation.

8. Stratification

Over the past decade the concept of vertical stratification of abundances in stellar atmospheres has been brought to the test by observations. In terms of interpreting CP star atmospheres by diffusion theory it is quite sensible to think that some (or all) elements may not be distributed homogeneously throughout the line formation region. In the presence of vertical stratification the abundance of an element may be different if determined from lines of different ionization stages or from lines originating from considerably different excitation potentials. In addition, the isotopic composition of an element may also vary with depth. Much work has been undertaken on stratification among the magnetic CP stars and the rapidly oscillating Ap stars (roAp). For the nmCP stars stratification has been suggested from observations involving i) ionization anomalies of Ga (Smith 1995), ii) abundance trends of Cr II lines in the wing of Hβ (Savanov & Kochukhov 1998, Savanov & Hubrig 2003), iii) He I line profiles (Dworetsky, this volume), and iv) weak emission lines (Sigut 2001a,b).

The arguements in favor of stratification from each of these techniques are weak. In all cases it is the modeling of the stellar atmospheres that is the most speculative aspect of the problem. What can be said is that the plane-parallel homogeneous model atmosphere does not serve to reproduce spectral observations in some cases. Another contention is the role of turbulence in the modeling. A depth independent turbulent velocity is the normal approach to modeling line profiles and equivalent widths. But this may not be a reasonable assumption, since turbulence profiles can lead to equivalent width variations.

Regarding the ionization anomaly for gallium, the case has been made for stratification by using high resolution optical spectra for higher excitation lines with *IUE* spectra used for resonance lines. The latter suffers from severe line blending (Nielsen 2002 and paper in preparation), which when accounted for reduces the magnitude of the anomaly. Mixing data of different quality requires caution in the claims of the results.

Ionization anomalies do exist, as noted above, when the spectrum is analysed with LTE computer codes. The application of NLTE codes is progressing but at a rather slow pace, in part due to a need for appropriate model atoms (particularly, collisional cross-sections) but also the result of momentum among astronomers to fend off changing their habits regarding the use of LTE codes. Claims of stratification from observations of ionization imbalance must for the present be tempered if the analysis was conducted in LTE.

9. Crossing boundaries

The study of CP stars has benefitted from earlier efforts to bring order to the perplexing variety of peculiarities exhibited by the stars of the upper main sequence. Through spectral classification we are able to identify common peculiarity traits and promote the study of certain groupings of stars in more depth. However, at some point the continuance of compartmentized studies alone becomes unhealthy because they are not equipped to answer questions of a more general nature, those questions posed regarding origins and evolution. The literature on the observations of CP stars is dominated by studies of specific stellar groups and does not promote thinking in terms of continuity across longer baselines of effective temperature, mass, or magnetic field properties.

Recent work is beginning to address connecting the nmCP star groupings to explain observed changes in spectral peculiarities over spectral type. Wahlgren & Dolk (1998) initiated a program to study chemical composition trends across the boundary separating the HgMn from the hot-Am stars near spectral type A0. It has since been realized that the abundances of certain elements (lanthanides) have a more continuous behavior than previously thought while other elements (Pt-group) display a rather discontinuous behavior across this boundary. Dolk (2002) undertook to extract from the literature on nmCP stars the abundance trends of 27 elements over the effective temperature range 7000 to 14000 K. From this one sees that the spread in abundance is large (> 0.5 dex) for most elements across the entire temperature range. This spread reflects both star-to-star variations and the systematics of different data sets and their analysis. Observed trends include deficiencies (He, C, N, O) or enhancements (Nd, Pr, Ce, Ba, Sr) at all temperatures as well as a tendency for abundance to increase (Mn, P, Ca, Sc, Ti, Y, Zr) or decrease (Ni, Al, S) with increasing temperature. These trends now serve as generalized constraints for theory, since the influence of stellar age is not taken into account.

The evolution of spectrum peculiarity in the vicinity of spectral type A0 has been taken up by Adelman *et al.* (2003). They have proposed that the hot-Am stars (spectral type A0-A2) have evolved from the stars at the cool end of the HgMn class. The mass range of interest is between 2.5 and 3.0 solar masses, where both HgMn and hot-Am star classes have members. This is an interesting idea that requires further theoretical consideration. Although not addressed by this group, an additional consideration should be whether isotopic anomalies and their temporal evolution are consistent with this suggestion. The known isotopic composition of mercury for the HgMn stars in this mass range display no trend with effective temperature or luminosity.

Much more work remains to be done before understanding the relationship between HgMn and hot-Am stars. Likewise, the relationships between the He-weak and HgMn stars as well as the hot-Am stars with the classical Am stars need to be explored. Still other boundaries remain to be crossed. The distinction between magnetic and nonmagnetic stars may be all too artificial and need to be readdressed in terms of continuity. And there may even be psychological barriers that must be crossed before agreement can be reached on the implementation of improved analysis tools. The many interesting problems posed by CP stars can best be solved by the expertise of many disciplines and their resolve to cooperate.

Acknowledgements

It is a pleasure to acknowledge financial support received from the Swedish Academy of Sciences.

References

Adelman, S.J., Adelman, A.S., & Pintado, O.I. 2003, *A&A* 397, 267
Adelman, S.J., Cowley, C. R., Leckrone, D. S., Roby. S. W., Wahlgren, G. M. 1993, *ApJ* 419, 276
Adelman, S.J., Gulliver, A. F., Kochukhov, O. P., Ryabchikova, T. A. 2002, *ApJ* 575, 449
Adelman, S.J., Proffitt, C. R., Wahlgren, G. M., Leckrone, D. S., Dolk, L. 2004, *ApJS* in press
Burkhart, C. & Coupry, M. F. 1997, *A&A* 318, 870
Bychkov, V.D., Bychkova, L. V. & Madej, J. 2003, *A&A* 407, 631
Castelli, F. & Hubrig, S. 2004, *A&A* 421, L1
Dolk, L. 2002, *PhD Thesis*, Lund Obervatory, Lund University
Dolk, L., Wahlgren, G.M. & Hubrig, S. 2003, *A&A* 402, 299
Evans, N.R. 1995, *ApJ* 445, 393
Evans, N.R., Vinko, J. & Wahlgren, G.M. 2000, *AJ* 120, 407
Evans, N.R., Carpenter, K., Rbinson, R. *et al.* 1999, *ApJ* 524, 379
Farthmann, M. *et al.* 1994, *A&A* 291, 919
Hubrig, S. & Castelli, F. 2001, *A&A* 375, 963
Hubrig, S. & Mathys, G. 1995, *Comments Astrophys.* 18, 167

Hubrig, S., Castelli, F. & Wahlgren, G.M. 1999, *A&A* 346, 139

Hui-Bon-Hoa, A., Burkhart, C. & Alecian, G. 1997, *A&A* 323, 901

Hui-Bon-Hoa, A. & Alecian, G. 1998, *A&A* 332, 224

Iliev, I. Kh., Fenovcik, M., Budaj, J., *et al.* 2005, *These Proceedings*, 301

Ilyin, I.V. 2000, *PhD Thesis*, Oulu University

Ivarsson, S. *et al.* 2004, *A&A* in press

Kurucz, R.L. 1995, *CD-ROM 23*

Landstreet, J. 1998, *A&A* 338, 1041

Lanz, T. & Mathys, G. 1993, *A&A* 280, 486

Leckrone, D.S., Proffitt, C. R., Wahlgren, G. M., *et al.* 1999, *AJ* 117, 1454

Mathys, G. & Hubrig, S. 1995, *A&A* 293, 810

Mathys, G. & Lanz, T. 1990, *A&A* 230, L21

Monier, R. & Richard, O. 2005, *These Proceedings*, AP2

Michaud, G., Reeves, H. & Charland, Y. 1974, *A&A* 37, 313

Nielsen, K., Karlsson, H. & Wahlgren, G.M. 2000, *A&A* 363, 815

Nielsen, K. 2002, *PhD Thesis*, Lund Observatory, Lund University

Preston, G.W. 1974, *ARA&A* 12, 257

Ryabchikova, T. 1998, *Contr. Astron. Obs. Skalnaté Pleso* 27, 319

Ryabchikova, T. A., Malanushenko, V. P., & Adelman, S. J. 1999, *A&A* 351, 963

Savanov, I.S. 1998, *Astronomy Reports* 42, 508

Savanov, I. & Hubrig, S. 2003, *A&A* 410, 299

Savanov, I.S. & Kochukhov, O.P. 1998, *AstL* 24, 516

Shorlin, S.L.S., Wade, G. A., Donati, J.-F. *et al.* 2002, *A&A* 392, 637

Sigut, T.A.A., Landstreet, J. D., Shorlin, S. L. S. 2000, *ApJ* 530, L89

Sigut, T.A.A. 2001a, *ApJ* 546, L115

Sigut, T.A.A. 2001b, *A&A* 377, L27

Sikström, C.-M. *et al.* 1999, *A&A* 343, 297

Smith, K.C. 1995, *A&A* 297, 237

Smith, K.C. 1996, *Ap&SS* 237, 77

Smith, K.C. 1997, *A&A* 319, 928

Stelzer, B., Huélamo, N., Hubrig, S. *et al.* 2003, *A&A* 407, 1067

Wahlgren, G.M. & Dolk, L. 1998, *Contr. Astron. Obs. Skalnaté Pleso* 27, 314

Wahlgren, G.M. & Evans, N.R. 1998, *A&A* 332, L33

Wahlgren, G.M. *et al.* 1993, in: M.M. Dworetsky, F. Castelli & R. Faraggiana (eds.), *Peculiar Versus Normal Phenomena in A-Type Stars*, IAU Coll. No. 138, ASP Conf. Ser. 44, 121

Wahlgren, G.M., Dolk, L., Kalus, G. *et al.* 2000, *ApJ* 539, 908

Wahlgren, G.M. & Hubrig, S. 2000, *A&A* 362, L13

Wahlgren, G.M. & Hubrig, S. 2001a, in: G. Mathys, S.K. Solanki & D.T. Wickramasinghe (eds.), ASP Conf. Ser. 248, p. 369

Wahlgren, G.M. & Hubrig, S. 2001b, ASP Conf. Ser. 248, p.365

Wahlgren, G.M. & Hubrig, S. 2004, *A&A* 418, 1073

Wahlgren, G.M., Ilyin, I. & Kochukhov, O. 2001, *BAAS* 33, 1506

Wahlgren, G.M., *et al.* 2002, in: K.G. Strassmeier & A. Washuettl, *1st Potsdam Thinkshop on Sunspots & Starspots, poster proceedings*, (Astrophysical Institute Potsdam), p. 87

Wahlgren, G.M., Hubrig, S. & Ivarsson, S. 2003, in: L.A. Balona, H.F. Henrichs & R. Medupe (eds.), *International Conference on Magnetic Fields in O, B, and A Stars* ASP Conf. Ser. 305 , p. 256

White, R.E., Vaughan, A. H., Jr., Preston, G. W., & Swings, J. P. 1976, *ApJ* 204, 131

Woolf, V.M. & Lambert, D.L. 1999, *ApJ* 521, 414

The A-Star Puzzle
Proceedings IAU Symposium No. 224, 2004   ⓒ 2004 International Astronomical Union
J. Zverko, J. Žižňovský, S.J. Adelman, & W.W. Weiss, eds.   DOI: 10.1017/S1743921304004685

A search for SB2 systems among selected Am binaries

I.Kh. Iliev[1,], M. Feňovčík[2], J. Budaj[3,4], J. Žižňovský[4], J. Zverko[4], I. Barzova[1] and I. Stateva[1]

[1]National Astronomical Observatory, Bulgarian Academy of Sciences,
P.O.Box 136, BG-4700 Smolyan, Bulgaria
e-mail: iliani@astro.bas.bg

[2]Pavol Jozef Šafárik University, 040 01 Košice, Slovak Republic

[3]Department of Astronomy and Astrophysics, Pennsylvania State University,
Davey Lab. 525, University Park, 16802 PA, USA

[4]Astronomical Institute, Slovak Academy of Sciences,
059 60 Tatranská Lomnica, Slovak Republic

Abstract. We report on the detection of secondary spectra in five spectroscopic binary systems: HD 434, HD 861, HD 108642, HD 178449, and HD 216608. High signal-to-noise high resolution spectroscopic observations were carried out at the Bulgarian NAO Rozhen as part of an extended project concerned mainly with Am stars in binary systems. Our knowledge about early type binaries has serious gaps. This is true especially when it is only based on older photographic techniques. We concluded that photographic data involving longer orbital periods (where Doppler shifts due to the orbital motion are comparable or even less than the rotational broadening of the spectral lines) and early type stars (that have only a few and usually broad lines) should be revisited or at least used with caution. We demonstrate that for the five systems how CCD observations made with 2-m class telescopes can discover the binary nature or secondary spectra of many currently unresolved SB1 systems. Important astrophysical information such as the atmospheric parameters and the mass ratios are used to unravel previous misinterpretations of the data leading often to spurious orbits.

Keywords. Stars: binaries: spectroscopic, stars: chemically peculiar, stars: individual: (HD 434, HD 861, HD 108642, HD 178449, HD 216608)

1. Introduction

Am stars are very often found in binary systems (North *et al.* 1998, Debernardi *et al.* 2000). They offer a unique possibility to study the role of tides and tidal interactions on the stellar hydrodynamics and the diffusion processes in stellar atmospheres. Recently we found indications (Budaj 1996, Budaj 1997, Iliev *et al.* 1998) that the observed Am abundance patterns may depend on the orbital elements of a binary system. It is more pronounced in systems with higher eccentricities and possibly also with longer orbital periods. Some years ago we started an observing project concerned especially with Am stars in binary systems. Its main goal is to collect sufficient high quality spectroscopic data to fulfill the rigid requirements of the spectrum synthesis procedures. The first results of this project have been already presented by Feňovčík *et al.* (2005). Soon after the start of the observations we found that spectra of some target stars exhibited clear signs of secondary components. Here we report on the new SB2 systems discovered among our selected Am binaries.

2. The observations

Our spectroscopic observations were made with the 2-m RCC telescope of the Bulgarian National Astronomical Observatory Rozhen. A Photometrics AT200 camera with a SITe S1003AB chip (1024×1024, $24\mu m$ pixels) was used in the Third camera of the coudé spectrograph to provide spectra in two 100 Å wide spectral regions centered at 6440 Å and 6720 Å with a resolving power $R \sim 32\,000$. A typical S/N ratio is about 300. Standard IRAF procedures were used for bias subtraction, flat-fielding and wavelength calibration. Hot, fast rotating stars were used for telluric lines removal. Continuum fitting and measurements of the radial velocities and equivalent widths were performed using the EQWREC2 code of Budaj & Komžík (2000).

Twenty-six Am binaries from Budaj (1997) were selected which: 1) are brighter than 7 th magnitude in V, 2) have declinations greater than $+10°$, and 3) have orbital periods between 10 and 180 days. This assures a full range of systems with original eccentricities which did not undergo circularization on the main sequence. We put no constraints on the rotational velocities and included an additional broad line "normal" star to test the spectrum synthesis procedure on highly rotating stars.

3. The results

3.1. *HD 434*

HD 434 (HIP 728, $m_V = 6.5$, $v \sin i = 60$ km s^{-1}) is known as SB1 system. A preliminary orbit has been reported by Hube & Gulliver (1985). Thirty-eight photographic spectra with a dispersion of about 15 Å mm^{-1} were used for this purpose. Later Sreedhar Rao & Abyhankar (1992) used 33 spectra with 33 Å mm^{-1} and determined that $P = 34.26$ days, $e = 0.475$, $\gamma = 2.6$ km s^{-1}, and $K = 24.1$ km s^{-1}. Their radial velocity curve, however, differs significantly in γ- and K-velocities from those of Hube & Gulliver (1985). Our recent observations (Iliev *et al.* 2001a, Budaj *et al.* 2003) discovered a pronounced secondary spectrum (Fig. 1). The combination of high eccentricity, high $v \sin i$, and low K-velocity results in the heavy blended spectrum (Sp. No. 1) seen during the most of the orbital period. Spectral lines of the secondary are isolated only in the very short phase interval during the maximum separation of the components (Sp. No. 2). We found a mass-ratio $q = M_1/M_2 = 1.19 \pm 0.06$, while $\gamma = +12.0$ km s^{-1}. With these values the K_1-velocity should be greater than 31 km s^{-1}. Finally, the Hube & Gulliver (1985) orbital elements satisfy our spectroscopic data better than those of Sreedhar Rao & Abyhankar (1992). But their γ-velocity is smaller than ours.

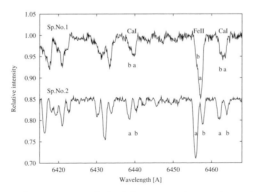

Figure 1. Two spectra of HD 434 are shifted for convenience along the intensity axis. Lines are marked with "a" for the primary, and with "b" for the secondary component.

3.2. *HD 861*

Although bright and close HD 861 (HIP 1063, $m_V = 6.6$, $v \sin i = 35$ km s^{-1}) is a rarely studied SB1 system. The orbital elements originated from Acker (1971): $P = 11.2153$ days, $e = 0.22$, $\gamma = -12.5$ km s^{-1}, and $K = 43.8$ km s^{-1}. Our observations (Budaj *et al.* 2004) reveal two systems of lines and evidence of orbital motion (Fig. 2). Sharp and weak details are seen only around the Ca I and the Fe I lines. Synthetic spectrum calculations for the primary spectrum show that no predicted theoretical lines can be identified with such details. Thus, the secondary component is substantially cooler, fainter, and less massive than the primary. Its rotation is much slower. Our radial velocity measurements lead to a mass-ratio $q \sim 2$.

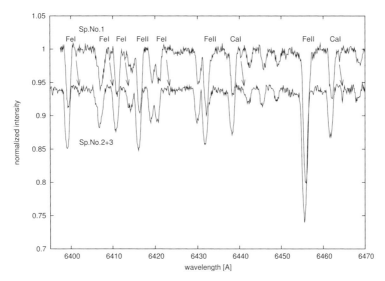

Figure 2. The two successive spectra of HD 861 are separated for convenience by 0.05. While the strong and well isolated lines of the primary are shifted blueward, weak and sharp lines apparently belonging to the secondary are shifted to the red. This is indicated by arrows.

3.3. *HD 108642*

HD 108642 (HIP 60890, $m_V = 6.5$) is a well-known SB1 system. The most recent orbital elements are by Abt & Willmarth (1999): $P = 11.7843$ days, $e = 0.0$, $\gamma = -0.7$ km s^{-1}, and $K = 41.14$ km s^{-1}. According to Abt & Morell (1995) $v \sin i = 13$ km s^{-1}, while Landstreet (1998) obtained $v \sin i < 4.5$ km s^{-1}. Magnetic field measurements by Shorlin *et al.* (2002) permit us to estimate as a by-product that the mass-ratio $M_1/M_2 = 1.9\pm0.1$ and $\gamma = -2 \pm 2$ km s^{-1} via polarimetric least-square deconvolution profiles. Two of our observations (Budaj *et al.* 2003) are presented in Fig. 3, where a small part of the spectrum of HD 108642 centered at the Ca I 6439 Å line is shown. For the mass-ratio we obtained $M_1/M_2 = 1.82\pm0.01$ which agrees with Shorlin *et al.* (2002). For the γ-velocity we found $\gamma = -0.4$ km s^{-1}, a value very close to that of Abt & Willmarth (1999) who were able to measure only the radial velocity of the primary.

3.4. *HD 216608*

The next star HD 216608 (HIP 113048) is the visual binary system ADS 16345AB. The V magnitudes of A and B components are 6.0 and 7.8, respectively. The brightest member HD 216608A is a SB1 star. Its Am characteristics were found by Walker (1966). Companion B is a F6V star that orbits the primary with a period of about 105 years and has a

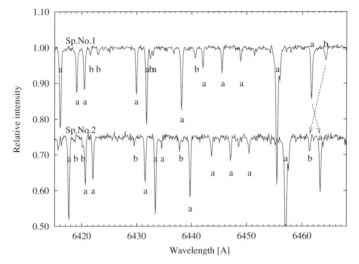

Figure 3. The spectrum of HD 108642 on two different nights. Its SB2 nature is demonstrated by the two line systems, "a" and "b", exchanging places between the first and the second night.

separation of less than one arcsecond. In addition there is an optical C companion, a faint 10.7 star, 28 arcseconds away. Data about the projected rotational velocity of the primary component are a bit controversial. While Walker (1966) measured 50 km s^{-1}, Abt & Morell (1995) determined 46 km s^{-1}, but Abt & Moyd (1973) reported 35 km s^{-1}. Orbital elements come from Abt & Levy (1985): $P = 24.1635$ days, $e = 0.2$, and $K = 10.1$ km s^{-1}. Photographic spectra with a dispersion of about 17 Å mm^{-1} were used.

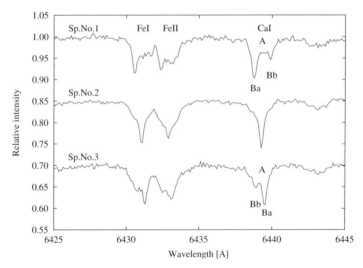

Figure 4. Three successive spectra of HD 216608. Note that while both sharp components marked with Ba and Bb apparently exchange their places due to orbital motion, the wide and shallow lines marked with A do not move.

Our observations of HD 216608 (Iliev *et al.* 2001b) disclose variable spectra that include three line systems. Fig. 4 illustrates how the spectrum changes on three successive nights. A "first sight" interpretation would assign the sharp moving details to the SB1 system HD 216608, while the broad lines could belong likely to the B companion. But the A component is brighter and hotter than the B component, and it is unlikely that its lines would be much sharper. Although well-determined these sharp lines make only a small contribution to the total equivalent width of the ternary blends. This is true especially for the iron lines. The most plausible explanation is that the sharp details belong to the visual B component which seems to be a newly discovered SB2 system HD 216608B. Thus, the wide lines obviously originate from the A component. Synthetic spectrum calculations confirm our suggestions. The mass-ratio $M_{Ba}/M_{Bb} = 1.4$, and the $v \sin i$ values of the A, Ba, and Bb components are 43, 9, and 5 $\mathrm{km\,s}^{-1}$, respectively. Sharp Ba and Bb lines are heavily blended with the broad lines of the primary, thus affecting all foregoing radial velocity measurements made at low resolution. Orbital elements of the system, previous mass estimates, and the SB1 nature of the HD 216608A have to be revisited as well.

3.5. *HD 178449*

17 Lyr (HD 178449, HIP 93917, $m_V = 5.2, v \sin i = 125$ $\mathrm{km\,s}^{-1}$) is a SB1 system, the primary component of ADS 12061. Component B is a 9.1 star 4 arcseconds away from A. The first orbital elements were by Abt & Levy (1976), but they were wrong and the data were analyzed by Dworetsky (1983). Abt & Morell (1995) classified HD 178449 as a mild Am star. It was in our initial target list mainly to test the synthetic spectrum analysis procedures at higher rotational velocities. During the observations we found it extremely interesting and observed it frequently. Very weak sharp absorption details in the bottom of many spectral lines are probably the most intriguing features in its spectrum (Fig. 5). Budaj & Iliev (2003) reported that both the broad and the sharp line systems did not change their radial velocities during the observations (22 months time span) within an error window of $1\sigma \sim 1.5$ $\mathrm{km\,s}^{-1}$. The sharp details are constantly shifted blueward by about 5 $\mathrm{km\,s}^{-1}$.

Three possibilities for the origin of these sharp details are: 1) interstellar, 2) shell (circumstellar), and 3) secondary companion. The first possibility can be eliminated as many sharp lines originate from excited energy levels. To check for a shell we made special observations in the spectral region containing the Ca II K, Ca II H, and Ti II 3913 Å lines. No shell patterns have been found. A further argument against the shell nature of these details is that HD 178449 is rather cool. A stellar origin of the details can be suggested as there are no sharp Fe II lines. Only Ca I and Fe I lines can be seen in Fig. 5. Synthetic spectrum calculations clearly show that the sharp additional spectrum can be reproduced fairly well if a $T_{\mathrm{eff}} = 5000\,\mathrm{K}, \log g = 4.5$, and solar abundance atmosphere for the secondary is assumed. This corresponds to a G-dwarf star which is about 4.2 fainter than the primary at 6450 Å. With this model we can conclude that the secondary would be 4.4 fainter in V, and 4.7 fainter at 3920 Å. It would be very difficult to detect such weak lines if the photographic plates are used. What if the secondary spectrum is produced by visual B component? It is 3.9 fainter and has small angular separation that probably could contaminate the spectrum of the primary. We think this is not the case. If we assume a circular orbit, then the orbital period could be at least 1200 years, but the largest radial velocity difference would be less than we observe. Assessing carefully all arguments we conclude that the weak sharp lines belong to the newly found spectroscopic component of the system.

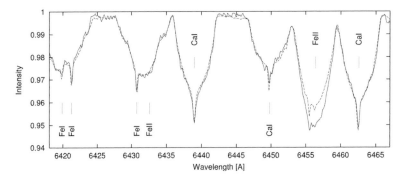

Figure 5. Observed (solid line) and computed (dashed line) spectra of HD 178449.

Acknowledgements

II and IS acknowledge the support provided by the SOC and the LOC of the Symposium. JB gratefully acknowledges grant support from Penn State University (NSF-NATO fellowship: NSF DGE-0312144) and partial support by the VEGA grant No. 3014 and the Science and Technology Assistance Agency under the contract No. 51-000802.

References

Abt, H.A., Levy, S.G. 1976, *ApJS*, 30, 273
Abt, H.A., Levy, S.G. 1985, *ApJS*, 59, 229
Abt, H.A., Morell, N.I. 1995, *ApJS*, 99, 135
Abt, H.A., Moyd, K.I. 1973, *ApJ*, 182, 809
Abt, H.A., Willmarth, D.W. 1999, *ApJ*, 521, 682
Acker, A. 1971, *A&A*, 14, 189
Budaj, J. 1996, *A&A*, 313, 523
Budaj, J. 1997, *A&A*, 326, 655
Budaj, J., Iliev, I.Kh. 2003, *MNRAS*, 346, 27
Budaj, J., Iliev, I.Kh., Barzova, I.S., Žižňovský, J., Zverko, J., Stateva, I. 2003, *IBVS*, 5423
Budaj, J., Iliev, I.Kh., Feňovčík, M., Barzova, I.S., Richards, M.T., Geordzheva, E. 2004, *IBVS*, 5509
Budaj, J., Komžík, R. 2000, `http://www.ta3.sk/~budaj/software`
Debernardi, Y., Mermilliod, J.-C., Carquillat, J.-M., Ginestet, N.J. 2000, *A&A*, 354, 881
Dworetsky, M.M. 1983, *MNRAS*, 203, 917
Feňovčík M., Budaj., I., Iliev, I., Richards, M.T., Barzova, I. 2005, *These Proceedings*, FP20
Hube, D.P., Gulliver, A.F. 1985, *JRASC*, 79, 49
Iliev, I.Kh., Budaj, J., Zverko, J., Barzova, I.S., Žižňovský, J. 1998, *A&AS*, 128, 497
Iliev, I.Kh., Budaj, J., Zverko, J., Žižňovský, J. 2001a, *IBVS*, 5051
Iliev, I.Kh., Budaj, J., Žižňovský, J., Zverko, J., Stateva, I., Geordzheva, E. 2001b, *IBVS*, 5199
Landstreet, J.D. 1998, *A&A*, 338, 1041
North, P., Ginestet, N., Carquillat, J.-M., J., Carrier, F., Udry, S. 1998, *CAOSP*, 27, 179
Shorlin, S.L.S., Wade, G.A., Donati, J.-F., Landstreet, J.D., Petit, P. *et al.* 2002, *A&A*, 392, 637
Sreedhar Rao, S., Abyhankar, K.D. 1992, *MNRAS*, 258, 819
Walker, E.N. 1966, *The Observatory*, 86, 154

Discussion

DWORETSKY: It is good to see such intensive detective work on stars that appeared to be well-determined. It shows how important these sorts of efforts can be. I expect to see many more such discoveries in the not-to-distant future.

The A-Star Puzzle
Proceedings IAU Symposium No. 224, 2004
J. Zverko, J. Žižňovský, S.J. Adelman, & W.W. Weiss, eds.
© 2004 International Astronomical Union
DOI: 10.1017/S1743921304004697

The variable light curves
of some mCP stars†

Diane M. Pyper[1]and Saul J. Adelman[2]

[1]Department of Physics, University of Nevada, Las Vegas, Las Vegas, NV 89154-4002, USA
[2]Department of Physics, The Citadel, 171 Moultrie Street, Charleston, SC 29409, USA‡

Abstract. The use of the Four College Automatic Photometric Telescope at the Fairborn Observatory in Southern Arizona with its ability to observe stars every clear night and for continuous runs has played a major role in our very successful cooperative program of Strömgren differential *uvby* photometry of magnetic CP (mCP) stars. Out of about 100 stars observed only a few have variable light curves. But more may be found when additional observations are made. We review the published theoretical basis and observations as well as present new results for CU Vir, V1093 Ori, MW Vul, and HR 7224. We interpret some observations as possibly requiring a more robust theory of the variability of mCP star light curves.

Keywords. Stars: chemically peculiar, stars: individual (CU Vir, V1093 Ori, HR 7224), stars: variables: other, techniques: photometric

1. Introduction

The light, magnetic and spectrum variations of the magnetic chemically peculiar (mCP) stars are generally accepted as being the result of observing spotted rotating stars. The large scale magnetic fields, which usually are primarily dipolar, are variable over the photosphere and produce changes in the local chemistry, via hydrodynamical process including diffusion, which in turn affect the radiated energy distribution. Shore & Adelman (1976) inspired by the changes in the line profiles of 56 Ari at a given phase interpreted these results as due to variations produced by the precession of the rotation axis. It could be the result of a forced precession due to a companion or to free-body precession caused a change in the moment of interia along the magnetic field axis when the magnetic and rotational axes are highly inclined to one another. The precessional periods were predicted to be about 6 years for mCP stars having rotational periods of less than one day. If rotation is the dominant perturbation, an analog of the Chandler wobble will result. Thus, we included searches for such effects in our long-term observation program using the Four College Automatic Photometric Telescope (FCAPT) at the Fairborn Observatory in Southern Arizona, which has completed its 14th year of operations.

Several observational consequences of precession are expected: 1). The timings of the maximum, the minimum, and of any identifiable portion of the light curves will experience periodic changes during the precessional cycle. 2). The observed value of $vs \sin i$ will be found to change, being smallest when the rotational pole is most nearly towards us. If the change of i is large enough, high dispersion spectra should be able to measure it and

† This research was supported by grant AST 0071260 from the US National Science Foundation

‡ Guest Investigator, Dominion Astrophysical Observatory, Herzberg Institute of Astrophysics, National Research Council of Canada, 5071 W. Saanich Road, Victoria, BC V9E 2E7, Canada

D.M. Pyper & S.J. Adelman

V1093 ORI - LONG RUNS

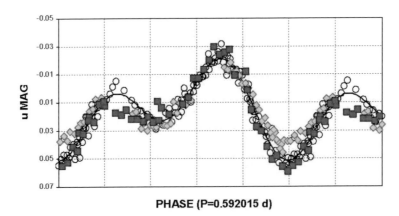

PHASE (P=0.592015 d)

Figure 1. Changes in the shape of the u light curve of V1093 Ori. Open circles represent data for 1996-97; the solid line is the fit for these data (Fourier series, 5 parameters). Diamonds represent data for 1995-96 and squares represent data for 2003-04. Each horizontal interval is 0.2 of the period.

provide information on the geometry. and 3). If the stars are rapidly rotating, they will be flattened ellipsoids of rotation with their polar diameters smaller than their equatorial diameters. Thus they will appear to be brighter when their polar axes point more nearly towards us.

Photometrc variations that can be attributed to precession have been definitely found for 108 Aqr (Adelman 1999), 20 Eri (Adelman 2000), 56 Ari (Adelman *et al.* 2001), MW Vul (Adelman & Young 2004), and V1093 Ori (Pyper, in preparation). In addition, we have found some surprising, even bizarre, variations in some other program mCP stars, such as the abrupt change in period in 1984 of the short-period mCP star CU Vir (Pyper *et al.* 1998) and the extraordinary switch from predominantly short (order 1 day) to longer (order 100 days) period variations of HR 7224 (Adelman 2004). We report on both published and recent results of this program. Variable light curves are found for only a small percentage of the some 100 stars we are observing and that the amplitudes of such variability as detected are of order 1%. In many cases the periods of mCP stars that we have derived are refinements of those found by earlier photometric studies. For some stars additional observations may reveal variable light curves. The use of one telescope and filter set has helped us in making comparisons between observations taken in different years.

2. Variable light curves just indicating precession

We have found four mCP stars which show variable light curves consistent with precession. For them we do not have observations over a sufficiently long time to discover the expected small increase in their rotation periods. FCAPT *uvby* data from 1995-2004 of V1093 Ori (HD 36313) show variations in the shape and amplitude of the light curves from year to year, as seen in Figure 1 for the u magnitude. Data for 2000-01 (not shown) display similar differences with more scatter. The v, b and y filter data show changes similar to u. All FCAPT data agree with the period of 0.592015 days determined by North (1984), making this the second shortest period mCP variable in our program.

MW Vul

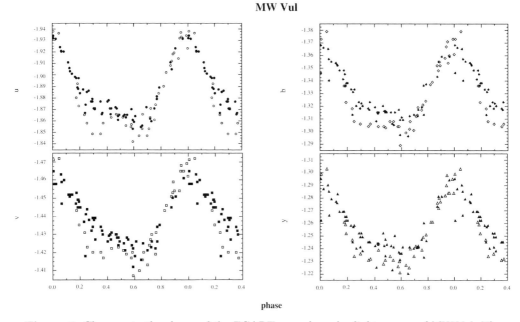

Figure 2. Changes in the shape of the FCAPT *u, v, b,* and *y* light curves of MW Vul. The solid and open symbols represent, respectively, observations made in 1990-91 and 2003-04.

Interestingly, the light curves of North are shifted slightly with respect to the FCAPT curves. A slightly longer period of 0.592035 days makes the two data sets agree, but this introduces shifts between successive years of the FCAPT observations. Thus we see an apparent phase shift between the 1984 and 1995-2004 data but no change in period which is likely to be due to the precessional period.

The variability of V1093 Ori is similar to that found by Adelman (2000) for 20 Eri (HD 22470) and by Adelman (1997) for 108 Aqr (HD 223640) with a periods of 1.92889 days and 3.735239 days, respectively. Small differences are found between data in the Strömgren *uvby* photometric system which were taken in different years.

Recently Adelman & Young (2004) found that MW Vul (HD 192913) whose period is 16.840 days (Adelman & Knox 1994) is also undergoing precession. There are clear differences in the shapes and amplitudes of the light curves for the same Strömgren passband in years 1 and 14 in Figure 2. Values for year 1 (1990-91) are solid symbols and those for year 14 (2003-04) are open symbols. Examination of more fragmentary data obtained in other years support the conclusion that MW Vul has variable light curves.

These four stars have rotational periods that range from the original expectation of Shore & Adelman (1976) to one that clearly violates it. Thus instead of just concentrating our efforts on the fastest rotating stars, we also need to look at those which are more slowly rotating. We need to try to determine when the slowly rotating stars which are now precessing begin this motion.

3. Two stars whose rotational periods have increased

Adelman (2002) found some evidence for the expectation that as mCP stars move away from the ZAMS their rotational velocities decrease. But only in the last few year have the first observed small period increases been well established for 56 Ari and CU Vir, two of the fastest rotating mCP stars.

CU VIR - O-C u/U (MIN)

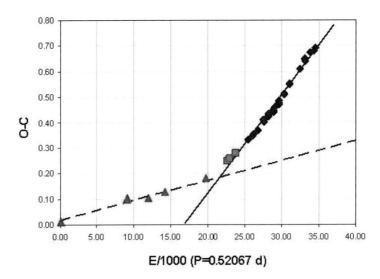

E/1000 (P=0.52067 d)

Figure 3. The period of CU Vir abruptly increased in approximately 1984 from 0.5206778
days (dashed line) to 0.52070854 days (solid line).

3.1. *56 Ari*

Adelman *et al.* (2001) used extensive sets of *U*BV and FCAPT *uvby* photometry as well
as spectroscopic data to investigate the light and equivalent width variations of 56 Ari
(HD 19832). Its rotational period is increasing at a rate of about 2 s per century. There
is evidence for a second period whose length is about 5 years which is attributed to
the precession of the axis of rotation. *U*BV (see, e.g. Fried & Adelman 2003) and *uvby*
photometry from different years show changes in the light curve shapes and amplitudes.

3.2. *CU Vir*

Based on spectroscopic data and FCAPT *U*BV and *uvby* data, Pyper *et al.* (1998) found
that the period of CU Vir (HD 124224) abruptly increased in approximately 1984 from
0.5206778 days (dashed line) to 0.52070854 days (solid line) in Figure 3. The latter period
is the average for 1987-1997. The data points on the upper O-C curve are as follows: red
triangles represent the JDmin of the *U* mags. for 1958-83, pink squares represent *U* mags
for 1987-89 and blue diamonds represent *u* mags. for 1991-2004 (Pyper *et al.* 1998).

The lower O-C curve in Figure 4 represents the *U* and *u* data from 1980-2004 based
on the period 0.5206778 days. The pink square represents *u* mags for 1980-83, green
triangles represent *U* mags for 1987-89, yellow circles represent *u* mags. for 1991-93 and
blue diamonds represent *u* mags. for 1994-2004. These data can be interpreted in two
ways, both indicating that the rotation of CU Vir is still slowing down.

1) CONTINUALLY CHANGING PERIOD: The parabolic fit (solid red line) corresponds to a period increase of P = 1.32E-06 d/yr, starting with P = 0.5206778 days in
1983. For this fit, R2 = 0.9959.

2) TWO CONSTANT PERIODS: A linear regression through the 1993-2004 data gives
almost as good a fit as the parabolic fit and indicates a constant period P = 0.5207125
days (solid black line; R2 = 0.9929), but only if the 1987-1992 data are fit to a different
constant period P = 0.5206999 days (dashed line; R2 = 0.9968).

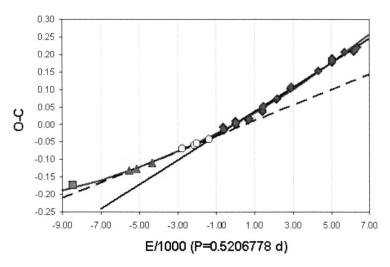

Figure 4. Another way to look at the *u* photometry of CU Vir.

Thus CU Vir is still showing period changes following its major slowdown in 1984. It has either experienced another discrete change in period or is continually slowing down. The former interpretation is perhaps preferable as it could be explained by a modification of the outer envelope of the star by its magnetic field, as suggested by Stepien (1998). A continual slowdown is more difficult to explain as there is no evidence for mass loss, unless there is a continual slowing effect by the magnetic field. CU Vir as the fastest rotating mCP star has the largest rotational deformation. Thus it may be undergoing a Chandler-like wobble.

4. HR 7224

Adelman (1997) obtained 154 sets of FCAPT differential Strömgren *uvby* observations in the 1993-94 and 1994-1995 observing season of HR 7224 (HD 177410). When a periodogram analysis was performed, a period of 1.123095 days was obtained. Figure 5 shows the *b* values plotted as a function of this period. The *u*, *v*, *b*, and *y* light curves are in phase with amplitudes of 0.05, 0.045, 0.035, and 0.03 mag., respectively. There is a narrow single maximum with a much broader miminium with two subminima. Winzer's (1974) 16 *V* magnitudes when transformed to *y* agree with those from Adelman (1997). Celestia 2000 (ESA 1998) gives a period of 1.12323 days from Hipparcos photometry. Adelman (1997) used Winzer's values to refine his period. and thus his result is based on a longer time period. But still the agreement is quite satisfactory.

In Spring 2003, a new set of FCAPT *uvby* observations was started to improve the period and to search for any rotational slowdown. As before the *u*, *v*, *b*, and *y* light curves are in phase, but soon it became apparent that there was something very different (Adelman 2004). At the end of observations for Fall 2003, a periodogram analysis indicated a period of 101 days. The amplitude of variability was about 0.21 mag. This was a result without presidence. Observations in Spring 2004 began to show a different picture. The simplist way to have a periodic light curve was to double the period to about 215 days and assume that the portion of the light curve now being seen was not previously

D.M. Pyper & S.J. Adelman

HR 7224

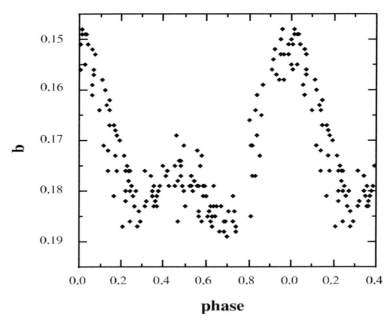

Figure 5. The b photometry of HR 7224 of Adelman (1997) plotted with a 1.123095 day period.

observed. The u, v, b, and y light curves are still in phase and have not too differing amplitudes which supports a geometric explanation.

A scenario which decreases the rotational period by such a large amount causes substantial theoretical difficulties. Fortunately we are saved from that fate. In June 2004, Adelman obtained two spectrograms of HR 7224 at the Dominion Astrophysical Observatory at $S/N = 200$ and a two pixel resolution of 0.072 Å. If the rotational rate had decreased, one would expect to see the spectrum of a sharp-lined star while if its rotation rate was around a day the star would be fast rotating. Observations of the Hβ region do not reveal any sharp-lines, the core of Hβ is rounded, and $v \sin i$ close to 100 km s^{-1}. The signature of a shell is missing. Mkritchian (private communication) has also obtained spectrograms, but his estimate of $v \sin i$ is smaller.

The 215 day period may be due to precession. We are now seeing a much larger portion of the surface than before when we watch long enough. If HR 7224 is still rapidly rotating, then there should be photometric evidence for this. Adelman had observed this star once a night. But in June 2004 he observed it continuously for about six hours for three nights. During the these runs, the amplitude of variability was about 0.04 mag. which is the previous rotational amplitude. The task now is to try to determine what is the rotational period.

But HR 7224 had another surprise. The mean light curve should have begun to go through a decline of 0.20 mag. if the dominant period was constant before this meeting began. But it was not to be (see Figure 6). The dynamical state of HR 7224 is evolving with the dominant period increasing. It is very important to measure the full amplitude for it may well decrease with an increasing precessional period. The FCAPT closed down the day this meeting began. Observations should resume in September.

HR7224-13&14

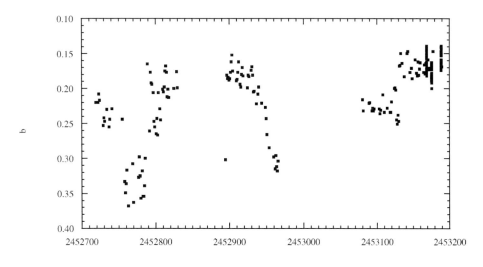

HJD

Figure 6. The *b* photometry of HR 7224 as a function of HJD beginning in Spring 2003.

What caused the changes in the dynamical state? There is a period of some 8 years without observations for HR 7224. The star was a well-behaved mCP star through the time the 1994-95 observations were taken. Sometime between then and Spring 2003 the precessional behavior was initiated.

HR 7224 is located near the Zero Age Main Sequence. It is unclear how to produce an internal mass inhomogeneity that would cause the star to start precessing. Increasing the magnetic field strength by a factor of 2 increases the difference in the moment of inertia between the magnetic and polar axes by a factor of 4 (Shore & Adelman 1976). Bohlender *et al.* (1993) failed to detect a significant magnetic field, Wade and colleagues will try again very shortly.

HR 7224 does not have any known close companions. Thus forced precession is not likely. It would probably involve a change in the orbit of a companion. Adelman's other current scenario is that it was hit by, captured, and now is slowly absorbing some object which survives for awhile and losses mass slowly to account for the changes we see. Possibilities include planets, black dwarfs, and the least massive stars.

If any of these possibilities are correct, HR 7224 poses interesting theoretical challenges. On the other hand, another explanation is still possible. HR 7224 has produced one surprise after another so we would not be surprised with new ones.

5. Conclusions

For all of the stars discussed in this paper, further photometric observations are desirable. Due to the type of coverage desired, the use of automated telescopes is the most feasible approach. Further our results to date show that it is desirable to obtain several sets of observations for 50 or more mCP stars to better define their light curves and to do a proper census of the class for light curve variability. Measurements of their $v \sin i$ values obtained throughout their apparent precessional cycles might yield important geometric information.

Acknowledgements

This research was supported by grant AST 0071260 and earlier grants from the US National Science Foundation. We appreciate the work of Louis J. Boyd in designing and maintaining the FCAPT. We acknowledge the roles of Drs. Robert J. Dukes and George McCook, the Principal Astronomers of the FCAPT, in the successful operation of this telescope.

References

Adelman, S. J. 1997 A&AS, 122, 249.
Adelman, S. J. 1999 Baltic Astronomy, 8, 369.
Adelman, S. J. 2000 A&AS, 146, 13.
Adelman, S. J. 2002 Baltic Astronomy, 11, 475.
Adelman, S. J. 2004 MNRAS, 351, 823.
Adelman. S. K. & Knox, J. R., Jr. 1994 A&AS, 103, 1.
Adelman, S. J. & Young, K. J. 2004, A&A, in press.
Adelman, S. J., Malanushenko, V., Ryabchikova, T. A. & Savanov, I. 2001 A&A, 375, 982.
Bohlender, D. A., Landstreet, J. D., & Thompson, I. B. 1993, A&A, 269. 355.
ESA 1997 The Hipparcos and Tycho Catalogues, SP-1200.
Fried, R. & Adelman, S. J. 2003 Journal of Astronomical Data 9, article 2.
North, P. 1984 A&AS, 55, 259.
Pyper, D. M., Ryabchikova, T., Malanushenko, V., Kuschnig, R., Plachinda, S. & Savanov, I. 1998 A&A, 339, 822.
Shore, S. N. & Adelman. S. J. 1976 ApJ, 209, 816.
Stepien, K. 1998 A&A, 337, 754.
Winzer, J. E. 1974 PhD Thesis University of Toronto.

Discussion

BREGER: The (O-C) curve of CU Vir resembles that of many pulsating stars. A very sudden change of period will not fit the observations; neither will a constant period change (no parabola). A possibility might be chaotic changes, since the curve shows structures beyond the main period jump. What are the sizes of the error bars?

ADELMAN: CU Vir has a magnetic field while δ Scuti stars do not. CU Vir is also not in the classical instability strip. The magnetic field helps stabilize the atmosphere. The mean periods of mCP stars increase with age from the ZAMS, chaotic processes are not likely. The error bars are about the same size as the symbols in this Figure.

MOSS: Any plausible, ever quite implausible(!), extrapolation of Ap stars surface fields to the interior predicts magnetic distortions that are much less than than the rotational, especially for a rotational period of $0(1)$ d.

ADELMAN: The magnetic field change the moment of interior along the magnetic axis relative to that perpendicular to it. The difference of the moments of interior along and perpendicular to the rotational axes need to be sufficiently large to come precession. The observations are consistent with precession in terms of phase shifts and magnitude differences at the same phase for the best determined case 56 Ari. What is needed for further proof is the observation of the change in $v \sin i$.

The A-Star Puzzle
Proceedings IAU Symposium No. 224, 2004
J. Zverko, J. Žižňovský, S.J. Adelman, & W.W. Weiss, eds.

© 2004 International Astronomical Union
DOI: 10.1017/S1743921304004703

NLTE ionization equilibrium of Nd II and Nd III in cool A and Ap stars

L.I. Mashonkina[1], T.A. Ryabchikova[1] and A.N. Ryabtsev[2]

[1]Institute of Astronomy, Russian Academy of Sciences, Pyatnitskaya 48, 109017 Moscow,
Russia
email: lima@inasan.ru

[2]Institute of Spectroscopy, Russian Academy of Sciences, 142190, Troitsk, Moscow region,
Russia

Abstract. The kinetic equilibrium of Nd II -Nd III in the atmospheres of A-type stars is investigated for the first time with a model atom containing 1651 levels of Nd II, 607 levels of Nd III, and the ground state of Nd IV. NLTE leads to an overionization of Nd II resulting in weakening the Nd II lines at mild neodymium overabundances relative to the solar Nd abundance ([Nd/H] < 2.5) and produces the opposite effect at the higher [Nd/H] values. NLTE abundance corrections grow with effective temperature and reach $\sim$0.6 dex at $T_{\rm eff}$ = 9500 K. The Nd III lines are strengthened compared with LTE. NLTE abundance corrections range between -0.3 dex and -0.2 dex for $T_{\rm eff}$ between 7500 K and 9500 K. Therefore NLTE effects may explain ionization discrepancies up to 0.8 – 0.9 dex, derived with the LTE approach.

NLTE effects are even larger for a stratified Nd abundance distribution compared with a homogeneous one resulting in positive NLTE abundance corrections up to 1.4 dex for the Nd II lines and negative ones as small as to -0.5 dex for the Nd III lines. The influence of uncertainty in the photoionization cross-sections on NLTE results is investigated. NLTE calculations were applied to Nd analyses in the atmospheres of roAp stars γ Equ and HD 24712.

Keywords. Atomic data, line: formation, stars: chemically peculiar, stars: abundances

1. Introduction

Abundance analyses of cool Ap stars revealed huge ionization discrepancies in Nd II–Nd III, which may reach 2 dex in the atmospheres of rapidly oscillating (roAp) stars (Ryabchikova *et al.* 2001). In a newer LTE analysis of one of them, γ Equ, Ryabchikova *et al.* (2002) interpreted the observed discrepancy using a stratified Nd distribution with the accumulation of the element above $\log \tau_{5000} = -8$. In upper atmospheric layers departures from LTE are expected. Thus non-local thermodynamical equilibrium (NLTE) line formation is used to obtain theoretical Nd II and Nd III line profiles and equivalent widths for a range of effective temperatures and Nd overabundances typical of the cool Ap stars.

2. NLTE calculations for Nd II - III

NLTE calculations were made with the DETAIL code using accelerated lambda iteration following the method described by Rybicki & Hummer (1991, 1992). DETAIL originally was created at Munich University by Butler & Giddings (1985) and modified later. In our NLTE calculations we use plane-parallel homogeneous model atmospheres computed with the MAFAGS code (Fuhrmann *et al.* 1997).

2.1. *Model atom*

The model atom includes 658 energy levels of Nd II from laboratory measurements (Martin *et al.* 1978, Blaise *et al.* 1984) and 993 levels of Nd II and 607 levels of Nd III predicted in this work. The predicted levels belong to quartet and sixtet terms of the $4f^4$ np (n = 7 - 11) and the $4f^3 5d$ 6p electronic configuration for Nd II and to triplet and quintet terms of the $4f^3$nl (nl = 4f, 5d, 6s) and $4f^2 5d^2$ electronic configuration for Nd III. The Nd II and Nd III spectra were calculated by using the RCN-RCG-RCE software package (Cowan 1981). The RCN subroutine determines the wave functions of the configurations by the Hartree-Fock single-configuration method with relativistic corrections. These wave functions are used to compute the average configuration energies, the Slater integrals of intra-configuration electrostatic interactions, the integrals of electric dipole transitions and inter-configuration electrostatic interactions, as well as spin-orbit interaction parameters. Using these quantities, the RCG subroutine computes the wavelengths and transition probabilities. By fitting the computed and experimental energy levels from (Blaise *et al.* 1984), the ab initio energy matrix parameters are modified and these parameters are again used by the RCG subroutine to compute a semiempirically adjusted spectrum. Calculated energy levels of the Nd II $4f^4 6p$ configuration agree with experimental values within several hundreds of cm^{-1}. Calculated gf values for the Nd II $4f^4 6s\,^6 I$ - $4f^4 6p$ transitions are greater by about 50% compared with new experimental data of Den Hartog *et al.* (2003). Similar accuracy is expected for the predictions of related values to the unknown $4f^4$np configurations.

Levels of the same parity with small energy differences were combined into single level. The final model atom includes 247 levels of Nd II, 68 levels of Nd III and the ground state of Nd IV.

For transitions between the measured Nd II energy levels, oscillator strengths f_{ij} are available in the Vienna Atomic Line Data Base (Kupka *et al.* 1999). Within 0.1 - 0.15 dex they are consistent with the new experimental data by Den Hartog *et al.* (2003). The remaining transitions f_{ij} have been calculated in this paper. No data on photoionization cross-sections for both Nd II and Nd III levels are available in the literature and we use hydrogenic cross-sections. For electron impact excitation we use the formula of van Regemorter (1962) for allowed transitions and that of Allen (1973) with $\Omega = 1$ for forbidden ones. Electron impact ionization cross-sections are computed according to Drawin (1961).

2.2. *Kinetic equilibrium calculations and NLTE effects*

NLTE calculations show that in the atmospheres of dwarfs with $T_{\rm eff}$ between 7500 K and 9500 K the ionization equilibrium of Nd II/Nd III deviates from the thermodynamical value. In line formation layers all the Nd II levels are underpopulated compared with the LTE populations while the Nd III levels are overpopulated at $T_{\rm eff} \leqslant 8000$ K (see Fig. 1, left panel for the LTE and NLTE total number densities of Nd II and Nd III in the model atmosphere with $T_{\rm eff} = 7700$ K) and keep their thermodynamical values for the higher effective temperatures. NLTE effects for the Nd II and Nd III lines are of opposite sign, and they are, therefore, important for the comparison of neodymium abundances deduced from these lines. In Fig. 1 right panel, the departure coefficients, $b_i = n_i^{\rm NLTE}/n_i^{\rm LTE}$ of the selected levels of Nd II and Nd III in the model atmosphere with $T_{\rm eff}/\log g/[{\rm M/H}] = 7700/4.2/0.1$ are shown as a function of continuum optical depth τ_{5000} refering to $\lambda 5000$. Here, $n_i^{\rm NLTE}$ and $n_i^{\rm LTE}$ are the kinetic equilibrium and thermal (Saha-Boltzmann) number densities, respectively. Everywhere in atmosphere [Nd/Fe] = 3 is adopted.

Overionization of Nd II in the atmospheric layers above $\log \tau_{5000} = 0$ is caused by superthermal radiation of a non-local origin near thresholds of $4f^4 6p$ levels with $E_{\rm exc}$ from

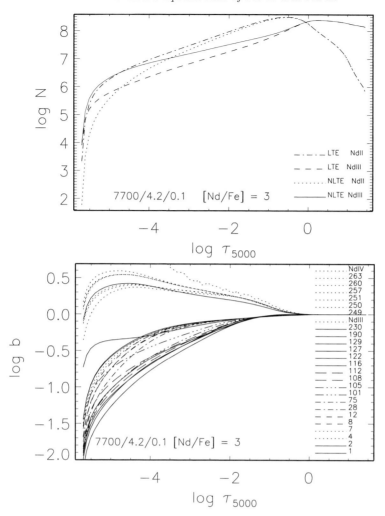

Figure 1. LTE and NLTE total number densities of Nd II and Nd III (top) and departure coefficients for selected levels of Nd II and Nd III (bottom) in the model atmosphere 7700/4.2/0.1. Everywhere in atmosphere [Nd/Fe] = 3.

3 eV to 4 eV (λ_{thr} = 1600 - 1850 Å). The Nd II lines are weakened compared with the LTE case at mild neodymium overabundances relative to the solar Nd abundance ([Nd/H] < 2.5) but strengthened at greater [Nd/H] values. The Nd III lines are always stronger than with LTE. In Table 1 for the model atmosphere 7700/4.2/0.1 which represents the roAp star γ Equ, theoretical NLTE and LTE equivalent widths are given for several lines of interest together with NLTE abundance corrections $\Delta_{\mathrm{NLTE}} = \log \varepsilon_{\mathrm{NLTE}}$ - $\log \varepsilon_{\mathrm{LTE}}$. As expected, departures from LTE for the Nd II and Nd III lines grow with T_{eff} (Fig. 2).

2.3. *NLTE effects for a stratified* Nd *abundance distribution*

For the model atmosphere 7700/4.2/0.1 NLTE calculations were performed assuming [Nd/H] = 4 in the atmospheric layers above $\log \tau_{5000} \simeq -3.6$ and a steep change of Nd overabundance to [Nd/H]= 0 in the layers below $\log \tau_{5000} \simeq -2.7$. The calculated departure coefficients for the selected levels of Nd II and Nd III are shown in Fig. 3 and

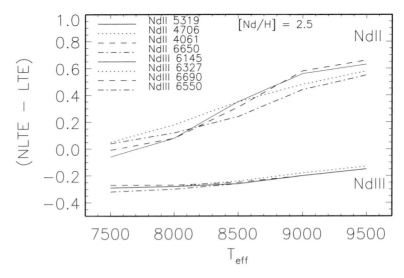

Figure 2. NLTE abundance corrections for the Nd II and Nd III spectral lines for different values of $T_{\rm eff}$. In all cases $\log g = 4$ and [Nd/H] $= 2.5$

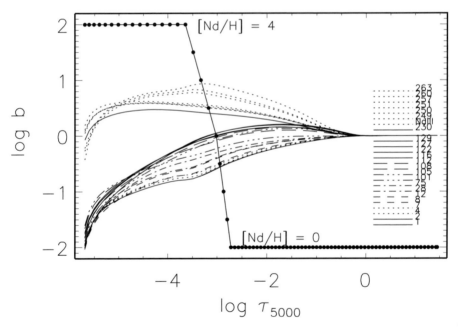

Figure 3. Departure coefficients for the selected levels of Nd II and Nd III in the model atmosphere 7700/4.2/0.1 with stratified Nd abundance distribution (the [Nd/H] ratio is shown by filled circles connected with the solid line).

theoretical NLTE and LTE equivalent widths as well as NLTE abundance corrections are given in Table 1 (columns "layer").

NLTE calculations show the overionization of Nd II and enhanced number densities of Nd III in line formation layers are similar to the homogeneous Nd abundance distribution. However, in contrast to Fig. 1, the Nd II low-excitation levels are strongly underpopulated above $\log \tau_{5000} = $ -2 and their departure coefficients are smaller compared with the levels of intermediate energy. The Nd II lines are weakened compared with the LTE case and NLTE abundance corrections range from 1.07 dex to 1.42 dex for different lines. The Nd III lines are strengthened compared with the LTE case and have approximately the same values as for the homogeneous Nd abundance distribution.

2.4. *Influence of uncertainties of atomic parameters on the final results*

A first test concerns photoionization cross-sections for the Nd II levels since kinetic equilibrium of Nd II depends strongly on radiative b-f transitions. Test NLTE calculations have been made for the stratified Nd abundance distribution by multiplying photoionization cross-sections of the Nd II levels by scaling factors of 100 and 0.01. In the first case the overionization on Nd II is amplified and Δ_{NLTE} increases by 0.09 dex - 0.14 dex for different Nd II lines. For the Nd III lines Δ_{NLTE} increases by only 0.02 dex. Reducing photoionization cross-sections by 100 times has much larger effect: Δ_{NLTE} decreases by 0.36 dex to 0.65 dex for different Nd II lines and by 0.10 dex to 0.14 dex in absolute value for the Nd III lines. However, we emphasize that even with such low photoionization cross-sections the ionization equilibrium Nd II/Nd III deviates significantly from thermodynamical value and a difference of neodymium abundances deduced from the Nd II and Nd III lines is reduced by about 1 dex if NLTE effects are taken into account.

For the homogeneous Nd abundance distribution NLTE calculations performed with increasing collision excitation cross-sections by a factor of 10 have show a small effect for the Nd II lines (Δ_{NLTE} have reduced by 0.02 dex to 0.05 dex) and a slightly larger one for the Nd III lines, at the level of 0.1 dex of Δ_{NLTE}.

Table 1. Theoretical NLTE and LTE and observed equivalent widths (in mÅ) of the Nd II and Nd III lines for γ Equ. Computations were made for two cases: [Nd/Fe] = 3 everywhere in atmosphere (columns "[Nd/Fe] = 3]") and enhanced Nd abundance with [Nd/Fe] = 4 in atmospheric layers outside $\log \tau_{5000} = -3.6$ (columns "layer"). i and j are level numbers according to the model atom.

λ, Å	E_{low}, eV	$\log gf$	$i - j$	[Nd/Fe] = 3			layer			W_{obs}
				W_{LTE}	W_{NLTE}	Δ_{NLTE}	W_{LTE}	W_{NLTE}	Δ_{NLTE}	
Nd II										
4706.54	0.00	-0.88	1 - 101	92	92	0.00	90	41	1.30	40
4811.34	0.06	-1.01	2 - 101	86	86	0.00	85	32	1.42	30
4061.08	0.47	0.55	7 - 122	127	128	-0.03	109	77	1.05	73
5319.82	0.55	-0.21	8 - 105	105	110	-0.12	103	53	1.31	47
5533.82	0.56	-1.23	8 - 104	67	67	0.00	63	12	1.39	9
5077.15	0.82	-1.04	12 - 114	62	62	-0.01	55	12	1.07	7
5399.09	0.93	-1.41	14 - 114	44	42	0.03	32	5	0.97	6
5033.51	1.14	-0.47	20 - 123	74	77	-0.07	69	22	1.05	24
6650.52	1.95	-0.17	75 - 129	66	71	-0.09	59	15	1.07	25
Nd III										
5294.10	0.00	-0.70	248 - 263	105	124	-0.42	106	127	-0.51	143
6550.23	0.00	-1.50	248 - 257	77	99	-0.40	86	112	-0.48	120
4796.49	0.14	-1.66	249 - 270	57	69	-0.27	53	70	-0.37	83
5633.55	0.14	-2.19	249 - 263	34	48	-0.26	31	55	-0.43	67
6327.26	0.14	-1.42	249 - 260	74	94	-0.37	81	104	-0.42	123
6145.07	0.30	-1.34	250 - 263	74	94	-0.36	78	99	-0.40	
5987.68	0.46	-1.27	251 - 266	67	87	-0.37	70	91	-0.38	108
6690.83	0.46	-2.36	251 - 263	20	32	-0.27	17	38	-0.46	53
5677.18	0.63	-1.43	252 - 271	54	68	-0.26	49	64	-0.27	90
5845.02	0.63	-1.18	252 - 270	63	82	-0.35	64	82	-0.33	110

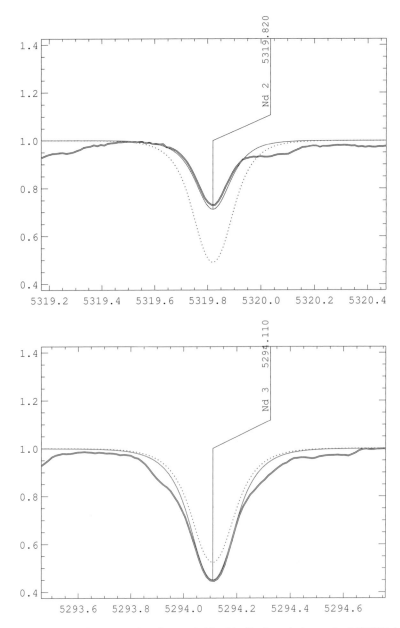

Figure 4. A comparison between the observed (double line) and theoretical NLTE (solid line) and LTE (dotted line) Nd II λ5319 (left panel) and Nd III λ5294 (right panel) line profiles calculated with the stratified Nd abundance distribution from Fig. 3.

3. Nd abundance distribution in the atmospheres of the roAp stars γ Equ and HD 24712

Ryabchikova *et al.* (2002) performed a stratification LTE analysis of Nd in the atmosphere of roAp star γ Equ and concluded that Nd should be concentrated above $\log \tau_{5000} = -8$ to fit the observed lines of Nd II–Nd III. This result was qualitatively consistent with the observed distribution of pulsational radial velocity (RV) amplitudes observed in spectral lines of different elements/ions. However, even with stratified Nd abundances the fit of line profiles was not good in many cases. New NLTE calculations were made of the Nd lines in γ Equ and in another roAp star HD 24712 (Sachkov *et al.* 2005). Equivalent widths in γ Equ were measured in the spectra described in Sachkov *et al.* (2004) and in Kochukov *et al.* (2004) and are given in the last column of Table 1. Fig. 4 shows a comparison between the observed and synthetic NLTE and LTE line profiles for chosen Nd II and Nd III lines calculated with the Nd abundance distribution from Fig. 3.

The results of Table 1 and Fig. 4 demonstrate a fairly good agreement between observations and calculations and provide strong evidence for Nd concentrations above $\log \tau_{5000} = -3$. This new Nd abundance distribution is more realistic than Ryabchikova *et al.* (2001) with LTE. Curiously, we also get good agreement both in equivalent widths and in line profile fit for another roAp star, HD 24712, using the same Nd abundance distribution, although this star is cooler than γ Equ (Sachkov *et al.* 2005).

Line formation depths were calculated and applied to the pulsational analysis of γ Equ atmosphere (Kochukhov 2005).

Acknowledgements

The authors thank Professor Thomas Gehren for providing the codes DETAIL and MAFAGS. All NLTE calculations were performed using the eridani computer of the Institute of Astronomy and Astrophysics of Munich University. MLI and RTA are partially supported by the Presidium RAS Programme "Nonstationary phenomena in astronomy", by RFBR grant number 04-02-16788 and by the Federal Programme "Astronomy".

References

Allen, C.W. 1973, *Astrophysical Quantities* (Athlone Press)

Blaise, J., Wyart, J.-F., Djerad, M.T., Ahmed, Z.B. 1984, *Phys. Scripta* **29**, 119

Butler, K., Giddings, J. 1985, *Newsletter on the analysis of astronomical spectra* No. 9, (University of London)

Cowan, R.D. 1981, *The Theory of Atomic Structure and Spectra* (Univ.of California Press. Berkely. California. USA)

Den Hartog, E.A., Lawler, J.E., Sneden, C., Cowan, J.J. 2003, *ApJS* **148**, 543

Drawin, H.-W. 1961, *Z.Physik* **164**, 513

Fuhrmann, K., Pfeiffer, M., Frank, C., Reetz, J., Gehren, T. 1997, *A&A* 323, 909

Kochukhov, O., Ryabchikova T., Piskunov, N. 2004, *A&A* 415, L13

Kochukhov, O. 2005, *These Proceedings*, 433

Kupka, F., Piskunov, N., Ryabchikova, T.A., Stempels, H.C., Weiss, W.W. 1999, *A&AS* 138, 119

Martin, W.C., Zalubas, R., Hagan, L. 1978, *Atomic energy levels - The Rare Earth Elements. NSRDS-NBS* 60, (U.S. Gov. Print. Off., Washington)

Ryabchikova, T.A., Savanov, I.S., Malanushenko, V.P., Kudryavtsev, D.O. 2001, *Astron. Rep.* 45, 382

Ryabchikova, T., Piskunov, N., Kochukhov, O., Tsymbal, V., Mittermayer P., Weiss, W.W. 2002, *A&A* 384, 545

Rybicki G.B., Hummer D.G. 1991, *A&A* 245, 171

Rybicki G.B., Hummer D.G. 1992, *A&A* 262, 209

Sachkov, M., Ryabchikova, T., Kochukhov, O., Weiss, W.W., Reegan, P., Landstreet, J.D. 2004, in: D.W.Kurtz & K.Pollard (eds.), *IAU Coll. 193 Variable Stars in the Local Group*, (in press)

Sachkov, M., Ryabchikova, T.,Ilyin, O., Kochukhov, O., Lueftinger, T. 2005, *These Proceedings*, GP2

van Regemorter H. 1962, *ApJ* **136**, 906

Discussion

G. MATHYS: Have you taken the hyperfine structure into account in your determinations of abundance of Nd? It is definitely seen as a difference between the separation of the σ_{b-} and π^- components, on the one hand, and of the σ_{r-} and π^- components, on the other hand, in strongly magnetic Ap stars with magnetically resolved lines.

T. RYABCHIKOVA: No, we have not. According to the laboratory analysis by Dolk *et al.* (2002, A&A 385, 111) the hyperfine structure of the two odd Nd isotopes, which represented 20% of the total terrestrial abundance, generally is not observable. Isotopic shifts are also small The heaviest ^{150}Nd isotope is only responsible for $\approx$ 6% of the terrestrial mixture. The observed asymmetry you have mentioned may be caused by line blending. The Nd III λ6145 line which you mean, I believe, is blended with weak Si I and Ca II lines. Both elements may be stratified, which increases the blending.

C. COWLEY: Are there some cases (stars) where the Nd II and Nd III LTE abundances do agree? And if so, does this mean that if we do the calculations for them in NLTE that the Nd II and Nd III will no longer agree?

T. RYABCHIKOVA: There are some cases where Nd II and Nd III seem to provide consistent results (mainly for effective temperatures above 8000 K), but the errors in the abundance determinations from Nd II lines are rather large, up to 0.5 dex, and may be of the order of NLTE effects.

N. PISKUNOV: Do you expect that the application of NLTE spectral synthesis to other rare-earth elements would bring down their formation depth as it did for Nd?

RYABCHIKOVA: Yes, we expect it.

The A-Star Puzzle
Proceedings IAU Symposium No. 224, 2004
J. Zverko, J. Žižňovský, S.J. Adelman, & W.W. Weiss, eds.

© 2004 International Astronomical Union
DOI: 10.1017/S1743921304004715

Panel discussion section F

CHAIR: D.W. Kurtz

SECTION ORGANIZER & KEY-NOTE SPEAKER: C.R. Cowley
INVITED SPEAKERS: T. Ryabchikova, G.M. Wahlgren
CONTRIBUTION SPEAKERS: I.Kh. Iliev, S.J. Adelman, T. Ryabchikova

Discussion

BALEGA: Is there any sense in performing a speckle interferometric survey to search for secondaries among the Ap stars? As a rule, the secondary companion will be always fainter and cooler than the primary star. Therefore, its input into the common spectrum will be negligible. A good example is 53 Cam with a magnitude difference between the components of 1.2.

ILIEV: Definitely "yes." Even if the secondary is 4 or 5 magnitudes fainter than the primary, its contribution to the overall spectrum can be crucial, as it was already shown for 17 Lyr (HD 178449). Since the presence of the secondary provides significant information about the entire system, there are no advantages in neglecting the companion.

RYABCHIKOVA: It depends on how much fainter is the secondary, because if it is some 5 - 10 times fainter in the violet it might be much brighter in the red and infrared regions which astronomer are now starting to analyze spectra. If we neglect the cool secondary we could make some errors in any analysis. I have an example when the difference was 2 magnitudes. α And A has $T_{\mathrm{eff}} = 13800\,\mathrm{K}$, but we clearly see the secondary lines ($T_{\mathrm{eff}} = 8000\,\mathrm{K}$) in the even not very red spectral region at $4950\,\text{Å}$. We used these secondary lines for accurate radial velocity determinations.

KURTZ: Comment. With your NLTE line profiles for Nd II and Nd III (Mashonkina *et al.* 2005, *These Proceedings*, 315) you have reduced the line strength substantially with NLTE comparing to LTE. That is going to change the diffusion calculation. Earlier Georges Alecian raised a problem regarding the small optical depths. It would be very interesting to know if one can perform a diffusion calculations that produces the kind of stratified layers of the Nd and Pr that Ryabchikova sees in these stars.

LEBLANC: Of course, if we have a sufficient atomic data it is possible to do calculations for these elements in NLTE using a code like PHOENIX. However, the addition of new elements to this code is nontrivial. We wish, in the not too distant future, to add more rare earth elements to our stratified model calculations.

GRIFFIN: Comment. The more we study A-type stars, the more we seem to be faced with samples of one. At the same time, Wahlgren is encouraging us to use the properties of stars to investigate parameters from a fresh perspective "across borders." With so many samples of one! This is calling for a more concerted effort from us all. Bear in mind that the more we look in detail at an A-star the more likely it is to prove abnormal in some aspect.

WEISS: I would like to recall a statement of Cowley that he made some years ago at another conference: "If a star is normal it depends on the spectral resolution one was using".

ADELMAN: There are some unique stars and we do not know if similar stars exist. If we have a complete sample down to some limiting magnitude, then estimates of the frequencies of the various types can be obtained. To find photometrically interesting stars often requires a large amount of observing on a small telescope, at least 200 good sets of differential measurements. Additional observations often are needed to follow up an initial discovery. We know only a small fraction of what could be learned as most published photometric sets of data contain 25 or so observations and as there are only a few active photometric observers of normal and peculiar A stars.

KURTZ: In Ryabchikova's first talk she brought up the problem of our discrepancy with the non-detection of magnetic variability and pulsation time scales in some roAp stars. Leone and I observed γ Equ with the Galileo telescope and claimed to have detected a 250 Gauss pulsation in the magnetic field strength in that star with a 12 min time scale. If this is real, as the pulsation mode moves around the star it will sample the magnetic field horizontally or vertically at different places. Further if this is true it gives a new three dimensional handle on the magnetic field structure. Kochukhov, Ryabchikova, and I observed that star using the 6-m telescope and we found absolutely nothing. Kochukhov and I have exchanged data. I find no variations in his data but I do find variations in our data. I have also re-reduced the data from the Galileo telescope using both the standard technique of circular polarization and the technique that Leone and I used originally; both of them gave the same resulting signal. We do not know if the star changed or there is something wrong with the data sets. So for the time being this all simply remains unclear. It may be true or it may not be true. We do not know where the problem is, we are searching for more ideas about where about to go: to a 10 m telescope or to another star.

RYABCHIKOVA: The first step should be some independent analyses of the two data sets and then we will have to decide.

KURTZ: I agree. It would be very good then to do.

The A-Star Puzzle
Proceedings IAU Symposium No. 224, 2004
J. Zverko, J. Žižňovský, S.J. Adelman, & W.W. Weiss, eds.

© 2004 International Astronomical Union
DOI: 10.1017/S1743921304004727

Pulsations of A stars

Luis A. Balona

South African Astronomical Observatory,
P.O. Box 9, Observatory 7935, Cape Town, South Africa
email: lab@saao.ac.za

Abstract. A review of recent work on pulsating A stars is presented. The types of pulsating stars are the roAp, δ Scuti, λ Bootis, Am stars, pre-Main Sequence pulsating A stars and α Cygni stars (pulsating A supergiants). Population II pulsating A stars such as SX Phe (blue stragglers) and RR Lyraes are also discussed. The emphasis is on the physics that can be derived from the study of stellar pulsations and suggestions for further progress.

Keywords. Stars: oscillations, stars: chemically peculiar

1. Introduction

The A-type stars present possibly the most interesting and challenging problems in stellar pulsation. They lie on the hot side of the Classical Instability Strip and include all evolutionary states from pre-Main Sequence to supergiants and also Population II pulsators such as RR Lyraes. Pulsating chemically peculiar and magnetic stars such as the roAp stars offer additional problems and opportunities.

In this paper I will discuss what can be learned from the pulsations in A-type stars, presenting the most recent results from the literature and examining what we have learned, or may learn from such studies.

2. The roAp stars

The roAp stars are the most complex pulsating stars. The observed pulsations are strongly influenced by the magnetic field and are affected by horizontal spatial chemical composition variations and vertical stratification. Furthermore, the pulsation frequencies are near the critical acoustic limit with wavelengths comparable with the pressure scale height in the atmosphere. As a result, it is not sufficient to use a simple one-layer approximation to model the line profile variations.

In the atmosphere the pulsations are, in fact, magneto-hydrodynamic waves with both longitudinal and transverse components. The eigenfunctions do not resemble a simple combination of normal p-mode eigenfunctions. Despite the very high pulsation frequency, when in normal stars one may expect the displacement to be almost entirely vertical, the horizontal displacement component is probably significant because of coupling with the magnetic field. In addition, the form of the eigenfunction probably varies across the photosphere due to the differing geometry of the magnetic field. As a result, I believe we are a long way from understanding and modeling the line profile variations. Nevertheless, these complexities present unique opportunities for understanding the interaction between pulsations and magnetic fields.

The first step in any asteroseismological investigation is to determine the mode of pulsation. Owing to the complex nature of the eigenfunctions, well-known photometric and spectroscopic methods applicable to other pulsating stars fail for roAp stars. In addition, the pulsation frequencies are modified quite significantly by the interaction with the magnetic field. As a result, the roAp stars are not suitable objects for asteroseismology on a detailed scale.

Recent high-resolution spectroscopic observations appear to show that the pulsations are apparently nonaxisymmetric (Kochukhov & Ryabchikova, 2001ab). Furthermore, the phase of the pulsations appears to vary from ion to ion, indicative of the small scale length and, possibly, a running wave component. These observations remain to be explained and, on first sight, are at odds with the very successful and generally accepted oblique pulsator model. It should be noted, however, that Shibahashi (2005) has proposed that the apparent nonaxisymmetric line profiles could possibly be understood as due to a shock wave in an axisymmetric mode. This is an interesting idea which deserves further study.

Bigot & Dziembowski (2002) introduced two main improvements in understanding the pulsations in magnetic stars. They treated the effect of the magnetic field with a non-pertubative theory and included the effects of the centrifugal distortion of the star. They found that rotation plays an important role, even in the slow rotators. As a result, the magnetic and pulsation axes are no longer aligned.

The mechanism responsible for exciting the oscillations in roAp stars is still unknown. Balmforth et al. (2000) suggested that the high magnetic field intensity at the poles will suppress convection. Freed of this damping, the star may pulsate in this region driven by the κ mechanism acting in the hydrogen ionization zone. Cunha (2002) estimated the theoretical edges of the instability strip using such models. Comparison with observations show that most roAp stars fall well within this strip, though mostly on the cooler side. Cunha suggests that the Ap stars within the instability strip which do not pulsate (the "noAp" stars) may, in fact, be pulsating with periods which are longer than expected. They may be undetected because of the observational technique normally used to discover roAp stars.

Cunha et al. (2003) attempted to constrain theoretical models which characterise the pulsations in HR 1217. They find that the interior chemical composition is more important than the modeling of convection and that lower abundances of heavy elements and larger abundance of helium lead to a closer match to the data.

Very recently, there has been several papers which examine the apparent variations in the magnetic field over a pulsational cycle. Leone & Kurtz (2003) found evidence of such an effect in the Nd III lines of γ Equ. Savanov et al. (2003) offer confirmation, but further observations by Kochukhov et al. (2004) do not show this effect. Even if such an effect is securely established, it will be difficult to model the line profiles for reasons already mentioned. As a result, it may not be possible to understand the physics involved in the process and, therefore, to make any firm deductions.

Further progress in our understanding of the roAp stars is likely to come from time resolved high-dispersion spectroscopy and from space observations. We need to understand the cause of the apparently nonaxisymmetric modes giving rise to the line profile variations and the variation of pulsation phase in the atmosphere. Space observations should uncover many new very low amplitude modes which could be very important for a full understanding of the interaction between pulsation and magnetic fields. An interesting possibility would be the detection of a chromosphere or corona around roAp stars. Because the pulsation frequency is so close to the acoustic limit, it is possible that the running wave component may deposit sufficient energy in the upper atmosphere to produce the necessary elevated temperature.

On the theoretical side, the main unsolved problem is the unknown excitation mechanism. The proposal by Balmforth et al. (2000) is promising, but requires further study.

3. δ Scuti stars

In many ways δ Scuti stars are probably the best candidates for asteroseismology, especially the hotter stars for which the effects of convection can be ignored. Unlike roAp stars, they have "simple" atmospheres and the pulsation frequencies are considerably higher than the rotational frequencies so that the eigenfunctions of the pressure modes are well approximated by simple spherical harmonics. The usefulness of these stars arises from the large number of modes that have been observed in a few cases. Breger *et al.* (1998) detected at least 24 significant frequencies of low degree in the A5 star FG Vir. Recent observations have more than doubled this number.

A large number of observed frequencies is an important requirement for asteroseismology as it constrains the solution. Of equal importance are the mode identifications and a model which is a good approximation to the star. Mantegazza & Poretti (2002) have recently obtained line profile observations of FG Vir which are in general agreement with the photometric mode identifications, but suffer from an uncertainty of ± 1 in the degree, ℓ. This uncertainty is, unfortunately, typical of line profile analyses, independently of the S/N ratio of the data, and negates much of the effort that is required to obtain these data. They do, however, provide valuable information on the inclination angle which in turn is necessary to apply rotational corrections to the frequencies. It has been shown that even a relatively small rotational velocity has a great impact on the expected frequencies leading to serious errors in matching the observed frequencies with models. Interactions in the eigenfunctions of closely spaced modes distort the mode identifications, especially in moderately and rapidly rotating stars.

Although the pulsation modes of most frequencies in FG Vir appear to be rather well identified from photometric techniques, a complete and unique astroseismic solution has still to be found, in spite of intensive efforts (Breger *et al.* 1999). One is beginning to suspect that there is something wrong with models of the "simple" envelopes of A-type stars. Templeton *et al.* (2001) find that changes to the stellar mass, the chemical composition and the convective core overshooting length change the observed pulsation spectrum significantly. It, therefore, seems that there are more unobservables than previously supposed. This stresses the importance of detecting as many frequencies as possible.

Most observations thus far have concentrated on modes of low degree which are, in fact, the most useful from an asteroseismological point of view. However, modes of high degree have been detected in some stars (e.g., τ Peg, Kennelly *et al.* 1998). These modes can only be detected if the star has a moderate to high projected rotational velocity. It is possible that many modes of high degree may become observable with MOST, although establishing the spherical harmonic degree may be an impossible task.

It seems to me that the modeling of pulsations in δ Scuti stars is well behind the observations. It would be interesting to explore the effects of changing the models in various ways to determine which physical processes lead to a better match with the data.

4. λ Bootis stars

The λ Boo stars are Population I late B- to early F-type stars with moderate to extreme surface underabundances of most Fe-peak elements and solar abundances of the lighter elements. Paunzen *et al.* (2002a) find that these stars comprise a homogeneous group containing stars at all phases of the Main Sequence evolution. This suggests that the process for depleting the Fe-peak elements works continuously during the Main Sequence phase.

In an investigation of the pulsations of λ Boo stars, Paunzen *et al.* (2002b) found that at least 70% of all λ Boo stars within the Classical Instability Strip pulsate and they do so with high overtone modes. These results indicate that there is probably no clear difference between λ Boo stars and δ Sct stars as far as pulsational properties are concerned. Thus it seems reasonable to suppose that the peculiar abundances are confined to the surface and do not significantly affect the pulsations. The challenge remains, of course, as to the mechanism whereby the chemical peculiarities come about.

Paunzen *et al.* (2003) observed several λ Boo stars in the near infrared with the object of detecting circumstellar material. They found evidence for an infrared excess in a quarter of the sample of stars. Gray & Corbally (2002) continued a program to search for CP A stars, particularly λ Boo stars, in open clusters. It now appears that no λ Boo stars occur in clusters. It would seem that some factor external to the star and related to membership in open clusters prevents the operation of the λ Boo phenomenon. They suggest that this may be a result of the photoevaporation of circumstellar material around these stars by UV radiation from massive cluster stars.

Paunzen (2005) has suggested that the peculiar abundances in λ Boo stars originate by accretion when the star passes through an interstellar dust cloud. There seems to be convincing evidence that this idea is correct.

5. Pulsations in Am stars

The classical Am stars are A-type stars whose spectra are characterized by an underabundance of Ca (and/or Sc) coupled with an overabundance of the Fe group and heavier elements with respect to normal stars of the same colour. Originally it was thought that Am stars do not pulsate because diffusion drains helium from the ionization zone which is responsible for driving the pulsations. This is consistent with the idea that diffusion is the mechanism for the line-strength anomalies of the Am and Ap stars.

Observations of several Am stars show that this simple model is not valid; pulsation and metallicism have been found to co-exit in classical Am stars, though pulsating Am stars are rare. Pulsations in evolved Am stars and marginal Am stars lying near the red edge of the instability strip can be understood within the context of diffusion theory in terms of driving by residual He II and the H ionization zone. Most of the observed pulsating Am stars seem to be of this class (e.g., HD 98851 and HD 102480, Joshi *et al.* 2003, HD 13079, Martinez *et al.* 1999). However, Am stars near the ZAMS are not expected to pulsate (Turcotte *et al.* 2000).

The complexities regarding the pulsations in Am stars was nicely summarized by Kurtz (2000). In a nutshell, the slowly rotating A-type stars appear as chemically peculiar because of diffusion and generally do not pulsate because the κ mechanism can no longer operate in the depleted He II ionization zone. However, evolved Am stars (the ρ Pup or δ Del stars) pulsate due to an evolutionary replenishment of some helium in the He II ionization zone. In marginal Am stars the He II ionization zone is not fully depleted, so low-amplitude pulsations can occur. In other words, diffusion in those Am stars which pulsate is assumed not to have completely drained helium from the He II ionization zone so that the κ mechanism can still operate, leading to low-amplitude pulsations.

To test these ideas, one needs a clearer understanding of the incidence of low amplitude pulsations in the A-star Main Sequence band. The role of photometric space observations of Am stars and apparently non-pulsating Main Sequence A-type stars is, therefore, of great importance for further progress.

6. Pulsations in pre-Main Sequence A-type stars

Intermediate mass pre-Main Sequence stars are expected to cross the instability strip on their way to the Main Sequence. The Herbig Ae stars are thought to be such objects as they are found in star forming regions and show the characteristics of infalling material. Although they normally exhibit significant irregular light variations, pulsation in these stars is not difficult to find because of the periodic nature and the short period. Catala (2003) has presented a nice review of pulsations in these stars.

Breger (1972) was the first to identify two pre-Main Sequence pulsators in the open cluster NGC 2264. The bright Herbig Ae star HR 5999 was found to be pulsating by Kurtz & Marang (1995) and Kurtz & Catala (2001). Strangely, the pulsations in this star do not seem to be detectable in high-dispersion spectroscopic observations (Balona, unpublished).

Marconi *et al.* (2000) discovered short-period variability in two Herbig Ae stars V351 Ori and HD 35929. Marconi *et al.* (2001) found that V351 Ori pulsates with at least four frequencies. A detailed photometric and spectroscopic analysis of V351 Orionis by Balona *et al.* (2002) shows that two frequencies are definitely present in the light curve and a further frequency is visible in the radial velocity variations. Furthermore, modes of high degree ($\ell \approx 8$) are sometimes visible.

Kurtz & Muller (2001) discovered pulsations in the light curve of the Herbig Ae star HD 142666, but could not detect pulsations in another Herbig Ae star, HD 142527, lying within the instability strip.

HD 104237 (A4V) is of particular interest. Donati *et al.* (1997) detected an unambiguous magnetic field (the first time detection of a magnetic field in a Herbig Ae star) and small amplitude radial velocity variations with a 37-min period. Recent results (Böhm *et al.* 2004) show that the star is a spectroscopic binary and pulsates with at least five periods. This star is a multiple object, there being at least 5 components (Feigelson *et al.* 2003) with four low-mass PMS companions.

Ripepi *et al.* (2002) obtained photometric time series observations of pre-Main Sequence stars in IC 348. They found one star, H254, which has a single mode with frequency 7.406 cycles d^{-1}. Ripepi *et al.* (2003) present multisite observations of V353 Ori in which 5 frequencies are detected. Pinheiro *et al.* (2003) detected two periods in V346 Ori.

Perhaps the most remarkable discovery in this regard is that of pulsations in the archetypal young A-type star β Pictoris. Koen (2003) discovered three frequencies in the light variations of β Pic. Subsequent high-resolution spectroscopic observations (Koen *et al.* 2003) revealed at least 18 frequencies, mostly of high degree ($4 < \ell < 10$). Comparisons with models suggest that the observed frequencies are compatible with those expected for a δ Scuti star with about the same temperature and luminosity.

The hotter A-type stars are not expected to be chromospherically active. The detection of strong emission lines of O VI and C III with FUSE in β Pic (Bouret *et al.* 2002) came as a complete surprise. The data are compatible with the presence of an active chromosphere in β Pic. Whether the presence of a chromosphere may be a result of energy transport by pulsation is an open question.

7. Pulsation in A supergiants

The variations in A supergiants are extremely complex and involve timescales from hours to years. Variations can occur in either the photosphere or the extended envelope that is associated with mass loss. The early work by Abt (1957), who observed the radial velocity of several A- and F-type supergiants for nearly every night for 30 consecutive nights, still provides an important benchmark. All the stars proved to be variable with characteristic periods in the range 4 - 30 days.

The most extensively studied A supergiant is α Cyg. Lucy (1976) performed a periodogram analysis of 144 velocities which suggest that the variability is due to the simultaneous excitation of many discrete pulsation modes with periods ranging from 6.9 to 100.8 d. The expected radial fundamental mode for α Cyg is about 14 d. Since that time, the only spectroscopic work of note on these stars is that of Kaufer *et al.* (1997). They obtained time series spectra of A-type and late B-type supergiants. They find complex cyclical variations of the radial velocities with typical velocity dispersion of about 3 km s^{-1}. Multiple periods both longer and shorter than the radial fundamental mode are present. Examination of the line profile variations show prograde traveling features. These could be g-modes of low degree ($\ell < 5$).

Adelman & Albayrak (1997) examined the *Hipparcos* database for variability of A0-A5 supergiants in the Bright Star Catalogue. They find that the amplitude increases with luminosity, but that no definite periods can be determined from the data other than those for the three A supergiants examined by the *Hipparcos* team. The periods given for these three stars are 6.36 d (HR 618), 1.33 d (HR 4169) and 2.36 d (HR 6825), all considerably shorter than the radial fundamental mode.

Because of the long periods and the large amount of data required to extract a periodic signal from the irregular variations which characterise these stars, very little work has been done. The most promising approach is a dedicated program of observations on an automatic photometric telescope.

8. SX Phe stars and pulsating blue stragglers

Blue stragglers are stars (mostly found in globular clusters) which are probably formed by stellar mergers from short-period binaries or collisions. These stars fall in a region of the colour-magnitude diagram consistent with having masses greater than the current cluster turnoff. The field star, SX Phe, is sometimes identified as a possible blue straggler. A remarkably high fraction of blue stragglers exhibit variations associated with binary interactions and pulsation. Some of them show two radial modes, which allows the mass to be constrained.

The most comprehensive investigation of pulsating blue stragglers is that of Gilliland *et al.* (1998). Observations of six SX Phe variables in the core of 47 Tuc were obtained using HST. Two of these stars show pulsations in the fundamental and first-overtone mode, two others in the forth and fifth radial overtone while the remaining two variables are multiperiodic variables. From these data they were able to estimate the masses which are well above the the turnoff mass in 47 Tuc and are consistent with the merger scenario.

9. RR Lyrae stars and the Blazhko effect

The nature of the Blazhko effect in RR Lyrae stars remains an unsolved problem. The effect is the slow periodic, or quasi-periodic, modulation of the amplitude and shape of the light curve. One possibility is that it is due to the modulation of the pulsations during a magnetic cycle. However, Chadid *et al.* (2004) found no evidence for the presence of a magnetic field in RR Lyrae, the brightest Blazhko star.

Other evidence which seems to exclude models involving rotational effects is that the Blazhko period does not remain constant. Jurcsik *et al.* (2002) find that the pulsation period of XZ Dra exhibits cyclic, but not strictly regular variations with a ≈ 7200 d period and that the Blazhko period seems to follow the observed period changes of the radial fundamental mode. LaCluyze *et al.* (2004) found the Blazhko period in XZ Cyg is anti-correlated with the primary period of XZ Cyg. These are important results because they exclude any explanation which requires that the Blazhko period be exactly equal or directly proportional to the rotation period of the star.

Recent attempts to explain the Blazhko effect have focused on resonance models, in which there is a nonlinear resonance between the dominant radial mode and a nonradial mode, or the oblique magnetic rotator. Nowakowski (2002) and Nowakowski & Dziembowski (2001) find that significant amplitude and phase modulation of the light curve might occur from a pulsation in which the pair of modes $\ell = 1, m = \pm 1$ are excited. The Blazhko period is then determined by the rotation frequency weighted with the Brunt-Vaisala frequency in the deepest part of the radiative envelope of the star. It is not clear whether this model can explain the correlation or anti-correlation of the Blazhko period with the fundamental radial mode.

The detection and identification of nonradial modes in RR Lyrae stars would be an important achievement for our understanding of the Blazhko effect. To this end, Kolenberg (2002) and Chadid *et al.* (1999) performed detailed frequency analysis of line profile variations in RR Lyr and obtained strong evidence for the presence of nonradial modes. Time resolved high dispersion spectra of RR Lyr stars seem to be a promising new tool for understanding the Blazhko effect.

10. Conclusions

The roAp stars present us with a large number of unsolved problems, of which the enigmatic line profile variations and phase shifts are just two examples. Progress is hampered because we do not yet fully understand the eigenfunction. Further observations of time-resolved line profile variations will undoubtedly assist in resolving this very difficult problem.

It seems to me that observations of pulsations in δ Scuti stars are well ahead of theory. We need to understand why it is so difficult to match the observed frequencies with those from models. Definite progress has recently been made in our understanding of the λ Boo stars (Paunzen 2005). It is likely that high precision photometry from space will enable very low amplitude Am stars to be detected and thus confirm our understanding of how pulsations are driven in these stars.

Detection of pulsations in pre-Main Sequence stars is necessary to map out the pre-MS instability strip, although we need to refine estimates of temperature and luminosity. Pulsations offer a possible mechanism of energy deposition in the upper atmosphere and may lead to the creation of a chromosphere. The evidence for chromospheres on pre-Main Sequence Vega-type stars and Herbig Ae stars is convincing and opens new lines of research.

There has been very little progress on observations on SX Phe stars, blue stragglers and supergiants. To some extent this can be attributed to the faintness of the former and the long-term observational effort in the latter.

Some progress is at last being made on the long-standing puzzle of the Blazhko effect in RR Lyrae stars. It is now clear that the Blazhko period is coupled in some way with the pulsation period. The challenge of detecting the nonradial modes responsible for the coupling is probably best addressed by high resolution spectroscopy.

References

Abt, H.A. 1957, *ApJ* 126, 138

Adelman, S.J. & Albayrak, B. 1997, *IBVS* 4541

Balmforth, N.J., Cunha, M.S., Dolez, N., Gough, D.O. & Vauclair, S. 2000, *MNRAS* 323, 362

Balona, L.A., Koen, C. & van Wyk, F. 2002, *MNRAS* 333, 923

Bigot, L. & Dziembowski, W.A. 2002, *A&A* 391 235

Böhm, T., Catala, C., Balona, L.A. & Carter, B. 2004, *A&A*, in press.

Bouret, J.-C., Deleuil, M., Lanz, T., Roberge, A., Lecavelier des Etangs, A. & Vidal-Madjar, A. 2002, *A&A* 390, 1049

Breger, M. 1972, *ApJ* 171, 539

Breger, M., Zima, W., Handler, G., *et al.* 1998, *A&A* 331, 271

Breger, M., Pamyatnykh, A.A., Pikall, H. & Garrido R. 1999, *A&A* 341, 151

Catala, C. 2003, *ApSS* 284, 53

Chadid, M., Kolenberg, K., Aerts, C. & Gillet, D. 1999, *A&A* 352, 201

Chadid, M., Wade, G.A., Shorlin, S.L.S. & Landsteet, J.D. 2004, *A&A* 413, 1087

Cunha, M.S. 2002, *MNRAS* 333, 47

Cunha, M.S., Fernandes, J.M.M.B. & Monteiro M.J.P.F.G. 2003, *MNRAS* 343, 831

Donati, J-F., Semel, M., Carter, B.D., Rees, D.E. & Cameron, A.C. 1997, *MNRAS* 291, 658

Feigelson, E.D., Lawson, W.A. & Garmire, G.P. 2003, *ApJ* 599, 1207

Gilliland, R.L., Bono, G., Edmonds, P.D., Caputo, F., Cassisi, S., Petro, L.D., Saha, A. & Shara, M.M. 1998, *ApJ* 507, 818

Gray, R.O. & Corbally, C.J. 2002, *AJ* 124, 989

Joshi, S., Girish, V., Sagar, R., Kurtz, D.W., Martinez, P., Kumar, B., Seetha, S., Ashoka, B.N. & Zhou, A. 2003, *MNRAS* 344, 431

Jurcsik, J., Benko, J.M. & Szeidl, B. 2002, *A&A* 396, 539

Kaufer, A., Stahl, O., Wolf, B., Fullerton, A.W., Gaeng, T., Gummersbach, C.A., Jankovics, I., Kovacs, J., Mandel, H., Peitz, J., Rivinius, T. & Szeifert T. 1997, *A&A* 320, 273

Kennelly, E.J., Brown, T.M., Kotak, R., Sigut, T.A.A., Horner, S.D., Korzennik, S.G., Nisenson, P., Noyes, R.W., Walker, A. & Yang S. 1998, *ApJ* 495, 440

Kochukhov, O. & Ryabchikova, T. 2001a, *A&A* 374, 615

Kochukhov, O. & Ryabchikova, T. 2001b, *A&A* 377, L22

Kochukhov, O., Ryabchikova, T. & Piskunov, N. 2004, *A&A* 415, 13

Koen, C. 2003, *MNRAS* 341, 1385

Koen, C., Balona, L.A., Khadaroo, K., Lane, I., Prinsloo, A., Smith, B. & Laney, C.D. 2003, *MNRAS* 344, 1250

Kolenberg, K. 2002, A spectroscopic study of the Blazhko effect in RR Lyrae, Ph.D. Thesis, University of Leuven

Kurtz, D.W. & Marang F. 1995, *MNRAS* 276, 191

Kurtz, D.W. 2000, in Delta Scuti and Related Stars, ed. M. Breger & M. H. Montgomery (San Francisco: ASP Vol 210), 287

Kurtz, D.W. & Catala C. 2001, *A&A* 369, 981

Kurtz, D.W. & Muller, M. 2001, *MNRAS* 325, 1341

LaCluyze, A., Smith, H.A., Gill, E.-M., Hedden, A., Kinemuchi, K., Rosas, A.M., Pritzl, B.J., Sharpee, B., Wilkinson, C., Robinson, K.W., Baldwin, M.E. & Samolyk, G. 2004, astro-ph/0401314

Leone, F. & Kurtz, D.W. 2003, *A&A* 407, 67

Lucy, L.B. 1976, *ApJ* 206, 499

Mantegazza, L. & Poretti, E. 2002, *A&A* 396, 911

Marconi, M., Ripepi, V., Alcala, E., Covino, E., Palla, F., & Terranegra L. 2000, *A&A* 355, L35.

Marconi, M., Ripepi, V., Bernabei, S., Palla, F., Alcala, E, Covino, E. & Terranegra, L. 2001, *A&A* 372, L21.

Martinez, P., Kurtz, D.W., Ashoka, B.N., Chaubey, U.S., Gupta, S.K., Leone, F., Catanzaro, G., Sagar, R., Raj, E., Seetha, S. & Kasturirangan, K. 1999, *MNRAS* 309, 871

Nowakowski, R.M. & Dziembowski, W.A. 2001, *Acta Astron.* 51, 5

Nowakowski, R. 2002, in Radial and Noradial Pulsations as Probes of Stellar Physics, ed. C. Aerts, T.R. Bedding & J. Christensen-Dalsgaard (San Francisco: ASP), 408

Paunzen, E., Iliev, I. Kh., Kamp, I. & Barzova, I.S. 2002a, *MNRAS* 336, 1030.

Paunzen, E., Handler, G., Weiss, W.W., Nesvacil, N., Hempel, A., Romero-Colmenero, E., Vuthela, F.F., Reegen, P., Shobbrook, R.R., Kilkenny, D. 2002b, *A&A* 392, 515

Paunzen, E., Kamp, I., Weiss, W.W. & Weisemeyer, H. 2003, *A&A* 404, 579

Paunzen, E. 2005, *These Proceedings*, 443

Pinheiro, F.J.G., Folha, D.F.M., Marconi, M., Ripepi, V., Palla, F., Monteiro, M.J.P.F.G. & Bernabei, S. 2003, *A&A* 399, 271

Ripepi, V., Palla, F., Marconi, M., Bernabei, S., Arellano Ferro, A., Terranegra, L. & Alcala, J.M. 2002, *A&A* 391, 587

Ripepi, V., Marconi, M., Bernabei, S., *et al.* 2003, *A&A* 408, 1047

Savanov, I., Musaev, F.A. & Bondar, A.V. 2003, *IBVS* 5468

Shibahashi, H. 2005, *These Proceedings*, GP17

Simon, T., Ayres, T.R., Redfield, S. & Linsky, J.L. 2002, *ApJ* 579, 800

Templeton, M., Basu, S. & Demarque, P. 2001, *ApJ* 563, 999

Turcotte, S., Richer, J., Michaud, G. & Christensen-Dalsgaard, J. 2000, *A&A* 360, 603

Discussion

SAIO: I would like to mention that δ Scuti type pulsations in roAp stars can be suppressed by a magnetic field larger than about 1 kG, according to a nonadiabatic analysis.

BALONA: That would certainly help us to understand why we do not find δ Sct pulsations in roAp stars.

BREGER: You are absolutely correct in concluding that refined models are most important for asteroseismology of δ Scuti stars. May I add that, in addition, observations supporting these refinements need to be made whenever possible.

BALONA: Yes, I agree. I believe we need to change the current models of A stars used in the pulsational analysis.

SHIBAHASHI: In the case of β Pic, are the low-degree modes found photometrically detected by spectroscopy?

BALONA: The answer is no, but I believe this is due to the low amplitudes of the modes. The low-degree modes cause long-scale line profile variations which are more difficult to detect.

SKODA: Could the high order line profile variations in β Pic be explained by periodic circumstellar obscuration rather than pulsation?

BALONA: The line profiles described here are detected in all photospheric lines and cannot be due to absorption by the circumstellar material.

KUPKA: I would like to point out that Simon *et al.* (2002) have presented FUSE observations of a number of A-type stars. They find signatures of chromospheric emission for stars cooler than 8300 K, but not for hotter stars.

BALONA: β Pictoris, with $T_{\mathrm{eff}} = 8200$ K , is in the hot end of the temperature range of the chromospherically active stars detected by Simon *et al.* (2002). My point is that the detection of a chromosphere in these stars is difficult to understand because we do not know how the very thin convective region can support magnetic activity. Bouret *et al.* (2002) show that their model of a thin region heated up to a few 10^5 K located close to the photosphere reproduces the observations in β Pic remarkably well. The challenge is to understand how such a chromosphere-transition region can be formed in these stars.

The A-Star Puzzle
Proceedings IAU Symposium No. 224, 2004
J. Zverko, J. Žižňovský, S.J. Adelman, & W.W. Weiss, eds.

© 2004 International Astronomical Union
DOI: 10.1017/S1743921304004739

δ Scuti and γ Doradus stars

Michel Breger

Institute of Astronomy, University of Vienna, Türkenschanzstr. 17, A-1180 Wien, Austria
email: michel.breger@univie.ac.at

Abstract. The paper emphasizes the connection between observations and theory in the Lower Instability Strip. It is argued that the δ Scuti and γ Doradus pulsators are very close relatives in the sense that the γ Doradus phenomenon can also be seen in the slightly hotter δ Scuti variables.

New developments and recent progress are reviewed. Arguments are given why all (or almost all) stars in the Lower Instability Strip are pulsators. Furthermore, the question of the so-called missing modes is answered as very extensive photometry campaigns have revealed a large number of pulsation modes with amplitudes less than one millimag. Arguments are given that the 500+ $\ell = 0$ to 2 modes predicted for δ Scuti stars are actually present.

The progress made in developing reliable methods of pulsation mode identification is briefly discussed. The discovery of many close modes with small amplitudes sets severe requirements concerning the length of future studies.

Keywords. Stars: variables: δ Scuti, stars: oscillations (including pulsations)

1. Introduction

δ Scuti stars are pulsators situated in the Classical Instability Strip on and above the Main Sequence. They pulsate with radial and nonradial pressure (p) as well as mixed pressure/gravity (p/g) modes. The excitation is due to the κ mechanism in the He II ionization zone.

γ Doradus stars are gravity (g) mode pulsators. In the H-R Diagram they overlap the cool δ Scuti stars and extend into the F stars. The details of the excitation are not entirely clear at this stage. Convective blocking has been proposed as a mechanism (Guzik *et al.* 2000, Dupret *et al.* 2004).

Despite their difference in the types of excited pulsation modes and the length of the pulsation periods, the two types of stars may be related. No known γ Doradus star has so far been found to show short-period δ Scuti pulsation. The converse may not be true: many cool δ Scuti stars may show the γ Doradus phenomenon, both from observations and theoretical models. This is illustrated in Fig. 1, where excitation as a function of frequency is shown for the $\ell = 2$ modes of the δ Scuti star FG Vir. The instability parameter, η, shows a second peak in the low-frequency region, viz., near $1.7 \, \mathrm{c\,d^{-1}}$. The peak remains negative, but varies with the convection treatment adopted. It is attractive to speculate that the peak can actually become positive so that gravity modes can be excited in a limited part of the low-frequency region.

These low-frequency peaks have been observed in some δ Scuti stars. Attempts to explain them in terms of other hypotheses such as combination modes have failed. An example is the star BI CMi (Breger *et al.* 2002). This star pulsates in the 4.8 to 13.0 $\mathrm{c\,d^{-1}}$ region, but also shows a clear peak at $1.662 \, \mathrm{c\,d^{-1}}$. Fig. 2 demonstrates this relative to the main comparison star, HD 66925. We note here that the peak is also seen relative to a second comparison star, HD 66829.

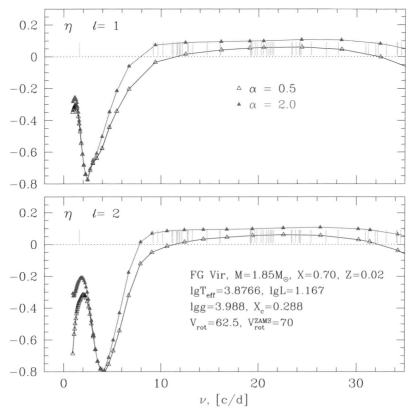

Figure 1. Pulsation excitation models calculated by A. A. Pamyatnykh for a typical evolved δ Scuti star. The parameter η denotes the amount of instability. The two curves were calculated with α values of 0.5 (bottom) and 2.0 (top). Instability occurs above the horizontal line and is in agreement with the observed frequencies of FG Vir. The local maximum at low frequencies may point towards an explanation of the observed low-frequency peaks in FG Vir and other δ Scuti variables.

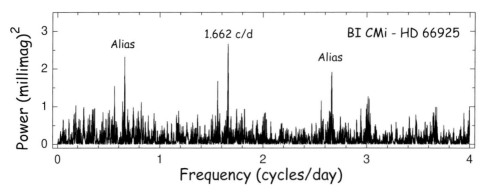

Figure 2. Power spectrum of the cool δ Scuti star BI CMi in the low-frequency region. The peak at 1.662 $c\,d^{-1}$ could be interpreted as a gravity mode, if the local excitation maximum near 1.7 $c\,d^{-1}$ shown by some models is strong enough.

2. What fraction of stars in the Lower Instability Strip are variable?

A number of photometric variability surveys of stars in and near the Lower Instability Strip have detected variability in about 25% - 33% of the stars. The results depend on the photometric precision (i.e., the amplitude threshold) and the width of the instability strip. A excellent recent discussion can be found in Poretti *et al.* (2003).

A much higher incidence of pulsation is obtained if one corrects for the observational limits of the surveys. If we examine the distribution of amplitudes of the δ Scuti variables found in surveys and extrapolate this to near-zero amplitude, all or maybe all the stars in the Lower Instability Strip should be pulsating.

This expectation is dramatically confirmed by new WIRE satellite measurements of the bright A star, Altair. These measurements are more accurate than ground-based photometry of equal observing time. (Later we will show that ground-based photometry can, and does detect, amplitudes of 0.2 mmag). Buzasi *et al.* (2004) show that Altair is variable with an amplitude of about 1 mmag. This confirms the expectation that most (or all) of the 'constant' stars in the Instability Strip are variable at small amplitudes.

What determines the amplitude of a δ Scuti or a γ Doradus star? The largest effect is due to stellar rotation (see Fig. 3): high amplitude requires low rotation. Often, but not always, these high-amplitude modes are radial. Stars rotating faster than ~30 km s^{-1} pulsate with a mixture of low-amplitude radial and nonradial modes. Why some of these amplitudes are as large as 0.02 mag. and some are as small 0.0002 mag. is not known at this time.

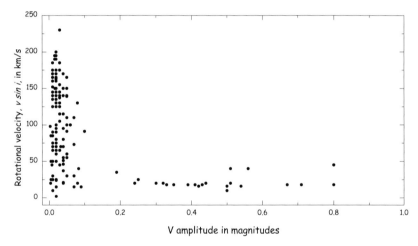

Figure 3. Relationship between observed photometric amplitude and measured rotational velocity, $v \sin i$. This diagram shows that high-amplitude pulsation is restricted to slow rotation.

3. Pulsation mode identifications

One of the most important prerequisites for the application of asteroseismology is the identification of the discovered frequencies with the stellar pulsation modes, i.e., the determination of pulsation quantum numbers n, ℓ, m. A variety of photometric and spectroscopic techniques are applied. In the last few years, there has been enormous progress in the development and checking of these techniques. Space limitations do not permit more details here, but examples can be found in Aerts *et al.* (2004) and Zima *et al.* (2004). Apart from mode identifications, spectroscopic line-profile analyses can also be used to determine the inclination of the rotation axis to the line of sight. Fig. 4, kindly supplied by W. Zima, demonstrates this for the star FG Vir.

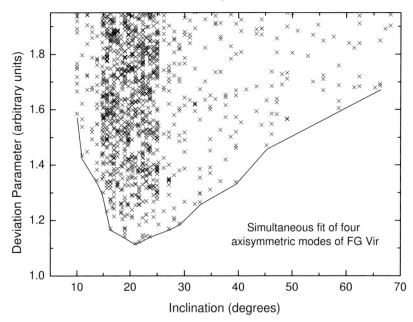

Figure 4. Line-profile variations can be used to also determine the inclination of the rotation axis to the line of sight. For the star FG Vir, the observed profile variations of four non-axisymmetric modes were fitted simultaneously with a common inclination. The diagram shows the quality of fits of more than 15000 simulations (some shown by crosses) to the observed line profiles of FG Vir. The best agreement is found for an angle of 20°.

4. Where are the missing modes?

δ Scuti star models predict pulsational instability in many radial and nonradial modes. The observed number of low-degree modes is much smaller than the number predicted by pulsation models. The problem of mode selection is most severe for post-Main-Sequence δ Scuti stars, which comprise about 40% of the observed δ Scuti stars. The theoretical frequency spectrum of unstable modes is very dense. Most modes are of mixed character as they behave like p-modes in the envelope and like g-modes in the interior. For example, a model of 4 CVn predicts 554 unstable modes of $\ell = 0$ to 2. Of these 6 are $\ell = 0$, 168 are $\ell = 1$, and 380 are $\ell = 2$. However, only 18 (and an additional 16 combination frequencies) were observed (Breger *et al.* 1999). The problem for other δ Scuti stars is similar.

Consequently, either the theoretical predictions or the observational techniques are imperfect (for a discussion, see Breger & Pamyathykh 2002). On the observational side it is, therefore, necessary to lower the observational threshold to search for the missing modes. The answer may be provided by the 1000+ hours of photometry obtained by the δ Scuti Network for FG Vir. The preliminary multifrequency solution already has identified 60 frequencies. The separate solutions for 2002 and 2003 are similar, so that we regard the results to be reliable.

The amplitude distribution is shown in Fig. 5. The diagram shows an almost exponential curve with decreasing amplitude. Since it is improbable that the star stops at exactly the observational threshold, considerably more modes should be present near 0.2 mmag.

This is supported by the power spectrum of the residuals after the 60-frequency solution is subtracted (Fig. 6). Excess power in the frequency regions, in which FG Vir is known to pulsate, indicates the existence of many additional modes.

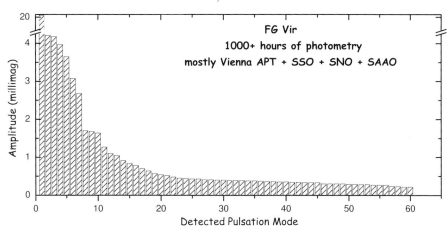

Figure 5. Photometric amplitudes of the pulsation modes of FG Vir. The detection of 60 modes is preliminary and based on 2002 and 2003 data of the δ Scuti Network. The exponential shape of the distribution suggests that below the observational limit 0.20 mmag considerably more modes are present.

While we have not detected all of the $\sim$500 predicted pulsation modes, the huge increase in detected modes with the low 0.20 mmag threshold strongly suggests that the predicted modes are not missing. While the question of mode selection may now be answered, the question of what determines the sizes of the amplitudes remains.

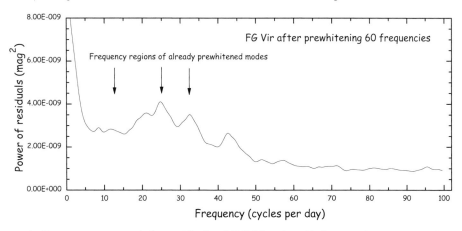

Figure 6. Power spectrum of the residuals of FG Vir after 60 frequencies were prewhitened. There exists excess power in two of the three main regions in which the prewhitened modes were found. This also suggests the presence of additional, so far undetected modes.

5. Problems that need to be considered

The recent developments in the study of these pulsating stars have shown a number of exciting discoveries as well as problems which need to be considered in future investigations from space or from the ground.

5.1. *Dense frequency spectrum requires exact model predictions*

An important tool to understand stellar structure is to compare the observed pulsation frequencies with those given by model calculations. As more frequencies as well as

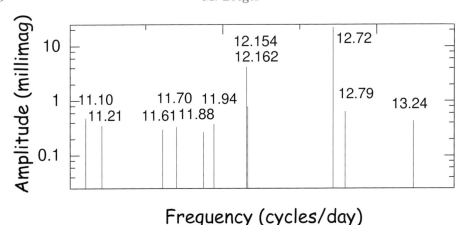

Frequency (cycles/day)

Figure 7. A very small frequency region for FG Vir, showing the dense observed frequency spectrum. This diagram demonstrates that pulsation models need to predict the pulsation frequencies to very high precision to enable unique comparisons with observations.

pulsation modes are identified, additional parameters (such as convective overshooting) can be finetuned in the models. However, the dense frequency spectrum observed leads to a uniqueness problem in the comparison. This can only be solved if the models can predict frequencies to much better than 1%.

Let us use FG Vir again as an example. Fig. 7 shows a small frequency region with the detected frequencies. Unpublished spectroscopic as well as photometric mode-identification techniques have identified three of these modes with different values of ℓ. We can identify the modes at 12.15, 12.72 and 12.79 c d^{-1} with $\ell = 0$, 1, and 2, respectively. The nature of the 12.16 c d^{-1} mode and the other frequencies is unknown.

The dense frequency spectrum indicates that the pulsation models need to make very accurate frequency predictions to avoid misidentifications!

5.2. *Extremely close frequency pairs*

The majority of well-studied δ Scuti stars show a number of very close frequencies with separations as small as 0.01 c d^{-1}. For the star BI CMi at least four close frequency pairs were discovered. It could be shown (Breger & Bischof 2002) from amplitude and phase arguments that these were independent modes, rather than artefacts of the amplitude variability of a single mode. This may be related to the Blazhko Effect observed in another group of pulsators, the RR Lyrae stars.

Another close frequency pair can also be seen in Fig. 7, where the 12.154 c d^{-1} and 12.162 c d^{-1} modes for FG Vir are shown.

The close frequencies still need to be explained theoretically. On the observational side, they demand very high frequency resolution in the measurements (from space or ground), requiring observing runs longer than 100 d.

5.3. *Beware of pulsation from companions*

Successful asteroseismology requires the matching between observed and predicted frequencies. It would be fatal if some of the observed frequencies would not belong to the star studied, but to a companion. In some cases, both components of a binary system pulsate, sometimes with similar periods, if their masses are similar. An example is the star θ^2 Tau shown below in Fig. 8 (Breger 2005).

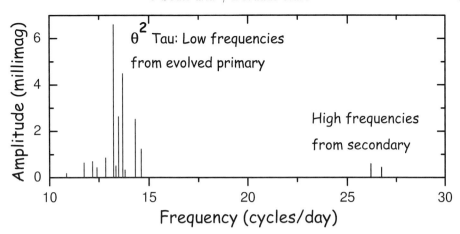

Figure 8. Example of a binary system, θ^2 Tau, in which both components pulsate. The low and the high frequency groups show opposite light-time (O-C) shifts, which agree with those predicted from the known orbits. This diagram demonstrates that before modelling pulsations, one needs to ensure that all of the observed modes originate in the same star.

5.4. *Rotational splitting is not equidistant*

Even for moderate stellar rotation, rotational splitting is not equidistant (e.g., see a discussion by Goupil *et al.* 2000). This departure from equidistant splitting is very complicated to compute and provides a powerful tool when observations are compared with various stellar models. While rotational splitting has been easily observed in g-mode pulsators such as white dwarfs, for main-sequence A stars the theoretical complexity is matched by the difficulty of obtaining mode identifications for a complete $\ell = 1$ triplet, or a $\ell = 2$ quintuplet.

A triplet has now been detected. Fig. 9 shows an observed $\ell = 1$ triplet for FG Vir. We confirm that rotational splitting is indeed nonequidistant.

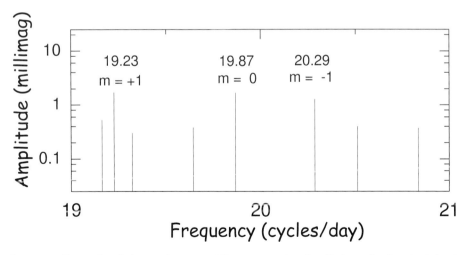

Figure 9. Example of observed non-equidistant rotational splitting of a $\ell = 1$ triplet in FG Vir.

Acknowledgements

This investigation has been supported by the Austrian Fonds zur Förderung der wissenschaftlichen Forschung, project number P17441-N02.

References

Aerts, C., Cuypers, J., De Cat, P., Dupret, M. A., De Ridder, J., Eyer, L., Scuflaire, R., Waelkens, C. 2004, *A&A*, **415**, 1079

Breger, M. 2005, *ASP Conf. Ser.*, in press (Granada Workshop)

Breger, M. & Bischof, K. M. 2002, *A&A* 385, 537

Breger, M. & Pamyatnykh, A. A. 2002, *ASP Conf. Ser.* 259, 388

Breger, M., Pamyatnykh, A. A., Pikall, H. & Garrido R. 1999, *A&A*, 341, 151

Breger, M., Garrido, R., Handler, G., Wood, M. A., Shobbrook, R. R., Bischof, K. M., Rodler, F., Gray, R. O., Stankov, A., Martinez, P., O'Donoghue, D. O., Szabo, R., Zima, W., Kaye, A. B., Barban, C., Heiter, U. 2002, *MNRAS*, 329, 531

Buzasi, D. L., Bruntt, H., Bedding, T. R., Retter, A., Kjeldsen, H., Preston, H. L., Mandeville, W. J., Catanzarite, J., Conrow, T., Laher, R. 2004, *ApJ*, in press

Dupret, M. A., Grigahcéne, A., Garrido, R., Gabriel, M., Scuflaire, R. 2004, *A&A*, 414,1081

Goupil, M.-J., Dziembowski, W. A., Pamyatnykh, A. A., Talon, S. 2000, *ASP Conf. Ser.*, 210, 267

Guzik, J. A., Kaye, A. B., Bradley, P. A., Cox, A. N., Neuforge, C. 2000, *ApJ*, 542, L57

Poretti, E., Garrido, R., Amado, P. J., Uytterhoeven, K., Handler, G., Alonso, R., Martín, S., Aerts, C., Catala, C., Goupil, M. J. 2003, *A&A*, **406**, 203

Zima, W., Kolenberg, K., Briquet, M. & Breger, M. 2003, *CoAst*, 144, 5

Discussion

MOEHLER: You mentioned the detection of variability at the 1 - 2 mmag level as an indication that indeed all stars in the variability strip vary. Are there any plans to study stars outside the instability strip at comparable precision to verify that they do not vary?

BREGER: Indeed, the space missions plan to investigate stars outside the classical instability strip. We already know that on both sides of the δ Scuti strip there are pulsators: the γ Doradus gravity-mode pulsators to the cool side and isolated variables to the hot side.

WEISS: Within the COROT Additional Programme it is planned to test the hot and cool borders of the instability strip. This is also an on-going project for MOST.

BALONA: The drop in the number of δ Scuti stars with rotation, as detected by photometry may be because the rapidly rotating stars pulsate with modes of high degree, which cannot be detected by photometry. The example of β Pic comes to mind.

The A-Star Puzzle
Proceedings IAU Symposium No. 224, 2004
J. Zverko, J. Žižňovský, S.J. Adelman, & W.W. Weiss, eds.

© 2004 International Astronomical Union
DOI: 10.1017/S1743921304004740

Some recent discoveries in roAp stars

D.W. Kurtz[1], V.G. Elkin[1], G. Mathys[2], J. Riley[1], M.S. Cunha[3], H. Shibahashi[4] and E. Kambe[5]

[1]Centre for Astrophysics, University of Central Lancashire, Preston PR1 2HE, UK
e-mail: dwkurtz@uclan.ac.uk

[2]European Southern Observatory, Casilla 19001, Santiago 19, Chile

[3]Centro de Astrofsica da Universidade do Porto, Rua das estrelas, 4150 Porto, Portugal

[4]Department of Astronomy, University of Tokyo, Tokyo 113-0033, Japan

[5]Department of Earth and Ocean Sciences, National Defence Academy, Yokosuka, Kanagawa 239-8686, Japan

Abstract. Research in roAp stars is being vigorously pursued, both theoretically and observationally by many groups. We report the discovery of a 21-min period, luminous roAp star, HD 116114. Longer periods for more luminous stars have been predicted theoretically and this is the first discovery of such a star. We discuss a model for the blue-to-red line profile variability observed in some roAp stars involving a shock wave high in the atmosphere of roAp stars, yet show that the Hα line in 33 Lib has the blue-to-red-to-blue line profile variability expected for subsonic dipolar pulsation concentrated towards the pulsation pole. Further we report for 33 Lib unprecedented observations of the amplitudes and phases of its principal mode at 2.015 mHz and its first harmonic of that at 4.030 mHz.

Keywords. Stars: chemically peculiar, stars: oscillations, stars: variables: other, stars: individual: (HD 116114, HD 137949)

1. Introduction

The roAp stars have been observed photometrically since their discovery by Kurtz (1982) over 20 years ago. Frequency analyses of their light curves have yielded rich asteroseismic information on the degrees of the pulsation modes, distortion of the modes from normal modes, magnetic geometries, and luminosities. The latter, in particular, are derived asteroseismically and agree well with Hipparcos luminosities (Matthews *et al.* 1999).

Theoretical work on the interaction of pulsation with both rotation and the magnetic field by Bigot & Dziembowski (2002) has presented a new look at the oblique pulsator model of these stars. They find that the pulsation axis is inclined to both the magnetic and the rotation axes, and the pulsation modes are complex combinations of spherical harmonics that result in modes that, in many cases, can be travelling waves looking similar to (but are not exactly) sectoral m-modes. Bigot & Kurtz (2004) have shown that the improved oblique pulsator model of Bigot & Dziembowski yields the rotational inclination and the magnetic obliquity for the roAp star HR 3831 that are in excellent agreement with those found from magnetic studies, whereas the old oblique pulsator model is in probable disagreement with the magnetic studies. Kochukhov (2005), however, finds very different rotational and magnetic inclinations for HR 3831 from the Doppler mapping of the pulsation mode. He suggests that the magnetic geometry needs to be reconsidered in this light, and his geometry is consistent with the original oblique pulsator model and the new model of Saio & Gautschy (2004), who find modes that are aligned with the magnetic axis and are distorted by the magnetic field so that they cannot be described by single spherical harmonics. They note that the horizontal motion can be comparable to the vertical motion for these modes. It is the unique geometry of the pulsation modes in roAp stars that allows us to examine their non-radial pulsation modes from varying aspects as can be done with no other type of star. At present there is a great deal of research, both observational and theoretical, on these stars, with some exciting conflicts that will be resolved in time.

The spectra of many roAp stars show a strong core-wing anomaly in the Hydrogen lines, particularly the Hα line (Cowley *et al.* 2001, Kochukhov *et al.* 2002). This indicates abnormal atmospheric structure, as does that consistent abundances for the second and third ionisation states of Rare Earth elements, particularly Nd and Pr, have not yet been found for these stars (Ryabchikova *et al.* 2002). Until atmospheric models can be found that solve these problems, caution is called for, but new high-resolution spectroscopic results for the roAp stars suggest vertical stratification of some ions, particularly the Rare Earths, and show the short vertical wavelengths of the pulsation modes. It must also be cautioned that there are known horizontal abundance variations with the concentration of the Rare Earth elements towards the magnetic poles. See, for example, studies of γ Equ (HD 201601) (Kochukhov & Ryabchikova 2001a), HD 166473 (Kurtz *et al.* 2003), α Cir (Kochukhov & Ryabchikova 2001b, Balona & Laney 2003, Baldry *et al.* 1999), HR 3831 (Kochukhov & Ryabchikova 2001b, Baldry & Bedding 2000), HR 1217 (Balona & Zima 2002, Sachkov *et al.* 2004) and 33 Lib (HD 137949) (Mkrtichian *et al.* 2003, Kurtz *et al.* 2004). The results of these studies are plausibly interpreted in terms of the chemical stratification of the elements in the atmospheres of these stars. In general, Fe lines originate near a radial node around optical depth $\tau = 1$ with little, or no radial velocity variations seen. The core of Hα forms higher in the atmosphere and shows radial velocity variations with amplitudes of hundreds of $m\,s^{-1}$. Lines from the first and second ionisation states of the Rare Earths Pr and Nd arise from a thin layer around optical depth $\tau = 10^{-3}$ and can show amplitudes of $km\,s^{-1}$.

The spectroscopic studies have begun a three-dimensional resolution of the pulsation modes, with the vertical stratification giving depth information, the rotation of the oblique mode providing information on the surface geometry of the modes, and, in the case of HR 3831, Doppler imaging of the pulsation mode giving two-dimensional geometrical information (Kochukhov 2005). While these spectroscopic tools are very powerful, they demand high spectroscopic resolution, high time resolution and high signal-to-noise (S/N), requirements that can only be met with large telescopes. It is thus not possible at present to study in detail the frequencies in roAp stars spectroscopically because of the need for high duty cycle data sets over extended periods of time. Photometric studies are still required to obtain asteroseismic frequency spectra.

2. roAp versus noAp stars: HD 116114

Hubrig *et al.* (2000) compared the positions of the roAp stars and noAp (nonoscillating Ap) stars in the HR diagram and concluded that the roAp stars are, on average, closer to the Main Sequence than the noAp stars. This has important implications for the still-uncertain driving mechanism for the roAp stars. There has been a question, however, about whether a selection effect could have caused this apparent separation of the roAp and noAp stars. Most of the 33 known roAp stars were discovered by Martinez and Kurtz photometrically in various studies. They preferentially searched for stars with strong Strömgren δm_1 indices and δc_1 indices that are negative, indicating strong line blanketing. For the coolest Ap stars a negative δc_1 index is often an indicator of an extremely peculiar star, and such stars are often found to be roAp stars. However, δc_1 increases with luminosity so that an extremely peculiar evolved cool Ap star may show a normal δc_1. The surveys of Martinez and Kurtz would have been much less likely to find roAp stars among such stars, so there is a selection effect in their discovery.

Cunha (2002) predicted an instability strip for roAp stars and found that the more evolved stars should pulsate with longer periods, in some cases in the range of $20-25$ min. Again, the high-speed photometric searches of Martinez and Kurtz could easily miss variability on this timescale because of confusion with sky transparency variations, since they usually searched for only $1-2$ hr per star when surveying for new roAp stars. To find longer periods of the order predicted by Cunha photometric searches should be for $3-4$ hr to try to separate the pulsation signal from low frequency sky transparency noise. This demands the most stable photometric conditions.

As it turns out, the discovery of a longer-period roAp star has come from a spectroscopic radial velocity study. Elkin *et al.* (2004) have found the cool Ap star HD 116114 to be pulsating with a period near 21 min, a value close to that predicted by Cunha (2002) for a star of the luminosity and temperature of HD 116114. This discovery found radial velocity variations with amplitudes of only tens of $\mathrm{m\,s^{-1}}$ in certain selected lines in the spectrum of HD 116114. Such precision is possible with the high resolution VLT UVES data obtained for this star. Fig. 1 shows an amplitude spectrum for 11 spectral lines with the pulsation signal clearly visible at $\nu = 0.79\,\mathrm{mHz}$ ($P = 21\,\mathrm{min}$).

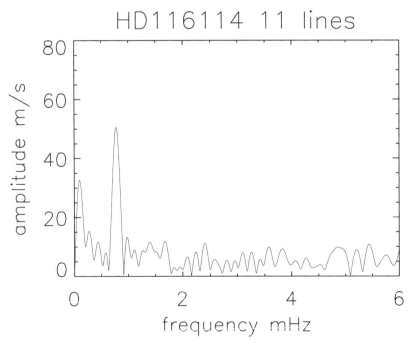

Figure 1. Amplitude spectrum of the radial velocity variations for 11 spectral lines in the Ap star HD 116114. The $50\,\mathrm{m\,s^{-1}}$ signal at $0.79\,\mathrm{mHz}$ ($P = 21\,\mathrm{min}$) is clear.

One implication of this discovery is that there may be more luminous roAp stars with longer periods as predicted by Cunha (2002). A search is now being undertaken to discover more of these stars and study them in some detail. If they are found in some numbers, then the luminosity difference between the roAp and noAp stars may disappear, leaving an understanding of why some Ap stars pulsate and some do not more mysterious.

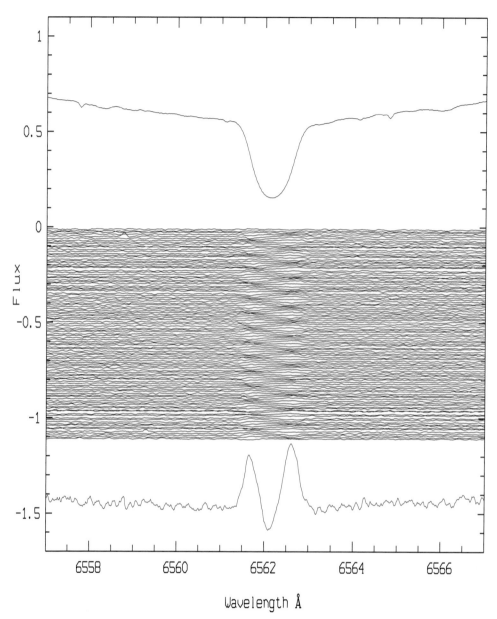

Figure 2. The Hα LPV in 33 Lib. The blue-to-red-to-blue pattern is as expected for subsonic pulsation for a dipole mode concentrated towards the pulsation pole, as is seen in model LPV in Fig. 3.

3. Line profile variability in roAp stars

Line profile variations (LPV) in γ Equ showing apparent blue-to-red travelling bumps were discovered by Kochukhov & Ryabchikova (2001a). Shibahashi *et al.* (2004) proposed that these line profile variations are a manifestation of a shock wave in the high atmosphere near the magnetic polar regions. This idea was vigorously discussed at IAUS224. Other stars with similar profile variations have now been discovered by several groups working on these stars.

An example is 33 Lib (see, e.g., Kurtz *et al.* 2004). For this star the LPV for Nd III $\lambda6145$, which forms high in the atmosphere of the star, has the same apparent blue-to-red bumps seen in γ Equ. The Hα line, however, shows a blue-to-red-to-blue, back-and-forth LPV as can be seen in Fig. 2. This is as expected for radial mode pulsation, or for dipole pulsation with the variation concentrated to the pulsation pole, where the velocities are subsonic. We show this in model LPV in Fig. 3 for a dipole concentrated towards the pulsation pole.

The important point is that the maximum speed must be faster than the atmospheric sound speed for the lines that show only blue-to-red LPV. The pulsation amplitude in velocity is expected to increase with the decrease of the density with height. So the amplitude in velocity naturally exceeds the sound speed at a certain level, and a shock wave is generated. When this shock wave propagates through the layer under consideration, the layer is suddenly kicked upward by the shock wave and then falls down after reaching the peak height, at which the speed of the layer is zero. This process repeats and the resultant variation in the Doppler shift is qualitatively in agreement with the observed one in Nd III lines of γ Equ and 33 Lib. The lack of variation in the standard deviation in the middle of each cycle occurs as the layer comes to a stop, then falls back.

This picture may also be consistent with the mode identifications from the photometric observations, which imply axisymmetric dipole modes whose symmetry axis coincides with the magnetic axis, both being inclined to the rotation axis of the star. If it is correct, then the following are expected: 1) The LPV should be sinusoidal for the spectral lines formed in layers where the wave motion is subsonic as seen for Hα in 33 Lib in Fig. 2, 2) The LPV should be sinusoidal for chemical elements that are not concentrated in the magnetic polar regions, so their pulsation velocities remain subsonic, and 3) The LPV should show monotonic blue-to-red motion only for the lines formed in the high atmosphere near the magnetic polar regions.

4. Atmospheric depth dependence of pulsation in 33 Lib

For the roAp stars that have been observed at high spectral resolution few generalities of the radial velocity behaviour, beyond those given above, can yet be made. Kurtz, Mathys and Elkin have now obtained high time resolution ($\leqslant 1$ min), high spectral resolution (R = 105 000) and high signal-to-noise ratio (50–600) spectra for two hours each for 15 known roAp stars (including HD 116114 which was discovered with this data set) and 6 other roAp candidates and sharp-lined A stars using the UVES spectrograph on UT2 (Kueyen) of the VLT on Paranal, Chile.

One of our targets, 33 Lib (HD 137949), is a bright ($V = 6.68$) roAp that is known photometrically to have a principal pulsation frequency of 2.015 mHz ($P = 8.27$ min) with a first harmonic frequency at 4.030 mHz (Kurtz 1991). We have analysed a set of 111 echelle spectra of 33 Lib obtained with the UVES spectrograph on UT2 (Kueyen) of the VLT. Each spectrum had an integration time of 40 s with the readout and overhead time between spectra of 25 s, giving a time resolution of 65 s. The typical signal-to-noise (S/N) ratio is about 200 and the time span of the observations is 2 hr.

33 Lib has three pulsation modes that can be seen in our 2-hour duration data set, the known 2.015-mHz mode and harmonic at 4.030 mHz, and a new, low-amplitude frequency at 1.769 mHz that we will not discuss here. The largest amplitudes found for the principal

SPH L=1 M= 0 AMP=0.4km/s C=4km/s Incl=0

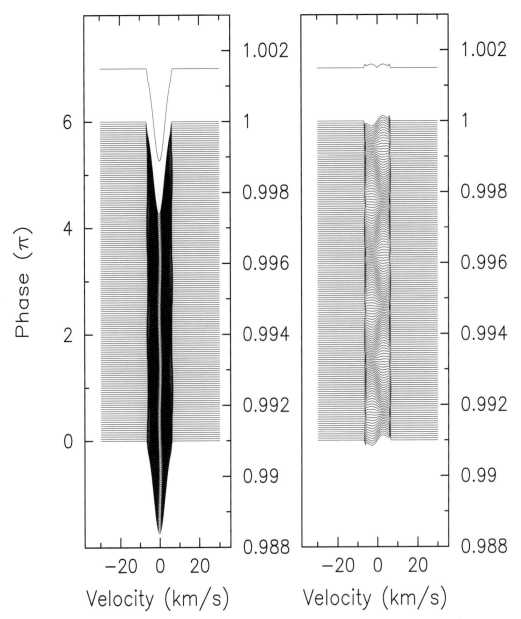

Figure 3. Theoretically expected LPV of an $\ell = 1$, $m = 0$ mode for a non-rotating star. The pulsation amplitude is assumed to be $0.4\,\mathrm{km\,s^{-1}}$ and the line profile is assumed to have an intrinsic line width corresponding to a velocity of $4\,\mathrm{km\,s^{-1}}$. The aspect angle between the pulsation axis and the line-of-sight is assumed to be $0°$. For each mode the LPV for one pulsation cycle are shown on the left panel, and the difference between the LPV and the average profile (shown top left) are displayed in the right panel. The LPV move from blue to red and back again, as is observed for the Hα line in 33 Lib as shown in Fig. 2.

pulsation frequency are for Hα and some lines of singly-ionised Rare Earth elements, Nd II, Ce II and Sm II with semi-amplitudes of $200 - 300 \, \mathrm{m \, s^{-1}}$. Lines of Nd III, Pr III and Eu II typically have amplitudes in the $120 - 200 \, \mathrm{m \, s^{-1}}$ range, while lines of Fe, Cr and Ca usually have amplitudes less than $40 \, \mathrm{m \, s^{-1}}$. Fig. 4 shows the amplitude spectra for Nd II λ5319.815 and Nd III λ6145.070 showing all three of these frequencies. The apparent independent behaviour of the 4.030-mHz peak led Kurtz *et al.* (2004) to suggest that it is an independent mode, rather than an harmonic of the 2.015-mHz peak. We are working on an interpretation of it as an harmonic, even so, this harmonic behaves in many ways like an independent mode.

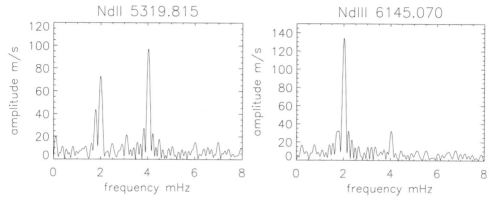

Figure 4. Left panel: The amplitude spectrum of the radial velocity variations for Nd II λ5319.815. The known principal frequency at 2.015 mHz and the new frequency at 1.769 mHz are resolved at a separation of 246 μHz. The highest amplitude mode is at the harmonic frequency 4.030 mHz. Note the absence of any second harmonic. Compare this with Nd III λ6145.070 in the right panel, showing the independence of the amplitudes of the 2.015-mHz and 4.030-mHz frequencies.

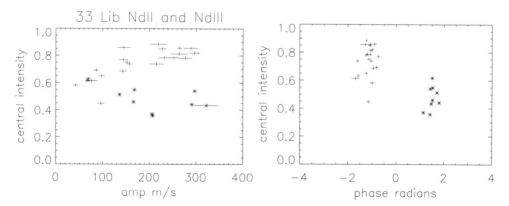

Figure 5. The amplitudes and phases of the Nd II and Nd III lines as a function of central line intensity normalised to 1 at the continuum. The smaller the value of the central intensity, the higher in the atmospheric layer for that ion the line forms. The left panel shows the amplitudes for the Nd II lines (plus signs) and Nd III lines (asterisks). Because of decreasing density and ionisation equilibrium, the Nd III lines form higher in the atmosphere than the Nd II lines. The two layers lie on opposite sides of a radial pulsation node with the Nd III amplitudes increasing outwards away from the node, and the Nd II amplitudes increasing inwards away from the node. The phases suggest standing waves which differ by nearly π radians.

Fig. 5 shows the amplitudes and phases determined by least squares for the 2.015-mHz mode for Nd II and Nd III. For each of these ions the phases do not appear to vary as a function of atmospheric depth. Those phases differ between the ions by 2.65 ± 0.31 radians, nearly in antiphase (π radians) as found by Mkrtichian *et al.* (2003). The new discovery is the clear increase in amplitude with atmospheric height for Nd III and the increase in amplitude with atmospheric depth for Nd II. This is completely consistent with Mkrtichian et al.'s suggestion that the lines of these two ions form on opposite sides of a pulsation node with Nd III above, and Nd II below the node.

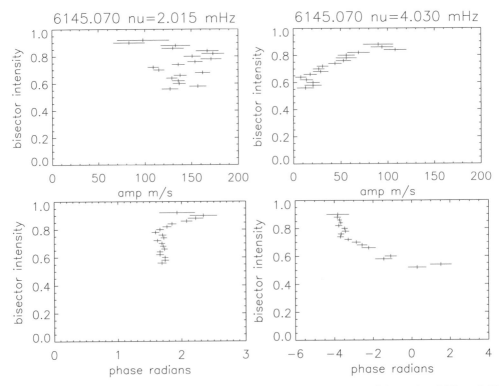

Figure 6. The amplitudes and phases for Nd III $\lambda6145.070$ at central intensity 0.56 to 0.92 in steps of 0.02 (where 1 is the continuum) for the 2.015-mHz mode on the left and for the 4.030-mHz harmonic on the right. The deeper the line depth (weaker the central intensity), the higher in the atmosphere the cut through the line is formed. The phases have been calculated using the function $\cos(2\pi ft + \phi)$, so the times of maxima occur earlier for larger phases.

The constancy of the phases for this mode suggests standing waves on both sides of the node, but there is an unprecedented new way (for any star other than the Sun) to examine the pulsation amplitudes and phases as a function of atmospheric height. We have done that by making cuts through the strong line Nd III $\lambda6145.070$ at central intensity 0.56 to 0.92 in steps of 0.02 (continuum = 1). We have measured the radial velocities of the line bisector at each of those depths, and then fitted by linear least squares the 2.015-mHz and 4.030-mHz frequencies to extract amplitudes, phases and their associated errors. The results are consistent with a constant radial velocity amplitude in the Nd III layer for the 2.015-mHz mode, although an increase with height is possible. There also seems to be a standing wave higher in the layer with a running wave travelling outwards from below. This is not what is expected of a purely acoustic mode. For that we expect a standing wave with an outwardly running wave above it where energy is being dissipated in the boundary layer. We do not know, this is the first observation of such behaviour, and theoretical studies (Saio & Gautschy 2004, Cunha & Gough 2000, Bigot *et al.* 2000) do not yet address it, but we suggest that the phase behaviour should not be interpreted as a running wave, rather as the result of the varying relation between the phases of the magnetic and acoustic parts of the magneto-acoustic mode in this high layer.

Much more interesting is the behaviour of the amplitudes and phases for the 4.030-mHz harmonic, shown in Fig. 6, for $\lambda6145.070$. For this frequency the amplitude decreases strongly with height, which is not expected for a standing wave with a node below, as is seen for the 2.015-mHz mode in the left panels of Fig. 6. This suggests another node near the top of the line-forming layer for the 4.030-mHz frequency. Whether this frequency is an harmonic or an independent mode, it provides new, independent constraints on the atmospheric structure of 33 Lib.

Fig. 6 shows that the phases of the 4.030-mHz harmonic go smoothly through at least π radian change from the top to the bottom of the layer in which this line forms, and even through a full 2π radians if the more uncertain phases from two cuts at bisector intensity 0.56 and 0.58 can be trusted. If the 4.030-mHz frequency is from an independent mode, then we suggest that this phase change is a result of interaction between the acoustic and magnetic components of the pulsation mode, although there are no detailed models yet to test this suggestion in the theories of Cunha & Gough (2000), Bigot *et al.* (2000) and Saio & Gautschy (2004). If it is an harmonic, then this phase behaviour is also yet to be modelled. In either case there is unprecedented depth information about the atmospheric structure, the radial part of the mode or modes and the interaction of the pulsation with the magnetic field.

References

Baldry, I.K., Bedding, T.R. 2000, MNRAS, 318, 341

Baldry, I.K., Viskum, M., Bedding, T.R., Kjeldsen, H., Frandsen, S. 1999, MNRAS, 302, 381

Balona, L.A., Laney, C.D. 2003, MNRAS, 344, 242

Balona, L.A., Zima, W. 2002, MNRAS, 336, 873

Bigot, L., Dziembowski, W.A. 2002, A&A, 391, 235

Bigot, L., Kurtz, D.W. 2004, A&A, submitted

Bigot, L., Provost, J., Berthomieu, G., Dziembowski, W.A., Goode, P.R. 2000, A&A, 356, 218

Cowley, C.R., Hubrig, S., Ryabchikova, T.A., Mathys, G., Piskunov, N., Mittermayer, P. 2001, A&A, 367, 939

Cunha, M.S. 2002, MNRAS, 333, 47

Cunha, M.S., Gough, D.O. 2000, MNRAS, 319, 1020

Elkin, V.G., Riley, J., Cunha, M.S., Kurtz, D.W., Mathys, G. 2004, in preparation

Hubrig, S., Kharchenko, N., Mathys, G., North, P. 2000, A&A, 355, 1031

Kochukhov, O., Bagnulo, S., Barklem, P.S. 2002, ApJ, 578, L75

Kochukhov, O., Ryabchikova, T. 2001a, A&A, 377, L22

Kochukhov, O., Ryabchikova, T. 2001b, A&A, 374, 615

Kochukhov, O. 2005, *These Proceedings*, GP20

Kurtz, D.W. 1982, MNRAS, 200, 807

Kurtz, D.W. 1991, MNRAS, 249, 468

Kurtz, D.W., Elkin, V.G., Mathys, G. 2003, MNRAS, 343, L5

Kurtz, D.W., Elkin, V.G., Mathys, G. 2004, submitted

Matthews, J.M., Kurtz, D.W., Martinez, P. 1999, ApJ, 511, 422

Mkrtichian, D.E., Hatzes, A.P., Kanaan, A. 2003, MNRAS, 345, 781

Ryabchikova, T., Piskunov, N., Kochukhov, O., Tsymbal, V., Mittermayer, P., Weiss, W.W. 2002, A&A, 384, 545

Sachkov, M., Ryabchikova, T., Kochukhov, O., Weiss, W.W., Reegen, P., Landstreet, J.D. 2004, ASP Conf. Ser., Vol. 310, *Variable Stars in the Local Group*, eds D.W. Kurtz & K.R. Pollard, San Francisco ASP, 208

Saio, H., Gautschy, A. 2004, MNRAS, 350, 485

Shibahashi, H., Kurtz, D., Kambe, E., Gough, D. 2004, ASP Conf. Ser., Vol. 310, *Variable Stars in the Local Group*, eds D.W. Kurtz & K.R. Pollard, San Francisco ASP, 287

Proceedings The A Star Puzzle
Proceedings IAU Symposium No. 224, 2004
J. Zverko, J. Žižňovský, S.J. Adelman, & W.W. Weiss, eds.

© 2004 International Astronomical Union
DOI: 10.1017/S1743921304004752

Pulsating pre-Main Sequence stars in young open clusters

Konstanze Zwintz[1], Marcella Marconi[2], Thomas Kallinger[1] and Werner W.Weiss[1]

[1]Institute of Astronomy, University of Vienna, Türkenschanzstrasse 17, A-1180 Vienna,
Austria
email: zwintz@astro.univie.ac.at

[2]INAF Astronomical Observatory of Capodimonte, Naples, Italy

Abstract. New pulsating pre-main sequence (PMS) stars have been discovered in the young open clusters IC 4996 and NGC 6383 using CCD time series photometry in Johnson B and V filters. As the cluster ages are both smaller than 10 million years, all members later than spectral type A0 are still contracting towards the ZAMS, hence providing ideal candidates for searches of pulsation. A dozen stars in NGC 6383 and 35 stars in IC 4996 lie within the boundaries of the classical instability region in the Hertzsprung-Russell (HR) diagram, but pulsation was detected for only two of them in each cluster.

For the well-studied cluster NGC 2264, the two already known PMS pulsating members, V 588 Mon and V 589 Mon, have been analysed using new data from a multi-site campaign. All data collected since their discovery in 1972 build the basis for the first measurements of period changes in PMS pulsators.

Keywords. Techniques: photometric, stars: pre–main-sequence, stars: variables: other, Galaxy: open clusters and associations: individual (NGC 6383, IC 4996)

1. Introduction

Pre-main sequence (PMS) stars are evolving from the birthline to the zero-age main sequence (ZAMS) in the HR-diagram. During this phase they are characterized by a high degree of activity. They display photometric and spectroscopic variabilities on time scales between a couple of minutes and several months. They sometimes strongly interact with the circumstellar environment, in which they are still embedded. Very often emission lines appear in their spectra and UV and/or IR excesses are present. As PMS stars cross the instability region while contracting towards the ZAMS, part of their activity is due to stellar pulsations. The time PMS stars spend in this phase is extremaly short. A 1.5 $M_\odot$ star crosses the instability region in 10^6 years while a 4 $M_\odot$ star needs only 80000 years. At present 24 PMS pulsators are known displaying periods between 18 minutes (Amado *et al.* 2004) and 6 hours.

2. Evolutionary stage

PMS stars differ from their more evolved counterparts of same temperature and luminosity only in their interior structure, while their envelope properties are quite similar. The determination of the evolutionary stage of a field star may be ambiguous as the evolutionary tracks for PMS and post-main sequence stars intersect each other several times (see Fig. 1, PMS and post-main sequence evolutionary tracks are taken from D'Antona & Mazzitelli (1994) and Breger & Pamyatnykh (1998), respectively).

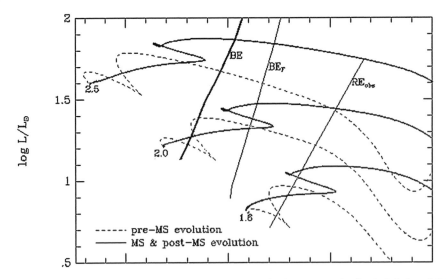

Figure 1. Intersecting pre- and post-main sequence evolutionary tracks for 1.6, 2.0 and 2.5 $M_\odot$ and the boundaries of the classical instability strip, where RE_{obs} denotes the empirical red edge, BE the blue edge for the radial overtones and BE_F the blue edge for the fundamental mode (Breger & Pamyatnykh 1998).

The study of young open clusters provides the possibility to reduce confusion with more evolved objects, as all members have the same age and distance. Clusters younger than a few 10^7 yr show a main sequence down to $\sim$ B9, while cluster members of later spectral types have not yet reached the ZAMS. Hence, time series observations of the young open clusters NGC 6383 and IC 4996 were carried out to increase the number of pulsating PMS stars. New data from the PMS pulsators V 588 Mon and V 589 Mon in NGC 2264 were obtained during a multi-site campaign in 2002, and the pulsational characteristics of the stars were investigated in detail.

3. NGC 6383

NGC 6383 belongs to the Sgr OB1 association ($\alpha_{2000} = 17^h 34.8^m$, $\delta_{2000} = -32° 34'$) and has an age of $\sim$1.7 Myr. Its main sequence reaches only to B9, so cluster members of spectral types A – F are still in their PMS phase, hence are ideal candidates to search for pulsation. CCD time series photometry in Johnson B and V filters was obtained with the 0.9-m telescope of Cerro Tololo Interamerican Observatory (CTIO), Chile, within 9 nights. Two new PMS pulsators have been detected, NGC 6383 170 and NGC 6383 198.

3.1. *NGC 6383 170*

For NGC 6383 170 ($V = 12.61$ mag), Thé *et al.* (1985) found H_α in emission and high excess radiation in the NIR, typical for Herbig Ae/Be stars. With a spectral type of A5 IIIp (van den Ancker *et al.* 2000), a confirmed membership in NGC 6383, and a position above the ZAMS, NGC 6383 170 is a newly discovered PMS pulsator (see Fig. 2). Five frequencies spanning a period range between 1.24 and 2.89 hours (see Table 1) have been detected (Zwintz *et al.* 2004).

Linear, nonadiabatic pulsation was calculated for radial modes of PMS models resulting in three possible solutions. No model reproduces all five frequencies simultaneously, but given the probable coexistence of radial and nonradial modes in these stars, not all frequencies need to correspond to radial pulsation. The model fitting the observed

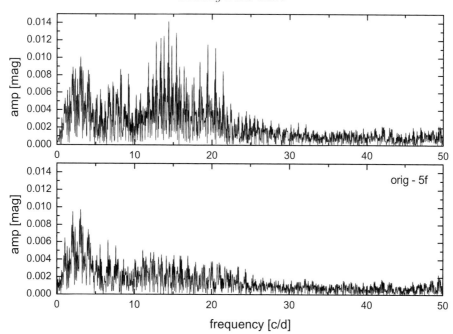

Figure 2. Amplitude spectra of NGC 6383 170 in the V filter; top: original data, bottom: after prewhitening the five significant frequencies.

frequencies best, gives a stellar mass of 2.5 $M_\odot$, $\log L/L_\odot = 1.68$, $T_{\text{eff}} = 8100$ K, and pulsation in third ($f1$) and fifth overtones ($f2$). The solution seems to be optimal, because it is closest to the parameters derived spectroscopically by van den Ancker *et al.* (2000) (see Zwintz et al. 2004).

3.2. *NGC 6383 198*

In the data of NGC 6383 198, only a single frequency of 19.024 d^{-1}, corresponding to a period of ~ 1.26 hours, is significant in both filters (see Fig. 3).

Calculations of linear, nonadiabatic, radial pulsation models were also performed. As no spectral classification is available for this star, the V and $(B - V)$ values were used to derive empirical ranges of luminosity and effective temperature based on the transformations given by Kenyon & Hartmann (1995). Given the cluster membership of NGC 6383 198, it seems reasonable that the star pulsates with a single frequency in the third radial overtone having 2.0 $M_\odot$, $\log L/L_\odot = 1.3$ and $T_{\text{eff}} = 7345$ K (Zwintz *et al.* 2004).

4. IC 4996

IC 4996 belongs to the Cygnus star forming region ($\alpha_{2000} = 20^h\,16^m\,34^s$, $\delta_{2000} = 37° 39' 54''$) and is ~ 7.5 Myr old. CCD time series photometry in Johnson B and V filters was obtained using the 1.5-m telescope at Sierra Nevada Observatory (OSN), Spain, within 10 nights. Two new PMS pulsators IC 4996 37 and IC 4996 40 were discovered.

4.1. *IC 4996 37*

IC 4996 37 ($V = 15.30$ mag) has a spectral type of A5 (Delgado *et al.* 1999) and pulsates with a single frequency corresponding to a period of 45.2 min (see Table 1) most probably in the 4th radial overtone mode (Marconi 2004).

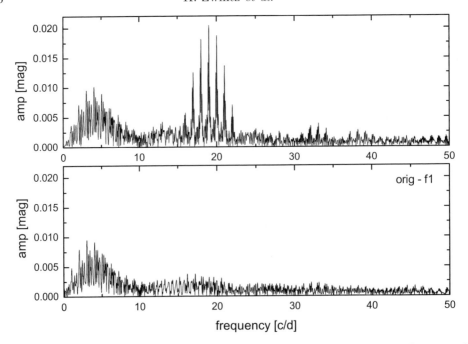

Figure 3. Amplitude spectra of NGC 6383 198 in the V filter; top: original data, bottom: after prewhitening the single significant frequency.

Table 1. Frequencies, amplitudes and phase shifts determined for the four new PMS pulsators NGC 6383 170, NGC 6383 198, IC 4996 37 and IC 4996 40.

star no		frequency $[\mathrm{d}^{-1}]$	V amp. [mmag]	B amp. [mmag]	Δphase
NGC 6383 170	$f1$	14.376	12.5	16.0	-0.243
	$f2$	19.436	11.3	14.9	-0.092
	$f3$	13.766	9.8	12.3	0.474
	$f4$	8.295	8.6	11.1	0.035
	$f5$	17.653	7.6	9.8	-0.050
NGC 6383 198	$f1$	19.024	20.8	26.4	0.114
IC 4996 37	$f1$	31.875	4.6	4.6	-0.200
IC 4996 40	$f1$	33.569	7.6	8.5	-0.138

4.2. IC 4996 40

IC 4996 40 ($V = 15.03$ mag) is of spectral class A4 (Delgado *et al.* 1999) and shows a single pulsation period of 42.9 min (see Table 1). Using linear, non-adiabatic, radial pulsation models, this period was identified as a 5th radial overtone mode (Marconi 2004).

5. NGC 2264

The well-known cluster NGC 2264 is ~ 8 Myr old and contains the first two detected PMS δ Scuti-like stars, V 588 Mon (NGC 2264 2, HD 261331) and V 589 Mon (NGC 2264 20, HD 261446) investigated by Breger (1972). A multi-site campaign was carried out to reinvestigate the pulsational properties of the two stars. This resulted in the detection of a much denser frequency spectra than available for any other PMS pulsator.

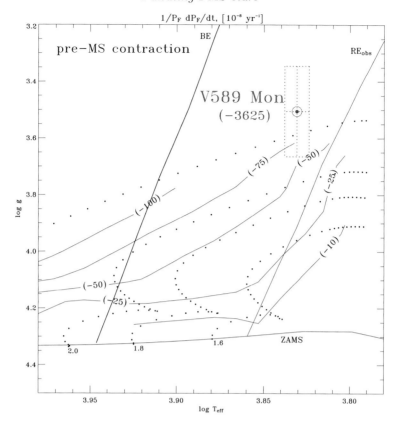

Figure 4. An HR diagram showing the ZAMS, PMS evolutionary tracks (dotted lines), the general blue edge (BE) and the empirical red edge (RE_{obs}) of the classical instability strip and lines of constant period change (solid lines). The position of V 589 Mon according to its T_{eff} and $\log g$ is marked (dot) including the errors (box). The star shows an expected period change of -75 to $-100 \cdot 10^{-8} \mathrm{yr}^{-1}$.

Using data from Breger (1972) together with observations from Peña (2002) performed in 1986, it is possible to detect evolutionary period changes in a PMS star for the first time. As an example, Fig. 4 (Pamyatnykh & Kallinger 2004) shows the position of V 589 Mon in the HR-diagram, also indicated are the PMS evolutionary tracks (dotted lines, D'Antona & Mazzitelli 1994), the ZAMS, the general blue edge (BE) and the empirical red edge (RE_{obs}) of the classical instability strip and lines of constant period change (solid lines). According to its position V 589 Mon is expected to show a period change of -75 to $-100 \cdot 10^{-8} \mathrm{yr}^{-1}$ within the given error bars. But our measurements yield a period change of $-3625 \cdot 10^{-8} \mathrm{yr}^{-1}$, indicating that stellar evolution in this phase is much faster than expected. More data are needed to investigate such period changes in greater detail.

6. Conclusions

From the 24 PMS pulsating stars now known, nine are members of young open clusters, while the others are Herbig Ae field stars. All PMS pulsators lie within the boundaries of the classical instability strip, but a gap seems to exist towards the red edge. It is essential to investigate whether this lack of stars is due to poor-number statistics, an overlap with the T Tauri domain or an evolutionary effect. A larger number of known PMS pulsators would help to settle this question, but also permit investigations to better define the boundaries of the PMS instability strip observationally and especially investigate a possible difference in the classical instability region.

Period changes due to evolutionary effects need to be studied in more detail to find whether stellar evolution is indeed a factor 40 faster than expected by theory.

Acknowledgements

This project was supported by the Austrian *Fonds zur Förderung der wissenschaftlichen Forschung* (P14984). Fourier Analysis was performed using the program PERIOD98 written by M. Sperl (1998). Use was made of the WEBDA database, operated at the Institute of Astronomy of the University of Lausanne.

References

Amado P.J., Moya A., Suárez J.C., Martin-Ruiz S., Garrido R., Rodriguez E., Catala C., Goupil M.J. 2004, *MNRAS*, 352, L11
Breger M. 1972, *ApJ*, 171, 539
D'Antona F., Mazzitelli I. 1994, *ApJS*, 90, 467
Kenyon S.J., Hartmann L. 1995, *ApJS*, 101, 117
Marconi M. 2004, *priv. communication*
Pamyatnykh A. & Kallinger T. 2004, *priv. communication*
Peña J. H., Peniche R., Cervantes F., Parrao L. 2002, *RMxAA*, 38, 31
Thé P.S., Hageman T., Westerlund B.E., Tjin A Djie H.R.E. 1985, *A&A*, 151, 391
van den Ancker M.E., Thé P.S., de Winter D. 2000, *A&A*, 362, 580
Zwintz K., Marconi M., Reegen P., Weiss W.W. 2004, *MNRAS*, in press

Discussion

NOELS: What is the error bar on the effective temperature? Could it be a way to populate the region of "missing" stars?

ZWINTZ: The accuracy in effective temperature for these stars is ± 150 K. Only with the discovery of cooler PMS pulsators can the lack of stars towards the red edge of the instability region be explained.

MICHEL: Do we have an idea of the distribution of rotational velocities for these objects?

ZWINTZ: Not for all of the stars rotational velocities and/or $v \sin i$ values are available, but generally young stars are believed to be fast rotators.

The A-Star Puzzle
Proceedings IAU Symposium No. 224, 2004   © 2004 International Astronomical Union
J. Zverko, J. Žižňovský, S.J. Adelman, & W.W. Weiss, eds.   DOI: 10.1017/S1743921304004764

Excitation of the oscillations in roAp stars: Magnetic fields, diffusion, and winds

M.S. Cunha[1], S. Théado[2] and S. Vauclair[3]

[1]Centro de Astrofísica da Universidade do Porto, Rua das Estrelas, 4150-762 Porto, Portugal
email: mcunha@astro.up.pt

[2] Institut d'Astrophysique et de Géophysique, Université de Liège, allée du Six Août 17, 4000 Liège, Belgium

[3] Observatoire Midi Pyrénées, 14, Avenue Edouard-Belin, 31400-Toulouse, France

Abstract. We discuss different physical aspects that may influence pulsation stability in rapidly oscillating Ap stars. We pay particular attention to the role of the magnetic field, atomic diffusion and winds and point out some of the important uncertainties involved in present studies of pulsation stability in these stars. We argue that presently the models that best reproduce the observations are those with convection suppressed, diffusion of helium, and no winds. With these models, in the region of the HR diagram where most roAp stars have been found, we predict the excitation of high order modes and the damping of low order modes, in accord with observations.

Keywords. Stars: chemically peculiar, stars: magnetic fields, stars: oscillations

1. Introduction

Since they were first discovered (Kurtz 1982), rapidly oscillating Ap (roAp) stars have challenged us in many different ways. Perhaps one of the most interesting questions raised soon after they were discovered was, 'why do these stars pulsate in the way they do?'

Even though the exact position of roAp stars in the HR diagram is still under debate, due, particularly, to the lack of good determinations of their effective temperatures, they are generally found in the Main Sequence part of the Classical Instability Strip, similar to the δ-Scuti stars. Thus, nonadiabatic calculations of linear oscillations performed with the original models of roAp stars predicted that their pulsations should have periods of few hours, rather than periods of few minutes, as observed by Don Kurtz. This problem, first discussed by Dolez & Gough (1982), is not yet fully resolved, despite having attracted the attention of many different investigators.

In this paper we will discuss how different aspects of the physics of roAp stars may play a role in determinng which oscillations are excited. This study follows those of Balmforth *et al.* (2000) (hereafter BCDGV) and Cunha (2002). A description of the codes used for calculating the equilibrium and pulsating models used throughout this paper can be seen in section 3.3 of BCDGV.

2. Building nonspherically symmetric models

Rapidly oscillating Ap stars are known to be patchy. This is mainly a result of the combination of magnetic fields and atomic diffusion. When carrying out theoretical studies of oscillations in these stars, one possible way to deal with the lack of spherical symmetry is to make use of the variational principal. Under given conditions, the variational principle can be used to calculate the frequencies of the oscillations appropriate to models composed of different angular regions, through a weighted average of the frequencies appropriate to different spherically symmetric models.

With this in mind, throughout the paper we will study spherically symmetric models only. To learn about the oscillation properties in models composed of different regions, it is enough to combine the correct spherically symmetric models in the way described in section 4 of BCDGV.

3. Some key ingredients for pulsation stability in roAp stars

To understand, simultaneously, the presence of high order modes and the absence of low order modes in roAp stars, it is important to focus on the properties that distinguish the latter from the δ-Scuti stars. In particular roAp stars are slow rotators and have strong magnetic fields. Consequently, very efficient diffusion processes, strongly influenced by the magnetic fields, take place in their surface layers. With this in mind, below we list some key ingredients that are likely to influence the pulsation stability in roAp stars.

    (i) Indirect effects of the magnetic field - interaction with convection;

    (ii) Chemical gradients - diffusion, winds;

    (iii) Direct effects of the magnetic field - interaction with pulsation;

    (iv) Wave reflection in the surface layers;

    (v) Etc.

3.1. *Indirect effects of the magnetic field - interaction with convection*

Strong magnetic fields influence convective motions. In a simple picture, in stars convective motions distort the magnetic field lines, and the Lorentz force resulting from the perturbation of the magnetic field may prevent convection from proceeding as it otherwise would. The interaction between magnetic fields and convection was studied in particular by Gough & Tayler (1966) and Moss & Tayler (1969). Moreover, the sufficient condition for suppression of convection derived by Gough & Tayler was tested by BCDGV for magnetic fields of magnitudes of a few kG, in models appropriate for roAp stars. These studies indicate that magnetic fields like those observed in roAp stars are sufficiently strong to influence envelop convection, and, possibly suppress it to a large extent, particularly in regions where the magnetic field is close to vertical. With this in mind BCDGV studied linear, nonadiabatic oscillations in models of roAp stars with envelop convection suppressed, and found that in these models high order modes, in the frequency range observed in roAp stars, were unstable. Moreover, they found that the excitation of these modes takes place in the region where hydrogen is ionized, while the region of second ionization of helium contributes to stabilize them (as suggested by Dziembowski & Goode (1996)).

With the goal of expanding the work of BCDGV, Cunha (2002) studied linear nonadiabatic oscillations in models with convection suppressed, spanning the region of the HR diagram where roAp stars are located. Fig. 1 shows the Theoretical Instability Strip (TIS) derived by Cunha (2002), together with those roAp stars for which Hipparcos parallaxes are available. Even though a general agreement was found between the observations and the TIS, a few discrepancies remain to be explained. In particular, a few stars are located beyond the theoretical red edge, the theoretical blue edge appears to be too hot, and, unlike what is observed, low order modes, with frequencies typical of those observed in δ-Scuti stars, are also predicted to be unstable.

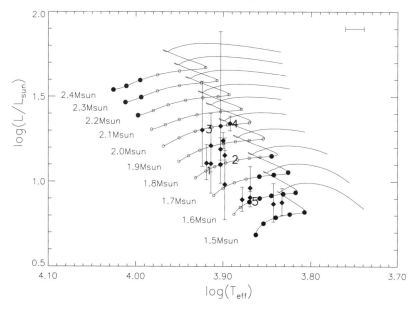

Figure 1. Theoretical Instability Strip (TIS) for roAp stars as derived by Cunha (2002). Modes like those observed in roAp stars are predicted to be unstable in the region filled with small open circles. roAp stars for which Hipparcos parallaxes are available are also shown. The numbers 1 to 5 correspond to models represented in Fig. 2 and Fig. 3

3.2. *Chemical gradients - diffusion, winds*

If the excitation of the oscillations observed in roAp stars depends strongly on nonadiabatic processes associated with the regions where hydrogen and helium are ionized, as suggested by the work of BCDGV, then the profiles of these two elements in the envelopes of roAp stars may play an important role in the excitation.

Theado *et al.* (in preparation) evolved models of roAp stars from the ZAMS, suppressing convection whenever a simplified criteria for suppression of convection (also tested by BCDGV) was satisfied. Since the intention was to test the influence that chemical profiles appropriate to different physical conditions would have in the excitation process, some of the models which were evolved accounted only for the diffusion of helium while others included the combined effect on helium of atomic diffusion and winds.

When no wind is present, helium simply sinks, and the regions where hydrogen and helium are ionized become very poor in helium and, consequently, rich in hydrogen. On the other hand, when a wind is present to compete with diffusion, an accumulation of helium takes place in the region where helium has its first ionization. Helium profiles were computed for models spanning the TIS of Cunha (2002), and the stability analysis of linear oscillations appropriate to these models was carried out.

In Fig. 2 we show the results of the stability analysis for models with different helium profiles. It is clear from the results that in the region of the HR diagram where these models are located, models with diffusion and no winds tend to show fewer unstable low order modes of the type observed in δ-Scuti stars, while showing unstable high order modes with frequencies typical of those observed in roAp stars. Since these models are located in the region where most roAp stars seem to be concentrated, this might be an indication that the levitation of helium by a wind is not a common phenomenon in roAp stars. The diffusion of helium, on the other hand, might be part of the solution for

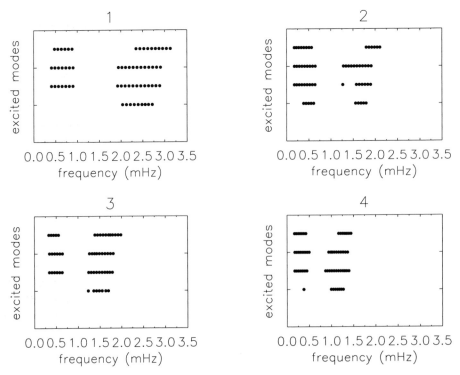

Figure 2. Each panel corresponds to a given location in the TIS (identified by the corresponding number in Fig. 1). In each panel results are given for four different models: homogeneous envelop (top), diffusion of helium plus stronger wind (second from top), diffusion of helium plus weaker wind (third from top) and diffusion of helium only (bottom). Winds were parametrized in terms of the helium outgoing flux, resulting in mass loss rates that vary slightly from model to model. The averages over the four panels are $\approx 5 \times 10^{-14} M_\odot \, \mathrm{yr}^{-1}$ for weaker winds, and $\approx 14 \times 10^{-14} M_\odot \, \mathrm{yr}^{-1}$ for stronger winds. Each filled circle corresponds to the frequency of an unstable mode.

the original mismatch between the predictions of BCDGV and Cunha (2002), and the observations, concerning the excitation of low order modes in roAp stars.

Another mismatch between the TIS and the observations is the position of the theoretical red edge, which appears to be too hot. With this in mind tests with different helium profiles where also carried out for models around the theoretical red edge. However, the position of the theoretical red edge was found to be unchanged, regardless of the helium profile considered.

3.3. ... etc.

Still with the problem of the red edge in mind, let us jump to the end of the list given at the start of this section.

The surface layers of roAp stars are rather complicate and often badly modeled. It is, therefore, important to study the sensitivity of the results to those physical aspects that are not yet well known, or not yet well modeled. Following Cunha (1999) and BCDGV, we have explored different $T - \tau$ relations in the atmosphere, different external boundary conditions in the pulsating code, and different extents of the atmosphere. We paid particular attention to models of lower masses and effective temperatures, to check the robustness of the theoretical red edge.

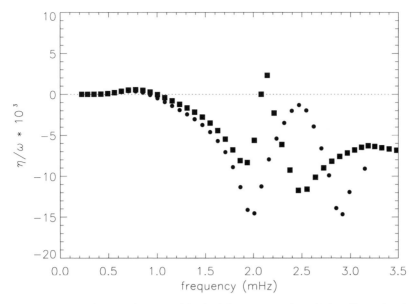

Figure 3. Relative growth rates for a model which location is identified in Fig. 1 by the number 5. The circles and squares show the growth rates for models with diffusion and no wind, and atmospheres which extend to optical depths of $\tau = 10^{-4}$ and $\tau = 10^{-6}$, respectively. The modes are excited if their growth rates are positive.

Fig. 3 shows the relative growth rates for oscillations in two models of $M = 1.6\,M_\odot$, which atmospheres extend to different optical depths. When the atmosphere is extended to smaller optical depths, the growth rates change in a systematic way. The curve is compressed and lift, shifting the largest growth rates to smaller frequencies. Moreover, due to the general increase of the growth rates, some high order modes which are found to be stable in the first model, become unstable in the model with an extended atmosphere.

We emphasize that far from the edges of the TIS, the results are rather robust, in the sense that high frequency modes are excited under all conditions explored. The exact frequency of the modes and, to a less extent, the number of modes excited is, however, influenced by the details of the model. Consequently, the fact that near the edges the energy balance for some high frequency modes results in predicted growth rates that are close to zero, implies that the uncertainty in the minimum optical depth can change the position of the edges of the TIS.

3.4. *Direct effects of the magnetic field - interaction with pulsation*

Oscillations of the kind observed in roAp stars are expected to perturb the magnetic field. The magnetic field will thus react back, influencing the oscillations.

Assuming the observed oscillations are global acoustic waves deep in the interior of the star, then, in the region where the magnetic pressure and the gas pressure are comparable, they must have a magnetoacoustic nature. As the wave propagates down to the region where the gas pressure dominates, it decouples into an (essentially) acoustic and a (essentially) magnetic wave. If, as suggested by Roberts & Soward (1983), the magnetic wave is assumed to dissipate as it propagates towards the interior of the star, due to the rapid increase in its wavenumber, the acoustic wave is constantly losing energy through the coupling with the magnetic field in the surface layers.

While studying linear, adiabatic oscillations in the presence of a magnetic field, Cunha & Gough (2000) found that the relative amount of energy lost through the coupling mentioned above, changes in a cyclic way with the frequency of the modes. Moreover, according to their results, at some frequencies the associated negative growth rates can be comparable to the positive growth rates obtained by BCDGV for the same high frequency oscillations. This means that the direct effect of the magnetic field on the energy balance that determines whether a given mode is excited, can compete with the nonadiabatic effects, and must be taken into account in a consistent way. Therefore, linear, nonadiabatic calculations which include the direct effect of the magnetic field are needed, to improve our understanding of the excitation mechanism in roAp stars. Saio (private communication) mentioned that preliminary results indicate that the coupling mentioned above can also contribute to stabilizing low order modes in roAp stars.

3.5. *Wave reflection in the surface layers*

The amount of energy lost in the surface layers of the star through wave transmission, and consequent dissipation, is also an essential ingredient of the energy balance that determines whether a given mode is excited in a given roAp star. Unfortunately, the models used so far to study wave reflection in the surface layers of roAp stars did not include a magnetic field. Models used for studying the magnetic field effect on pulsations, on the other hand, tend to use boundary conditions for mode reflection/transmission which are rather artificial. The understanding of the mechanism by which waves are reflected is thus one of our present research priorities.

4. Conclusions

The excitation of oscillations results from a subtle balance between energy losses and gains. Thus, the study of all possible energy 'sources' and 'sinks' is absolutely necessary, if the problem of the excitation mechanism in roAp stars is to be fully understood.

In this paper we have discussed different effects that we believe will determine the excitation of the high order modes observed in roAp stars, as well as to the damping of the low order modes which, so far, have never been found. We have discussed some of the uncertainties involved in the calculations and, last, but certainly not least, we have argued that we still have a lot of work in front of us, before we can claim that the question 'why do these stars pulsate in the way they do?' has been fully answered.

Acknowledgements

MC is supported by FCT-Portugal, through the grants BPD/8338/2002 and POCTI/FNU/43658/2001.

References

Balmforth, N.J., Cunha, M.S., Dolez, N., Gough, D.O. & Vauclair, S., 2000, *MNRAS*, 323, 362 (BCDGV)

Cunha, M.S., 1999, PhD thesis, University of Cambridge, UK

Cunha, M.S., 2002, *MNRAS*, 333, 47

Cunha, M.S. & Gough, D. O., 2000, *MNRAS*, 319, 1020

Dolez, N. & Gough, D. O., 1982, in: J. P. Cox & C.J. Hansen (eds.), *Pulsations in Classical and Cataclysmic Variable stars*, (JILA, Bouder, CO), p. 248

Dziembowski, W.A. & Goode, P.R., 1996, *ApJ*, 458, 338

Gough, D. O. & Tayler, R.J., 1966, *MNRAS*, 133,85

Kurtz, D. W., 1982, *MNRAS*, 200, 807

Moss, D.L. & Tayler, R.J., 1969, *MNRAS*, 145, 217

Roberts, P. H. & Soward, A., 1983, *MNRAS*, 205, 1171

Discussion

R. TRAMPEDACH: Which convection theory do you use to address the interaction between convection and p-modes?

M. S. CUNHA: In the models in which convection was not suppressed, we have used a nonlocal generalization of a time-dependent, mixing-length theory (Gough, 1977, ApJ 214, 196, Balmforth, 1992, MNRAS 255, 603).

R. TRAMPEDACH: In that case I would like to point out that the Gough *et al.* convection model suffers from the same unphysical large gradient of the turbulent pressure at the top of the convective zone as do MLT models. This results in a sharp feature in the gas pressure and temperature. I do not know how this affects the modes and their interactions with convection, but the effect is most likely significant.

T. RYABCHIKOVA: Maybe abundance gradients of metals, observed and calculated in roAp stars below $\log \tau = -1$ influence the proposed pulsation excitation mechanism?

M. S. CUNHA: I believe they will have no direct effect in the excitation, because of their low abundances, when compared with hydrogen. They might have a small indirect effect, if they have a strong impact in the structure of the atmosphere.

The A-Star Puzzle
Proceedings IAU Symposium No. 224, 2004
J. Zverko, J. Žižňovský, S.J. Adelman, & W.W. Weiss, eds.
© 2004 International Astronomical Union
DOI: 10.1017/S1743921304004776

New approaches to solve the old Blazhko puzzle in RR Lyrae stars

K. Kolenberg

[1]Institute for Astronomy, University of Vienna, Türkenschanzstrasse 17, A - 1180 Wien,
Austria
email: kolenberg@astro.univie.ac.at

Abstract. Almost a century after its discovery, the phenomenon of amplitude and/or phase modulation, observed in a large percentage of the RR Lyrae stars, still lacks a widely accepted theoretical understanding. Recent attempts to theoretically explain the effect focus on two alternatives, the magnetic models and the resonances models, both involving the presence of nonradial pulsation components.

We present the *'Blazhko project'*, a large international collaboration focused on understanding the Blazhko effect. The aim of the *'Blazhko project'* is to combine spectroscopic and photometric data from a sample of well-selected Blazhko and nonBlazhko stars, to reveal decisive information on the physical mechanism responsible for the modulation.

Keywords. Stars: oscillations; stars: variables: RR Lyrae, stars: horizontal-branch

1. RR Lyrae stars and the Blazhko effect

The study of RR Lyrae stars is a particularly active field of research, with numerous present and future applications. These pulsating variables have mean periods of less than a day, and show brightness variations of the order of a magnitude. Until not so long ago, RR Lyrae stars were considered to be prototypes of purely radial pulsators.

The most intriguing subclass of RR Lyrae stars consists of stars showing the Blazhko effect, the phenomenon of amplitude and/or phase modulation, which was named after one of its discoverers (Blazhko 1907). The lightcurve of a Blazhko star periodically changes its shape over a timescale of typically tens to hundreds of days (Figure 1). There seems to be a lack of correlation between the pulsation period and the Blazhko period.

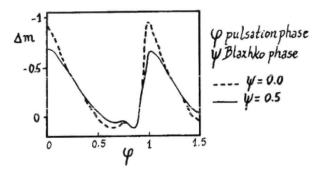

Figure 1. Illustration of the light curve changes of a Blazhko star.

The estimated incidence rate of Blazhko variables among the galactic RRab stars (fundamental mode pulsators) is about 20-30.% (Szeidl 1988, Moskalik & Poretti 2002). For the RRc Blazhko stars (first overtone pulsators) this rate is less than 5%. In the

Large Magellanic Cloud the incidence rate for RRab stars is only half as large, which is probably a metallicity effect (Alcock *et al.* 2003).

2. Past and present studies of the Blazhko effect

The Blazhko effect has been the frequent subject of photographic and photometric studies (e.g., Szeidl & Kollath 2000). Rigorous frequency analyses of photometric data for Northern field Blazhko stars were carried out by, e.g., Borkowski (1980), Kovacs (1995), and Smith *et al.* (2003), just to mention a few. For RR Lyr, the brightest Blazhko star, important spectroscopic studies were carried out by Struve & Blaauw (1948) and Preston *et al.* (1965).

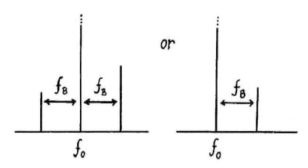

Figure 2. Triplet or doublet in the frequency spectrum of a Blazhko star.

In the past decade systematic studies of accurate CCD data of globular clusters and the Magellanic Clouds (e.g., MACHO, OGLE) have cast a new light upon the study of RR Lyrae variability (Moskalik & Poretti 2003, Alcock *et al.* 2000, 2003), and have yielded important statistics on the phenomenology of the Blazhko effect. The frequency spectra of light and radial velocity curves of RR Lyrae Blazhko stars exhibit either a doublet structure or an equally-spaced triplet structure around the main pulsation frequency and its harmonics, with a small frequency separation corresponding to the Blazhko frequency (Figure 2). The observed period ratios are 0.95-1.05, which excludes the possibility of another radial mode being excited. A large majority of the Blazhko stars have a larger modulation peak at the higher (rather than the lower) frequency side of the main pulsation component. The Blazhko effect is not found among the long period RRab stars (Smith 1981). Blazhko stars at their greatest light amplitude fall approximately on the curve of amplitude versus period as defined by stars with regular light curves. This indicates that the Blazhko effect tends to reduce the maximum light level.

Period changes are a common feature in RR Lyrae stars, and also occur in Blazhko stars (Smith 1995, Szeidl & Kollath 2000, LaCluyzé *et al.* 2002). The observed period variability is too fast to be of evolutionary nature. In some stars the Blazhko effect ceased. Some well-studied field Blazhko stars are reported to display, besides their Blazhko cycles, also very long periods of the order of years. RR Lyrae, for example, shows a cycle of about 4 years, at the end of which the strength of the modulation suddenly decreases, and a phase shift of about 10 days occurs in the Blazhko cycle. This phenomenon is still unexplained, though it has reminded some investigators of the solar magnetic cycle.

3. Explanations for the Blazhko effect in RR Lyrae stars

The most plausible hypotheses to explain the phenomenon focus on two types of models, both involving nonradial pulsation components (Figure 3).

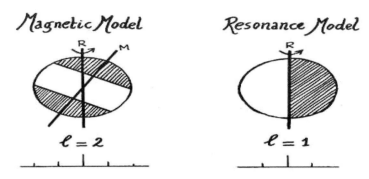

Figure 3. The two prevailing models for the Blazhko effect.

3.1. *The resonance models*

The resonance models are based on a (nonlinear) resonance between the radial fundamental mode and a nonradial mode. In these models the dipole ($\ell = 1$) modes have the highest probability to be nonlinearly excited (Cox 1993, Van Hoolst *et al.* 1998). Nowakowski & Dziembowski 2001) predict significant amplitude and phase modulation in the case of excitation of a rotationally split $m = \pm 1$ pair. The modulation period is determined by the rotation rate (currently unknown) and the Brunt-Väisälä frequency in the deepest part of the radiative interior. Peterson (1996) measured the line-widths via cross-correlation for 27 RR Lyrae stars and obtained an upper limit for $v \sin i$ of 10 km s^{-1}.

3.2. *The magnetic models*

The magnetic models, like the simple oblique pulsator model for roAp stars (Kurtz 1982), suppose that Blazhko stars have a magnetic field inclined to the stellar rotation axis (Cousens 1983, Shibahashi & Takata 1995). The main radial mode is deformed by the magnetic field to have an additional quadrupole component ($\ell = 2$), whose symmetry axis coincides with the magnetic axis. Due to the star's rotation our view of the pulsation components changes, causing the observed amplitude modulation. Shibahashi & Takata (1995) predicted a quintuplet structure in the frequency spectrum, but also showed that the quintuplet looks almost like a triplet for certain geometrical configurations. Depending on which of the side components we then observe, the Blazhko period is supposed to be equal to the rotation period or half of the rotation period. According to the latest model, a magnetic field of about 1 kG is needed in this model for the amplitude modulation to be observable. Babcock (1958) and Romanov *et al.* (1994) reported a variable magnetic field in RR Lyr with a strength up to 1.5 kG, whereas Preston (1967) and Chadid *et al.* (2001, 2004) contradict these measurements.

Despite the progress made in the last decade both in observations and modelling, a basic physical understanding of the Blazhko phenomenon is still lacking. Unsolved questions involve the modulation components of unequal amplitudes, the deviations from strict amplitude/phase modulation, the different incidence rates for fundamental mode and first overtone pulsators, the role of metallicity, the role of convective turbulence, etc. As all current models for explaining the Blazhko effect are based on the presence of nonradial components, their detection and identification is of utmost importance for understanding the mechanism behind the amplitude modulation.

4. New developments in line profile analysis

Up to now most observational studies of Blazhko stars were based on photometric data. High-resolution line profiles offer much better diagnostics to find and identify non-radial oscillation components in pulsating stars. A few years ago the first line profile study aiming at an identification of the nonradial mode(s) was carried out by Kolenberg *et al.* (2003). It was based on a set of 669 high-resolution ($R = 40000$) spectra of RR Lyr, obtained with the spectrograph ELODIE attached to the 1.93-m telescope at the Observatoire de Haute-Provence in France (Chadid *et al.* 1999). A detailed study of the variations of the Fe II λ4923.921 led to a clear detection of nonradial pulsation components in the star. By means of an adapted version of the moment method (Balona 1986, Aerts 1996, Kolenberg 2002), the detected nonradial modes were identified as nonaxisymmetric (with respect to the rotation axis) modes of degree $\ell \leqslant 3$. The incomplete coverage of the data over the Blazhko cycle hampered a more precise identification. As this kind of analysis is a first essential step towards a decisive confrontation between the theoretical models and the observations, more spectroscopic data, additional data, better spread over the Blazhko cycle, and similar data sets of additional (Blazhko) RR Lyrae stars are highly desired.

5. The Blazhko project

The Blazhko project is a large international collaboration, set up to join efforts in obtaining a better understanding of the Blazhko phenomenon in RR Lyrae stars. The project was founded in Vienna and started its activities in the autumn of 2003.

5.1. *Methodology*

The starting point for improving the modelling is an extensive database for a limited sample of field RR Lyr Blazhko and nonBlazhko stars. We selected a few RRab Blazhko stars, in the Northern and in the Southern hemisphere, and one RRc Blazhko target. Important also is the inclusion of a few well-selected, nonmodulated RR Lyrae stars in the target list, of which similar data are being gained, to be compared with the Blazhko stars.

The required data set consists of high-resolution ($R \geqslant 40000$), high-S/N ($\geqslant 100$) spectroscopic data evenly spread over the Blazhko cycle for the target stars. A few very detailed snapshots ($S/N \geqslant 200$) are being (and will be) obtained with telescopes of the 8-m class, and will help to distinguish between different nonradial modes. We carried out a simulation study to find out how much high-resolution data are minimally needed to be able to disentangle the modes, and what would be their optimal time spread. Additional radial velocities over a longer time base can be obtained with smaller telescopes and provide essential information to interpret the line profile variations. Finally, photometric data gathered over a time base of at least a year, are needed to ensure the required frequency resolution. For the interpretation of the data we will use the available spectroscopic identification methods, as well as combined techniques presently being developed in Vienna (see Zima *et al.* 2004).

5.2. *Present status*

At the time of writing the project has about 25 collaborators, observational and theoretical astronomers, in both hemispheres.

In 2004, an extensive spectroscopic and photometric campaign was dedicated to RR Lyrae, gathering observations from 10 different observatories. For the Southern hemisphere, a sample of relatively unstudied Blazhko stars is being sampled photometrically, with the aim of determing Blazhko periods and selecting the best targets for a spectroscopic study.

From the theoretical side, line profile variations are being studied for stars pulsating in a radial mode, modulated by nonradial components with close frequencies. These experiments also serve to determine the optimal time spread for the observations, to make efficient use of telescope time. Theoretical studies may also reveal how light curves are influenced at maximum/minimum light by the interaction of different modes. Finally, spectroscopic mode identification methods and their applicability are being tested in a Vienna-Leuven collaboration (Zima *et al.* 2004)

The status of the Blazhko project can be checked on a new website dedicated to the collaboration: `http://www.astro.univie.ac.at/~dsn`.

Acknowledgements

I would like to acknowledge the International Astronomical Union and the SOC for awarding me a IAU grant to participate in this Symposium. I also thank for useful suggestions and comments on the topic given during the conference, and for the positive reactions, which in the meantime have given rise to fruitful collaboration. This project is supported by the Austrian Research Foundation (Wissenschafstfonds, FWF grant P17097-N02).

References

Aerts, C. 1996, *A&A*, 314, 115

Alcock, C., Allsman, R., Alves, D. R., Axelrod, T., Becker, A., Bennett, D., Clement, C., Cook, K. H., Drake, A., Freeman, K., & the MACHO collaboration 2000, *ApJ* 542, 257

Alcock, C., Alves, D. R., Becker, A., Bennett, D., Cook, K. H., Drake, A., Freeman, K., Geha, M., Griest, K., Kovács, G., & the MACHO collaboration 2003, *ApJ* 542, 257

Babcock, H.W. 1958, *ApJS*, 3, 141

Balona, L.A. 1986, *MNRAS*, 219, 111

Blazhko, H.W. 1907, *Astron. Nachr.* 175, 325

Borkowski, K.J. 1980 *Acta Astron.* 30, 393

Chadid, M., Kolenberg, K., Aerts, C., Gillet, D. 1999, *A&A* 352, 201

Chadid, M., Wade, G.A., Shorlin, S.L.S., Landstreet, J.D. 1999, *A&A* 413, 1087

Cousens, A. 1983, *MNRAS*, 203, 1171

Cox, A.N. 1993, *Proc. IAU Coll.* 139, 409

Kolenberg, K. 2002, PhD Thesis, Leuven University, Belgium, `http://www.ster.kuleuven.ac.be/pub/kolenberg_phd/`

Kolenberg, K., Aerts, C., Fokin, A., Dziembowski, W., Chadid, M., Gillet, D. 2003, *ASP Conf. Ser.* 259, 396

Kovacs, G. 1995, *A&A*, 295, 693

Kovacs, G. 2002, *ASP Conf. Ser.* 259, 396

Kurtz, D.W. 1982, *MNRAS* 200, 807

LaCluyzé, A., Smith. H.A., Gil, E.-M., Hedden, A., Kinemuchi, K. Rosas, A. M., Pritzl, B. J., Sharpee, B., Robinson, K. W., *et al.* 2002, *ASP Conf. Ser.* 259, 416

Moskalik, P. & Poretti, E. 2003, *A&A* 398, 213

Nowakowski, R.M. & Dziembowski, W.A. 2001, *Acta Astron.*, 51, 5

Peterson, R.C., Carney, B.W., Latham, D.W. 1996, *ApJ*, 465, 47

Preston, G.W., Smak, J., Paczyński, B. 1965, *ApJS*, 12, 99

Shibahashi, H., Takata, M. 1995, *ASP Conf. Ser.* 83, 42

Smith. H.A. 1981, *PASP*, 93, 721

Smith. H.A. 1995, *RR Lyrae stars*, Cambridge University Press

Smith, H.A.,0 Church, J.A., Fournier, J., Lisle, J., Gay, P., Kolenberg, K., Carney, B.W., Dick, I., Peterson, R.C., Hakes, B. 1981, *PASP*, 115, 43

Struve, O., Blaauw, A. 1948, *ApJ*, 93, 721

Szeidl, B., Kollath, Z. 2000, *ASP Conf. Ser.* 203, 281

Van Hoolst, T., Dziembowski, W.D., Kawaler, S.D. 1998, *MNRAS*, 297, 536

Zima, W., Kolenberg, K., Briquet, M., Breger, M. 2004, *Comm. in Asteroseismology (CoAst)* 144, 5

Discussion

SHIBAHASHI: I suppose that, when you say something about the azimuthal order m, you are implicitly assuming 'm' with respect to the rotation axis. Since the magnetic model deals with the axisymmetric component with respect to the magnetic axis being oblique to the rotation axis, this nonradial component is expressed in terms of a sum of nonaxisymmetric components with respect to the rotation axis. So, in this respect, your observational claim of the presence of nonaxisymmetric components (with respect to the rotation axis) does not contradict the magnetic model.

KOLENBERG: Indeed, in the version of the moment method we used it is assumed that the pulsation axis coincides with the rotation axis. We used the slow-rotation approximation.

MICHEL: Is there any known correlation between the amplitude (or existence) of the blazhko effect and the projected rotational velocity ($v \sin i$)? If yes, this could be an argument in favor of a given type of nonradial mode possibly 'interfering' with the radial mode.

KOLENBERG: Projected rotational velocities of RR Lyr stars are not well known/studied (see text). Your suggestion is definitely something to explore in the future.

PRESTON: Blazhko periods projected back to the main sequence predict rotational velocities of many tens of km s^{-1}, but to my knowledge no Population II turnoff stars possess measured rotation.

PRZYBILLA: Does the stellar energy distribution vary over the Blazhko cycle, i.e., does the mean energy output of the star change between pulsation cycles in the Blazhko minimum/maximum? Is there some compensation in different photometric passbands?

KOLENBERG: *If* the effective temperature range changes during the Blazhko cycle (e.g., if the RR Lyrae gets brighter at maximum because it gets hotter), then we might expect the wavelengths at which the energy is emitted to change slightly during the Blazhko cycle. I do not know whether this change has been observed and studied, and what is its size.

PRESTON: It barely changes.

WADE: You mentioned our article (Chadid *et al.* 2004) on the results of our magnetic field measurements of RR Lyrae, which led to the conclusion that "RR Lyrae does not show any magnetic field over the course of the 4-year cycle". I want to underscore that the Chadid *et al.* study, because it was extended over 4 years, explicitly ruled out the idea of a magnetic cycle of this duration (and rebutted, therefore, any argument that we observed the star at 'magnetic minumum' or somesuch).

The A-Star Puzzle
Proceedings IAU Symposium No. 224, 2004
J. Zverko, J. Žižňovský, S.J. Adelman, & W.W. Weiss, eds.
© 2004 International Astronomical Union
DOI: 10.1017/S1743921304004788

Panel discussion section G

Chair: **H. Shibahashi**

Section organizer & key-note speaker: L.A. Balona
Invited speakers: M. Breger, D.W. Kurtz
Contribution speakers: K. Zwintz, M.S. Cunha, K. Kolenberg

Discussion

1. roAp stars

Noels: How is convection suppressed in model computations?

Cunha: We do this in two different ways. One is just to make the mixing parameter smaller and smaller. As it approaches zero we are, in effect, suppressing convection. The other way is simply to use the equation for radiative energy transport throughout the envelope. The results are very similar.

Reegen: I can think of two reasons for the phase lags in the line bisector analysis of roAp stars. Firstly, each bisector refers to a different layer and it is possible that the rotational velocity is different for different layers. The result is a phase lag at the boundaries of each layer. Secondly, the speed of sound is different at the boundaries of the layers and the eigenfrequencies no longer match. Instead, there is a periodic lag in the region where you have a standing wave and this causes the phase lags.

Kurtz: That may be true. What I did not talk about today and what is definitely important in this connection is the effect of horizontal stratification. Oleg Kochukhov is working on this aspect and trying to understand what we see in the line profiles. I think that in some stars the vertical stratification of neodymium is a very good explanation for the phase lags, but that may not be true in other roAp stars.

Noels: Is it possible that driving in roAp stars may be due to iron? As Georges Michaud has shown, iron can accumulate in certain layers due to diffusion and it may be possible that this could act as a driving mechanism in a certain mass range.

Cunha: I actually tried this idea, but it does not seem to make much difference for the roAp stars which have high-frequency oscillations, but I did not try the same thing for the low-frequency δ Scuti pulsators.

Kurtz: What do you need to get driving from iron at about $200000\,\mathrm{K}$? I suspect that for the roAp stars the iron layer is too deep to cause much driving of the roAp oscillations because these have high amplitudes only near the surface, so you need a mechanism which operates close to the surface.

Shibahashi: To cause driving the thermal timescale must closely correspond to the pulsational timescale. For the iron ionization zone, the thermal timescale is much longer than the periods observed in the roAp stars.

CUNHA: Yes that is true, although I do not think that by itself is the reason why the iron ionization zone cannot drive high-frequency oscillations. I do not believe that you have to have the same kind of timescale. What you do need is that the thermal timescale in the ionization region must be comparable *or longer* than the pulsation period. You cannot have a shorter thermal timescale, however. The argument that the timescale has to be of the same order as the period is based on the change of the profile of Γ_1, and that is not perfectly symmetric. The argument goes that on one side driving occurs and on the other side there is damping, and that it is basically symmetric. So if the timescale is much larger than the period of the oscillations, the pulsations are almost adiabatic and driving is compensated almost exactly by the damping. I think the real problem is the very small amplitudes when you get down to the region of iron bump, so there would be no way to excite the oscillations.

MICHEL: The destabilizing effect should be active also for the δ Scuti stars. Should we therefore not see short period pulsations similar to those in the roAp stars in the δ Scuti stars?

CUNHA: That is exactly what I was saying, that the thermal timescale has to be of the same order or longer. If it is smaller, then there is no driving. In the δ Scuti stars, whose periods are longer than those of the roAp stars, the thermal timescale of the hydrogen layer (which we believe to be the main driving region) has a smaller timescale than the periods of the δ Scuti stars, but is comparable to that of the periods of roAp stars. That means that we should not see δ Scuti periods in roAp stars due to the H ionization zone.

MICHEL: Yes, I understand that this is true for the periods observed in the δ Scuti stars, but what I was asking is why we do not see roAp pulsations (shorter periods) in the δ Scuti stars?

CUNHA: That is because convection dampens the driving of such oscillations in the δ Scuti stars. Remember that we had to introduce the concept of suppressing convection in the roAp stars by their strong magnetic fields to get driving due to hydrogen ionization. This does not work for the δ Scuti stars.

VAUCLAIR: The iron convective zone is able to destabilize stars in the sdB region (hot end of the horizontal branch), but probably not cooler stars (cf. S. Chapinet's PhD thesis).

SHIBAHASHI: To explain the strange LPV found in the Nd III and the Pr III lines of γ Equ, I am suggesting that those LPV are due to a shock wave generated in the upper atmosphere somewhere between the photosphere and the line forming layer of both Nd III and Pr III. I would like to ask the experts if the possible presence of a shock wave is allowed in the framework of the diffusion hypothesis.

MICHAUD: The question is how much mixing the shocks may cause. In so far as they are, for instance, in a region of vertical magnetic field one can imagine they can be modeled by adding to a one dimensional random walk the effects of gravity and of the radiative acceleration. It would widen the layer created by diffusion but not necessarily eliminate it. So shock waves would not necessarily eliminate the layering caused by atomic diffusion.

MONIER: How do you think we would see evidence of shock waves in the spectrum of an Ap star, and on what timescale?

SHIBAHASHI: So far the evidence is in just the line profile variations of Nd III and Pr III lines during the pulsational cycle. There is a monotonic shift from blue to red, a sudden jump from red to blue, and again a monotonic shift from blue to red. Other lines, such as Hα and Nd II, show more-or-less sinusoidal line profile variations, unlike those of Nd III. We believe Nd III to be formed very high in the atmosphere, which is exactly the region where you may expect to see a shock.

MONIER: Why do you not expect that in some cases the shock waves would dampen and completely decrease to zero amplitude?

SHIBAHASHI: We have not yet carried out any computations, so my proposal is just a phenomenological model.

KOCHUKHOV: I think that a comment on the observational aspect of your model is in order. Do we really see a shock in the line profile variability? A shock means a very rapid change in the line profiles as observed, for example, in other types of pulsating stars. I agree that your model potentially explains the blue-to-red moving feature. But currently it seems that your model also predicts a very rapid change in the line profiles and it is not clear if this change is consistent with the observational data. In the observations we see fairly smooth changes of all characteristics of the line profile: line widths, radial velocity, line asymmetry, etc. This aspect needs to be explored in greater detail.

KOLENBERG: I am also very interested in knowing what the variations of the moments of the line profile looks like if there is a shock wave. This is something that should be checked to see if it is consistent with the data.

SHIBAHASHI: At this moment, the shock wave model is just an idea which has been proposed in an attempt to explain a very puzzling phenomenon. The details of the theoretical line profile variations should be dependent on some free parameters. So far I have not attempted to see if the computed line profiles match the observations in detail, but I think that the model is promising for explaining the observed line profile variations in some of the roAp stars.

2. δ Scuti/PMS stars

NOELS: Are there examples of Pre-Main Sequence (PMS) δ Scuti stars located in exactly the same part of the H-R diagram as the evolved δ Scuti stars? If so, are there any observational differences in the frequency spectra?

BREGER: The PMS and the evolved δ Scuti stars are in the same part of the H-R Diagram. For every PMS δ Scuti star you can find at least one evolved analog with a similar temperature and luminosity. Nevertheless, there are still differences in the rotational velocity and the metallicity. These differences may have a larger effect on the pulsations than the difference in evolutionary status. More detailed studies of PMS stars to examine their pulsational behaviour are important.

ZWINTZ: The first problem is that we do not have spectra of all the pulsators we have detected. We have short runs, of two weeks mostly, or one week, or just a couple of nights, which is not good enough to detect all the frequencies. We just say "here is another point and there is another point". So we need a longer time series using multi-site observations or perhaps data from space. If we had a frequency spectrum sufficiently dense, we could compare the evolved stars with the young stars; but at this stage this is not possible. There should be differences. Theory predicts differences which originate in the interior of the star. There is probably some observational hint that Herbig Ae stars, which are characterized by emission lines, interact strongly with their environment. So there are some observational indications, but not enough to be convincing.

BALONA: It is extremely difficult to get accurate effective temperatures and luminosities for PMS stars. These stars usually have emission lines, are obscured by dust and are generally too far for Hipparcos parallaxes to be used. So you cannot directly determine their positions in the H-R diagram with any degree of accuracy.

ŠKODA: We have heard that in β Pic there are a large number of modes with high spherical harmonic numbers which appear to be unique to the star. We know that there is a circumstellar disk around beta β Pic. Is it not possible that the line profile variations are not due to pulsation but to periodic obscuration arising from circumstellar matter orbiting the star?

BALONA: In β Pic there is, indeed, nonperiodic obscuration of circumstellar material seen in the H and K lines of Ca II and which is generally attributed to infalling bodies. However, the periodic line profile variations seen in β Pic occur in lines that are formed in the photosphere and are not affected by the circumstellar medium. Therefore, the variations cannot be due to obscuration by the circumstellar medium.

KURTZ: Michaud showed just a few years ago from his models that iron ionization could possibly drive g-modes in Am stars. This has not yet been tested observationally. Are there any observations that can say whether g-modes exist in Am stars?

BREGER: This depends on which of the modes you mean. The observations can exclude relatively short periods around, say, 14 hours. The observational limits become poorer for much longer periods.

DASZYNSKA-DASZKIEWICZ: Breger showed the instability parameter versus frequency. For δ Scuti models we get a peak of instability for the low frequency g-mode range, but not enough to cause driving of pulsations. In this case we can attribute this to our lack of understanding of the interaction of convection and pulsation. So my comment is that the opacities used in the models may still need to be increased.

BREGER: This was an excellent comment, but I must add that the models do not include the effects (if any) of convective blocking.

DASZYNSKA-DASZKIEWICZ: I think that the δ Scuti and the β Cephei stars still have problems with the opacity tables.

BREGER: The groups have very different temperatures, but both show longer-period analogs with gravity modes, as recent discoveries have shown.

DASZYNSKA-DASZKIEWICZ: In theory we should get peaks in both type of stars.

3. The Blazhko effect

KURTZ: Stars showing the Blazhko effect have mostly similar behaviour, but it can range from nearly pure amplitude modulation to nearly pure phase modulation for some C-type RR Lyrae stars. Some of the modulation envelopes are nearly symmetrical such that as the peak amplitude increases, the minimum amplitude decreases. In others, the minimum amplitude changes much less than the maximum amplitude. It is difficult to see how the resonance or magnetic models can match the wide range of behaviour we see in the Blazhko stars.

KOLENBERG: Yes, I agree. But both models need to be developed. Also, we cannot exclude the possibility of a combination of both effects.

BREGER: In the resonance model, a combination of a radial and a single nonradial mode cannot explain the observed variations. However, what happens when we have four of five close modes? We will probably see amplitude and phase variations in varying degrees. This is seen in δ Scuti stars, for example.

KOLENBERG: Yes, it can definitely explained by more than one mode. Observations are under way to detect these modes using high-resolution spectroscopy.

BREGER: Although the predicted frequency region with excited pulsation modes may be quite large, a nonradially pulsating star seems to pick several small frequency subregions, in which an especially large number of pulsation modes are excited and observed. To my knowledge, this has not yet been explained theoretically. Could this be related to the resonance model of the Blazhko Effect, which requires many close modes?

PRESTON: Well I am on my angular momentum kick. I would like to make a comment on Kolenberg's talk. In the case of the magnetic model, the Blazhko periods suggest that the MS progenitors had rotational velocities of many tens of kilometers per second. But to my knowledge there has never been a measurable rotational velocity of a Population II MS star.

BUDOVIČOVÁ: How long are the Blazhko periods?

KOLENBERG: Typically it is tens to hundreds of days. The shortest Blazhko period is about 11 d and the longest about 500 days. There seems to be no correlation between the period of radial pulsation and the Blazhko period.

The A-Star Puzzle
Proceedings IAU Symposium No. 224, 2004   © 2004 International Astronomical Union
J. Zverko, J. Žižňovský, S.J. Adelman, & W.W. Weiss, eds.   DOI: 10.1017/S174392130400479X

Evolved A-type stars

M. Parthasarathy

Indian Institute of Astrophysics, Bangalore - 560034, India
email: partha@iiap.res.in

Abstract. A review of various types of evolved A stars is presented.

Keywords. Stars: abundances, stars: evolution, stars: post-AGB, binaries: general, stars: mass loss, circumstellar-matter

1. Introduction

Low and intermediate mass stars and massive stars during their post-Main Sequence evolution go through the evolved A-type star stages. The evolved A-type stars are located in the H-R diagram on the horizontal branch (HB), post-HB, post-asymptotic giant branch (post-AGB), and also in the region occupied by supergiants. Detailed multi-wavelength studies of various types of evolved A-type stars enables us to understand the nucleosynthesis, mixing, diffusion, mass-loss, and similar processes, as well as the evolutionary sequence/links with their cooler and hotter analogues.

2. A-type post-AGB stars

Post-AGB stars (proto-planetary nebulae (PPNe)) are objects that have recently evolved off the AGB (after the termination of the AGB phase of evolution of low and intermediate mass stars) but have not yet reached high enough temperatures to photoionize their circumstellar envelopes. This evolutionary stage is very short-lived depending on the core mass (Schönberner 1983, Blöcker 1995). The low and intermediate mass stars evolve through the post-AGB phase and planetary nebulae (PNe) phase and end as white dwarfs. During the transition from the tip of the AGB to the PN phase they appear as A-type post-AGB stars for a short period.

That some bright A and F supergiants have far-infrared (IRAS) colours similar to PNe and detached cold circumstellar dust shells similar to PNe central stars was first pointed out by Parthasarathy & Pottasch (1986). They concluded that these stars are not massive supergiants, but low mass stars in the post-AGB stage of evolution. From further analyses and multi-wavelength observations of IRAS sources with far-IR colours similar to PNe, several post-AGB stars were discovered (Parthasarathy & Pottasch 1986, 1989, Lamers *et al.* 1986, Hrivnak *et al.* 1989, Kwok 1993 and references therein, Van Winckel 2003 and references therein). They have supergiant spectral types from cool K, G, F, A types to OB types, forming an evolutionary sequence in the transition region from the tip of the AGB to early stages of PN (Parthasarathy 1993a, Parthasarathy *et al.* 1993, 1995). Some are high galactic latitude A and F supergiants with low metallicities and high radial velocities. It is now believed that many high galactic latitude A and F type supergiants are old low-mass stars in the post-AGB stage of evolution (Parthasarathy & Pottasch 1986, Lamers *et al.* 1986, Bond & Luck 1987). The AGB phase of evolution of these stars was terminated only recently by the superwind type of severe mass-loss which removed most of the outer envelope. Because of the very thin extended outer envelope that remained around the core these post-AGB stars mimic the spectra of supergiants.

2.1. *Very metal-poor post-AGB A-supergiants*

Recent studies have established the existence of a class of post-AGB A-supergiants with extremely low iron-group abundances (HR 4049 ([Fe/H] = -4.8, Waelkens *et al.* 1991, Takeda *et al.* 2002), HD 44179 (central star of Red Rectangle) ([Fe/H] = -3.1, Waelkens *et al.* 1996), HD 52961 ([Fe/H] = -4.8, Waelkens *et al.* 1991), BD +39° 4926 ([Fe/H] = -2.9). Their photospheric abundances are similar to the gas-phase abundances of the interstellar medium. The refractory elements Fe, Mg, Si, Al, Ti, Ca, etc are depleted and the abundances of volatile elements C, N, O, S, and Zn do not show such extreme depletions. The photospheric abundances of these stars correlate with the grain condensation temperatures (Bond 1991, Van Winckel *et al.* 1992, Parthasarathy *et al.* 1992). Van Winckel *et al.* (1992), Takeda *et al.* (2002), and Waelkens *et al.* (1996) found the Zn abundances in the extremely Fe-poor post-AGB stars HD 52961, HR 4049 and HD 44179 to be : HD 52961 ([Zn/H] = -1.5, [Fe/H] = -4.8), HR 4049 ([Zn/H] = -1.3, [Fe/H] = -4.7), and HD 44179 ([Zn/H] = -0.6, [Fe/H] = -3.3). S and Zn have very low condensation temperatures and therefore do not easily condense into dust grains. From Figure 6 and Table 2 in Takeda *et al.* (2002) it is clear that the abundance of Zn traces the original metallicity of stars that show the depletion of refractory elements.

The nearly normal abundances of Zn, S, C, N, O and the extreme depletions of refractory elements Fe, Mg, Si, Al, Ti, Ca, etc. in the very metal-poor A-type post-AGB supergiants indicate that the depletion of the refractory elements is the result of chemical fractination. The presence of circumstellar dust shells/disks around these stars lends strong support to the idea that the depleted refractory elements are in circumstellar dust grains. Bond (1991) suggested that there is a selective removal of the refractory elements from the photosphere through grain formation and mass-loss. It is still not understood how the dust and the gas are separated resulting in the remarkable peculiarity in the photospheric abundances of very metal-poor A-type post-AGB stars.

The processes that produce photospheres with depleted refractory elements in the λ Boo stars and in the A-type post-AGB stars described above may be the same (Parthasarathy 1994).

2.2. *Binarity of very metal-poor A-type post-AGB stars*

There is conclusive evidence that all the known very metal-poor post-AGB A and F supergiants (HR 4049, HD 44179, HD 52961, HD 46703, BD +39° 4926, and HD 213985) with depleted refractory elements in their photospheric abundances are single-lined spectroscopic binaries with orbital periods of several hundred days (Van Winckel *et al.* 1995). Waters *et al.* (1992) suggested that the gas-dust separation needed to account for the depletion pattern and photospheric abundance peculiarities, the presence of binary companions (Van Winckel *et al.* 1995), and dust disks (Cohen *et al.* 2004, Van Winckel *et al.* 2000) may be a necessary condition. For HR 4049 Monier & Parthasarathy (1999) found that its UV energy distribution is very deficient compared to that of the standard A6 Ib star. An ATLAS 9 model ($T_{\rm eff}$ = 7500 K, $\log g$ = 2.0 and [Fe/H] = -5.0) best fits the optical continuum. The UV-flux deficiency of HR 4049 scales as λ^{-1} (Monier & Parthasarathy 1999, Waelkens *et al.* 1995). The UV and optical light variations of HR 4049 are found to occur in phase as a result of variable circumstellar extinction (Monier & Parthasarathy 1999, Waelkens *et al.* 1991). The high galactic latitude (b = 57°) post-AGB star HD 213985 (A2 Ia) also shows variable optical and UV circumstellar extinction (Waelkens *et al.* 1987, 1995, Ambika & Parthasarathy 2004, Whitelock *et al.* 1989). Further study of the optical and UV variability of the A-type post-AGB stars HR 4049 and HD 213985 and similar stars may enable us to understand the nature of the circumstellar dust around the evolved A-type stars and the depletion of refractory elements. However, there are several A and F type post-AGB stars with dust disks (Sivarani 2000), but without any depletion of refractory elements in their chemical composition. For example HD 101584 which is a late A or early F post-AGB supergiant (Sivarani *et al.* 1999) with a hot and cold circumstellar dust disk and bipolar outflow does not show any depletion of refractory elements in its photospheric spectrum (Sivarani 2000, Sivarani *et al.* 1999).

The A-type post-AGB binary HD 44179 (Red Rectangle) (Cohen *et al.* 2004) is associated with a bipolar protoplanetary nebula with biconical appearance known as the Red Rectangle. High resolution HST images reveal a dusty disk and complex structures. Many are unique to this object. The Red Rectangle and the central A-type post-AGB star are unique spectroscopically. Its spectrum shows a plethora of unidentified optical, molecular bands in emission (Van Winckel *et al.* 2002) superimposed on a continuous Extended Red Emission (ERE) (Cohen *et al.* 1975, Witt & Boroson 1990). The carrier of the ERE is still under investigation (Witt & Vijh 2004). Schultheis *et al.* (2002) found two or three A and B stars with mid-IR excesses whose optical spectra indicate the presence of ERE. Further study of these stars is needed. These stars may be similar to HD 44179. HD 44179 (Red Rectangle) is one of the brightest known sources of extended red emission as well as of unidentified infrared (UIR) band emission which is attributed to PAHs. Recently Vijh *et al.* (2004) discovered a band of blue luminescence in the Red Rectangle nebula. They suggest that the spectrum of this newly discovered blue luminescence is most likely fluorescence from small neutral polycyclic aromatic hydrocarbon (PAH) molecules. Some other A and F-type post-AGB stars also show PAHs and dusty disks. Some of the best examples are HD 44179, CRL 2688, Hen 401, and Hen 1475 (Cohen *et al.* 2004, Sahai 2001, Sahai *et al.* 1999, Waelkens *et al.* 1996, Parthasarathy *et al.* 2001a).

3. F-type post-AGB stars with overabundance of carbon and s-process elements

Post-AGB stars which have gone through the carbon-star stage on the AGB are expected to show the products of the triple-alpha process, CN and ON cycling and the s-process on the surface as a result of thermally pulsing AGB evolution and the third dredge-up. Chemical composition analyses of several slightly metal-poor ([Fe/H] = -0.5) F-type post-AGB stars with 21 μm IR emission features show that they are all overabundant in carbon and s-process elements (Parthasarathy *et al.* 1992, Van Winckel 2003, Reddy *et al.* 1997, 2002, Reyniers *et al.* 2004). These post-AGB stars show a clear relation between neutron irradiation and the s-process overabundances based on the ratio of barium peak elements (heavy s-process elements) to strontiumr-peak elements (light s-process elements).

From a comparison with the predictions of the recent AGB models, Reyniers *et al.* (2004) suggest that the observed spread in nucleosynthesis efficiency is intrinsic and indicates that different ^{13}C pockets are needed for stars with comparable mass and metallicity to explain their abundances.

Several post-AGB stars lack overabundances of s-process elements. Their photospheres and circumstellar dust shell are found to be oxygen rich indicating the absence of efficient third dredge-up in these stars. All the post-AGB stars with 21 μm emission are found to show high C/O ratios and overabundances of s-process elements. These stars have evolved through carbon star stage on the AGB (Parthasarathy 1999, 2000).

Some of the most s-process enriched F-type post-AGB stars are IRAS 05341+0852 ([Fe/H] = -0.7, ([s-process/Fe] = + 2.2, Reddy *et al.* (1997), and IRAS 06530-0213 ([Fe/H] = -0.5, [s-process/Fe] = +2.06, Reyniers *et al.* (2004). The optical spectra of these two stars are completely dominated by atomic transitions of s-process elements. For low-metallicity stars ([Fe/H] < -1) the s-process nucleosynthesis can lead to large overabundances of lead (Pb) with respect to other s-process elements such as barium (Ba). Low-metallicity ([Fe/H] < -1) A and F type post-AGB stars with overabundances of carbon and s-process elements are expected to show overabundances of lead (Pb). Discovery and study of such stars will enable us to further understand the s-process nucleosynthesis as well as the lead (Pb) rich binary Blue Metal-Poor stars (BMP stars) (Preston & Sneden 2000, 2005, Sneden *et al.* 2003). The binary companions (now unseen) of the lead rich Blue Metal-Poor stars seem to have gone through a post-AGB stage similar to the F-type post-AGB stars with overabundances of carbon and s-process elements. The Binary Blue Metal-Poor stars seems to have acquired their overabundances of carbon and s-process elements including lead (Pb) via mass transfer during the AGB and post-AGB evolution of their companion stars (Preston & Sneden 2005, Sneden *et al.* 2003, Sivarani *et al.* 2004).

4. Hot post-AGB stars

The A-type post-AGB stars evolve towards the left in the HR-diagram at constant luminosity into hotter post-AGB stars. As a result of photoionisation of the surrounding circumstellar dust and gas shell by the hot post-AGB star, they eventually appear as planetary nebulae with hot post-AGB central stars. From the analysis of IRAS data and spectra of IRAS sources with far-IR colours similar to post-AGB stars and PNe several new hot post-AGB stars have been identified (Parthasarathy *et al.* 2000a, Parthasarathy *et al.* 2001b, Parthasarathy & Pottasch 1989). Gauba & Parthasarathy (2003) studied the UV(IUE) spectra of several high galactic latitude hot post-AGB stars. In many cases, the UV(IUE) spectra suggested partial obscuration of the hot central stars due to circumstellar dust. The circumstellar extinction law in these cases was found to be linear in λ^{-1} suggesting an upper limit of $\sim$200 Å as the radii of the small grains. Carbon-rich and oxygen-rich features were identified in the ISO spectra of these objects (Gauba & Parthasarathy 2004). Absorption lines of C II, N II, O II, Al III, Si III and Fe III and emission lines of H I, He I, C II, N I, [N II], O I, [O I], Mg II, Al II, Si II, V I, Mn I, Fe III, [Fe II], and [Cr II] have been identified in the high resolution spectra of some hot post-AGB stars (Parthasarathy *et al.* 2000b, Klochkova *et al.* 2002, Gauba *et al.* 2004). P-Cygni profiles of H_α indicate ongoing post-AGB mass loss in these stars. The hot post-AGB star, IRAS01005+7910 (Klochkova *et al.* 2002) was found to be carbon-rich. Gauba *et al.* (2004) estimated C/O $\sim$ 0.78 for IRAS13266-5551.

Several hot post-AGB stars have also been identified in the high galactic latitudes and the halo of our Galaxy on the basis of their derived atmospheric parameters and chemical composition (Conlon *et al.* 1991, 1994, McCausland *et al.* 1992, Moehler & Heber 1998, Mooney *et al.* 2002 and references therein). (These stars appear to be hotter analogues of A and F - type post-AGB stars). These stars seem to have evolved from the G, F, A - type post-AGB stars in that order. However, the abundance patterns of hot post-AGB stars found in the halo of our Galaxy show significant differences when compared with the abundance pattern of G, F, A - type post-AGB stars. Most significant are the underabundances of carbon in the halo hot post-AGB stars. Such carbon underabundances are also observed in the hot-post AGB stars in the globular clusters. Thus it is likely that the hot post-AGB stars with underabundances of carbon left the AGB before undergoing the third dredge-up.

The observed abundance patterns of post-AGB stars shows the following groups : (1) Post-AGB A and F supergiants with nearly normal abundance of Fe and slightly overabundant CNO and oxygen-rich circumstellar dust shells. The high galactic latitude F-supergiant HD 161796 (Parthasarathy & Pottasch 1986) is a good example of this group, (2) A-type post-AGB supergiants showing extreme underabundance of Fe and other refractory elements, and nearly normal abundances of CNO, S and Zn. Stars belonging to this group are HR 4049, HD 52961, and HD 44179 (Van Winckel *et al.* 1992, 1995, Van Winckel 2003, Takeda *et al.* 2002), (3) F-type post-AGB supergiants are carbon-rich and overabundant in s-process elements indicating that they have gone through the third dredge-up during their AGB stage of evolution. IRAS 05341+0852 is a good example of such a post-AGB star (Reddy *et al.* 1997), and (4) The high galactic latitude low gravity hot post-AGB supergiants show underabundances of Fe and significant underabundances of carbon indicating that these are old disc or halo stars in the post-AGB stage of evolution (McCausland *et al.* 1992, Parthasarathy *et al.* 2000b, Mooney *et al.* 2002). Further studies are needed to understand how the AGB and post-AGB evolution of low and intermediate mass stars result in the above mentioned post-AGB abundance patterns.

5. Rapidly evolving A and B type post-AGB stars

From multiwavelength studies of IRAS sources with far-IR (IRAS) colours similar to PNe few rapidly evolving A and B type post-AGB stars have been discovered (Parthasarathy *et al.* 1993, 1995, 2000b, 2001b, Bobrowsky *et al.* 1998, Gauba & Parthasarathy 2003). The best examples of rapidly evolving A and B type post-AGB stars are Henize 1357 = IRAS 17119-5926 = SAO 244567 = Stingray Nebula and SAO 85766 = IRAS 18062+2410. Less than 100 years ago they were A type post-AGB stars and now they are OB central stars of low excitation planetary nebulae.

5.1. *Henize 1357 = Stingray Nebula*

Hen 1357 was the first object that was shown to have evolved from a B1 type post-AGB supergiant into a PN in the extremely short timescale of 20 years (Parthasarathy *et al.* 1993, 1995, Bobrowsky *et al.* 1998). The UV (IUE) spectra of Hen 1357 (see Figure 1 in Parthasarathy *et al.* 2001b), shows the rapid evolution of the central star from 1988 to 1996. The P-Cygni profiles of NV(1240 Å) and C IV(1550 Å) in the spectra taken in 1988, had indicated a terminal wind velocity of -3500 km s^{-1}. The wind appears to have completely ceased by 1994 (Fiebelman 1995, Parthasarathy *et al.* 1995). We find that the object has faded by a factor of 3 in the UV, from 1988 to 1996. Then, if the luminosity were to remain the same, the temperature of the central star should have increased from 37000 K to 50000 K in the same period. The fading of the UV continuum and the termination of the stellar wind (see Figure 1 in Parthasarathy *et al.* 2001b) suggest that the nuclear fuel is almost extinct as a result of the post-AGB mass loss. Alternatively, the fading could be due an episode of high mass-loss and it is possible that the shell hydrogen burning has not stopped or stopped temporarily. The fading suggests that the central star of Hen 1357 is rapidly evolving into a DA white dwarf.

The rapid evolution of Hen 1357 indicates that the fading is occurring much more rapidly than predicted, considering that it has taken 2700 years to go from the tip of the AGB to $T_{\rm eff} = 21000$ K ($T_{\rm eff}$ in 1971), and about 20 years to turn into a PN. If the nuclear shell burning has stopped, it has stopped at a much lower surface temperature than any of the post-AGB evolutionary models suggest. Using a distance of 5.6 kpc, Parthasarathy *et al.* (1993) estimated the luminosity of the central star to be $3000L_\odot$. The core mass luminosity relation then suggests that the core mass should be $0.55M_\odot$. A $0.6M_\odot$ core mass post-AGB star takes about 10,000 years to evolve from the tip of the AGB to the white-dwarf cooling track. The luminosity, the core mass, observed rapid evolution and fading of of Hen 1357 are not in agreement with the time scales of evolution of low or high mass post-AGB models. Recent HST images of Hen 1357 (Bobrowsky *et al.* 1998) showed the presence of a $1.67'' \times 0.92''$ nebula around the central star. They also revealed the presence of collimated outflows and a binary companion to the central star of Hen 1357. The object needs to be monitored in the UV, the optical and the IR during the next few decades to understand the birth and early evolution of PNe and their central stars.

5.2. *SAO85766*

SAO85766 appears to have rapidly evolved in the last 75 years (Arkhipova *et al.* 1999, Parthasarathy *et al.* 2000b) similar to Hen 1357 described above. It was classified as a high galactic latitude ($b = 20°$), post-AGB star having far infrared colors similar to PNe (Volk & Kwok 1989, Parthasarathy 1993b). In the HDE catalogue it was classified as an A5 star based on a plate taken in July 1924 (JD 2423971) (Arkhipova *et al.* 1999). Now it is a B1 supergiant with a low excitation nebula (Arkhipova *et al.* 1999, 2001a, Parthasarathy *et al.* 2000b). Its A5 spectral type in 1924 indicates that its T_{eff} in 1924 was 8500 K. Now its B1 I type spectrum and an analysis of recent high resolution spectra yields $T_{\text{eff}} = 22000$ K. The UV colors of this star listed in the catalogue of stellar ultraviolet fluxes indicate that in 1973 its spectral type was around A5 I (Parthasarathy *et al.* 2000b). However, the spectral energy distribution of this star obtained in 1985-87 indicated that it is an early B-type star (B1 I). An analysis of its high resolution spectra revealed underabundances of carbon and metals, a high radial velocity, and the presence of low excitation nebular emission lines (Parthasarathy *et al.* 2000b). These characteristics coupled with its high galactic latitude, the presence of a PN type cold detached shell, and the variations observed in its spectrum suggest that SAO85766 has evolved rapidly during the last 75 years from a A5 type post-AGB star and it is now in early stages of the PN phase as a B1I type post-AGB star. The significant underabundance of carbon indicates that SAO 85766 left the AGB before undergoing third dredge-up (Parthasarathy *et al.* 2000b, Mooney *et al.* 2002, Ryans *et al.* 2003). SAO 85766 shows irregular rapid light variations of low amplitude due to pulsation (Arkhipova *et al.* 1999, 2001a). In the UV similar light variations are detected (Gauba & Parthasarathy 2003).

According to Blöcker (1995) a post-AGB star of 0.7 $M_{\odot}$ traverses the spectral interval from F type to B type in about 100 years. The rapid post-AGB evolution of Hen 1357 and SAO 85766 indicates that their core-masses are of the order of 0.7 $M_{\odot}$. About 100 years ago these two stars were A, F type post-AGB supergiants. Post-AGB mass-loss may have speeded their evolution into hot-post-AGB and PN stages.

5.3. *LS II +34 26 and LS IV -12 111*

LS II + 34 26 (IRAS20462+3416 = V1853 Cyg) was classified as a massive B1.5 Ia - Iabe supergiant by Turner & Drilling (1984). They estimated its distance to be 17.8 kpc by assuming that it is a massive population I B supergiant located in the outer edge of the Galaxy. Parthasarathy (1993a) discovered it to be an IRAS source with far-IR (IRAS) colours similar to A and F-type post-AGB supergiants and PNe. Based on its far infrared colors, global energy distribution, low gravity, and high galactic latitude, Parthasarathy (1993a) suggested that LS II +34 26 (IRAS20462+3416) is a low-mass post-AGB B-type supergiant surrounded by a cold detached dust envelope located at a distance of 3 to 4.6 kpc, rather than a massive population B supergiant. The B supergiant spectrum of LS II + 34 26 shows an evolutionary sequence of post-AGB supergiants from cooler (G,F,A) to hotter (B and O) types. The nebular emission lines in the optical spectrum of LS II+34 26 indicated that it is rapidly evolving into a very low excitation PN (Parthasarathy 1994,

Garcia-Lario *et al.* 1997). Such nebular emission lines seems to absent in the spectrum of LS II + 34 26 obtained by Turner & Drilling in 1977. Differences in the spectrograms obtained by Turner & Drilling in 1977 and 1981 and the recent spectra of LS II +34 26 (Parthasarathy 1994, Garcia-Lario *et al.* 1997, Arkhipova *et al.* 2001b) suggest that it is rapidly evolving into the PN stage. A multiwavelength study by García-Lario *et al.* (1997) confirmed that the central star is surrounded by a very low excitation and compact nebula (Ueta *et al.* 2000). The presence of P-Cygni profiles indicate ongoing post-AGB mass-loss (Garcia-Lario *et al.* 1997, Arkhipova *et al.* 2001b). Like Hen 1357 and SAO 85766, LS II +34 26 seems to be another rapidly evolving hot post-AGB star. Recently Arkhipova *et al.* (2001b) made photometric and spectroscopic observations of LSII +34 26 and found rapid irregular low amplitude light variations similar to that of SAO 85766 (Arkhipova *et al.* 1999, 2001a).

LS IV -12 111 (IRAS 19590-1249) was classified as a high galactic latitude hot post-AGB star based on its IRAS data and high galactic latitude (Parthasarathy 1990a, 1993b). Analysis of its high resolution spectrum by McCausland *et al.* (1992) and Conlon *et al.* (1993) confirmed that it is a hot post-AGB star rapidly evolving into a low excitation planetary nebula. McCausland *et al.* (1992) derived chemical compositions of several high latitude hot post-AGB stars. In contrast to the cooler post-AGB stars, LS IV -12 111 showed a severe carbon depletion (Ryans *et al.* 2003). Similar carbon depletions have been reported in other hot post-AGB stars at high galactic latitudes. These include LS II+34 26 (García Lario *et al.* 1997), PG1323-086, PG1704+222 (Moehler & Heber 1998) and SAO85766 (Parthasarathy *et al.* 2000b).

6. Final helium shell flash objects

Some post-AGB stars which are entering the white dwarf cooling track (pre-white dwarfs) with enough helium in their envelopes may experience a final helium shell flash (Schönberner 1979, Iben 1982). As a result of this late thermal pulse the very hot post-AGB star returns to the AGB. The transition is characterized by an increase in luminosity and a drop in effective temperature in a very short time (of the order of 5 to 15 years). During this transition the photosphere becomes very rich in helium, C, N, O and s-process elements. Herwig (2001) has shown that the evolutionary time scale of very late thermal pulse stars strongly depends on the value chosen for the convective mixing efficiency parameter. The evolutionary paths have a double loop structure (see Figures 1, 2, and 3 in Lawlor & MacDonald 2002). The double loop in the HR diagram is a result of the penetration of the He-flash driven convection zone into H-rich layers. During this evolutionary phase the star once again very rapidly passes through the effective temperature and luminosity region occupied by the A and F type post-AGB stars.

Sakurai's object (V4334 Sgr) (Duerbeck *et al.* 2000), FG Sge (Blöcker & Schönberner 1997), and V605 Aql (Harrison 1996) are understood to be the objects that experienced very late thermal pulse (final helium shell flash). All three have increased in luminosity, decreased in temperature, and show evidence for significant changes in chemical composition. From detailed study and monitoring of these stars, we may be able to test stellar evolution and nucleosynthesis models of the very late thermal pulse which occurs in about 10% of the central stars of planetary nebulae (Iben & MacDonald 1995). V4334 Sgr (Sakurai's object) evolved very rapidly from a pre-white dwarf stage to the AGB in less than six years (Duerbeck *et al.* 2000). Lawlor & MacDonald (2002) indicate that each of the three observed objects (Sakurai's object, V605 Aql, and FG Sge) represent one of the three crossings of the HR diagram from hot to cool (Sakurai's object), from cool to hot (V 605 Aql) and then back to cool giant (FG Sge). During these crossings of the HR diagram these stars appear to have spent a very brief time as A-F post-AGB stars for the second time or third time. These three objects may once again pass through the A-F post-AGB stars stage within the next 10 to 50 years. Lawlor & MacDonald 2002) find that the critical aspect of the double loop evolution (their Figures 1, 2, and 3) is the requirement that the convective mixing efficiency be much lower than predicted by standard mixing length theory.

A very late thermal pulse (final helium shell flash) is also used to explain the [WC] and [WC]-PG 1159 central stars of planetary nebulae (Acker & Neiner 2003, Parthasarathy & 1998). The F-type post-AGB stars (see Section 3) with overabundances of carbon and s-process elements as they evolve towards higher temperatures may become [WC] Central stars of planetary nebulae (Parthasarathy 1999).

7. Unusual eruptive and nova-like variables

Recentlya few unusual eruptive variables have been discovered whose evolutionary stages and progenitors are not clear. These are V838 Monocerotis, V4332 Sagittari, and V445 Puppis. It is unclear if they are experiencing late thermal pulses (final helium shell flash), or nova like outbursts and if they are evolved massive stars or low mass stars belonging to a new class of objects.

V838 Mon was first reported to be in outburst in 2002 and became very bright ($V =$ 6.7) and blue and became a very red object. The light of the outbursts illuminated the circumstellar matter, resulting in spectacular HST images of the expanding light echo (Bond *et al.* 2003). It now has cooled and shows a late M spectral type to become the first supergiant classified as type L. The observations indicate that V838 Mon has evolved to a cool temperature of the order of 2500 K (Banerjee & Ashok 2002) within 130 days after the outburst. Recent studies of the circumstellar envelope and infrared data indicate that V83 Mon underwent two or three prior outbursts in the past. It may be a post-AGB star or an evolved giant going through repeated outbursts and mass-loss episodes. The progenitor of V838 Mon may be an A-type star.

Two other stars that had nova-like outbursts, brightening and rapid cooling similar to V838 Mon are V4332 Sagittari (Martini *et al.* 1999, Munari *et al.* 2002, Banerjee *et al.* 2003) and V445 Puppis (Ashok & Banerjee 2003 and references therein). The pre and post-outburst flux distribution of V445 Puppis appears to correspond to that of F and A stars (Ashok & Banerjee 2003). The nova-like outbursts of V838 Mon, V4332 Sagittari and V445 Puppis may or may not be related to late thermal pulse/final helium shell flash of post-AGB stars. These stars may be close binaries with evolved companions and accretion and mass-transfer events. Further spectroscopic and photometric monitoring and detailed chemical composition analyses of V838 Mon, V 4332 Sagittari and V 445 Puppis and similar objects are needed to understand if they are related to the post-AGB evolution of single low or intermediate mass stars.

8. A-type post-HB and post-AGB stars in globular clusters and related UV-bright stars in globular clusters

There are several types of evolved A-type stars in globular clusters including blue straggler stars, interacting binaries, A and B type horizontal branch (HB) stars (Moehler 2005), A-type post-HB and post-AGB stars (Ambika *et al.* 2004, Jasniewicz *et al.* 2004, Mooney *et al.* 2004). In earlier surveys some were classified as UV-bright stars (Zinn *et al.* 1972). The term UV-bright stars was introduced by Zinn *et al.* (1972) for stars in globular clusters that are above the horizontal branch (HB) and bluer than red giants. Further studies showed that the UV-bright stars in globular clusters include suprahorizontal branch stars, post-AGB stars, post-early-AGB stars and AGB-manque stars (de Boer 1985, 1987, Dorman *et al.* 1993, Moehler 2001). The list of UV-bright stars in globular clusters prepared by Zinn *et al.* (1972), Harris *et al.* (1983), and de Boer (1987) contain several A-type post-AGB, post-HB and post-early-AGB (PEAGB) stars.

Chemical composition studies of B, A, and F type UV-bright stars in globular clusters reveal that some of them are post-AGB A and F stars showing evidence for third dredgeup and overabundance of carbon and s-process elements for example ROA 24 in ω Cen (Gonzalez & Wallerstein 1994), while some show severe carbon deficiency indicating that they left the AGB before the third dredgeup occurred for example ZNG 1 in M10, ROA 5701 in ω Cen and Barnard 29 in M13 (Moehler *et al.* 1998, Jasniewicz *et al.* 2004, Mooney *et al.* 2004).

Recently Ambika *et al.* (2004), Jasniewicz *et al.* (2004), and Mooney *et al.* (2004) derived chemical composition of few A and B type evolved stars in several globular clusters. Some are in the post-HB and the post-AGB stages of evolution. The abundance pattern in a few A-type post-HB stars indicates the presence of diffusion and radiative levitation process, already found in blue-HB (BHB) stars in globular clusters (Moehler 2005). Chemical composition studies of several types of evolved stars and UV-bright stars in globular clusters are needed to further understand their evolutionary sequences and links and abundance peculiarities.

9. Blue stragglers

Globular clusters contain exotic evolved A and B stars that cannot be explained by canonical stellar evolutionary models, for example, the blue straggler stars. Blue stragglers are present in all the globular clusters. The explanations for their formation involve mass transfer in and/or the merger of a binary star system, or collisions between the stars. All these mechanisms may be at work in the globular clusters resulting in the formation of blue stragglers. Recent Hubble Space Telescope observations of a large sample of globular clusters reveal that every globular cluster contains between 40 and 400 blue stragglers (Piotto *et al.* 2002). The population does not correlate with either stellar collision rate (as would be expected if all blue stragglers were formed via collisions) or total mass (as would be expected if all blue stragglers were formed via the unhindered evolution of a subset of the stellar population) (Davies *et al.* 2004). The observed blue straggler frequency in globular clusters can be the result of the combination of the two formation processes, i.e., collisions or mergers in a crowded place (dynamical blue stragglers) and the evolution of isolated binaries (primordial blue stragglers). The comparison of the observed blue straggler distribution in 47 Tuc (Mapelli *et al.* 2004) with the simulated distribution supports the hypothesis that internal blue stragglers result from stellar collisions, while external (outside of cluster 20 core radii) are exclusively generated by mass transfer in primordial binaries (where are evolved in isolation experiencing mass transfer). Recent observations of a few blue stragglers in globular clusters indicate the presence of circumstellar shells around them (De Marco *et al.* 2004).

In wide binaries blue straggler stars may result when the primary evolves off the Main Sequence transferring mass on to the Main Sequence secondary. The blue stragglers and blue metal-poor binary stars (Preston & Sneden 2005) in the galactic halo may have resulted from such a process (Sneden *et al.* 2003).

A number of blue stragglers in old stellar systems are found to be photometrically variable (Mateo 1993). Some are pulsating blue stragglers and others eclipsing binary blue stragglers. Detailed spectroscopic and photometric study of photometrically variable blue stragglers (pulsating and eclipsing) can provide the physical properties of blue stragglers and can help us to understand the formation of blue stragglers in globular clusters, in old open clusters, and in the field. Some photometrically variable blue stragglers are dwarf Cepheids, or SX Phoenicis variables. The reason for the relative lack of large-amplitude pulsating blue stragglers in old open clusters is unclear.

10. Close-binary systems with A-type components

There are many Algol-type close binaries in which the primary stars are of A-spectral type. The Algol-type systems are semi-detached binaries. The cool subgiant/giant secondaries have filled their Roche lobes, which still show evidence of mass-transfer and mass-loss. The present A-type primaries are not in thermal equilibrium. The rapidly-oscillating (pulsating) and mass-accreting A-type components of Algol-type eclipsing binary systems form a small group (Mkrtichian et $al.$ 2004). These are Y Cam, AB Cas, RZ Cas, R CMa, TW Dra, AS Eri, RX Hya, and AB Per. Except for R CMa whose primary is a F1V star in all other above mentioned systems the pulsating primary components are A-stars. The observed pulsation periods range from 22 minutes to 282 minutes. The A-type stars in Algol-type close-binaries are important for asteroseismology as they permit asteroseismic estimates of accretion rates (Mkrtichian et $al.$ 2004) and for understanding the mass-loss, the mass-transfer and the evolution.

Recently Mkrtichian et $al.$ (2004) studied the frequency spectrum of the rapidly-oscillating mass-accreting A3 V component of AS Eri. AS Eri is a semi-detached eclipsing binary with an orbital period of 2.66 days. The A3V (1.92 $M_\odot$, 1.8 $R_\odot$) primary is well within its Roche lobe. The cool very low mass secondary (0.207 $M_\odot$, 2.25 $R_\odot$ subgiant fills its Roche lobe. AS Eri evolved to the present configuration with a substantial loss of angular momentum. There are rapid pulsations in the A3 V star of AS Eri with a frequency of 59.03 d^{-1} and second and third oscillation modes with frequencies 62.5 d^{-1} and 61.67 d^{-1}, respectively. These modes are related to the 5-6 overtone oscillations and are among the shortest period excited in nonmagnetic A-F stars. AS Eri is a very low mass ratio Algol type system. The mass of evolved secondary is only 0.207 $M_\odot$. The totally eclipsing Algol-type systems S Cnc and DN Ori with A-type primary components and with very low mass evolved secondaries are similar to AS Eri. The A stars in S Cnc, DN Ori and in other very low mass ratio Algol systems may reveal rapid pulsations similar to that observed in the A-type star in AS Eri. Study of the characteristics of rapid pulsations of A-type stars in Algol-type systems (in particular very low mass ratio Algol systems) may enable us to understand the mass-accretion process and evolution of Algol-type close-binary stars.

There are a few peculiar Algol-type systems such as V356 Sgr. It is a 8.9 day period semidetached system consisting of a B4V primary and a giant A2 II secondary. The A2 II secondary has filled its Roche lobe and is now burning hydrogen in a shell. From an analysis of the ultraviolet spectrum of the A2 II star obtained during the total eclipse Polidan (1988) finds that the evolved A2 II component is extremely underabundant in carbon. The cool evolved secondary components in Algol-type systems were also found to show underabundances of carbon (Parthasarathy et $al.$ 1983, Parthasarathy 1990b). Further studies of the A2 II star in V356 Sgr and its pulsation characteristics may enable us to further understand the accretion and evolution of V356 Sgr and similar close-binary stars. High speed photometric monitoring of V 356 Sgr during the totality may reveal the rapid pulsations of the A2 II secondary.

υ Sgr, HD 30353 (KS Per), LSS 4300 and CPD-58 2721 (LSS 1922) form a small group of hydrogen poor, helium rich close binaries with A-type components. The primary components in these systems are hydrogen-poor and helium rich late B and A-type supergiants. The ultraviolet spectra of v Sgr and HD 30353 indicate that their secondary components may be late evolved OB stars. UV high resolution spectra of these two peculiar binaries with hydrogen-poor A-type supergiants show violet shifted N V, C IV, Si IV, C II, Al III, and Mg II stellar wind profiles (Parthasarathy *et al.* 1986, 1990) indicating a stellar wind and mass-loss similar to that observed in OB stars. The IRAS far-infrared fluxes also show evidence for the presence of warm and cold circumstellar dust shells (Parthasarathy 1990b). Further studies of v Sgr, HD 30353, LSS 4300 and LS 1992 are needed to understand their peculiar chemical composition, mass-loss rates and evolution. It is also important to search for the rapid pulsations of the A-type primary components of v Sgr and HD 30353 (KS Per).

Another important class of close-binaries some of which contain an evolved A or a Main Sequence A star are planetary nebulae with binary central stars. There are more than 16 planetary nebulae with close-binary nuclei (Bond 2000) with periods of 2.7 hours to 16 days. These are so close that the nebulae must have been ejected through common-envelope interactions (Bond 2000 and references therein). Further studies of planetary nebulae with binary nuclei with A and F type components will enable us to derive accurate distances and luminosities of central stars, as the Bolometric Corrections of A and F type stars are very small. Also detailed study of such objects may enable us to understand the formation of jets, bipolar structures, shaping and evolution of planetary nebulae, and common envelope evolution (Bond 2000).

11. Stars related to massive A supergiants

Massive stars of 8 to 40 $M_\odot$ during their post-Main Sequence evolution go through the A supergiant stage. For more detail see Przybilla (2005). Some luminous blue variables (LBVs) are related to massive A supergiants.

LBVs include Hubble-Sandage variables, P Cygni stars, and S Doradus stars. All are hot massive stars and show A supergiant type spectra during the periods of maximum brightness. LBVs show variations in brightness by 1 to 2 magnitudes on time scales of a few decades. Occasionally, they erupt and increase in their brightness by more than 3 magnitudes. During the periods of maximum brightness they resemble spectral type A supergiants. For more details on LBVs see Nota & Lamers (1997a). The LBVs are the very short lived phase of the evolution of very massive stars.

There are about 10 LBVs known in the Milky Way Galaxy, about six in the LMC, about two in the SMC, and four or five in M31 and M33. Some of the well studied LBVs that show A supergiant spectrum during the period of maximum brightness are AG Car, HR Car, S Dor, R 71, and R 127. Most LBVs have nebulae and circumstellar dust shells that are the result of the ongoing large scale mass-loss and outbursts. HR Car is associated with circumstellar filamentary nebula. From a recent set of high resolution coronographic images Notal *et al.* (1997b) confirm that the nebula around HR Car is truly bipolar and reminiscent of the η Carinae Nebula. The polarization measurements of HR Car also indicate bipolar geometry (Parthasarathy *et al.* 2000c).

Recently Clark & Negueuela (2004) discovered W 243 in the galactic starburst cluster Westerlund 1 to be a luminous blue variable. They find the spectral type of W 243 to be A2 Ia with a rich emission line spectrum indicating high mass loss rate.

Waters *et al.* (1997) discuss the solid state features in the ISO-SWS spectra of some LBVs. They find evidence for crystalline forms silicates with a composition similar to those seen in red supergiants, AGB and post-AGB stars. They suggest that the dust was formed in a high-density, probably also high temperature environment, during the red supergiant phase of LBVs or in an ejection during a very red LBV phase. Observationally, there is strong evidence that very massive stars do not evolve to become red supergiants (Langer 1993) or if they do they spend a very short time as a yellow or red supergiant. There are few supergiants of this type with circumstellar dust shells. Some authors classified these stars with supergiant spectra as low mass stars in post-AGB stage of evolution while others as massive supergiants. The best known examples are HD 168625 (IRAS 18184-1623) (classified as A-type hypergiant by Morgan *et al.* 1955 and also as a post-AGB A supergiant (Parthasarathy 1989)) and HD 179821 (IRAS 19114+0002) (classified as a massive yellow supergiant (Jura *et al.* 2001) and also as post-AGB star (Pottasch & Parthasarathy 1988)).

Sterken *et al.* (1999) classified HD 168625 as a variable A-type hypergiant or marginally dormant LBV. On the basis of spatially resolved nebula around HD 168625 Hutsemekers *et al.* (1994) classified it as an LBV. From the velocities and chemical composition derived from spatially resolved spectroscopy Nota *et al.* (1996) confirm that the nebula is associated with the central star HD 168625 and it is nitrogen rich. It was also classified as a post-AGB candidate on the basis of far-infrared (IRAS) colours and detached cold circumstellar dust shell. Its far IR colours are similar to that of post-AB stars and proto-planetary nebulae (Parthasarathy 1989). Sivarani (2000) (see Garcia Lario *et al.* 2001) found that the equivalent widths of the spectral lines in the high resolution spectrum of HD 168625 are variable. She also finds that the star is nitrogen rich in agreement with the nitrogen-rich nebula, indicating that the nebula may have been ejected after the star became nitrogen-rich during the CNO cycle. The distance and luminosity derived from the Hipparcos parallax seem to be inconsistent with the previous distance estimates (Nota *et al.* 1996, Hutsmekers *et al.* 1994, Sivarani 2000, Garcia-Lario *et al.* 2001). The evolutionary status of HD 168625 is unclear, it may be a LBV or a post-AGB star (Sivarani 2000) with a core mass of the order of 0.8 to 1.0 $M_\odot$. The overabundance of nitrogen is similar to that observed in type I planetary nebulae.

HD 179821 ($V = 8.4$, G5Ia, b=4.96) (AFGL 2423 = IRAS 19114 + 0002) continues to be the subject of much debate as to whether it is a nearby (distance 1 kpc) post-AGB star (Pottasch & Parthasarathy 1988) or a distant (distance 6 kpc) high mass (30 $M_\odot$) post-red supergiant near the Humphreys-Davidson limit that will become a supernova within the next 100000 years (Jura *et al.* 2001). Detailed spectroscopic abundance analyses and atmospheric parameters of HD 179821 (Reddy & Hrivnak 1999, Thevenin *et al.* 2000) have failed to resolve the issue. Jura *et al.* (2001) argue that HD 179821 is a hypergiant at a distance of 6 kpc from the Sun. HD 179821 is a strong IRAS source with 25 micron flux of 650 Jy (Pottasch & Parthasarathy 1988). The extended dust shell scatters light at optical and near-IR wavelengths (Ueta et al. 2000, Jura & Werner 1999). Thevenin *et al.* (2000) on the basis of a LTE analysis of its spectra find $T_{eff} = 5660$ K, $\log g = -1.0$ and [Fe/H] = -0.5 and underabundances of carbon and the s-process element zirconium indicating that HD 179821 has not gone through the third dredge-up. They find its abundance pattern to be similar to that of the carbon-poor halo planetary nebula DDDM-1. They conclude that it is a low mass, slightly metal-poor star in the post AGB stage of evolution (Thevenin *et al.* 2000). However, Jura *et al.* (2001) on the basis of detailed high resolution mapping of the J = 1-0 CO emission from the circumstellar shell around HD 179821 conclude that it is a highly evolved massive supergiant and that it may explode as a supernova within the next 100000 years.

If it is a massive star evolved from A-type LBV supergiant phase to the present yellow/red hypergiant phase, significant variations in brightness and spectrum are expected. However there is no observational evidence for any major changes in brightness, colour and temperature and spectral type during the past 100 years (Arkhipova *et al.* 2001c, Jura *et al.* 2001). Based on ten years of photometric monitoring of HD 179821, Arkhipova *et al.* (2001c) found that it exhibits semiregular light variations with amplitudes of 0.25, 0.15, 0.10 magnitudes in the UBV bands respectively. The variability seems to be due to pulsation similar that observed in other post-AGB A and F supergiants. The characteristics of HD 179821 appears to be different from that of the yellow hypergiant star IRC +10420 whose spectral type is F8 Ia, although there are indications of a recent increase in temperature, up to A5 type (Klochkova *et al.* 1997). Accurate distance determinations in the next one or two decades may enable us to understand the evolutionary status of HD 179821 and HD 168625 and other similar supergiants.

References

Acker A., Neiner C. 2003 *A&A*, 403, 659

Ambika S., Parthasarathy M., Aoki W. 2004 *A&A* 417, 293

Ambika S., Parthasarathy M. 2004 *Preprint*

Arkhipova V.P., Ikonnikova N.P., Noskova R.I. 1999 *Ast. Lett.* 25, 25

Arkhipova V.P., Klochkova V.G., Sokol G.V. 2001a *Ast. Lett.* 27, 99

Arkhipova V.P., Ikonnikova N.P., Noskova R.I., *et al.* 2001b *Ast. Lett.* 27, 719

Arkhipova V.P., Ikonnikova R.I., Noskova R.I., *et al.* 2001c *Ast. Lett.* 27, 156

Ashok N.M., Banerjee D.P.K 2003 *A&A* 409, 1007

Banerjee D.P.K., Ashok N.M. 2002 *A&A* 395, 161

Banerjee D.P.K., Varricatt W.P., Ashok N.M., *et al.* 2003 *ApJ* 598, L31

Blöcker T. 1995 *A&A* 299, 755

Blöcker T., Schönberner D. 1997 *A&A* 324, 991

Bobrowsky M., Sahu K.C., Parthasarathy M., Garcia-Lario P. 1998 *Nature* 392, 469

Bond H.E. 1991 *IAU Symposium* 145, 341

Bond H.E. 2000 *ASP Conf. Ser.* 199, 115

Bond H.E. 2003 *Nature* 422, 405

Bond H.E., Luck R.E. 1987 *ApJ* 312, 203

Clark J.S., Negueruela I. 2004 *A&A* 413, L15

Cohen M., *et al.* 1975 *ApJ* 196, 179

Cohen M., Van Winckel H., Bond H.E., *et al.* 2004 *AJ* 127, 2362

Conlon E.S., Dufton P.L., Keenan F.P. 1991 *MNRAS* 248, 820

Conlon E.S., Dufton P.L., McCausland R.J.H., *et al.* 1993 *ApJ* 408, 593

Conlon E.S., Dufton P.L., Keenan F.P. 1994 *A&A* 290, 897

Davies M.B., Piotto G., de Angeli F. 2004 *MNRAS* 348, 129

de Boer K.S. 1985 *A&A* 142, 321

de Boer K.S. 1987 *IAU Colloq.* 95, 95

De Marco O., Lanz T., Ouellette A., *et al.* 2004 *ApJ* 606, L151

Dorman B., Rood T., O'Connell R.W. 1993 *ApJ* 419, 596

Duerbeck H.W., Liller W., Sterken C., *et al.* 2000 *AJ* 119, 2360

Fiebelman W.A 1995 *ApJ* 443, 245

Gauba G., Parthasarathy M. 2003 *A&A* 407, 1007

Gauba G., Parthasarathy M. 2004 *A&A* 417, 201

Gauba G., Parthasarathy M., Reddy, B.E. 2004 *A&A* in press

Garcia-Lario P., Parthasarathy M., De Martino D., *et al.* 1997 *A&A* 326, 1103

Garcia-Lario P., Sivarani M., Parthasarathy M., *et al.* 2001 *in Post-AGB objects, eds. Szczerba & Gorny* Kluwer, 309

Gonzalez G., Wallerstein G. 1992 *MNRAS* 254, 343

Harris H.C., Nemec J.M., Hesser J. 1983 *PASP* 95, 256

Harrison T.E. 1996 *PASP* 108, 1112

Herwig F. 2001 *ApJ* 554, L71

Hrivnak B.J., Kwok S., Volk K.M. 1989 *ApJ* 346, 265

Hutsemekers D., Van Drom E., Gosset E. 1994 *A&A* 290, 906

Jasniewicz G., de Laverny P., Parthasarathy M., *et al.* 2004 *A&A* 423, 353

Iben I.Jr. 1982 *ApJ* 260, 821

Iben I.Jr., MacDonald J. 1995 *White Dwarfs* ed. D.Koester & K.Werner (Springer), p48

Jura M., Werner M.W. 1999 *ApJ* 525, 113

Jura M., Velusamy T., Werner M.W. 2001 *ApJ* 556, 408

Klochkova V.G., Chentsov E.L., Panchuk V.E., 1997 *MNRAS* 292, 19

Klochkova V.G., Yushkin M.V., Miroshnichenko A.S., *et al.* 2002 *A&A* 392, 143

Kwok S. 1993 *ARA&A* 31, 63

Lamers H.J.G.L.M., Waters L.B.F.M., *et al.* 1986 *A&A* 154, L20

Langer N. 1993 *ASP. Conf. Ser.* 35, 159

Lawlor T.M., MacDonald J. 2002 *ASP Conf. Ser.* 279, 193

Mapelli M., Sigurdsson S., Monica C., *et al.* 2004 *Apj* 605, L29

Martini P., Wagner R., Tomaney A. 1999 *AJ* 118, 1034

Mateo M. 1993 *ASP Conf. Ser.* 53, 74

McCausland R.J.H., Conlon E.S., Dufton P.L. *et al.* 1992 *ApJ* 394, 298

Mkrtichian D.E., Kusakin A.V., Rodriguez E., *et al.* 2004 *A&A* 419, 1015

Moehler S. 2001 *PASP* 113, 1162

Moehler S. 2005, *These Proceedings*, 395

Moehler S., Heber U. 1998 *A&A* 335, 985

Moehler S., Heber M., Lemke M., *et al.* 1998 *A&A* 339, 537

Moehler S., Sweigart A.V., Landsman W.B., *et al.* 1999 *A&A* 346, L1

Monier R., Parthasarathy M. 1999 *A&A* 341, 117

Mooney C.J., Rolleston W.R.J., Keenan F.P., *et al.* 2002 *MNRAS* 337, 851

Mooney C.J., Rolleston W.R.J., Keenan F.P., *et al.* 2004 *A&A* (in press)

Morgan W.W., Code A.D., Whitford A.E. 1955 *ApJS* 2, 41

Munari U., Henden A., Kiyota S., *et al.* 2002 *A&A* 389, L51

Nota A., Pasquali A., Clampin M., *et al.* 1996 *ApJ* 473, 946

Nota A., Lamers H.J.G.L.M 1997a *ASP Conf. Ser.* 120,

Nota A., Smith L., Pasquali A., *et al.* 1997b *ApJ* 486, 338

Parthasarathy M. 1989 *IAU Colloquium* 106, 384, eds. H.R.Johnson and B.Zuckerman, CUP

Parthasarathy M. 1990a *IAU Symp.* 145, poster papers: p119, eds. Michaud, Tutukov, Bergevin

Parthasarathy M. 1990b *BASI* 18, 261

Parthasarathy M. 1993a *ApJ* 414, L109

Parthasarathy M. 1993b *ASP Conf.Ser.* 45, 173

Parthasarathy M. 1994 *ASP Conf.Ser.* 60, 261

Parthasarathy M. 1999 *IAU Symp.* 191, 475

Parthasarathy M. 2000 *IAU Symp.* 177, 225

Parthasarathy M., Pottasch S.R. 1986 *A&A* 154, L16

Parthasarathy M., Pottasch S.R. 1989 *A&A* 225, 521

Parthasarathy M., Lambert D.L., Tomkin J. 1983 *MNRAS* 203, 1063

Parthasarathy M., Cornachin M., Hack M. 1986 *A&A* 166, 237

Parthasarathy M., Hack M., Tektunali G. 1990 *A&A* 230, 136

Parthasarathy M., Garcia-Lario P., Pottasch S.R. 1992 *A&A* 264, 159

Parthasarathy M., Garcia-Lario P., Pottasch S.R., *et al.* 1993 *A&A* 267, L19

Parthasarathy M., Garcia-Lario P., de Martino D., *et al.* 1995 *A&A* 300, L25

Parthasarathy M., Acker A., Stenholm B. 1998 *A&A* 329, L9

Parthasarathy M., Vijapurkar J., Drilling J.S. 2000a *A&AS* 145, 269

Parthasarathy M., Garcia-Lario P., Sivarani T., *et al.* 2000b *A&A* 357, 241

Parthasarathy M., Jain S.K., Bhatt H.C 2000c *A&A* 355, 221

Parthasarathy M., Garcia-Lario P., Gauba G., *et al.* 2001a *A&A* 376, 941

Parthasarathy M., Gauba G., Fujii T., Nakada Y. 2001b *in Post-AGB objects as a phase of stellar evolution*, 29, (eds Szczerba & Gorny :Kluwer)

Piotto G., *et al.* 2002 *A&A* 391, 945

Polidan R.S. 1988 *Decade of UV Astronomy with IUE* ESA SP-281 Vol 1, 205

Pottasch S.R., Parthasarathy M. 1988 *A&A* 192, 182

Preston G.W., Sneden C. 2000 *AJ* 120, 1014

Preston G.W., Sneden C. 2005, *These Proceedings*, 403

Przybilla, N. 2005, *These Proceedings*, 411

Reddy B.E., Hrivnak B.J. 1999 *AJ* 117, 1834

Reddy B.E., Parthasarathy M., Gonzalez G., Bakker E.J. 1997 *A&A* 328, 331

Reddy B.E., Lambert D.L., Gonzalez G., Yong D. 2002 *ApJ* 564, 482

Reyniers M. 2002 *Ph.d Thesis*

Reyniers M., Van Winckel H., Gallino R., Straniero O. 2004 *A&A* 417, 269

Ryans R.S.I., Dufton P.L., Mooney C.J., *et al.* 2003 *A&A* 401, 1119

Sahai R. 2001 *in Post-AGB objects as a phase of stellar evolution*, p 53, (eds. Szczerba & Gorny: Kluwer)

Sahai R., Bujarrabal V., Zijlstra A. 1999 *ApJ* 518, L115

Schönberner D. 1979 *A&A* 79, 108

Schönberner D. 1983 *ApJ* 272, 708

Schultheis M., Parthasarathy M., Omont A., *et al.* 2002, *A&A* 386, 899

Sivarani T. 2000 *Ph.D thesis* Bangalore University

Sivarani T., Bonifacio P., Molaro P., *et al.* 2004 *A&A* 413, 1073

Sivarani T., Parthasarathy M., Garcia-Lario P., *et al.* 1999 *A&AS* 137, 505

Sneden C., Preston G.W., Cowan J.J. 2003 *ApJ* 592, 504

Sterken C., Arentoft T., Duerbeck H.W. 1999 *A&A* 349, 532

Takeda Y., Parthasarathy M., Aoki W., *et al.* 2002 *PASJ* 54, 765

Thevenin F., Parthasarathy M., Jasniewicz G. 2000 *A&A* 359, 138

Turner D.G., Drilling J.S. 1984 *PASP* 96, 292

Ueta T., Meixner M., Bobrowsky M. 2000 *ApJ* 528, 861

Van Winckel H. 2003, *ARA&A* 41, 391

Van Winckel H., Cohen M., Gull T.R. 2002 *A&A* 390, 147

Van Winckel H., Mathis J.S., Waelkens C. 1992, *Nature* 356, 500

Van winckel H., Waelkens C., Waters L.B.F.M. 1995, *A&A* 293, L25

Van Winckel H., Waelkens C., Waters L.B.F.M. 2000, *IAU Symposium* 177, 285

Venn K.A. 1997, *ASP Conf. Ser.* 120, 95

Vijh U.P., Witt A.N., Gordon K.D. 2004, *ApJ* 606, L65

Volk K.M., Kwok S. 1989, *ApJ* 342, 345

Waelkens C., Waters L.B.F.M., Cassatella A., *et al.* 1987, *A&A* 181, L5

Waelkens C., Van Winckel H., Bogaert E., *et al.* 1991, *A&A* 251, 495

Waelkens C., Van Winckel H., Waters L.B.F.M., *et al.* 1996, *A&A* 314, L17

Waelkens C., Waters L.B.F.M., Van Winckel H. 1995, *ApSS* 224, 357

Waters L.B.F.M., Trams N.R., Waelkens C. 1992, *A&A* 262, L37

Waters L.B.F.M., Morris P.W., Voors R.H.M., *et al.* 1997, *ASP Conf.Ser.* 120, 326

Whitelock P.A., Menzies J.W., Catchpole R.M., *et al. et al.* 1989, *MNRAS* 241, 393

Witt A.N., Boroson T.A. 1990, *ApJ* 355, 182

Witt A.N., Vijh U.P. 2004, *ASP Conf.Ser.* 309

Zinn R.J., Newell E.B., Gibson J.B. 1972, *A&A*, 18, 390

The A-Star Puzzle
Proceedings IAU Symposium No. 224, 2004
J. Zverko, J. Žižňovský, S.J. Adelman, & W.W. Weiss, eds.

© 2004 International Astronomical Union
DOI: 10.1017/S1743921304004806

Horizontal branch A- and B-type stars in globular clusters

Sabine Möhler

Institut für theoretische Physik und Astrophysik, Christian-Albrechts-Universität zu Kiel,
Olshausenstraße 40, D 24118 Kiel, Germany
email: moehler@astrophysik.uni-kiel.de

Abstract. Globular clusters offer ideal laboratories to test the predictions of stellar evolution. When doing so with spectroscopic analyses during the 1990s, however, the parameters we derived for hot horizontal branch stars deviated systematically from theoretical predictions. The parameters of cooler, A-type horizontal branch stars, on the other hand, were consistent with evolutionary theories. In 1999, two groups independently suggested that diffusion effects might cause these deviations, which we verified subsequently. I will discuss these observations and analyses and their consequences for interpreting observations of hot horizontal branch stars.

Keywords. Diffusion, stars: atmospheres, stars: evolution, stars: horizontal branch, Galaxy: globular clusters: general

1. Historical background

Globular clusters are densely packed, gravitationally bound systems of several thousand to about one million stars. The dimensions of the globular clusters are small compared to their distance from us. Half of the light is generally emitted within a radius of less than 10 pc, whereas the closest globular cluster has a distance of 2 kpc and 90% lie more than 5 kpc away. We can thus safely assume that all stars within a globular cluster lie at the same distance from us. With ages in the order of 10^{10} years globular clusters are among the oldest objects in our Galaxy. As they formed stars only once in the beginning and the duration of that star formation episode is short compared to the current age of the globular clusters, the stars within one globular cluster are essentially coeval. In addition all stars within one globular cluster (with few exceptions) show the same initial abundance pattern (which may differ from one cluster to another). Globular clusters are thus the closest approximation to a physicists' laboratory in astronomy.

The horizontal branch, which is the topic of this article, was discovered by ten Bruggencate (1927), when he used Shapley's (1915) data on M 3 and other clusters to plot magnitude versus colour (replacing luminosity and spectral type in the Hertzsprung-Russell diagram) and thus produced the first colour-magnitude diagrams ("Farbenhelligkeitsdiagramme"). In these colour-magnitude diagrams (CMDs) ten Bruggencate noted the presence of a red giant branch that became bluer towards fainter magnitudes, in agreement with Shapley (1915). In addition, however, he saw a horizontal branch ("Horizontaler Ast") that parted from the red giant branch and extended far to the blue at constant brightness. As more CMDs of globular clusters were obtained it became obvious that the relative numbers of red and blue horizontal branch stars (i.e., the horizontal branch morphology) varied quite considerably between individual clusters, with some clusters showing extensions of the blue horizontal branch (so-called "blue tails") towards bluer colours and fainter visual magnitudes, i.e., towards hotter temperatures†.

† The change in slope of the horizontal branch towards higher temperatures is caused by

Sandage & Wallerstein (1960) noted a correlation between the metal abundance and the horizontal branch morphology seen in globular cluster CMDs, the horizontal branch (**HB**) became bluer with decreasing metallicity.

About 25 years after the discovery of the horizontal branch Hoyle & Schwarzschild (1955) were the first to identify horizontal branch stars with post-red giant branch stars that burn helium in the central regions of their cores. Faulkner (1966) managed for the first time to compute zero age horizontal branch (**ZAHB**) models that qualitatively reproduced the observed trend of HB morphology with metallicity without taking into account any mass loss but assuming a rather high helium abundance of $Y = 0.35$. Iben & Rood (1970), however, found that they could *"... account for the observed spread in colour along the horizontal branch by accepting that there is also a spread in stellar mass along this branch, bluer stars being less massive (on the average) and less luminous than redder stars."* Comparing HB models to observed globular cluster CMDs Rood (1973) found that an HB that *"... is made up of stars with the same core mass and slightly varying total mass, produces theoretical c-m diagrams very similar to those observed. ... A mass loss of perhaps 0.2 $M_\odot$ with a random dispersion of several hundredths of a solar mass is required somewhere along the giant branch."* The assumption of mass loss on the red giant branch diminished the need for very high helium abundances.

Thus our current understanding sees horizontal branch stars as stars that burn helium in a core of about 0.5 $M_\odot$ and hydrogen in a shell and evolve to the asymptotic giant branch, when the helium in the core is exhausted (for a review on HB evolution see Sweigart 1994). The more massive the hydrogen envelope e.g. the cooler is the resulting star at a given metallicity†. The masses of the hydrogen envelopes vary from $\leqslant 0.02$ $M_\odot$ at the hot end of the horizontal branch (about 30000 K) to $0.3 - 0.4$ $M_\odot$ for the cool HB stars at about 4000–5000 K (depending on metallicity, e.g., Dorman *et al.* 1993). The stable red and blue HB stars are separated by the variable RR Lyrae range at about 6500–7500 K. This article deals with blue HB stars, which at effective temperatures of about 8000 K to 20000 K show spectra rather similar (at moderate resolution) to main sequence stars of spectral types A and B and are therefore called HBA and HBB stars. In the field of the Milky Way such stars are often denominated by FHB (field HB star) and used as tracers for halo structure.

2. Atmospheric parameters

Already early studies of HBA and HBB stars in globular clusters showed discrepancies between observational results and theoretical expectations: Graham & Doremus (1966) mentioned that the comparison of $(c_1)_0$ vs. $(b - y)_0$ for 50 blue HB stars in NGC 6397 with models from Mihalas (1966) indicated low surface gravities and a mean mass of 0.3 $M_\odot$ (0.4 $M_\odot$) for solar (negligible) helium abundance, assuming $(m - M)_0 = 12.0$ and $E_{B-V} = 0.16$. Later spectroscopic analyses of HB stars (see cited papers for details) in globular clusters reproduced this effect (cf. Fig. 1): Crocker *et al.* (1988, M 3, M 5, M 15, M 92, NGC 288), de Boer *et al.* (1995, NGC 6397), Moehler *et al.* (1995, 1997a, M 15), and Moehler *et al.* (1997b, NGC 6752).

The zero age HB (ZAHB) in Fig. 1 marks the position where the HB stars have settled down and started to quietly burn helium in their cores. The terminal age HB

---

the decreasing sensitivity of $B - V$ to temperature on one hand and by the increasing Bolometric Correction for hotter stars (i.e., the maximum of stellar flux is radiated at ever shorter wavelengths for increasing temperatures, making stars fainter at V) on the other hand.

† Due to the higher opacities in their envelopes metal-rich HB stars are cooler than metal-poor ones with the same envelope mass.

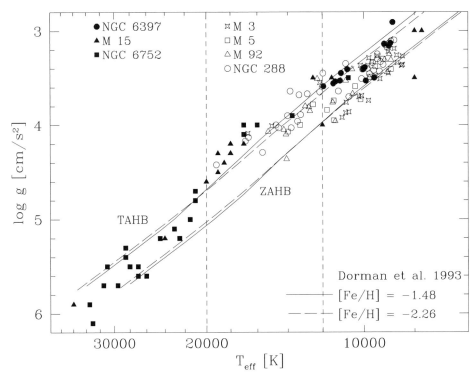

Figure 1. The results of Crocker *et al.* (1988, M 3, M 5, M 92, NGC 288), de Boer *et al.* (1995, NGC 6397), Moehler *et al.* (1995, 1997a, M 15), and Moehler *et al.* (1997b, NGC 6752) compared with evolutionary tracks from Dorman *et al.* (1993). ZAHB and TAHB stand respectively for the zero age and the terminal age HB (see text for details). The short-dashed lines mark the regions of low $\log g$ (see text for details).

(TAHB) is defined by helium exhaustion in the core of the HB star ($Y_C < 0.0001$). For temperatures between 12000 K and 20000 K the observed positions in the ($\log g, T_{\rm eff}$)-diagram fall mostly above the ZAHB and in some cases even above the TAHB. This agrees with the finding of Saffer *et al.* (1997) that field HBB stars show a larger scatter away from the ZAHB in $T_{\rm eff}, \log g$ than the hotter subdwarf B stars with $T_{\rm eff} > 20000$ K. Knowing the atmospheric parameters of the stars and the distances to the globular clusters we can determine masses for the stars (cf. Moehler *et al.* 1995, 1997b, de Boer *et al.* 1995). While the stars in M 3, M 5, and NGC 6752 have mean masses consistent with the canonical values, the hot HB stars in all other clusters have masses that are significantly lower than predicted by canonical HB evolution even for temperatures cooler than 12000 K.

Also some UV observations suggest discrepancies between theoretical expectations and observational results: The IUE (International Ultraviolet Explorer) and HUT (Hopkins Ultraviolet Telescope) spectra of M 79 (Altner & Matilsky 1993, Dixon *et al.* 1996) suggest lower than expected gravities and higher than expected metallicities for hot HB stars (but see Vink *et al.* 1999, who do not need low surface gravities to fit the HUT data). Hill *et al.* (1996) find from Ultraviolet Imaging Telescope (UIT) photometry of M 79 that stars bluer than $m_{152} - m_{249} = -0.2$ lie above the ZAHB, whereas cooler stars scatter around the ZAHB. UIT data of M 13 (Parise *et al.* 1998) find a lack of stars close to the ZAHB at a colour (temperature) range similar to the low $\log g$ range shown in Fig. 1. Landsman *et al.* (1996) on the other hand find good agreement between UIT photometry of blue stars in NGC 6752 and a standard ZAHB.

3. Atmospheric abundances

It has been realized early on that the blue HB and blue tail stars in globular clusters show weaker helium lines than field main sequence B stars of similar temperatures: Searle & Rodgers (1966, NGC 6397); Greenstein & Münch (1966, M 5, M 13, M 92); Sargent (1967, M 13, M 15, M 92). Greenstein *et al.* (1967) already suggested diffusion to explain this He deficiency. Michaud *et al.* (1983) performed the first theoretical study of diffusion effects in hot horizontal branch stars. Using the evolutionary tracks of Sweigart & Gross (1976) they found for the metal-poor models that *"in most of each envelope, the radiative acceleration on all elements* (i.e., C, N, O, Ca, Fe) *is much larger than gravity which is not the case in main-sequence stars."* The elements are thus pushed towards the surface of the star. Turbulence affects the different elements to varying extent, but generally reduces the overabundances†. Models without turbulence and/or mass loss (which may reduce the effects of diffusion) predict stronger He depletions than observed. A weak stellar wind could alleviate this discrepancy (Heber 1986, Michaud *et al.* 1989, Fontaine & Chayer 1997, and Unglaub & Bues 1998 discuss this effect, albeit for hotter stars). The extent of the predicted abundance variations varies with effective temperature, from none for HB stars cooler than about 5800 ± 500 K (due to the very long diffusion timescales) to $2 - 4$ dex in the hotter stars (the hottest model has $T_{\rm eff} = 20700$ K) and also depends on the element considered.

Observations of blue HB and blue tail stars in globular clusters support the idea of diffusion being active above a certain temperature. Abundance analyses of blue HB stars cooler than 11000 K to 12000 K in general show no deviations from the globular cluster abundances derived from red giants, while for hotter stars departures from the general globular cluster abundances are found, e.g., iron enrichment to solar or even super-solar values and strong helium depletion (see Moehler 2001 for references and more details). This agrees with the finding of Altner & Matilsky (1993) and Vink *et al.* (1999) that solar metallicity model atmospheres are required to fit the UV spectra of M 79.

All this evidence supports the suggestion of Grundahl *et al.* (1999) that the onset of diffusion in stellar atmospheres may play a role in explaining the jump along the HB towards brighter u magnitudes at effective temperatures of about 11500 K. This jump in $u, u - y$ is seen in all CMDs of globular clusters that have Strömgren photometry of sufficient quality. The effective temperature of the jump is roughly the same for all clusters, irrespective of metallicity, central density, concentration or mixing evidence, and coincides roughly with the onset of the "low $\log g$ problem" seen in Fig. 1 at $T_{\rm eff} \approx$ 11000 K to 12000 K. This in turn coincides with the region where surface convection zones due to hydrogen and He I ionization disappear in HB stars (Sweigart 2002).

Radiative levitation of heavy elements decreases the far-UV flux and by backwarming increases the flux in u. Grundahl *et al.* (1999) show that the use of metal-rich atmospheres ([Fe/H] = +0.5 for scaled-solar ATLAS9 Kurucz model atmospheres with $\log \epsilon_{\rm Fe, \odot} =$ 7.60) improves the agreement between observed data and theoretical ZAHB in the $u, u - y$-CMD at effective temperatures between 11500 K and 20000 K, but it worsens

---

† Michaud (1982) and Charbonneau & Michaud (1988) showed that meridional circulation can prevent gravitational settling and that the limiting rotational velocity decreases with decreasing $\log g$. Behr *et al.* (2000b) note that two of the HB stars hotter than 10000 K show higher rotational velocities and much smaller abundance deviations than other stars of similar temperature.

the agreement between theory and observation for hotter stars in the Strömgren CMD of NGC 6752 (see their Fig. 8). Thus diffusion may either not be as important in the hotter stars or the effects may be diminished by a weak stellar wind. The gap at $(B - V)_0 \approx 0$ discussed by Caloi (1999) is not directly related to the u-jump as it corresponds to an effective temperature of about 9000 K and is also not seen in every cluster (which would be expected if it were due to an atmospheric phenomenon). The gap at $T_{\text{eff}} \approx 13000$ K seen in the $c_1, b - y$ diagram of field horizontal branch stars (Newell 1973, Newell & Graham 1976) may be related to the u-jump as the c_1 index contains u.

The abundance distribution within a stellar atmosphere influences the temperature stratification and thereby the line profiles and the flux distribution of the emergent spectrum. A deviation in atmospheric abundances of HB stars from the cluster metallicity due to diffusion would thus affect their line profiles and flux distribution. Model atmospheres calculated for the cluster metallicity may then yield wrong results for effective temperatures and surface gravities when compared to observed spectra of HB stars. This effect could explain at least part of the observed discrepancies (see Sect. 5 for more details). Self-consistent model atmospheres taking into account the effects of gravitational settling and radiative levitation are, however, quite costly in CPU time and have started to appear only quite recently for hot stars (Dreizler & Wolff 1999, Hui-Bon-Hoa *et al.* 2000).

4. Rotational velocities

Peterson (1983, 1985a, 1985b) found from high-resolution spectroscopic studies that clusters with bluer HB morphologies show higher rotation velocities among their HB stars. However, the analysis of Peterson *et al.* (1995) shows that while the stars in M 13 (which has a long blue tail) rotate on average faster than those in M 3 (which has only a short blue HB), the stars in NGC 288 and M 13 show *slower* rotation velocities at *higher* temperatures. These results are consistent with those reported by Behr *et al.* (2000a), who determined rotational velocities for stars as hot as 19000 K in M 13. They found that stars hotter than about 11000 K have significantly lower rotational velocities than cooler stars and that the change in mean rotational velocity may coincide with the gap seen along the blue HB of M 13 (see Ferraro *et al.* 1998 for an extensive discussion of gaps). Also the results of Cohen & McCarthy (1997, M 92) and Behr *et al.* (2000b, M15) show that HB stars cooler than ≈ 11000 K to 12000 K in general rotate faster than hotter stars.

The studies by Behr (2003a) and Recio *et al.* (2002, 2004), which both consider several clusters, confirm the trends mentioned above. Below the diffusion threshold about 20% to 30% of the blue HB stars show rotation velocities of $v \sin i \approx 20 \ldots 30 \, \text{km s}^{-1}$, whereas there are no fast rotators among the hotter stars. From observations of field HB stars Behr (2003b) and Carney *et al.* (2003) note that red and cool blue HB stars show similar distributions of rotational velocities (after accounting for the larger radii of the red HB stars), whereas field RR Lyrae stars show no evidence for rotation.

Sills & Pinsonneault (2000) studied theoretical models for the rotation of HB stars and find that the observed rotation of cool blue HB stars in M 13 can be explained if their red giant precursors have rapidly rotating cores and differential rotation in their convective envelopes and if angular momentum is redistributed from the rapidly rotating core to the envelope on the horizontal branch. If, however, turn-off stars rotate with less than $4 \, \text{km s}^{-1}$, rapidly rotating cores in the Main-Sequence stars (violating helioseismological results for the Sun) or an additional source of angular momentum on the red giant branch (e.g., mass transfer in close binaries or capture of a planet as described by Soker & Harpaz 2000) are required to explain the rotation of blue HB stars. Talon & Charbonnel (2004) speculate that such differential rotation could be understood within the framework of internal gravitational waves. The change in rotation rates towards higher temperatures is not predicted by the models of Sills & Pinsonneault (2000) but could be understood as a result of gravitational settling of helium, which creates a mean molecular weight gradient, that then inhibits angular momentum transport to the surface of the star. Sweigart (2002) suggests that the weak stellar wind invoked to reconcile observed abundances in hot HB stars with diffusion calculations (cf. Sect. 3) could also carry away angular momentum from the surface layers and thus reduce the rotational velocities of these stars.

Soker & Harpaz (2000) argue that the distribution of rotational velocities along the HB can be explained by the spin-up of their progenitors due to interactions with low-mass companions, predominantly gas-giant planets, in some cases also brown dwarfs or low-mass Main-Sequence stars (especially for the very hot HB stars). The slower rotation of the hotter stars in their scenario is explained by mass loss *on* the HB, which is accompanied by efficient angular momentum loss. This scenario, however, does not explain the sudden change in rotational velocities and the coincidence of this change with the onset of radiative levitation.

5. Where do we stand?

Analysis of a larger sample of hot HB stars in NGC 6752 (Moehler *et al.* 2000) showed that the use of model atmospheres with solar or super-solar abundances removes much of the deviation from canonical tracks in effective temperature, surface gravity and mass for hot HB stars discussed in Sect. 2. However, some discrepancies remain, indicating that the low $\log g$, low mass problem cannot be completely solved by scaled-solar metal-rich atmospheres (which *do* reproduce the *u*-jump reported by Grundahl *et al.* 1999). As Michaud *et al.* (1983) noted diffusion will not necessarily enhance all heavy elements by the same amount and the effects of diffusion vary with effective temperature. Elements that were originally very rare may be enhanced even stronger than iron. The question of whether diffusion can fully explain the "low gravity" problem cannot be answered without detailed abundance analyses to determine the actual abundances and subsequent analyses using model atmospheres with non-scaled solar abundances (like ATLAS12, Kurucz 1992). Additional caution is also recommended by the results of Moehler *et al.* (2003) on M 13, where even the use of the metal-rich model atmospheres does not eliminate the problem of the low masses or the low gravities for effective temperatures between 12000 K and 16000 K. In that cluster we also found significant disagreement between atmospheric parameters derived from Strömgren photometry and from Balmer line fitting for stars cooler than about 9000 K, for which we found no explanation.

Still unexplained are also the low masses found for cool blue HB stars (which are not affected by diffusion) in, e.g., NGC 6397 and M 92. For those stars a longer distance scale to globular clusters would reduce the discrepancies (Moehler 1999). Such a longer distance scale has been suggested by several authors using HIPPARCOS results for metal-poor field subdwarfs to determine the distances to globular clusters by fitting their main sequence with the local subdwarfs. One should note, however, that de Boer *et al.* (1997), report that HIPPARCOS parallaxes for field HBA stars still yield masses significantly below the canonical mass expected for these objects. Carretta *et al.* (2000) present an extensive and excellent discussion of various globular cluster distance determinations and the zoo of biases that affect them, while Gratton *et al.* (2003) concentrate on the error budget of distances from main sequence fitting in their paper.

The problem of the different rotation velocities for cool and hot HB stars, however, remains rather stubbornly opposed to any attempted solution.

Acknowledgements

I would like to thank my colleagues and collaborators Drs. A.V. Sweigart, W.B. Landsman, M. Lemke, U. Heber, and B.B. Behr for many fruitful discussions and helpful comments.

References

Altner B., Matilsky T.A., 1993, ApJ 410, 116
Behr, B.B. 2003a, ApJS 149, 67
Behr, B.B. 2003b, ApJS 149, 101
Behr B.B., Djorgovski S.G., Cohen J.G., *et al.* , 2000a, ApJ 528, 849
Behr B.B., Cohen J.G., McCarthy J.K., 2000b, ApJ 531, L37
Caloi V., 1999, A&A 343, 904
Carney, B.W., Latham, D.W., Stefanik, R.P., Laird, J.B., Morse, J.A. 2003, AJ 125, 293
Carretta E., Gratton R., Clementini G., Fusi Pecci F., 2000, ApJ 533, 215
Charbonneau P., Michaud G., 1988, ApJ 327, 809
Cohen J.G., McCarthy J.K., 1997, AJ 113, 1353
Crocker D.A., Rood R.T., O'Connell R.W., 1988, ApJ 332, 236
de Boer K.S., Schmidt J.H.K., Heber U., 1995, A&A 303, 95
de Boer K.S., Tucholke H.-J., Schmidt J.H.K., 1997, A&A 317, L23
Dixon W.V., Davidsen A.F., Dorman B., Ferguson H.C., 1996, AJ 111, 1936
Dorman B., Rood, R.T., O'Connell, W.O., 1993, ApJ 419, 596
Dreizler S., Wolff B., 1999, A&A 348, 189
Faulkner J., 1966, ApJ 144, 978
Ferraro F.R., Paltrinieri B., Fusi Pecci F., Dorman B., Rood R.T., 1998, ApJ 500, 311
Fontaine G., Chayer P., 1997, in *The 3rd Conf. on Faint Blue Stars*, eds. A.G.D. Philip, J. Liebert & R.A. Saffer (L. Davis Press, Schenectady), p. 169
Graham J.A., Doremus C., 1966, AJ 73, 226
Gratton R.G., Bragaglia A., Carretta E., Clementini G., Desidera S., Grundahl F., Lucatello S., 2003, A&A 408, 529
Greenstein G.S., Münch G., 1966, ApJ 146, 518
Greenstein G.S., Truran J.W., Cameron A.G.W., 1967, Nature 213, 871
Grundahl F., Catelan M., Landsman W.B., Stetson P.B., Andersen M., 1999, ApJ 524, 242
Heber U., 1986, A&A 155, 33
Hill R.S., Cheng K.-P., Smith E.P., *et al.* , 1996, AJ 112, 601
Hoyle F., Schwarzschild M., 1955, ApJS 2, 1
Hui-Bon-Hoa A., LeBlanc F., Hauschildt P., 2000, ApJ 535, L43
Iben I.Jr., Rood R.T., 1970, ApJ 161, 587
Kurucz R.L., 1992, in *The Stellar Populations of Galaxies*, eds. B. Barbuy & A. Renzini, IAU Symp. 149 (Kluwer:Dordrecht), 225
Landsman W.B., Sweigart A.V., Bohlin R.C., *et al.* , 1996, ApJ 472, L93
Michaud G., 1982, ApJ 258, 349
Michaud G., Vauclair G., Vauclair S., 1983, ApJ 267, 256
Michaud G., Bergeron P., Heber U., Wesemael F., 1989, ApJ 338, 417
Mihalas D.M., 1966, ApJS 13, 1
Moehler S., 1999, Reviews in Modern Astronomy, ed. R. Schielicke, Vol. 12, p. 281
Moehler S., 2001, PASP 113, 1162
Moehler S., Heber U., de Boer K.S., 1995, A&A 294, 65
Moehler S., Heber U., Durrell P., 1997a, A&A 317, L83
Moehler S., Heber U., Rupprecht G., 1997b, A&A 319, 109
Moehler S., Sweigart A.V., Landsman W., Heber U., 2000, A&A 360, 120
Moehler S., Landsman W.B., Sweigart A.V., Grundahl F., 2003, A&A 405, 135
Newell E.B., 1973, ApJS 26, 37
Newell E.B., Graham J.A., 1976, ApJ 204, 804

Parise R.A., Bohlin R.C., Neff S.G., *et al.* , 1998, ApJ 501, L67

Peterson R.C., 1983, ApJ 275, 737

Peterson R.C., 1985a, ApJ 289, 320

Peterson R.C., 1985b, ApJ 294, L35

Peterson R.C., Rood R.T., Crocker D.A., 1995, ApJ 453, 214

Recio-Blanco, A., Piotto, G., Aparicio, A., Renzini, A. 2002, ApJ 527, L71

Recio-Blanco, A., Piotto, G., Aparicio, A., Renzini, A. 2004, A&A 417, 597

Rood R.T., 1973, ApJ 184, 815

Saffer R.A., Keenan F.P., Hambly N.C., Dufton P.L., Liebert J., 1997, ApJ 491, 172

Sandage A.R., Wallerstein G., 1960, ApJ 131, 598

Sargent W.L.W., 1967, ApJ 148, L147

Searle L., Rodgers A.W., 1966, ApJ 143, 809

Shapley H., 1915, Contr. Mt. Wilson 116

Sills A., Pinsonneault M.H., 2000, ApJ 540, 489

Soker N., Harpaz A., 2000, MNRAS 317, 861

Sweigart A.V., 1994, in *Hot Stars in the Galactic Halo*, eds. S. Adelman, A. Upgren, C.J. Adelman, (Cambridge University Press, Cambridge), p. 17

Sweigart A.V., 2002, Highlights of Astronomy, Vol. 12, ed. H. Rickman (ASP), p. 292

Sweigart A.V., Gross P.G., 1976, ApJS 32, 367

Talon S., Charbonnel C., 2004, A&A 418, 1051

ten Bruggencate P., 1927, *Sternhaufen* (Julius Springer Verlag, Berlin)

Unglaub K., Bues I., 1998, A&A 338, 75

Vink J.S., Heap S.R., Sweigart A.V., Lanz T., Hubeny I., 1999, A&A 345, 109

Discussion

FREYTAG: What are "real" abundances in contrast to "diffusion" abundances? Can they still be reliably defined?

MOEHLER: By real abundances I mean the abundances, which the star would show if its atmosphere were not affected by diffusion (like the abundances derived from red giants in globular clusters). Those cannot be derived if the spectrum *is* affected by diffusion

PRESTON: Have you compared the abundances of hot HB stars with those derived from red giants, which are affected in a different way by diffusion processes?

MOEHLER: Red giants should show no diffusion effects in their atmospheres due to their deep convection zones. HB stars with diffusion usually show strong enhancements of heavy elements compared to red giants in the same cluster.

CORBALLY: Some dozen years ago I was working with Richard Gray on the field horizontal branch stars identified by Dave Philip. These were all slightly cooler than those you have shown, i.e., truly A-type horizontal branch stars. We found that they lay near or slightly above the Main Sequence, indeed they were metal-poor as you said, and that also from their Balmer line profiles at medium resolution were probably helium-rich. Could you comment on the last possibility?

MOEHLER: Not really, as I find this a very puzzling result, which cannot be easily understood in the terms of diffusion.

The A-Star Puzzle
Proceedings IAU Symposium No. 224, 2004
J. Zverko, J. Žižňovský, S.J. Adelman, & W.W. Weiss, eds.

© 2004 International Astronomical Union
DOI: 10.1017/S1743921304004818

Blue metal-poor stars

George W. Preston[1] and Christopher Sneden[2]

[1]Carnegie Observatories, 813 Santa Barbara Street, Pasadena, CA 91101, USA
email: gwp@ociw.edu

[2]Department of Astronomy and McDonald Observatory, University of Texas,
Austin, TX 78712, USA
email: chris@verdi.as.utexas.edu

Abstract. We review the discovery of blue metal-poor (BMP) stars and the resolution of this population into blue stragglers and intermediate-age Main-Sequence stars by use of binary fractions. We show that the specific frequencies of blue stragglers in the halo field and in globular clusters differ by an order of magnitude. We attribute this difference to the different modes of production of these two populations. We report carbon and s-process enrichment among very metal-poor field blue stragglers and discuss how this result can be used to further resolve field blue stragglers into groups formed during RGB and AGB evolution of their erstwhile primary companions.

Keywords. Binaries: spectroscopic, blue stragglers, stars: chemically peculiar, stars: abundances

1. Introduction: Identification of BMP stars

In an isolated very old Milky Way Galactic halo there should be no Main Sequence stars of spectral type A-F. All these stars should have long-since evolved to become members of the white dwarf sequence. But hundreds of such objects that are collectively called blue metal-poor (hereafter BMP) stars have been discovered to date in Galactic halo photometric surveys, too many to be ignored in attempts to understand the origin and evolution of our Galaxy. These BMP stars comprise a laboratory for the investigations of stellar astrophysics in two very different environments: (1) in interacting halo binaries and (2) in (probably) dwarf satellite galaxies of the Milky Way. Numerous BMP stars were identified (Preston *et al.* 1994) among the metal-poor candidates and A-type stars of the HK survey (Preston *et al.* 1991) by their locations blueward of all globular cluster (hereafter GC) turnoffs in the $(U - B)$ vs $(B - V)$ plane, see Figure 1 of Preston *et al.* (1994). That paper employed a blue cutoff at $B - V = 0.15$ because the $U - B$ blanketing effect provides little or no discrimination at bluer colors (higher temperatures). Hotter BMP stars may exist, but we do not know how to identify them in the halo field.

The BMP population possesses the photometric characteristics of the blue straggler (hereafter BS) families in globular clusters (Fusi Pecci *et al.* 1992), but it differs from them in important respects. The specific frequency of halo BMPs with [Fe/H] < -1, reckoned as the number of BMPs per horizontal branch star, exceeds the similarly defined specific frequency of GC BSs by a factor of 15 (Preston & Sneden 2000, hereafter PS2000), so we know that these two samples of stars cannot have been drawn from the same parent population. Further understanding of the BMP phenomenon has come with spectroscopic investigations, and in this paper we review the results of two related high resolution studies by our group.

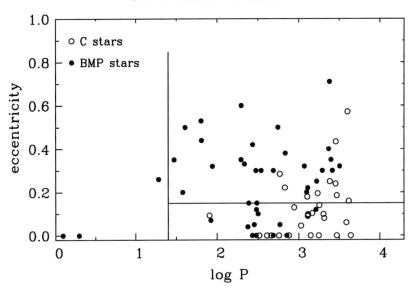

Figure 1. Correlation of binary orbital periods and eccentricities for BMP stars (PS2000), and in "C stars" (subgiant CH stars, McClure 1997; CH giants and barium stars, McClure & Woodsworth 1990). This figure is an adaptation of the lower two panels of Figure 19 in PS2000. The vertical line drawn at $P = 25$ days ($\log P = 1.4$) indicates the approximate transition period between short-period binaries that almost always have low-eccentricity orbits and long-period binaries that do not have strongly correlated periods and eccentricities. The horizontal line drawn at (an arbitrary) eccentricity $e = 0.15$ represents an approximate lower limit to the eccentricities of binaries that have not undergone forced orbital circularization from their companions. For clarity in this figure, we have plotted only mean parameters for those few BMP stars with 2-3 alias periods listed by PS2000.

2. RV-constant and binary BMP stars

PS2000 conducted a radial velocity survey of 62 BMP stars, using high-resolution $R \equiv \lambda/\Delta\lambda \simeq 30000$), low signal-to-noise ($S/N \sim 10$) echelle spectra obtained with the Las Campanas du Pont Telescope. Between 10 and 70 individual spectra (on average $\simeq$20) were gathered for each program star. Their velocity analysis revealed that $\simeq$60% of BMPs are in spectroscopic binaries with predominantly long periods. This binary frequency greatly exceeds the value of 15% found for spectroscopic binaries with orbital periods less than 4000 days in both the Galactic halo (Latham *et al.* 1998) and disk (Duquennoy & Mayor 1991). Additionally, the fraction of double-lined spectroscopic binaries among BMPs is abnormally low (1 of 42), their mass functions are small, and an unusually large fraction of their orbits have low eccentricities. We illustrate these unexpected results in Figure 1, which compares the periods and eccentricities of BMPs to some known classes of carbon-rich binaries (McClure 1997, McClure & Woodsworth 1990). Both groups of stars contain a large fraction of stars that have nearly circular orbits ($e < 0.15$) in spite of being in widely separated, long-period orbits ($P > 25$ days). This distribution of eccentricities is very different from those of normal disk and halo binary stars (Duquennoy & Mayor 1991, Latham *et al.* 1988, 1992), for which long-period, low-eccentricity orbits are rare.

Noting that the incidence of binaries among the BSs in M67 is very high, >60% (Latham & Milone 1996), and supposing that the BMPs are a mixture of BSs and a field population of intermediate-age stars with a normal (15%) binary frequency, PS2000 calculated that 60% of the BMPs are field BSs, most of which were created by mass transfer during the evolution of their erstwhile primary companions (McCrea 1964). The specific frequency of the BS component of the BMPs exceeds the GC value by a factor of 10. We attribute this high value in the field to formation of BSs by mass transfer among long-period binaries which dominate the period distribution of field binaries (Duquennoy & Mayor 1991). Formation of BSs by collisions and disruption of binaries by stellar encounters simply cannot occur in the very low-density halo. In GCs the specific frequency of BSs is low because preponderant wide binaries with long periods are destroyed by encounters in dense stellar systems: BS production is limited to the merger of the small population of short period survivors, an argument first made in a study of NGC 5466 by Mateo *et al.* (1990) and recently confirmed for 47 Tuc by Mapelli *et al.* (2004). A small percentage ($\simeq$10%) of primordial halo binaries with initial periods less than $\simeq$5 days (Duquennoy & Mayor 1991) merged (Vilhu 1982) to form more massive BSs, most or all of which left the Main Sequence long ago, so virtually all field blue stragglers are located in binaries, the products of mass transfer from erstwhile primaries that are now their degenerate companions. The binary BS population is contaminated by a small ($\simeq$10%) contribution due to the normal binaries that we presume accompany the intermediate-age field population.

We suggest that the single (RV-constant) BMP stars are the identifiable blue tips of intermediate-age populations captured from systems like the Carina dwarf galaxy (Smecker-Hane *et al.* 1994). Preston *et al.* (1994) estimated that such captured populations comprise perhaps 4–10% of the present halo. Subsequently, Unavane *et al.* (1996) drew a similar conclusion in a broader discussion of Galactic accretion.

3. Chemical compositions of binary and RV-constant BMP stars

Further insight into the two subclasses of BMP stars can be gained from their chemical compositions. PS2000 co-added the $\simeq$10-70 individual echelle spectra of each of their BMP stars to produce moderate S/N ($\sim$30-40), adequate to derive [Fe/H]† metallicities and relative [X/Fe] abundance ratios for: (a) Fe-peak elements Sc, Cr, and Mn; (b) light α-elements Mg, Ca, and Ti; and (c) neutron-capture (*n*-capture) elements Sr and Ba. Their observations included a sample of 62 stars, 19 of which rotate too fast ($V_e \sin i > 25\,\mathrm{km\,s^{-1}}$) to permit accurate determination of much more than their overall metallicity. The rapidly-rotating stars will not be considered further here. In Figure 2 we summarize the abundances for the remaining 43 stars, distinguishing the binaries from RV-constant stars via different symbols.

In the top panel of Figure 2 we show the mean abundance of the α elements Mg, Ca, and Ti for each star. For almost all stars, the "normal" Galactic halo-star enhancement of [α/Fe] $\simeq$ +0.3 ±0.1 appears in BMP stars.‡ No obvious separation is seen between the α-element abundances of RV-constant and binary BMP stars. In the middle panel, the mean of Fe-peak elements Sc and Cr are displayed. The abundances of this element group in BMP stars are consistent with the values in other kinds of halo stars, and there is no difference between the two BMP subclasses. This statement extends also to Mn, which is not included in the Fe-peak means because it was not as well determined in the PS2000 study, with only unmeaningful upper limits recorded for many BMP stars.

The *n*-capture abundance means <[Sr/Fe],[Bi/Fe]> shown in the bottom panel of Figure 2 are more interesting. At lowest metallicities, [Fe/H] $<$ −2, the RV-constant BMP

---

† For elements A and B, standard abundance notation is used: $[A/B] \equiv \log_{10}(N_A/N_B)_{star} - \log_{10}(N_A/N_B)_\odot$

‡ The one clear exception is the star CS 22966-043 ([Fe/H] = −1.96, [α/Fe] = −0.13 in the top panel of Figure 2. This star is an example of the apparently rare subclass of α-poor halo stars, discussed in detail by Ivans *et al.* (2003).

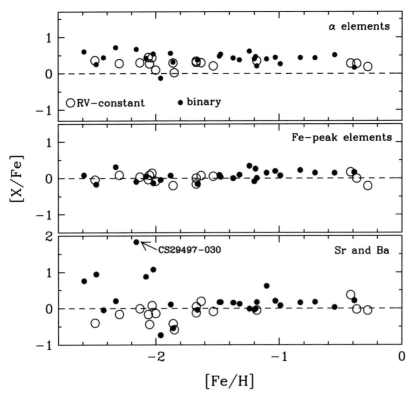

Figure 2. Mean abundances of different element groups in BMP stars derived from the survey data of PS2000. Abundances for their 43 slowly-to-moderately rotating stars ($V_e \sin i <$ 25 km s^{-1}) are plotted, but not for their 11 more rapidly rotating stars (for which abundances are less certain). Different symbols are used to distinguish BMP binary stars and those with constant radial velocities. In the upper panel the ordinate is the mean abundance of three α elements: $<$[Mg/Fe],[Mg/Fe],[Ti/Fe]$>$. In the middle panel the means are of two Fe-peak elements: $<$[Sc/Fe],[Cr/Fe]$>$. In the bottom panel the means are of two n-capture elements: $<$[Sr/Fe],[Bi/Fe]$>$. Attention is called to the star CS 29497-030 because it exhibits nearly a two-order-of-magnitude enhancement of these n-capture elements.

stars have no enhancements or perhaps deficiencies of the n-capture elements, while five out of the seven BMP binaries have very large enhancements of these elements. Sneden *et al.* (2003) obtained higher S/N spectra for a sample of 10 BMP stars drawn from the PS2000 survey, five in each subclass. Of these, three binaries and three RV-constant BMP stars were very metal-poor, [Fe/H] $\sim$ −2.1. Their study included the BMP binary CS 29497-030, which has the largest Sr and Ba abundance mean of the stars in Figure 2. The principle results of their abundance analysis can be summarized by inspection of the spectra displayed in Figure 3. In this figure we contrast the spectrum of CS 29497-030 to that resulting from the co-addition of the individual spectra of the three very metal-poor RV-constant stars. Although CS 29497-030 is the most extreme case, chosen deliberately for display purposes, almost all of the following conclusions apply to each of the three very metal-poor binaries.

(1) BMP binaries are carbon-rich. In Figure 3's top panel, we show some C I lines in CS 29497-030 that are absent in the RV-constant stars; the other two binaries also have detectable, albeit weaker C I lines. Additionally, the CH G-band can also be seen in all three binaries, but not in the RV-constant stars. In the BMP (T_{eff}, log g, [Fe/H]) regime, C-containing species should be completely undetectable. Noting that (high excitation) C

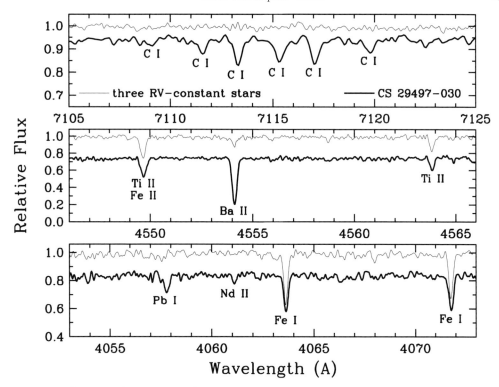

Figure 3. Spectra of selected small wavelength regions in the C-rich, *n*-capture-rich BMP binary star CS 29497-030, contrasted with co-added spectra of three RV-constant stars of comparable metallicity ([Fe/H] $\sim$ −2). This figure is adapted from Figures 4 and 6 of Sneden *et al.* (2003). We emphasize that the mean RV-constant spectra have a complete absence of detectable C I lines (top panel), a very weak Ba II line (middle panel), and no sign of Pb I or Nd II (bottom panel), while those features are prominent in CS 29497-030 (and in the other two BMP binaries investigated by Sneden *et al.* 2003).

I naturally strengthens and CH weakens with increasing $T_{\rm eff}$, clearly a pure abundance effect is at work. Carbon is enhanced by orders of magnitude in BMP binaries. *(2) BMP binaries are n-capture-rich.* In the middle panel of Figure 3 we contrast the Ba II resonance line strength between CS 29497-030 and the RV-constant stars. While the PS2000 Ba abundance estimates were based on this single feature, the more detailed Sneden *et al.* (2003) investigation included 3-4 Ba II lines and other *(3) Lead is present in BMP binary star CS 29497-030.* In the bottom panel of Figure 3 we show the spectrum of the Pb I λ4057 line; clearly this feature is strong (there are few stars with detectable Pb I in their spectra). *(4) The n-capture-element enhancements in BMP binaries are due to the s-process.* Although Sr II and Ba II lines are extremely strong in the BMP binary spectra, lines of Eu II (not shown in Figure 3) are very weak or undetectable. Rapid *n*-capture nucleosynthesis (the *r*-process) produces Eu much more efficiently than Ba, hence the lack of any Eu II detections immediately suggests that slow *n*-capture synthesis (the *s*-process) has been at work.

Sneden *et al.* (2003) determined abundances for nine *n*-capture elements in CS 29497-030 ([Fe/H] = −2.16), finding extremely large overabundances for most of these elements. For example, they derived [Ba/Fe] =+2.45 and [Pb/Fe] = +3.75, meaning that this very metal-poor star has a larger surface Ba content and more than a factor of 10 more Pb than does the Sun. Other Pb-rich stars have been discovered in recent years (e.g.,

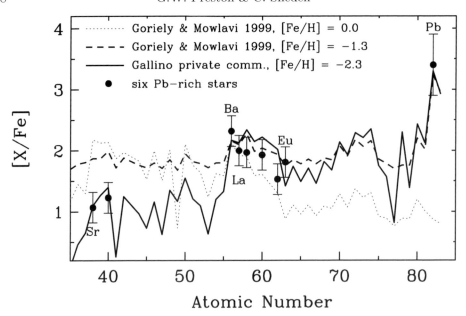

Figure 4. Reproduction of Figure 8 of Sneden *et al.* (2003): normalized mean *n*-capture abundances of six Pb-rich metal-poor stars (CS 22183-015, Johnson & Bolte 2002; CS 29526-110, Aoki *et al.* 2002; HD 187861 and HD 196944, Van Eck *et al.* 2003; HE 0024-2523, Lucatello *et al.* 2003; CS 29497-030, Sneden *et al.* 2003), and theoretical *s*-process predictions for nucleosynthesis environments of three different metallicities (Goriely & Mowlavi 2000, Gallino 2003). See the discussion in Sneden *et al.* (2003) for a detailed description on formation of the mean observed abundances.

Johnson & Bolte 2002, Aoki *et al.* 2002, Van Eck *et al.* 2003, Lucatello *et al.* 2003), these other stars are most often evolved red giants. The Pb-rich stars have various [Ba/Pb] ratios, but the general *n*-capture abundance patterns are very similar in all of them. In Figure 4 we show Figure 8 of Sneden *et al.* (2003) that compares six of the stars with largest Pb abundances to several *s*-process synthesis predictions. One of the *s*-process theoretical curves was computed by Goriely & Mowlavi (2000) for a solar-metallicity fusion environment, one was produced by these same authors for a moderately metal-poor ([Fe/H] = −1.3) environment, and one is an unpublished prediction calculated by R. Gallino (private communication) for a very metal-poor ([Fe/H] = −1.3) regime. Clearly the best match is with the Gallino *s*-process predications, because these were made with the lowest assumed metallicity (comparable to that of CS 29497-030). The explanation, originally due to Clayton (1988) and developed quantitatively by, e.g., Travaglio *et al.* (1999, and references therein), suggests that as the metallicity decreases in an *s*-process synthesis zone, the ratio of free neutrons to available seed nuclei increases. This naturally drives the *s*-process neutrons to attach to fewer target nuclei, pushing the resultant abundance distribution to heavier nuclei. For sufficiently low metallicities, an *s*-process piles up at Pb (and Bi), the heaviest stable nuclei in the Periodic Table.

4. Putting it all together

Statistical inferences from the HK survey, and recent radial velocity and abundance investigations all confirm that field BMP stars comprise two distinct halo subpopulations. The capture origin of the RV-constant stars remains as the only viable explanation for their presence in the halo, because we cannot imagine how these metal-poor stars could have been produced recently within our Galaxy.

The origin of the BMP BS binary component seems more clear. All evidence points such BMP stars being the end product of post Main Sequence mass transfer. The stars seen today originally possessed less than or about the mass of an ordinary 12-15 Gyr metal-poor halo turnoff star. Their companions were probably much more massive, up to perhaps several solar masses (we cannot be more precise on the basis of present spectroscopic data, though some abundance details do depend on the mass of the primordial primary that undergoes helium shell burning). As the erstwhile primaries evolved, their outer envelopes expanded until they overflowed their Roche lobes and transferred mass to their companions, which in response moved to a brighter and bluer portion of the Main Sequence; the original primaries are now white dwarfs. Transfer could occur during RGB or AGB evolution. The existence of C and *s*-process abundance enrichments in some (most?) of the most metal-poor BMP binaries guarantees that often this event occurred during the AGB "third dredge-up" evolutionary phase. The BMPs with long orbital periods present challenges for theoretical interpretation. Systems with periods greater than about 300 days or more should complete RGB evolution without Roche-lobe overflow. Therefore, we expect that the preponderance of the stars with periods greater than some cutoff value (not precise because of the unknown initial primary mass) should show AGB peculiarities to some degree. The expectations are clouded because AGB superwinds offer a second path to mass transfer. The amount of mass that can be transferred by winds in widely separated pairs is small, but then the contaminated convective layers of the present rather warm BMP primaries are shallow. The future of this topic rests on better theoretical calculations that connect plausible mass transfer to better abundance estimates based on superior spectra.

On the purely observational side, orbital parameters are not in a satisfactory state. The orbital periods derived in PS2000 are plagued by alias problems associated with annual trips to Las Campanas Observatory. Furthermore, a number of stars in our sample exhibit low-amplitude velocity variations on time scales comparable to and in some cases exceeding the time duration of our survey. Therefore a proper discussion of mass transfer in such wide systems must await additional velocity data that will lead to definitive orbits.

The wherewithal for improved BMP spectroscopy is now greatly enhanced with the completion of a number of new, large Southern Hemisphere telescope facilities (Gemini, Magellan, SOAR, VLT). We hope that some time at these facilities will be devoted to the issues raised in this presentation.

Acknowledgements

Some of the studies reviewed here occurred while C. S. was a Visiting Scientist at the Carnegie Observatories; their hospitality and financial support are gratefully acknowledged. This research has been supported in part by NSF grants AST-9987162 and AST-0307495 to C. S.

References

Aoki, W., Ryan, S.G., Norris, J.E., Beers, T.C., Ando, H., & Tsangarides, S. 2002, *ApJ* 580, 1149

Clayton, D.D. 1988, *MNRAS* 234, 1

Duquennoy, A., & Mayor, M. 1991, *A&A* 248, 485

Fusi Pecci, F., Ferraro, F.R., Corsi, C.E., Cacciari, C., & Buonanno, R. 1992, *AJ* 104, 1831

Gallino, R. 2003, *private communication*

Goriely, S. & Mowlavi, N. 2000, *A&A* 362, 599

Ivans, I.I., Sneden, C., James, C.R., Preston, G.W., Fulbright, J.P., Höflich, P.A., Carney, B.W., & Wheeler, J.C. 2003, *ApJ* 592, 906

Johnson, J.A. & Bolte, M.B. 2002, *ApJ* 579, L87

Latham, D.W., & Milone, A.E. 1996, ASP Conf. Ser. 90, The Origins, Evolution, and Destinies of Binary Stars in Clusters, ed. E. F. Milone & J.-C. Mermilliod (San Francisco: Astr. Soc. Pac.), 385

Latham, D.W., Mazeh, T., Carney, B.W., McCrosky, R.E., Stefanik, R.P., & Davis, R.J. 1988, *AJ* 96, 567

Latham, D. W., *et al.* 1992, *AJ* 104, 774

Latham, D.W., Stefanik, R.P., Mazeh, T., Goldberg, D., Torres, G., & Carney, B.W. 1998, ASP Conf. Ser., 154, Cool Stars, Stellar Systems, and the Sun, ed. R. A. Donahue & J. A. Bookbinder (San Francisco: Astr. Soc. Pac.) 2129

Lucatello, S., Gratton, R., Cohen, J.G., Beers, T.C., Christlieb, N., Carretta, E., & Ramírez, S. 2003, *AJ* 125, 875

Mapelli, M., Sigurdsson, S., Colpi, M., Ferraro, F.R., Possenti, A., Rood, R.T., Sills, A., & Beccari, G. 2004, *ApJ* 605, L29

Mateo, M., Harris, H.C., Nemec, J. & Olszewski, E.W. 1990, *AJ* 100, 469

McClure, R.D. 1997, *PASP* 109, 536

McClure, R.D. & Woodsworth, A.W. 1990, *ApJ* 352, 709

McCrea, W.H. 1964, *MNRAS* 128, 147

Preston, G.W., Beers, T.C., & Shectman, S.A. 1994, *AJ* 108, 538

Preston, G.W., Shectman, S.A., & Beers, T.C. 1991, *ApJ.S* 76, 1001

Preston, G.W. & Sneden, C. 2000, *AJ* 120, 1014

Smecker-Hane, T.A., Stetson, P.B., Hesser, J.E., & Lehnert, M.D. 1994, *AJ* 108, 507

Sneden, C., Preston, G.W., & Cowan, J.J. 2003, *ApJ* 592, 504

Travaglio, C., Galli, D., Gallino, R., Busso, M., Ferrini, F., & Straniero, O. 1999, *ApJ* 521, 691

Unavane, M., Wyse, R.F.G., & Gilmore, G. 1996, *MNRAS* 278, 727

Van Eck, S., Goriely, S., Jorissen, A., & Plez, B. 2001, *Nature* 412, 793

Vilhu, O. 1982, *A&A* 109, 17

Discussion

LANDSTREET: Your abundance plots of [X/H] *vs.* [Fe/H] show a fairly uniform distribution of stars down to [Fe/H] $\approx$ −2.6 where the sample ends abruptly. Why does the sample stop there, rather than continuing to lower metallicities?

PRESTON: Stars of lower metallicity are simply hard to find. There is no other reason for the lower bound to our Fe abundances.

WEISS: Can you please comment on the errors of the individual factors you were using in estimating the BMP incidence and how errors would propagate?

PRESTON: I would estimate that the errors in the specific frequencies of field and globular cluster blue stragglers are both typically 10 to 20 per cent − far smaller than the order-of-magnitude difference between the frequencies of the two groups.

MÖHLER: Why do you compare the blue straggler frequency among the BMP stars to the BSS frequency in globular clusters? Wouldn't open clusters be more appropriate, considering the dynamical conditions?

PRESTON: The abundances, spatial distributuions and kinematical characteristics of the BMP stars have a classical halo signature. This is why we use the globular clusters as reference systems. The BS's in M 67, however, appear to have the binary frequencies that we have adopted for our BS population.

The A-Star Puzzle
Proceedings IAU Symposium No. 224, 2004
J. Zverko, J. Žižňovský, S.J. Adelman, & W.W. Weiss, eds.
© 2004 International Astronomical Union
DOI: 10.1017/S174392130400482X

A supergiants

N. Przybilla[1], F. Bresolin[2], K. Butler[3], A. Kaufer[4], R.P. Kudritzki[2] and K.A. Venn[5]

[1]Dr. Remeis-Sternwarte Bamberg, Sternwartstrasse 7, D-96049 Bamberg, Germany
email: przybilla@sternwarte.uni-erlangen.de

[2]Institute for Astronomy, 2680 Woodlawn Drive, Honolulu, HI 96822, USA

[3]Universitäts-Sternwarte München, Scheinerstrasse 1, D-81679 München, Germany

[4]European Southern Observatory, Alonso de Cordova 3107, Casilla 19001, Santiago 19, Chile

[5]Macalester College, 1600 Grand Avenue, Saint Paul, MN 55105, USA

Abstract. A-type supergiants are the primary targets for the quantitative spectroscopy of individual stars in nearby galaxies because of their intrinsic brightness. An overview is given on the non-LTE techniques required for their analysis. Applications concentrate on placing observational constraints on evolutionary models for massive stars and their host galaxies by detailed abundance analyses. Results from high-resolution studies of A-type supergiants in Local Group galaxies and from intermediate-resolution multi-object spectroscopy of supergiants far beyond the Local Group are summarised.

Keywords. Stars: abundances, early-type, evolution, supergiants; galaxies: abundances

1. Introduction

Massive stars of $\sim 8-40\,M_\odot$ cross the A-star regime of the HRD during their post-Main Sequence evolution. Being supergiants at that time they are characterised by extended atmospheres, stellar radii measuring several tens to a few hundred $R_\odot$, and immense luminosities, on the order of 10^4 to several $10^5\,L_\odot$. The enormous intrinsic brightness, in coincidence with low Bolometric Corrections, makes BA-type supergiants the primary targets for the young field of extragalactic stellar astronomy. Using 8m-class telescopes these objects become accessible to high-resolution spectroscopy in the galaxies of the Local Group, and to medium-resolution spectroscopy out to distances of several Mpc.

This allows observational constraints to be placed on stellar evolution in a variety of galactic environments, in particular, on the effects of metallicity and rotation on stellar mass loss and the efficiency of chemical mixing. Moreover, important contributions can be made for the study of the galactochemical evolution of the host galaxies through the determination of present-day abundance patterns and gradients. BA-type supergiants can help us to verify classical studies of nebulae and extend the elemental inventory to iron-group and s- & r-process species. Finally, they can act as extragalactic distance indicators, via application of the wind momentum–luminosity and flux-weighted gravity–luminosity relationships (WLR: Kudritzki & Puls 2000; FGLR: Kudritzki *et al.* 2003).

To exploit the full potential, a few complications have to be overcome in model atmosphere analyses. The high luminosities drive stellar winds, to be solved in a hydrodynamical approach, and low atmospheric densities facilitate departures from LTE, which require a simultaneous solution of radiative transfer and statistical equilibrium (e.g., Kubát & Korčáková 2005, Krtička & Kubát 2005). In the following, we will concentrate on the latter aspect, as this allows us to draw important conclusions for studies of 'normal'

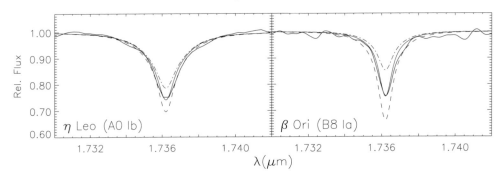

Figure 1. The observed H I $n = 4$–11 transition (thick line) in two supergiants is well reproduced by the recommended H I non-LTE model atom of Przybilla & Butler (2004, thin line), which accounts for accurate electron collision data from quantum-mechanical *ab-initio* computations for excitation processes. For comparison, synthetic spectra from computations using the Johnson (1972) approximation (dashed) and assuming LTE (dashed-dotted line) are also shown. The computations are performed for stellar parameters derived from the analysis of the visual spectra (η Leo: $T_{\text{eff}} = 9600$ K, $\log g$, = 2.00; β Ori: $T_{\text{eff}} = 12000$ K, $\log g = 1.75$). An analogous comparison in Main Sequence stars like Vega indicates that non-LTE departures are underestimated when using the Mihalas *et al.* (1975) approximation for electron collision rates.

A-stars as well, before discussing recent highlights from the quantitative spectroscopy of extragalactic A-type supergiants.

2. Quantitative spectroscopy of A-type supergiants

A-type supergiants were rediscovered as tools for astrophysics in the seminal work of Venn (1995), where modern LTE model atmosphere techniques were shown to suffice for their quantitative analysis using the spectroscopic indicators, ionization equilibria and Stark-broadened hydrogen lines, for the stellar parameter determination (see, e.g., Tanriverdi *et al.* 2005, Yüce 2005). In the following, the focus shifted from bright, though typically less luminous (LC Ib and Iab) Galactic objects towards supergiants in other galaxies of the Local Group, where only the more luminous stars are accessible to high-resolution spectroscopy using the currently available telescopes. Finally, when meeting the challenge of quantitative spectroscopy of supergiants beyond the Local Group only objects near the the Eddington limit are accessible with present-day instrumentation.

The progress on the observational side initiated a reinvestigation of the analyses techniques. Classical line-blanketed LTE atmospheres still turn out to be the best choice for studies of high-luminosity objects at present, however only in combination with massive non-LTE line-formation for the modelling of the photospheric spectrum, i.e., a hybrid non-LTE approach (Przybilla 2002). This is facilitated by the fact that the main atmospheric constituents (H, He) and the important metal opacities stay close to detailed equilibrium. In the following we discuss the major results from these investigations.

We begin with the most basic element, hydrogen, for which the effects of non-LTE departures in early-type stars were investigated more than three decades ago (e.g., Auer & Mihalas 1969a,b). Surprisingly, present-day modelling of the hydrogen spectrum fails in reproducing the observed Paschen, Brackett and Pfund lines, both in LTE and non-LTE, though good agreement is obtained for the Balmer lines. All early-type stars are affected, with the discrepancies reaching a maximum in the A-type supergiants, where a mismatch in the line strengths by factors up to 2–3 are found, see Fig. 1. The spectral features in the visual and IR are consistently reproduced only when commonly-used approximation

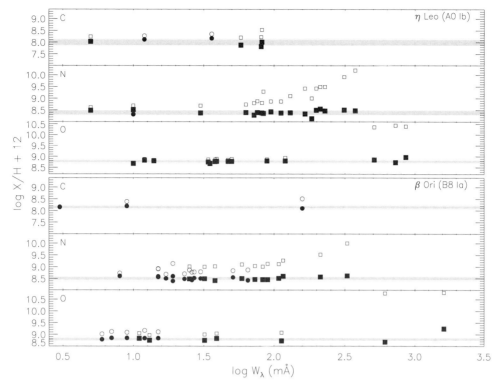

Figure 2. Elemental abundances in two Galactic supergiants from individual spectral lines of CNO plotted as a function of equivalent width: non-LTE (solid) and LTE results (open symbols) for neutral (boxes) and singly-ionized species. The grey bands cover the 1σ-uncertainty ranges around the mean values. Proper non-LTE calculations reduce the line-to-line scatter and remove systematic trends. Note that even weak lines can show considerable departures from LTE.

formulae for the evaluation of electron-collision excitation rates are dropped in favour of exact data from quantum-mechanical *ab-initio* computations (Przybilla & Butler 2004). This is because of a strong sensitivity of the line source-function to non-LTE departures in the Rayleigh-Jeans limit, resulting in an amplification of non-LTE effects in the IR.

Carbon, nitrogen and oxygen are the most abundant metals. All the observed spectral lines in the visual/near-IR originate from high-excitation, (quasi-)metastable levels that favour departures from LTE. Comprehensive non-LTE model atoms, accounting for more sophisticated atomic data than previously possible (Przybilla *et al.* 2000, 2001, Przybilla & Butler 2001) allow us to reproduce the entire observed CNO spectra to an unprecedented degree of accuracy, see Fig. 2. In particular, non-LTE abundance analyses remove systematic trends that trouble the LTE approach, and help to reduce the statistical uncertainties from the line-to-line scatter of typically ~ 0.2 dex in the literature down to better than 0.1 dex. Contrary to common assumptions, significant non-LTE abundance corrections by ~ 0.3 dex can occur even in the weak line limit.

The next step is to broaden our discussion towards a comprehensive study of the entire spectra of BA-type supergiants. Examples are shown in Fig. 3, where results from non-LTE and LTE abundance analyses of primarily weak lines are compared to the solar standard. From this we conclude that the non-LTE analysis reveals a striking similarity of the Galactic supergiant abundance patterns to the solar abundance distribution. This is also found for the M31 object, at slightly higher average metallicity. Fewer chemical

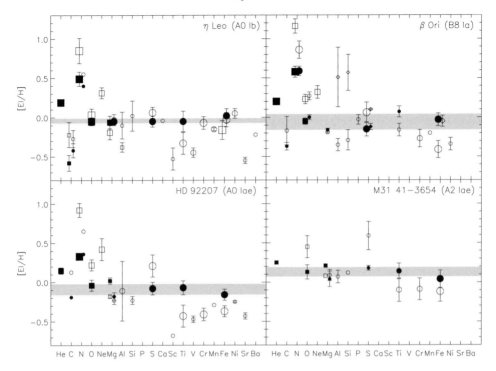

Figure 3. Abundance patterns for four Galactic and M31 supergiants, relative to the solar standard (Grevesse & Sauval 1998) on a logarithmic scale. Symbol designation as in Fig. 2, with the addition of double-ionized species (diamonds). The symbol size codes the number of spectral lines analysed. Error bars represent 1σ-uncertainties from the line-to-line scatter and the grey shaded areas mark the deduced stellar metallicity within 1σ-errors. The non-LTE computations reveal a striking similarity to the solar abundance distribution, except for the light elements which have been affected by mixing with nuclear-processed matter.

species are accessible in this case because of a more restricted spectral coverage and lower S/N of the observations. The light elements He, C and N show marked deviations which are interpreted as mixing of the atmospheric layers with nuclear-processed matter, qualitatively in good agreement with the predictions of the most recent models of massive star evolution (e.g., Maeder & Meynet 2000). Note in particular that non-LTE calculations can bring several ionization equilibria simultaneously into agreement, thus putting very tight constraints on the stellar parameters. LTE analyses on the other hand produce a large scatter of the individual elemental abundances, and result in increased uncertainties for the different species. They can even suggest α-enhancement for the more luminous objects, i.e., apparent overabundances of the α-elements, in coincidence with underabundant iron-group elements. This occurs because of selective non-LTE effects, which favour non-LTE line-strengthening in the α-elements in analogy to CNO. On the other hand, iron-group elements experience non-LTE line-weakening, because they are characterised by a plethora of energetically-close levels easily coupled via collisions that are collectively subject to non-LTE overionization. Ignoring these non-LTE effects can introduce systematic errors to abundance analyses of more luminous supergiants of typically $\sim$0.3 dex. Though the astrophysically most important elements are covered by our non-LTE computations, an extension to other chemical species is desirable. However, in many cases a lack of the required atomic data prevents such efforts at present.

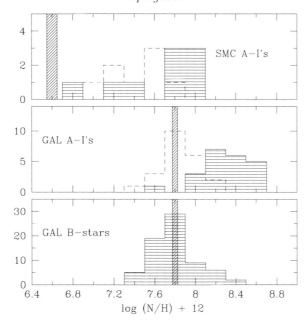

Figure 4. Histogram of nitrogen abundances in Galactic and SMC A-type supergiants, as obtained from an improved non-LTE reanalysis (Venn & Przybilla 2003) of the Venn (1999, dashed line) data, and of Galactic B-stars (Gies & Lambert 1992, Cunha & Lambert 1994), their Main-Sequence progenitors. The initial solar and SMC ISM nitrogen abundances are indicated.

Finally, we like to emphasise that using A-type supergiants as testbeds for the study of non-LTE effects offers a unique opportunity to improve stellar analysis techniques for other classes of stars in general. Due to the universality of atomic properties, sets of reference atomic data should be compiled and verified under the most extreme conditions.

3. A-type supergiants in the Local Group

The different galactic environments of the individual Local Group members offer unique opportunities to study the influence of metallicity on the evolution of massive stars. A comparison of Galactic and SMC A-type supergiant abundances (Venn 1999, Venn & Przybilla 2003) suggests that most of the objects have undergone substantial mixing with CN-cycled material, see Fig. 4. The efficiency of rotational mixing appears to correlate with metallicity since the SMC stars (at 0.2×solar metallicity) show larger nitrogen enrichments. This is again in good qualitative agreement with the predictions of the latest models of stellar evolution. The effect is a consequence of the reduced metal-line opacity which gives more compact objects and lower mass-loss rates, such that angular momentum losses are considerably reduced, which enables the mixing mechanisms (meridional circulation, shear instabilities) to retain their efficiency. A predicted correlation of mixing efficiency with stellar mass is not verified in this study. Note however, that the sample objects in the SMC are on average more massive than the Galactic supergiants.

The emission spectra of nebulae have been the primary sources for chemical abundances in extragalactic systems beyond the Magellanic Clouds until recently. Despite a widespread use for extragalactic applications, abundance determination techniques for H II regions (and planetary nebulae) are still subject to a number of inherent problems (see, e.g., Stasińska 2004). Abundances from individual stars in these galaxies open the opportunity to verify such studies with independent and well understood indicators.

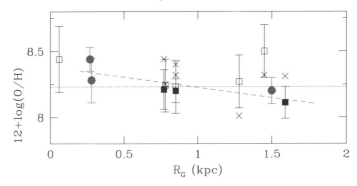

Figure 5. Stellar (Venn *et al.* 2001, filled circles) and nebular oxygen abundances as a function of galactocentric distance in NGC 6822 suggest a galactic abundance gradient of ~ -0.1 dex/kpc (linear least-squares fit: dashed line). H II region abundances are from Pagel *et al.* (1980, filled squares: abundances derived from O II and O III lines, hollow squares: from O II alone). The mean nebular abundance is indicated by the dotted line. Results from a redetermination of the nebular abundances by Pilyugin (2001) are also displayed (St. Andrews crosses). Note that Pilyugin's mean abundance is $\sim$0.1 dex higher.

Results from a comparison of nebular and stellar oxygen abundances in the dIrr galaxy NGC 6822 are displayed in Fig. 5. Both match within their mutual uncertainties in this low-metallicity case. Unexpectedly comes an indication of an abundance gradient, if a sub-set consisting of the stellar and the more reliable nebular data (i.e., observations showing both, O II and O III) is considered, which, if confirmed by additional measurements, could provide strong constraints on mixing timescales for galactochemical evolution.

While nebulae can provide abundances for a variety of light elements, iron-group and s- & r-process elements are accessible only in stars. These can provide constraints on other important parameters for galaxy evolution, as they trace nucleosynthesis sites complementary to SNe II, which are the main producers of the α-elements. Metal-poor dwarf irregular galaxies have attracted particular attention recently, as they can be understood as nearby analogues of the basic building-blocks for hierarchical galaxy formation in the early universe. Spectra of A-type supergiants in the dIrr galaxies SMC, NGC 6822, WLM, Sextans A and GR8 have been obtained with VLT/UVES and Keck/HIRES (Venn 1999, Venn *et al.* 2001, 2003, Kaufer *et al.* 2003), requiring exposures of several hours each. At metallicities down to $\sim$0.05×solar these objects are among the most metal-poor massive stars analysed so far. Mean α-element abundance (an average of O, Mg and Si) to iron abundance ratios for the sample stars as a function of metallicity, [Fe/H], are compared to Galactic disk and halo stars of similar metallicity in Fig. 6. The [α/Fe] ratios turn out to be roughly solar in these systems, indicating a similar contribution of SNe Ia and II to the chemical evolution as for the young Galactic star population, thus lacking the α-enhancement characteristic of old star populations. Globally lower star formation rates than in the solar neighbourhood lead to the lower present-day [Fe/H] of the dIrr galaxies.

4. ... and beyond

The Local Group provides us with a dozen star-forming galaxies, among those three giant spirals, where detailed high-resolution studies of individual luminous stars, primarily BA-type supergiants, are feasible. However, this impressive laboratory is still insufficient for a comprehensive study of galaxy formation and evolution. The step beyond the Local Group has to be taken to investigate all the actively star-forming systems along the Hubble sequence in clusters and other groups of galaxies, and in the field population.

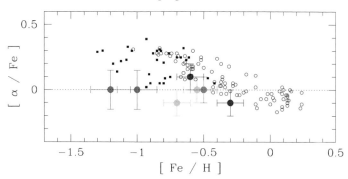

Figure 6. Stellar [α/Fe] versus [Fe/H] in dwarf irregular galaxies (solid circles), from lowest to highest metallicity: Sex A, GR8, WLM, SMC (average), WLM (2nd object), NGC 6822 (average of 3 objects), LMC (average). For comparison, Galactic disk stars (open circles, Edvardsson *et al.* 1993) and metal-rich halo stars (filled squares, Nissen & Schuster 1997) are also shown.

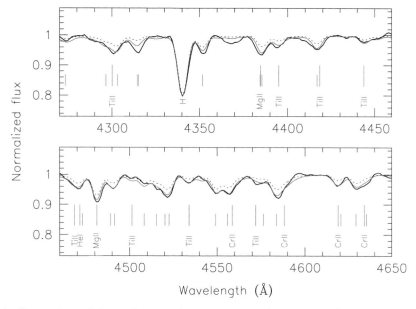

Figure 7. Comparison of the medium-resolution spectrum (thick full line) of a NGC 3621 A-type supergiant ($V = 20.47$ mag) with model predictions ($T_{\mathrm{eff}} = 9000$ K, $\log g = 1.05$) at 0.5 (thin line) and 0.2× solar metallicity (dotted line). Line identifications are provided (Fe II lines are indicated by the shorter marks). A total of 19 objects in the range $V \approx 20 - 22$ mag were observed using the multi-object capability of FORS1 on the VLT, of which 10 could be identified as supergiants, requiring a total integration time of 10.7 hr.

We have initiated a project to investigate the blue supergiant populations in nearby galaxies beyond the Local Group, using the FORS multi-object spectrograph on the VLT. Quantitative spectroscopy of individual stars out to distances of ~ 7 Mpc has been performed for the first time, estimating stellar parameters, chemical abundances, reddening, extinction and stellar wind properties. The first target was the field galaxy NGC 3621 (Bresolin *et al.* 2001). Medium-resolution spectra require spectrum synthesis techniques for the modelling of the entire spectral range to facilitate a closer analysis. While detailed abundance analyses are hampered by line blending, estimates (within roughly 0.2 dex) are still possible, taking full advantage of the methods presented in Sect. 2, as demonstrated in Fig. 7. The spectrum of the object investigated is well fitted with a chemical composition comparable to that of the LMC, i.e., $\sim$0.5 × solar, while a significantly lower metallicity, like that of the SMC can confidently be ruled out.

In the second target, the Sculptor Group galaxy NGC 300, four FORS fields have been observed, yielding spectra of 62 blue supergiants. The first steps to constrain the NGC 300 elemental abundance gradients and the internal reddening have been undertaken (Bresolin *et al.* 2002, Urbaneja *et al.* 2003). This will allow us, in addition to the topics already mentioned, to address the influence of two major sources of systematic uncertainty on the Cepheid period-luminosity relation.

References

Auer, L.H. & Mihalas, D. 1969a, *ApJ* 156, 157
Auer, L.H. & Mihalas, D. 1969b, *ApJ* 156, 681
Bresolin, F., Kudritzki, R.P., Méndez, R.H., Przybilla, N. 2001, *ApJ* 548, L159
Bresolin, F., Gieren, W., Kudritzki, R.P., Pietrzyński, G., Przybilla, N. 2002, *ApJ* 567, 277
Cunha, K. & Lambert, D.L. 1994, *ApJ* 426, 170
Edvardsson, B., Andersen, J., Gustafsson, B., *et al.* 1993, *A&A* 275, 101
Gies, D.R. & Lambert, D.L. 1992, *ApJ* 387, 673
Grevesse, N. & Sauval, A.J. 1998, *Space Sci. Rev.* 85, 161
Johnson, L.C. 1972, *ApJ* 174, 227
Kaufer, A., Venn, K.A., Tolstoy, E., Pinte, C., Kudritzki, R.P. 2004, *AJ* 127, 2723
Kubát, J. & Korčáková, D. 2005, *These Proceedings*, 13
Krtička , J. & Kubát, J. 2005, *These Proceedings*, 23
Kudritzki, R.P., Bresolin, F., Przybilla, N. 2003, *ApJ* 582, L83
Kudritzki, R.P. & Puls, J. 2000, *ARA&A* 38, 613
Maeder, A. & Meynet, G. 2000, *ARA&A* 38, 143
Mihalas, D., Heasley, J.N., Auer, L.H. 1975, *A Non-LTE Model Stellar Atmosphere Computer Program*, NCAR-TN/STR 104
Nissen, P.E. & Schuster, W.J. 1997, *A&A* 326, 751
Pagel, B.E.J., Edmunds, M.G., Smith, G. 1980, *MNRAS* 193, 219
Pilyugin, L.S. 2001, *A&A* 374, 412
Przybilla, N. 2002, PhD Thesis, Ludwig-Maximilians-University, Munich
Przybilla, N., Butler, K., Becker, S.R., Kudritzki, R.P., Venn, K.A. 2000, *A&A* 359, 1085
Przybilla, N., Butler, K., Kudritzki, R.P. 2001, *A&A* 379, 936
Przybilla, N. & Butler, K. 2001, *A&A* 379, 955
Przybilla, N. & Butler, K. 2004, *ApJ* 609, 1181
Stasińska, G. 2004, in: C. Esteban *et al.* (eds.), *Cosmochemistry* (Cambridge: CUP), p. 115
Tanriiverdi, T., Adelman, S. J., Albayrak, B. 2005, *These Proceedings*, HP2
Urbaneja, M.A., Herrero, A., Bresolin, F., *et al.* 2003, *ApJ* 584, L73
Venn, K.A. 1995, *ApJS* 99, 659
Venn, K.A. 1999, *ApJ* 518, 405
Venn, K.A. & Przybilla, N. 2003, in: C. Charbonnel, D. Schaerer & G. Meynet (eds.), *CNO in the Universe*, ASP Conf. Ser. (San Francisco: ASP), vol. 304, p. 20
Venn, K.A., Lennon, D.J., Kaufer, A., *et al.* 2001, *ApJ* 547, 765
Venn, K.A., Tolstoy, E., Kaufer, A., *et al.* 2003, *AJ* 126, 1326
Yüce, K. 2005, *These Proceedings*, HP3

Discussion

DWORETSKY: What values did you derive for the microturbulent velocity parameter?

PRZYBILLA: The microturbulent velocities apparently correlate with luminosity class. At LC Ib typical values are $\sim$4 km s^{-1}, increasing up to $\sim$10 km s^{-1} for objects close to the Eddington limit. The microturbulent velocities remain sub-sonic in all cases.

The A-Star Puzzle
Proceedings IAU Symposium No. 224, 2004
J. Zverko, J. Žižňovský, S.J. Adelman, & W.W. Weiss, eds.

© 2004 International Astronomical Union
DOI: 10.1017/S1743921304004831

Panel discussion section H

CHAIR: **S.J. Adelman**

SECTION ORGANIZER & KEY-NOTE SPEAKER: M. Parthasarathy
INVITED SPEAKERS: S. Möhler, G.W. Preston, N. Przybilla

Discussion

DWORETSKY: In the blue metal poor stars shown by you (from C. Sneden) you had a lot of Pb. Were there any signs of Hg, Mn, or Y, typical signatures of diffusion? Or is nuclear processing a sufficient explanation without diffusion?

PRESTON: Yttrium is enhanced, but is, along with Strontium and Zirconium, one of the "light" s-process elements, no Hg is seen and Mn is typically deficient, a common feature of metal poor halo stars. The abundnce signature of the BMP binaries is unabigously that of the AGB helium shell-burning phase of stellar evolution.

MICHAUD: It has been suggested that the integrated light of all objects is dominated by HHB stars spectra. It was also suggested that the spectra of some galaxies were dominated by A supergiants. Could you reconcile the two?

MOEHLER: By "old" I understand older than 10 Gigayears. In such objects the ultraviolet flux is dominated by hot Horizontal Branch stars with a minor contribution from hot post-AGB stars (minor due to their short life times).

PRZYBILLA: Massive stars spend only a short fraction - of the order of a few 10^4 years - of their life time as A-type supergiants during later evolution stages. It is therefore implausible to assume that the integrated light properties of galaxies can be dominated by these objects.

TANRIVERDI: To calculate systematic errors, the effect of T_{eff}, $\log g$, etc. are independent of each other? What is the reason for that? If it is considered dependent, what causes in result?

PRZYBILLA: The practice for estimating systematic uncertainties in stellar studies is changing parameters one at a time and in investigating the effect of this on the analysis. A formal (mean square) error may be calculated from the individual uncertainties. However, one has to bear in mind that many of the parameters are NOT independent, such that this formal systematic error has to be viewed as an estimate of the real uncertainties only. Thorough error analysis is VERY time-consuming: Monte Carlo simulations for error propagation require many hundred model runs, and with a single NLTE computation taking several CPU hours, like in the case of our Fe II model ion, we would rather perish than publish ...

GREVESSE: Comment on Non LTE - results. Your failure, when using data from the literature, and then success, when you recomputed yourself more accurate cross-sections, in the NLTE computations on Hydrogen, are a very good lessons and should be kept in mind by all the users of NLTE codes!

COWLEY: In the case of the H alpha profile of Arcturus were you able to make your nonLTE fit with reasonable value of the microturbulence? I could never fit the H depth in LTE, but one could fit the width. However, to fit the width an uncomfortably large microturbulence was required. Can you now feel confident of the non-LTE fits for stars like Arcturus so H alpha could be used as a temperature diagnostic?

PRZYBILLA: Note: This aims at the finding from a spin-off paper of results presented here (Przybilla & Butler, 2004, ApJ 610, L61).) Increasing the microturbulence velocity to supersonic values can improve the fit quality of the H line width in Arcturus. This however, stands in contrast to the results from the standard microturbulence indicators and, more serious, is hard to justify on physical grounds. The synopsis from that work is that the observed Balmer lines of K-giants cannot be reproduced by present-day standard model atmospheres. The failure to model the spectral lines of the most abundant and most basic (in terms of atomic physics) element indicates that all future conclusions from the analysis of such objects have to be viewed with caution.

YUCE: Did you look at the evolutions of your supergiants with LTE calculations? How was differences between LTE and NLTE calculations?

PRZYBILLA: NLTE & LTE stellar parameter determinations will give similar values for luminosity classes Ib to Iab, thus resulting in similar positions in the HRD. Proper NLTE calculations however, help to reduce the uncertainties. At higher luminosities NLTE computations become mandatory, if the same degree of accuracy in the parameter determination is aspired. More drastic are the effects on abundance determinations for the elements He, C, N and O as tracers for mixing of nuclear processed matter into the stellar atmosphere. Here, the differences between NLTE an LTE abundances can amount to a factor of 2 in weak lines to a factor of 100 in strong line analyses.

PISKUNOV: Are you familiar with the work by Paul Barklem of Hydrogen line profiles and broadening? There are several issues which are relevant to the previous discussion, for example for solar type stars there is broadening which changes the wings quite much in particular in lower members of Balmer series and also Paschen series?

PRZYBILLA: We have reproduced the solar Balmer and Paschen lines using the Stehle tables for Stark broadening and the Barklem formalism for the resonance broadening plus a model atmosphere by the Avrett group, and have studied also several other commonly used model atmospheres. Using this one arrives at the conclusion that one needs a chromosphere plus NLTE calculations to derive the physical run of the Hydrogen populations in the Sun. One can also reproduce the observed Balmer and Paschen lines rather accurately by LTE model atmospheres and NLTE line formation, but the physical solution is different: the line-formation depths in the latter case do not match the observation.

PISKUNOV: What Paul Olsson did rather recently for A stars he demonstrated that occupational probability algorythms were very important for correctly reproducing Balmer jump do you do that?

PRZYBILLA: We do not account for it explicitly, but we introduce an empirical correction for this by restricting the model atoms to the number of levels actually observed, i.e. the classical Inglis-Teller limit. This is important.

ADELMAN: You mentioned that the depth dependent microturbulence is necessary?

PRZYBILLA: No depth dependent microturbulence.

ADELMAN: So why the people get different micturbulence in different atomic species?

PRZYBILLA: We derive a single value for the microturbulent velocity from all the species in A-type supergiants. In the previous discussion on Arcturus this point did arise as a possibility to bring the hydrogen lines and the metal lines simultaneously into match, using microturbulence as a fudge factor to a greater extent than usual. To my knowledge the only use of different microturbulences for different species in A-type supergiants occurred in the seventies when quantitative analyses of supergiants were accomplished for the first time, due to less-accurate atomic data. I am not aware that modern studies still derive this. But I would suggest that this is because of unaccounted NLTE effects. The weak and strong lines of the typical microturbulence indicators (e.g. Ti and Fe) show different sensitivities to NLTE departures, such that the NLTE analysis may result in a single value while in LTE discrepancies are found.

SKODA: You have shown us mostly the IR spectra of A supergiants. How do they look like in UV and (if is there any activity observed) in X-ray or radio bands?

PRZYBILLA: The UV spectra of A type supergiants are dominated by symmetric absorption lines, like in the visual. Only the resonance lines of the most abundant species indicate a presence of a stellar wind (in the more luminous objects), in form of saturated "black" absorption troughs. P Cygni profiles are not observed. A radio excess is observed in some stars. I am not aware of any X-ray observations of A-type stars.

SKODA: You have stressed the influence of winds in spectra (of blue supergiants). Did you try to incorporate a model of stellar wind (even a simple one) into your calculations?

PRZYBILLA: We have used NLTE unified model atmospheres for successfully modelling the wind affected $H\alpha$ - $H\gamma$ lines in A type supergiants (Kudritzki *et al.* 1999, A&A 350, 970). Our present calculations for the photospheric absorption line spectra do not account for mass outflow. These lines appear to be basically unaffected by the stellar wind as indicated by preliminary results from a comparison with Hillier's CMFGEN code for β Ori.

BALEGA: Do you have any ideas about origin of fast rotators below 11000 K in globular clusters?

MÖHLER: Sills & Pinnsanneault (2000) suggest that red giants with rapidly rotating cores and differentially rotating envelopes could explain the fast rotators, if angular momentum is transported from the core to the surface on the horizontal branch. It is however, unclear, whether such fast rotating cores are possible. Soker and collaborators suggested a spin-up of the red giants precursor by a stellar companion or a planet entering the red giant's envelope as possible explanation for fast rotators. There are no observations so far to verify this hypothesis.

The A-Star Puzzle
Proceedings IAU Symposium No. 224, 2004
J. Zverko, J. Žižňovský, S.J. Adelman, & W.W. Weiss, eds.

© 2004 International Astronomical Union
DOI: 10.1017/S1743921304004843

A stars as physics laboratories

John D. Landstreet

University of Western Ontario, Department of Physics & Astronomy, London, Ontario,
Canada N6A 3K7
email: jlandstr@astro.uwo.ca

Abstract. Stars in various parts of the HR diagram often have atmospheres in which the departure from the simplest kind of plane-parallel model is largely dominated by a single physical effect. For example, massive stars and giants exhibit symptoms of strong winds and lower Main Sequence stars are very strongly influenced by the presence of deep and energetic envelope convection. Main Sequence A stars, in contrast, appear to display the competing effects of several physical effects of comparable magnitude. The effects which can be detected by observation include large and relatively simple magnetic fields, strong surface convection, pulsation (often in multiple modes), diffusion of specific species under the competing influences of gravity and radiative acceleration, and (more indirectly) internal turbulent mixing, weak winds, and non-thermal heating. This situation makes these stars extremely useful as laboratories to explore and to understand the physics of these various phenomena, and how these effects interact with each other. This review will summarize some of the interconnections that are gradually being understood and emphasize some of the major remaining problems.

Keywords. Convection, diffusion, turbulence, stars: atmospheres, (stars:) Hertzsprung-Russell diagram, stars: magnetic fields, stars: rotation, stars: winds, outflows

1. What are the A stars?

It is interesting to start our discussion by recalling some general characteristics of A stars. On the Main Sequence, A stars have effective temperatures between about 10000 and 7000 K. Popper (1980) lists a number of A stars in EB systems. They range in mass from 2.6 to 1.6 $M_\odot$. The actual mass range may be somewhat larger. From the models of Schaller *et al.* (1992), stars as massive as 3.5 $M_\odot$ evolve to effective temperatures less than 10000 K before leaving the Main Sequence. If we take the actual mass range of stars that spend some part of their Main Sequence lives as A stars as between about 3 and 1.6 $M_\odot$, the Main Sequence lifetimes range from 6×10^8 to 2×10^9 yr, and the luminosities range from about 40 to 10 $L_\odot$ on the Main Sequence.

It is easily forgotten that the effective temperature of a star evolves significantly during its Main Sequence lifetime, decreasing by about 30% from the ZAMS to the TAMS. Thus a star with a mass of a little more than 3 $M_\odot$ will be an A star only near the end of its Main Sequence life, while a star of 1.6 or 1.7 $M_\odot$ will become an F star during the Main Sequence stage.

Main Sequence A stars cover the transition region in the HR diagram between stars with Sun-like evolution to the giant branch with a strong increase in luminosity up to about 10^3 $L_\odot$ at the helium flash, and stars massive enough for evolution to the helium flash to occur at roughly constant luminosity, as is the case already around 4 or 5 $M_\odot$ (e.g., Schaller *et al.* 1992).

The surface convection zone (H I – H II, He I – He II and He II – He III) extends to a depth where the temperature is of order 50000 to 150000 K, and involves about 10^{-8} or more of the stellar mass.

423

For many of the problems that interest us, it is important to consider the progenitors of the A stars, the Herbig Ae stars. These stars have similar masses to the Main Sequence A stars, and also similar effective temperatures, but are often several times more luminous than the Main Sequence stars of the same T_{eff}. They are typically located in the galaxy in regions of recent star formation, along with the lower mass T Tauri stars. With increasingly powerful telescopes and the ability to observe in new wavelength regimes such as X-rays, we are now rapidly accumulating valuable information about these stars, and developing reasonably comprehensive (but still mostly one-dimensional) models both of the underlying star and its active chromosphere, disk and wind (e.g., Catala 2003).

A giants and supergiants, of course, originate as more massive stars, above 4 or 5 $M_\odot$. For such stars the A star phase may be reached as the star evolves rapidly to the red giant state after reaching the Schönberg-Chandrasekhar limit. In this case, the star is an A star for only $10^5 - 10^6$ yr.

2. What characteristics make A stars useful as physics laboratories?

An extremely interesting process in its own right, and a powerful probe of other physical processes in the stellar interior, is the microscopic diffusion of trace elements under the combined influence of gravity and radiative acceleration. This process can be a very valuable probe of other physical processes if competing processes do not occur with much higher velocities.

The characteristic time scale of this process is easily estimated. If we take the typical speed of a trace ion to be the thermal speed $v_{\text{th}} \sim (2k_{\text{B}}T/Am_{\text{u}})^{1/2}$, the collision time is of order $t_{\text{coll}} \sim 1/(n_A v_{\text{th}}\sigma)$. If we assume that the trace ion is accelerated between collisions relative to the dominant H medium by gravity and radiation with some net acceleration g_{e} (typically of order 10^4 cm s^{-2} in Main Sequence A stars), an estimate of the diffusion velocity is $v_{\text{dif}} \sim g_{\text{e}} t_{\text{coll}}/2$. Near $\tau = 1$, the number density of scatterers is of order 10^{14} cm^{-3}, and the diffusion speed is a small fraction of 1 cm s^{-1}.

In contrast, in O and early B stars, winds can occur with mass loss rates as large as 10^{-8} $M_\odot$ yr^{-1}. If this is assumed uniform over the star, an estimate of the vertical velocity of the wind is given by $\dot{M} \approx 4\pi R^2 n_A v_{\text{wind}}$. A mass loss rate of 10^{-8} $M_\odot$ yr^{-1} corresponds to velocity of some m s^{-1}. This velocity is much larger than the typical diffusion velocities, and as a result, large surface abundance variations are not able to develop. Thus chemical anomalies are generally ineffective as tracers of internal physical processes in hot stars.

A different situation occurs in the cool stars (mid F and later), where a deep convection zone is present. Again the velocities (which can rise to a significant fraction of the speed of sound, thus up to a couple of km s^{-1} even in the atmosphere) are far larger than the diffusion velocities, so the convective region is extremely well-mixed. Diffusion can occur into and out of this region at the bottom, but because the mixed reservoir is massive, at most rather small changes in the surface abundances occur. Again, abundance anomalies are not very useful as tracers of interior processes.

In contrast, the envelope convection in A and late B Main Sequence stars occurs only in a very shallow layer of tiny mass, and winds (if any) are apparently quite weak. Thus no large competing velocity fields prevent the development of abundance anomalies at the stellar surface. These can serve as valuable tracers of processes that modify or compete with diffusion in the stellar interior. This is perhaps the fundamental reason that middle Main Sequence stars ("tepid stars") are particularly useful as stellar physics laboratories.

A further important advantage of A stars, in contrast to more massive Main Sequence stars, is that we are able to observe the later stages of the pre-Main Sequence phase

(Herbig Ae stars). We may hope that in the near future we will be able to trace some of the interesting phenomena discussed below from this early evolutionary stage.

Finally, there is also an important observational characteristic of A stars that offers significant practical advantages. These stars are hot enough not to have molecules in their spectra, making spectral analysis enormously simpler than for G, K and M stars, but they are cool enough to have relatively rich line spectra in the easily accessible visible spectral window. This characteristic is not fundamental, but it certainly is valuable as we consider using A stars as laboratories to study stellar physics.

3. Chemical peculiarities: diffusion as a physical process

For later discussion, it is useful to recall the basic characteristics of trace element diffusion (Michaud *et al.* 1976). (1) Under the action of gravity alone, all ions heavier than H would slowly settle towards the interior of the star. (2) The outward flow of radiation in the star exerts an acceleration on atoms and ions. This acceleration is largest for ions of locally low abundance, and for ions with a rich array of low-lying energy levels. The acceleration *per ion* diminishes with increasing abundance and for ions in noble gas states. (3) The *net* acceleration at one level on a particular ion may be upwards or downwards. This acceleration will vary with depth and will *evolve in time* as ions diffuse from one level to another. (4) Consequences of internal diffusion are visible at the surfaces of many A stars, and may even include pronounced vertical stratification within the atmosphere. (5) Diffusion is modified by competing velocity fields inside the star and consequences of this competition may become visible at the surface (e.g., Richer *et al.* 2000). (6) The upper boundary conditions (presence and nature of a stellar wind, possible accretion) also modify atmospheric abundances (Vauclair 1975, Babel 1992). (7) Observationally we see that diffusion near the surface is profoundly modified by the presence of a strong magnetic field. A magnetic field strongly inhibits horizontal mixing and thus makes possible horizontal variations in composition. High in the atmosphere (but not elsewhere) ions are forced to drift along field lines. However, the basic mechanism leading to large surface inhomogeneities remains mysterious (Babel 1992). (8) Note that trace ions may diffuse upwards or downwards through a convective region. They will be homogeneously mixed in the convective zone, but may be added to or removed from such layers at their boundaries.

4. Chemical peculiarities: diffusion as a probe

Up to now it has proven remarkably difficult to get the results of diffusion computations to resemble closely the surface abundance patterns observed in various types of tepid stars. It seems likely that this is due to that diffusion competes with mixing processes in the stellar interior, and may be strongly influenced by the (very uncertain) upper boundary conditions at the top of the atmosphere. This sensitivity to competing processes makes diffusion, as observed in the atmospheric chemistry of various stars, potentially a very important probe of such invisible processes.

A very interesting aspect of the A stars is that one very important competing process, envelope convection, varies dramatically from the most massive to least massive A stars in the depth of the layer to which it operates. Thus in a limited mass range we have, in principle, the possibility of using the lower boundary of the convective zone as a kind of movable probe of the effects of diffusion at various depths.

A very important recent development, for testing the time dependence expected for the results of diffusion against observed abundance patterns in stars of know ages, has been the great improvement in astrometric data (from the Hipparcos, Tycho, and Tycho-2 data) for cluster stars, whose ages may be determined much more securely than the ages of most field stars. These data have greatly clarified which stars found in the fields of clusters and associations are actually physical members of these groups, and thus have relatively well-defined initial chemical composition and ages. Cluster stars are being exploited by Hui-Bon-Hoa & Alecian (1998) to study the evolution of anomalous abundances, and thus the effects of diffusion, in Am stars, and by Bagnulo and collaborators (still unpublished) to study the evolution of magnetic fields through the Main Sequence phase.

A very nice example of using diffusion to probe other physical effects is Sylvie Vauclair's (1975) hypothesis for the origin of He-strong magnetic B stars. These are stars having T_{eff} around 20000 K in which He is greatly overabundant in the atmosphere, in contrast to all cooler magnetic stars in which He is underabundant, almost certainly due to the downward diffusion of this element, which has little support from radiation. The occurrence of He-strong stars appears to be due to the presence of a weak stellar wind of order of $10^{-12} M_{\odot} \mathrm{yr}^{-1}$. This wind exerts a strong frictional drag on the ionized He below the atmosphere, lifting it into the atmosphere. At this level, the He ions become neutral, greatly reducing their interaction with the wind, so above this level the He atoms are *not* dragged along with the wind, but collect in the atmosphere. The beauty of this hypothesis is that it reveals the presence of an otherwise (currently) undetectable wind.

If winds of similar mass loss rate occurred in late B magnetic stars, they would be expected to lead to overabundances of Ne and/or O at the surface (Landstreet *et al.* 1998). Since these are not observed, either the winds do not occur with the necessary minimum mass loss rate, or they are not turbulent enough to be well-mixed. (Note that in any case winds much stronger than $10^{-12} M_{\odot} \mathrm{yr}^{-1}$ would erase most atmospheric chemical anomalies as they do in hot stars, so the occurrence of such winds seems to be excluded.)

Another nice example of using diffusion to probe invisible processes is offered by Richer *et al.* (2000) who have carried out computations of the expected abundance anomalies in Am stars. They find that if diffusion is allowed to occur in their models without competition, the surface anomalies are considerably larger than are observed in actual Am stars. Much better agreement with observation is obtained by assuming that meridional circulation currents are turbulent, and are able to mix somewhat (though not to completely homogenize) the outer layers down to a level including about $10^{-4} M_{\odot}$. Again we have an example of using surface abundances to explore otherwise invisible internal mixing processes.

As a final example, Charbonneau & Michaud (1988) have proposed that the occurrence of the HgMn peculiarities in late B stars is limited to stars with equatorial rotation velocities of less than about 75 $\mathrm{km \, s^{-1}}$ because of rapidly increasing competition with mixing due to meridional circulation currents. Diffusion thus appears to probe meridional circulation.

5. Atmospheric velocity fields

The widely used mixing length model of convection is really only an order-of-magnitude estimate of the effects of convection. In situations in which convection operates "efficiently", - in which the density and the velocity of the convecting gas are high enough to transport almost all the outward flux of energy and the temperature gradient is forced to the local adiabatic gradient, mixing length theory provides an adequate framework for computing stellar structure. However, in low density regions such as the outer envelopes of stars, convection is inefficient, and one is obliged to compute the flux carried by both convection and radiation. In this situation the inaccuracy of the mixing length model is an important limitation on the accuracy of stellar models.

In principle this situation can be improved by numerical computation of the structure of a convecting layer from physical first principles. However, such computations are limited in scope by the enormous memory requirements, by the long thermal relaxation times of inefficiently convecting gas, and by the necessity to model sub-grid scale dissipation. For the near future, such computations will not provide a solution to the need for a better model of convection, but very interesting results have been presented at this meeting by Freytag (2005) and by Trampedach (2005).

Several improved models of convection which are computationally tractable have been developed in recent years (Canuto & Mazzitelli 1992, Kupka & Montgomery 2002), and are beginning to be applied to the computation of the structure of inefficient convective regions. Testing the predictions of these models is an important part of this process. A stars are potentially an important laboratory for such tests, since in such stars an inefficient convection zone reaches into the atmosphere, where in sufficiently slowly rotating stars its velocity field is directly observable in deviations of line profiles from those predicted by simple models such as isotropic microturbulence (Landstreet 1998).

One intriguing result is already available from such studies, which have been carried out for some years for lower Main Sequence stars. In cool stars, the profiles of spectral lines are observed to have an extended long-wavelength wing, which is interpreted as a symptom of a convective pattern of slowly rising flow over a large fraction of the surface, together with a rapidly descending return flow over a smaller surface fraction (e.g., Gray 1989, Dravins 1990). Such a flow pattern is consistent with the results of numerical simulations, and detailed observations of the solar surface. In A stars, the extended wing is on the *short* wavelength side of the spectral line, suggesting that the convective structure is quite different from that of cooler stars, possibly with rapid updrafts on a small fraction of the surface and slower downdrafts over much of the rest, as is the case in the Earth's atmosphere. Furthermore, the velocities inferred from the study of A stars suggests that the speed of convective flows in the atmospheres of such stars are significantly larger than in the cool star case. The results for A stars are presently in serious disagreement with the numerical hydrodynamical computations of Freytag (2005) and Trampedach (2005), both of whom find that A star convection is structurally similar to that of the Sun.

6. Magnetic fields

Another observable physical effect in many A and B stars is the presence of a global magnetic field of considerable strength. This meeting has been dedicated to the memory of the discoverer of these fields, Horace Babcock, and of Vera Lvovna Khokhlova, a great pioneer in modelling the surfaces of magnetic stars. But on a much happier note, we have been very fortunate to have the second great pioneer of magnetic field observations, George Preston, here to participate in the meeting with us.

The magnetic fields of A and B stars dramatically affect the stars in which they occur, both in the nature of their atmospheres and in global properties. In the atmospheres, the magnetic stars have highly distinctive abundance anomalies which vary systematically with effective temperature, but also considerably from star to star among apparently rather similar stars. Furthermore, some of the stars have substantial abundance variations over the surface, and there is strong evidence for vertical stratification (both in magnetic and non-magnetic stars) (see reviews by Mathys (2005) and by Ryabchikova (2005)).

Two extremely interesting global effects of these fields occur as well. One important effect is that most (but not all) magnetic Ap stars have only about 0.1 of the angular momentum per unit mass found in normal A stars, while a few have less than 10^{-4} of the normal specific angular momentum. Since only a very small fraction of magnetic Ap stars are in close binary systems, this must be due to the magnetic field (see Mathys 2003). The other effect is that some of the coolest magnetic Ap stars pulsate, usually in a number of non-radial modes, with periods in the range of 4 to 15 min (see the review by Kurtz *et al.* 2005).

These stars provide us with a really valuable laboratory to observe the interactions of a strong magnetic field with a large-scale plasma, which in the cooler Ap stars is probably unstable to convection according to the Schwarzschild criterion. These tepid magnetic stars have two great advantages over the cooler magnetic stars as laboratories. First, that a range of field strengths, strong enough to be easily observable, are available in the numerous magnetic Ap stars. And second that the field is relatively homogeneous over the whole surface, making it far easier to study than that of unresolved magnetically active cool stars whose spectra are largely dominated by the hotter, unmagnetized plasma.

The list of unsolved problems concerning such stars is impressively long. (1) It is still uncertain how these stars lose so much angular momentum, although Stępień (2000) has proposed a very plausible semi-empirical theory involving magnetic coupling during the pre-Main Sequence phase to an accretion disk and to a wind. (2) Moss (2001) has carried out a number of illuminating model calculations of fossil field evolution during the Main Sequence phase, but it is not yet possible to link these closely to the observed fields, or to test the results observationally. In fact, it is still not completely clear that these fields are really fossils, although this is the general view. (3) On the basis of similar birth rates and the correct relative field strengths, it appears that the fields in Main Sequence Ap stars may be the progenitors of megagauss fields in magnetic white dwarfs (Angel *et al.* 1981). However, no calculations are available to really support this idea. (4) Detailed computations of atomic diffusion (e.g., Babel & Michaud 1991) have not yet succeeded in reproducing the observed abundance patterns in any magnetic Ap, and have been even less successful in explaining the origin of the abundance patches. One of the main uncertainties here is whether a stellar wind is present, and if so whether it is well-mixed or separated (i.e., consists mainly of ions driven out by radiation, without dragging along unsupported ions such as H and He). A wind could greatly modify the elements expected to accumulate in the atmosphere. Babel (1992) has suggested that magnetic control of a wind could account for the occurrence of abundance patches. (5) Finally, it seems very likely that the structure of the atmosphere of a magnetic Ap star, in which the magnetic field leads to significant forces and the abundances are probably quite non-uniform vertically, is very poorly described by the atmosphere models currently used to model observed spectra. The consequences of this uncertainty are still unexplored (see e.g., Kochukhov *et al.* 2002).

Although magnetic A stars are potentially an extremely valuable laboratory for stellar magnetohydrodynamics, much work still needs to be done in the lab before it will be fully functional.

7. Rotation and braking

Rotation of peculiar A stars has recently been reviewed by Mathys (2003, see also other contributions in this volume). The evolution of angular momentum under various circumstances is another area where A stars (together with the Herbig Ae stars) can provide an important laboratory. One of the great values of this particular laboratory is that a wide range of parameters is found. Among the A stars we have examples of stars which have some of the highest stellar values of specific angular momenta (of order

3×10^{17} cm^2 s^{-1}, corresponding to $v \approx 300$ km s^{-1} or $P \approx 0.5$ d), and some of the very lowest (a few magnetic Ap stars have rotation periods of decades or more). The observed values of specific angular momenta cover at a range of almost 10^5 from highest to lowest.

On one hand, this means that we can readily explore the effects *of* rotation, for example in producing mixing through meridional circulation that may compete with diffusion. Among the nonmagnetic A stars a wide range of periods is available. It has been argued (Charbonneau & Michaud 1988) that the restriction of the HgMn peculiarity to stars with $v \sin i < 75$ km s^{-1} is a consequence of the rapidly increased efficiency of mixing above this rotation rate.

On the other hand, it is also possible to use the A stars to look at the effects of other physical mechanisms *on* rotation and its evolution. It appears (North 1998) that the magnetic stars have been slowed during the pre-Main Sequence phase. Stępień (2000) has explained how magnetic interaction with a disk and wind could achieve this. Stępień & Landstreet (2002) have discussed how the same physics might be extended to explain how the longest periods might be produced, and why most of the shorter-period magnetic Ap stars have a large angle between the axis of their roughly dipolar structure and the rotation axis, while in long-period magnetic Aps stars this angle is usually small (Landstreet & Mathys 2000).

Similarly, a large fraction of HgMn and Am stars are in close binary systems, some of which do not show stellar rotation synchronized with the orbital period. These systems should furnish valuable information about angular momentum exchange in close binaries.

One very intriguing recent result is observational evidence that the rotation periods of some magnetic Ap stars are note precisely constant for a period of several decades (see Pyper and Adelman 2005). The interpretation of this phenomenon is still quite uncertain.

8. Pulsation

Among the A stars, three different kinds of pulsation are recognized. The pulsation properties of the pulsating A stars are potentially extremely important in the context of using such stars as laboratories. As pulsation modes are identified and compared to models of these stars, the pulsations will furnish enormously valuable constraints on the internal structure of various types of A stars.

The best known pulsation mode among the A stars is the δ Scuti type of pulsation. δ Sct variables are late A Main Sequence stars. Some pre-Main Sequence Herbig Ae stars are also δ Sct variables. These stars pulsate with periods of the order of 1 or 2 hours, and show both light and line profile variations. In these stars up to roughly 30 periods are observed by long multi-site ground-based observing campaigns (e.g., Breger *et al.* 2002). The observed pulsations are primarily low degree l, low overtone n p-modes (that is, the restoring force is primarily due to gas pressure, so that the oscillations resemble sound waves). This type of pulsation is, roughly, an extension of the classical Cepheid Instability Strip to the Main Sequence (Unno *et al.* 1989).

Some of the coolest magnetic Ap stars also exhibit short-period (4 to 15 min) light variations, typically with several frequencies that may come and go. Such stars are known as roAp (rapidly oscillating Ap) stars. Recently line profile variations, usually with one of the photometric periods, have also been detected in some of these stars, although often only in a few lines, particularly those of doubly ionized rare earths. A rich array of phenomena connect the pulsations to the magnetic fields. For example, the pulsations seem to be at least approximately aligned with the axis of the roughly dipolar field. These pulsations are low-l, high-n p-modes. The driving mechanism for the pulsations has not yet been securely identified (e.g., Shibahashi 2003).

A recent addition to the classes of pulsating A stars is the γ Doradus stars (Kaye et al. 1999, Mathias et al. 2004). These are multi-periodic variables showing variations with periods of the order of 0.4 to 3 days. Variations are seen in both light and spectral lines profiles. The γ Dor variables range roughly from A7 to F5 and are close to the Main Sequence. In this case, the occurrence of multiple long periods clearly points to non-radial g-mode pulsations (that is, the restoring force is mainly due to buoyancy forces rather than to gas pressure).

Because these classes of pulsators all are known to show multiple periods of variation, the number of constraints on internal structure available in principle is interestingly large (although we will not have anything like the astonishingly detailed information emerging about the interior of the Sun from helioseismology any time in the foreseeable future). However, to date it has proven very difficult to identify the observed frequencies with model pulsation frequencies (particularly for the δ Sct stars), so this potential is still largely unrealized.

What is needed observationally is a still larger number of observed frequencies. Such data are obtained from the ground only by organizing large multi-site photometric or spectroscopic campaigns involving several observatories and observing runs lasting days or weeks. Campaigns of this sort have produced some really remarkable data. For a few objects, the situation is shortly going to become remarkably better due to observations from photometric satellites such as MOST (launched and working), COROT and MONS (both to be launched in the near future), from which it should be possible to detect substantially weaker pulsation modes (a few micromagnitudes) with greatly improved frequency resolution and freedom from aliases.

Theoretically, the pulsation models depend on a large number of parameters, mass, age, metallicity, rotation, magnetic field, and very large model grids, together with algorithms for searching such grids for pulsation frequency sets approaching observed sets, will be needed. Indeed, for the roAp stars the situation is even worse, as one does not yet understand the mechanism that aligns the pulsations near the magnetic axis, or the mechanism that selects only certain modes to pulsate with observable amplitudes, or why some cool magnetic Ap stars pulsate and others do not.

Our understanding of the pulsating A stars is presently frustratingly limited, but this is a field in which rapid observational and theoretical progress is occurring. It is a field which offers really important rewards as models account for more and more of what is observed. We can look forward to having an increasingly detailed view of the interior of some of the A stars, which will make it possible in turn to study effects of invisible physics. In the future asteroseismology may make it possible to constrain the size of the internal magnetic field, to study the internal distribution of angular velocity, and to constrain the spatial variation of chemical abundances. These are such important potential rewards that we may be almost sure that progress in this field will continue at a rapid pace (see Christensen-Dalsgaard 2003).

9. Conclusion

It is clear that the A-type stars that are the subject of this meeting offer a really wide range of possibilities for use as laboratories to study stellar physics.

Acknowledgements

This review has benefited greatly from ongoing discussion and collaborations with Drs. S. Bagnulo, G. Mathys, and G. Wade.

References

Angel, J. R. P, Borra, E. F., Landstreet, J. D. 1981 *ApJS*, 45, 457

Babel, J. 1992 *A&A*, 258, 449

Babel, J., Michaud, G. 1991 *ApJ*, 366, 560

Breger, M. *et al.* 2002 *MNRAS*, 329, 531

Canuto, V., Mazzitelli, I. 1992 *ApJ*, 389, 724

Catala, C. In *Magnetism and Activity of the Sun and Stars* (ed. J. Arnaud and N. Meunier), p. 325. EDP Sciences

Charbonneau, P., Michaud, G. 1988 *ApJ*, 327, 809

Christensen-Dalsgaard, J. 2003 *A&SS*, 284, 277

Dravins, D. 1990 *A&A*, 228, 218

Freytag, B. 2005, *These Proceedings*, 139

Gray, D. F. 1989 *PASP*, 101, 832

Hui-Bon-Hoa, A., Alecian, G. 1998 *A&A*, 332, 224

Kaye, A. B., Handler, G., Krisciunas, K., Poretti, E., Zerbi, F. M. 1999 *PASP*, 111, 840

Kochukhov, O., Bagnulo, S., Barklem, P. S. 2002 *ApJ*, 578, L75

Kupka, F., Montgomery, M. H. 2002 *MNRAS*, 330, L6

Kurtz, D. W., Elkin, V. G., Mathys, G., Riley, J., Cunha, M.S., SDhibahashim H., Kambe, E. 2005, *These Proceedings*, 343

Landstreet, J. D. 1998 *A&A*, 338, 1041

Landstreet, J. D., Dolez, N., Vauclair, S. 1998 *A&A*, 333, 977

Landstreet, J. D., Mathys, G. *A&A*, 359, 213

Mathias, P. *et al.* 2004 *A&A*, 417, 189

Mathys, G. 2003 In *Stellar Rotation: Proceedings of IAU Symposium No. 215* (ed. A. Maeder & P. Eenens), EDP Sciences, in press

Mathys, G. 2005, *These Proceedings*, 225

Michaud, G., Charland, Y., Vauclair, S., Vauclair, G. *ApJ*, 210, 445

Moss, D. 2001 In *Magnetic Fields across the Hertzsprung-Russell Diagram* (eds. G. Mathys, S. K. Solanki, D. T. Wickramasinghe), p. 305. ASP Conference Series, Vol. 248

North, P. 1998 *A&A*, 334, 181

Popper, D. M 1980, *ARA&A*, 18, 115

Pyper, D. M., Adelman, S. J. 2005, *These Proceedings*, 307

Richer, J., Michaud, G., Turcotte, S. 2000 *ApJ*, 529, 338

Ryabchikova, T. 2005, *These Proceedings*, 283

Schaller, G., Schaerer, D., Meynet, G., Maeder, A. 1992 *AA&S*, 96, 269

Shibahashi, H. 2003 In *Internation Conference on Magnetic Fields in O, B and A Stars* (eds. L. Balona, H. F. Henrichs, R. Medupe), p. 55. ASP Conference Series, Vol. 305

Stępień, K. 2000 *A&A*, 353, 227

Stępień, K., Landstreet, J. D. 2002 *A&A*, 384, 554

Trampedach, R. 2005, *These Proceedings*, 155

Unno, W., Osaki, Y., Ando, H., Saio, H., Shibahashi, H. *Nonradial Oscillations of Stars*, Univ. of Tokyo Press

Vauclair, S. 1975 *A&A*, 45, 233

Discussion

FREYTAG: What steps do you recommend to extract the information buried in the complex line profiles you showed [profiles of the very sharp-lined Am star HD 108642, from Landstreet (1998, *A&A*, 338, 1041)]?

LANDSTREET: I have personally been trying to model these data using simple parameterized velocity fields, for example specifying vertical upward and downward flows over specified fractions of each integration grid area in my spectrum synthesis code. When I do this, I find that I can get line profiles that resemble those of strong lines in HD 108642 by assuming that the microturbulent velocity increases with height in the atmosphere, along with global vertical flows of the order of 8 km s^{-1} upward over 20% of the area, and downward at 2 km s^{-1} over the remaining area. Thus it appears to me that the data are suggesting a flow pattern which is opposite to that of the Sun, with rapid downdrafts in smaller regions and slow updrafts over large areas. But these results are still quite preliminary.

BALONA: In the pre-Main Sequence phase, a magnetic field inclined to the rotational axis by an angle β will experience a torque by coupling to the circumstellar disk, which will tend to further increase the angle of inclination of the magnetic axis, β. One may, therefore, expect to find a relationship between β and the time spent in the pre-Main Sequence phase, i.e., between β and mass. Is this confirmed by observations?

LANDSTREET: The higher-mass magnetic Bp stars spend of order 10^7 yr as PMS stars. They seem to never be able to slow to rotation periods longer than a couple of weeks, and generally seem to have a large angle between their magnetic and rotation axes. In contrast, the cool magnetic Ap stars, which spend more like 10^8 yr as PMS objects, sometimes exhibit very long rotation periods, of order years or more, and these very slow rotators seem to usually have their field and rotation axes parallel to each other. Perhaps this is the relationship that you are suggesting.

RYABCHIKOVA: Could you comment on the possibility that an A stars may lose a substantial part of its angular momentum while on the Main Sequence?

LANDSTREET: North (1984, A&A, 141, 328) showed from cluster data that Ap Si stars do not appear to lose angular momentum while on the Main Sequence. For cooler magnetic Ap stars, which are (mysteriously) largely absent from clusters, the only strong argument I know of that little or no angular momentum is lost on the Main Sequence is the absence of evidence for circumstellar material that could carry off that angular momentum.

RYABCHIKOVA: If we observe a weakly magnetic peculiar star near or beyond the TAMS, and this star has a very small rotational velocity, does this mean that the star arrived on the Main Sequence with slow rotation?

LANDSTREET: From my comment immediately above, I would be inclined to think that most or all of the angular momentum loss for a magnetic star occurs during the PMS phase. However, for an individual star, one cannot rule out the possibility of interaction during the star's life with an interstellar cloud which could be given some of the angular momentum.

The A-Star Puzzle
Proceedings IAU Symposium No. 224, 2004
J. Zverko, J. Žižňovský, S.J. Adelman, & W.W. Weiss, eds.

© 2004 International Astronomical Union
DOI: 10.1017/S1743921304004855

Diffusion and magnetic field effects
on stellar surfaces

Oleg Kochukhov

Institut für Astronomie, Universität Wien, Türkenschanzstraße 17, 1180 Wien, Austria
email: kochukhov@astro.univie.ac.at

Abstract. Chemically peculiar stars are ideal astrophysical laboratories for furnishing a wide range of theoretical and observational aspects of our knowledge of the physical processes in stars. Recent dramatic improvements in the quality of the observational data and refinements of the modelling techniques led to an emergence of a new branch of stellar astrophysics which is focused on the reconstruction and the understanding of the origin of the three-dimensional structures in stellar surface layers. In this contribution I present an overview of recent results of the detailed modelling of the chemical nonuniformities, the magnetic and the pulsation velocity fields in the atmospheres of A stars. New Doppler imaging analyses of the magnetic field and the chemical inhomogeneities reveal an unexpected complexity of the surface formations and suggest that nonmagnetic phenomena play an important role in shaping the geometry of chemical spots. Consideration of the line profile shapes observed at high spectral and time resolution has made it possible to probe the radial dependence of the chemical abundances and the pulsation characteristics of cool pulsating Ap stars. An extension of Doppler mapping to the reconstruction of non-radial stellar oscillation structure delivers a solution of the long-standing problem of the pulsational geometry of roAp stars and helps to elucidate the interrelation between the pulsations, the magnetic field and the stellar rotation.

Keywords. Stars: magnetic fields, stars: chemically peculiar, stars: oscillations, stars: spots, stars: individual (α And, α^2 CVn, α Cir, 53 Cam, HR 3831)

1. Introduction

The investigation of stellar magnetism and its influence on the stellar surfaces and the envelopes is a quickly maturing research area. In the past different observational aspects of A stars, such as the stellar rotation, the atmospheric structure, the magnetic field and the chemical inhomogeneities induced by radiative diffusion, were considered separately, using simple phenomenological models. In contrast, quite a few recent observations and modelling efforts established a very close and sometimes surprising connection between the different phenomena observed in the peculiar A stars. These important developments emphasize the necessity for self-consistent models based on elaborate theoretical foundations and suggest that we are witnessing the emergence of a new branch of stellar astrophysics focused on the comprehensive analysis of the origin and the evolution of 3-D formations in stellar atmospheres.

In this contribution I highlight the results of some recent analyses of the complex interrelations between the magnetic field, the chemical structures and the non-radial pulsation in the atmospheres of CP stars. I focus on the studies which concentrated on the detailed modelling of the shapes and the variability of stellar line profiles.

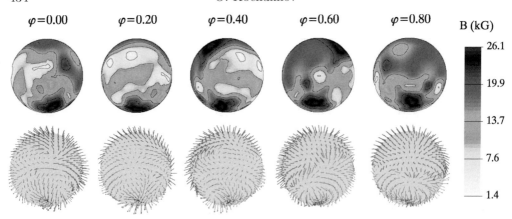

Figure 1. Distribution of the magnetic field strength (upper panel) and the field orientation (lower panel) over the surface of 53 Cam as derived from the Stokes $IQUV$ profiles of three Fe II lines. The star is shown at five equidistant rotation phases as indicated at the top of the figure. The aspect corresponds to the inclination angle $i = 123°$ and a vertically oriented rotation axis.

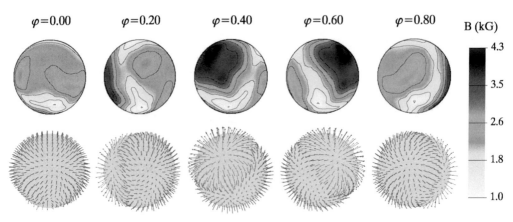

Figure 2. Same as Fig. 1 for the maps of the magnetic field strength (upper panel) and field orientation (lower panel) derived for α^2 CVn using the Stokes $IQUV$ profiles of two lines of Fe II and one line of Cr II. The inclination angle of the stellar rotation axis is $i = 130°$.

2. Magnetic field topologies

For many decades studies of the line polarization in the magnetic stars were limited to the Stokes I and V spectra and relied on a statistical multi-line analysis carried out under a number of restrictive approximations (e.g. Mathys 1991). This situation changed drastically a few years ago with the possibility of acquiring high-resolution spectra in all four Stokes parameters. The most important contribution came from the MuSiCoS spectropolarimeter which was used to obtain the first systematic 4 Stokes parameter observations of the magnetic CP stars (Wade *et al.* 2000). To match this improvement in the quality of the data, several groups have developed computer codes for realistic synthesis of the stellar polarization spectra (Wade *et al.* 2001). A more challenging problem of the magnetic inversion, a reconstruction of the stellar magnetic geometry from the Stokes parameter observations, was addressed by Piskunov & Kochukhov (2002). The magnetic Doppler imaging code that we have developed makes it possible to derive magnetic maps self-consistently with the distributions of the chemical elements. Most importantly, in mapping the magnetic geometry we do not use a multipolar field model and, thus, attempt to reconstruct the vector field distribution making no *a priori* assumptions.

The first inversion of the stellar high-resolution spectra in all 4 Stokes parameters was carried out for the well-known magnetic CP star 53 Cam (Kochukhov *et al.* 2004a). Bagnulo *et al.* (2001) demonstrated that the previous multipolar models of the magnetic field in this star were unable to reproduce the polarization profiles observed with MuSiCoS. Even a direct fit of a general dipole plus quadrupole superposition to the Stokes parameter observations fails to match the low amplitude of the linear polarization signal. The failure of such a simple magnetic distribution suggests that the magnetic topology is considerably more complicated. Kochukhov *et al.* (2004a) obtained a satisfactory model of the field in 53 Cam using magnetic Doppler imaging. With this method the magnetic map has been reconstructed directly from the Stokes $IQUV$ parameters, without relying on a low-order multipolar description of the field. This model successfully reproduces the Stokes parameter spectra and the other magnetic observables of 53 Cam.

The resulting map of the field strength (Fig. 1) appears to be strikingly complex. However, the overall distribution of the field orientation is still dominated by just two large regions of inward and outward directed field. What are the physical effects which gave rise to such a field distribution and how typical is it for other magnetic stars? At the moment only one additional star, α^2 CVn, has been analysed in all 4 Stokes parameters (Kochukhov *et al.* , in preparation). When we applied the same inversion technique as was employed for 53 Cam, a somewhat less structured magnetic map was found sufficient to fit the Stokes spectra of α^2 CVn. The respective spherical maps of the magnetic field strength and orientation are illustrated in Fig. 2. A comparative numerical analysis of the magnetic maps derived for the two stars confirms the impression that the magnetic field in α^2 CVn is simpler. Thus, we find that some CP stars possess a much more complicated field geometry than commonly believed. At the same time, there also seems to exist a large intrinsic scatter in the level of field complexity among different magnetic stars.

3. Surface chemical structures

In contrast to the Stokes parameter inversion, abundance Doppler mapping is a well-known tool, which has been applied to spotted Ap stars for many years. Recent progress in this field stems from improvements in the numerical methods and in the quality of observational material. These two factors allow one to construct more accurate chemical maps for a larger number of elements. Recently Kochukhov *et al.* (2004b) completed a very detailed abundance Doppler imaging study of the roAp star HR 3831. Our analysis is the most thorough examination of the surface chemical patterns for an Ap star. Maps of 17 chemical elements were recovered. Some of these distributions are based on the exceedingly high quality spectra collected by Kochukhov & Ryabchikova (2001b) ($R =$ 123 000, $S/N > 500$, 36 rotation phases), which enabled us to explore the surface spot pattern down to the fundamental resolution limit of Doppler imaging.

Fig. 3 shows an example of the chemical maps reconstructed for HR 3831. Only for a few elements do we find an obvious correlation with the magnetic geometry. For instance, Li seems to be accumulated at the poles of a dipolar field and O is enhanced at the magnetic equator. These nicely symmetric distributions are exceptional. Almost all other elements, for instance iron (Fig. 3), are distributed in a rather complex manner and do not follow the symmetry of the dominant dipolar magnetic field component of HR 3831.

Ba has one of the most high-contrast distributions. This element is concentrated in a series of very small spots which are found at the magnetic poles and at the intersection of the magnetic and the rotation equators. The geometry of the surface chemical patterns is

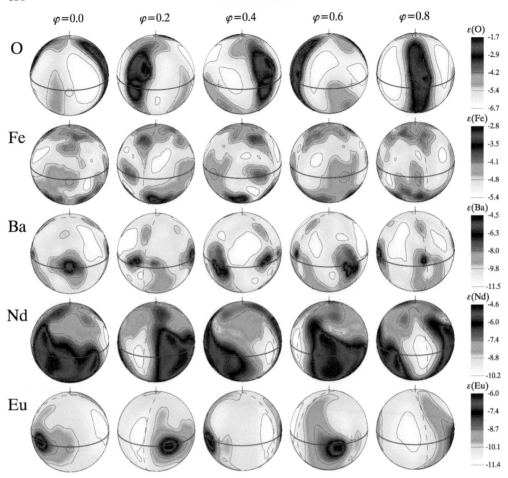

Figure 3. Surface distributions of O, Fe, Ba, Nd and Eu reconstructed for the roAp star HR 3831. The star is shown at five equidistant rotation phases at the inclination angle $i = 68°$ and a vertically oriented axis of rotation. In the spherical plots darker areas correspond to higher element concentration. The thick solid line shows the location of the rotational equator, whereas the dashed line, plus sign and circle correspond, respectively, to the magnetic equator and poles of the dipolar magnetic field in HR 3831.

extremely diverse. Even elements with similar atomic weights often show vastly dissimilar maps. It was commonly believed that all rare-earth elements have approximately similar surface distributions and are concentrated at the magnetic poles. This is certainly not true for HR 3831. Eu shows small spots, but they are noticeably displaced from the magnetic poles. On the other hand, Nd avoids the magnetic equator, but is present elsewhere at the surface. These results demonstrate the complexity of the diffusion in Ap stars and discard a naïve view of the chemical spots and rings symmetric with respect to magnetic field. Abundance distributions are clearly affected by a range of poorly understood phenomena, which are not directly related to magnetic field.

More compelling evidence for the nonmagnetic modulation of the radiative diffusion has emerged with the discovery of surface inhomogeneities in the brightest Hg-Mn binary star α And. Hg-Mn stars are known to be nonmagnetic (Shorlin et al. 2002), and α And is not an exception from this rule. Recent spectropolarimetric observations by Wade et al. (2004) find no longitudinal field above 50 G and establish an upper limit of the global field at about 300 G, which is roughly an equipartition field for α And. Yet, Adelman *et al.* (2002) reported convincing observations of the variability of the resonance Hg II $\lambda3984$ line and reconstructed the surface distribution of mercury. The concentration of this element was found to increase from the rotation poles towards the equator. A clear trend of abundance versus latitude may be connected with the change of surface gravity and resulting latitudinal modulation of the diffusion processes due to a fairly rapid rotation of α And. This does not explain the origin of the longitudinal abundance variation, which is observed as a series of 4 spots of Hg overabundance at or near the equator. Perhaps, the longitudinal structure is caused by the dynamic tides induced by the secondary star in the α And system, or it may be a signature of an undetected weak magnetic field. In any case, the surprising α And phenomenon shows how incomplete is our understanding of diffusion even when we deal with the brightest nonmagnetic stars.

4. Spectroscopic signatures of non-radial pulsation

The most outstanding characteristic of the spectroscopic pulsational variability in magnetic Ap stars is the diversity of pulsational changes observed in lines of different elements. Pulsations are clearly detected in lines of singly and doubly ionized rare-earth elements (REE) but typically do not show up in other lines (e.g., Kochukhov & Ryabchikova 2001a). Time-resolved radial velocity (RV) analyses demonstrated differences of almost a factor of 100 in amplitude and up to 180° in phase between the variability of different lines. Nothing like that is observed in normal Main Sequence pulsating stars. It was anticipated that the surface abundance spots may be responsible for this diversity of pulsational behaviour. However, Doppler mapping of the roAp star HR 3831 (Kochukhov *et al.* 2004b) proved this hypothesis to be wrong. The chemical elements showing maximal pulsational variability are not distributed over the surface in any special way.

Looking for an explanation of the puzzling roAp pulsation signature, we have established a clear relation between the oscillations and the stratification of the chemical abundances (Ryabchikova *et al.* 2002). All lines with high pulsation amplitude belong to REEs, which are affected by an extreme stratification and are probably formed high in the atmospheres of roAp stars. On the other hand, light and iron-peak elements tend to be concentrated in deep layers and show little pulsational variability. Thus, chemical stratification acts as a spatial filter for observations of the pulsational fluctuations and allow us to resolve the vertical structure of *p*-modes. This is done with the pulsation tomography technique, which is a method to study depth dependences of the pulsation properties using lines formed at different depth. A key point here is that the line formation depths in roAp stars are strongly modified by the chemical gradients. Therefore, the derivation of the vertical chemical profiles is a crucial ingredient of the roAp pulsational analysis.

Fig. 4 demonstrates an example of the application of the pulsation tomography method to the roAp star γ Equ. The left panel shows the pulsation RV amplitude at the formation depth of the selected spectral lines and the RV bisector measurements at different levels of the Hα core. A recent analysis by Mashonkina *et al.* (2005) indicated that NLTE effects become significant for the REEs concentrated in the upper atmospheric layers. They found that the bottom of the REE cloud and the formation depth of the corresponding lines is shifted to about $\log \tau_{5000} = -4$ when NLTE effects are accounted for. This leaves us with a very clear picture of a rapid increase of the pulsation amplitude towards the upper atmosphere of roAp stars. This behaviour has no theoretical explanation so far.

Interpretation of the pulsation phases is a more difficult task since it crucially depends on the accuracy that we can achieve in modelling the difference between individual REE

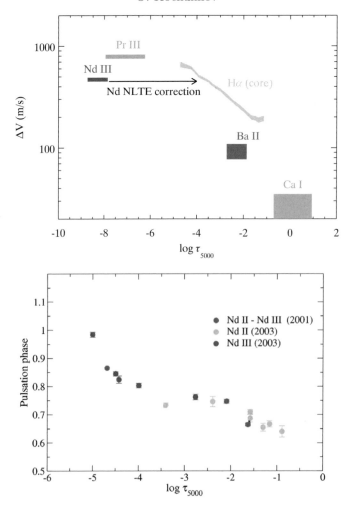

Figure 4. Pulsation tomography reconstruction of the vertical pulsation mode structure in the roAp star γ Equ. Left panel: pulsation amplitude as a function of optical depth derived from the RV variation of metal lines and bisector measurements in the Hα core. Right panel: pulsation phase as a function of the formation depth of Nd II and Nd III lines calculated taking into account NLTE effects and the vertical stratification of Nd.

lines. It appears that a NLTE analysis is compulsory for revealing the variation of pulsation phase with depth. The right panel in Fig. 4 shows results of the calculation for neodymium in γ Equ (Mashonkina *et al.* 2005). The relative phase of the RV variation is plotted as a function of formation depth calculated taking into account stratification and NLTE effects. There is a convincing evidence for the depth-dependence of pulsation phase. This means that in the REE-rich atmospheric layers of γ Equ we observe a running pulsation wave propagating upwards.

5. Surface mapping of stellar non-radial pulsations

The vertical diffusion has a very prominent effect on the oscillations and enables us to recover the vertical mode structure. This information has to be complemented by the horizontal picture to construct a comprehensive 3-D map of the pulsational disturbances and to understand the interaction between the global magnetic field and the non-radial

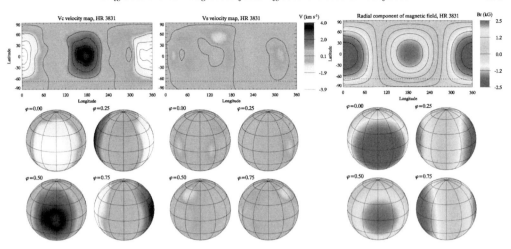

Figure 5. The first Doppler image of a stellar non-radial pulsation velocity field. Left and center: rectangular and spherical projections of the pulsation velocity amplitude maps derived for HR 3831. Right: distribution of the vertical component of magnetic field in HR 3831.

pulsations. This is a classical problem for roAp research. However, the traditional photometric time-series observations are not informative enough to tell us any details beyond a rather unspecific inference that the pulsations are probably oblique and are roughly described by axisymmetric dipolar modes (e.g., Kurtz 1990). Spectroscopic modelling is the key to resolving the horizontal mode structure and verifying conflicting theoretical predictions of the distortion of pulsation modes by the magnetic field and the rotation.

A recent breakthrough in understanding the roAp pulsation geometry has been achieved with the help of Doppler mapping of the pulsation velocity field. I developed a new Doppler imaging code (Kochukhov 2004, 2005) to recover the non-radial pulsation structure directly from the variability of line profiles. The new inversion procedure takes into account surface chemical inhomogeneities and does not assume a spherical harmonic basis for pulsation. To describe a time-dependent pulsation field, I represent each component of the velocity vector with the two constant surface maps, e.g., $V_r(t, \theta, \varphi) = V_r^C(\theta, \varphi) \cos(\omega t) + V_r^S(\theta, \varphi) \sin(\omega t)$. This is equivalent to mapping distribution of both pulsation amplitude and phase over the stellar surface.

The new pulsation mapping technique was applied to the high-resolution time series observations of the Nd III λ6145 line in the spectrum of HR 3831. It was possible to obtain a satisfactory fit to the profile variability with only the vertical pulsation displacement. Fig. 5 shows the rectangular and the spherical projections of the first Doppler image of the stellar pulsation velocity and compares the pulsation and the magnetic field maps. These surface images show that the pulsations are clearly inclined with respect to the rotation axis. This is the first direct evidence for oblique pulsations in stars. Furthermore, the obliquity angle is close to $90°$, the pulsations are almost axisymmetric and the pulsation axis coincides with the axis of the dipolar magnetic component. This is, essentially, the standard oblique pulsator model of Kurtz (1982) suggested long time ago. However, pulsation mapping derives this structure directly from the observations, whereas previously it was adopted *ad hoc*.

A closer look at the pulsation maps reveals some nontrivial and significant deviations from a pure oblique dipolar mode geometry. Pulsations are found to be strongly confined to the magnetic poles. Harmonic expansion in the magnetic reference frame demonstrates that the pulsation structure of HR 3831 can be roughly reproduced with a superposition of $\ell = 1$ and $\ell = 3$ axisymmetric components, in a very good agreement with the recent theoretical prediction by Saio & Gautschy (2004). On the other hand, the nonaxisymmetric $\ell = 1$ components are not needed for the description of pulsations in HR 3831. The revised oblique pulsator model of Bigot & Dziembowski (2002) predicted such components as a result of the influence of stellar rotation. These authors also argued that pulsations are not aligned with magnetic field. Neither prediction is confirmed, hence the model of Bigot & Dziembowski (2002) is not applicable to HR 3831. This can be taken as an indication that the magnetic field has a much stronger effect on the mode geometry compared with the stellar rotation.

Acknowledgements

I would like to acknowledge the Lise Meitner fellowship granted by the Austrian Science Fund (project No. M757-N02).

References

Adelman, S. J., Gulliver, A. F., Kochukhov, O. P., & Ryabchikova, T. A. 2002, *ApJ*, 575, 449
Bagnulo, S., Wade, G. A., Donati, J.-F., *et al.* 2001, *A&A*, 369, 889
Bigot, L., & Dziembowski, W. A. 2002, *A&A*, 391, 235
Kochukhov, O., & Ryabchikova, T. 2001a, *A&A*, 374, 615
Kochukhov, O., & Ryabchikova, T. 2001b, *A&A*, 377, L22
Kochukhov, O., Piskunov, N., Ilyin, I., Ilyina, S., & Tuominen, I. 2002, *A&A*, 389, 420
Kochukhov, O. 2004, *A&A* in press
Kochukhov, O., Bagnulo, S., Wade, G. A., *et al.* 2004a, *A&A*, 414, 613
Kochukhov, O., Drake, N. A., Piskunov, N., & de la Reza, R. 2004b, *A&A* in press
Kochukhov, O. 2005, *These Proceedings*, GP20
Kurtz, D. W. 1982, *MNRAS*, 200, 807
Kurtz, D. W. 1990, *ARA&A*, 28, 607
Mashonkina, L., Ryabchikova, T., & Ryabtsev, A. 2005, *These Proceedings*, 315
Mathys, G. 1991, *A&AS*, 89, 121
Piskunov, N., & Kochukhov, O. 2002, *A&A*, 381, 736
Ryabchikova, T., Piskunov, N., Kochukhov, O., *et al.* 2002, *A&A*, 384, 545
Saio, H., & Gautschy, A. 2004, *MNRAS*, 350, 485
Shorlin, S. L. S., Wade, G. A., Donati, J.-F., *et al.* 2002, *A&A* , 392, 637
Wade, G. A., Donati, J.-F., Landstreet, J. D., & Shorlin, S. L. S. 2000, *MNRAS*, 313, 823
Wade, G. A., Bagnulo, S., Kochukhov, O., *et al.* 2001, *A&A*, 374, 265
Wade, G. A., Abecassis, M., Auriere, M., *et al.* 2004, in *Magnetic Stars 2003*, eds. Yu. V. Glagolevskij and I.I. Romanyuk, in press

Discussion

RYABCHIKOVA: Do you think that the more complex topology of the magnetic field in HR 3831, compared to that of 53 Cam, may explain the diversity of abundance maps?

KOCHUKHOV: There is no indication that the field in HR 3831 is substantially more complicated than a dipole. On the other hand, we clearly see that a complex field can coexist with a fairly simple abundance structures in 53 Cam, whereas small-scale spots in HR 3831 seem to have no counterpart in the magnetic geometry. Thus, a relation between magnetic and chemical structures is not as straightforward as it was commonly believed.

MATHYS: How do you deal with the isotopic effect in mapping Hg on the surface of α And?

KOCHUKHOV: In mapping the Hg distribution on α And we recovered the surface distribution of the isotope mixture using the classical q-parameterization. We found no evidence for large surface variations of the isotope composition.

ALECIAN: Is there any indication of changes of the magnetic topology with time? In other words, have you observed the same star after one or two years to see if the reconstructed structures are the same?

KOCHUKHOV: The first magnetic maps of Ap stars derived from all 4 Stokes parameters were based on a unique observational material collected over the time-span of 3-4 years. We had no chance to collect two or more independent datasets for any of the targets. Nevertheless, we are certainly planning to reobserve 53 Cam to check the stability of the small-scale magnetic structures detected with our Stokes inversion technique.

MICHAUD: When we made a model for 53 Cam, Babel and I could reproduce the time averaged abundances observed in that star with no arbitrary parameters, using only the observed magnetic field and the known stellar properties. However, Babel had to introduce a mass loss at the magnetic poles to reproduce approximately the then observed spots on 53 Cam. I do not know how this should be changed by the much more precise magnetic and abundance maps now available.

KOCHUKHOV: The abundance distributions of 53 Cam derived with new techniques do not differ considerably from the earlier maps. The major change is in the field structure. However, despite the complexity of the field, it is still dominated by the two large regions of inward and outward directed magnetic flux, which is probably the most relevant for confining the mass loss. Thus, it may not be necessary to modify the theoretical diffusion model dramatically. But it is certainly worth continuing the work you have initiated with Babel and consider these diffusion calculations in the context of the new surface maps.

The A-Star Puzzle
Proceedings IAU Symposium No. 224, 2004   © 2004 International Astronomical Union
J. Zverko, J. Žižňovský, S.J. Adelman, & W.W. Weiss, eds.   DOI: 10.1017/S1743921304004867

The λ Bootis stars

Ernst Paunzen

Institute for Astronomy, University Vienna, Türkenschanzstr. 17, A-1180 Vienna, Austria
email: Ernst.Paunzen@univie.ac.at

Abstract. A comprehensive overview of the λ Bootis stars is presented. This small group (only a maximum of 2% of all objects in the relevant spectral domain) are Population I late B to early F-type stars, with moderate to extreme (up to a factor 100) surface underabundances of most Fe-peak elements and solar abundances of lighter elements (C, N, O, and S). They form a separate group among the classical chemically peculiar objects of the upper Main Sequence as their underabundances are quite outstanding. The basic membership criteria as well as the theories to explain the λ Bootis phenomenon are covered. New observations including detailed elemental (surface) abundances, tests for pulsational instability, fluxes in the IR region, and measurements from the Hipparcos satellite permit one to make a statistical sound analysis of the group properties. Details of the evolutionary status, abundance pattern and tests for developed theories are presented.

Keywords. Stars: abundances, chemically peculiar, fundamental parameters, oscillations

1. Introduction

λ Bootis stars are Population I late B to early F-type stars, with moderate to extreme (up to a factor 100) surface underabundances of most Fe-peak elements and solar abundances of lighter elements (C, N, O, and S). Only a maximum of about 2% of all objects (about 50 stars in total) in the relevant spectral domain are believed to be λ Bootis type stars as referenced by Paunzen *et al.* (2001) and Gray & Corbally (2002). Although the number of members is small, the group is an excellent test field for various astrophysical theories such as accretion, diffusion, mass-loss and asteroseismology.

2. Classification resolution spectroscopy and membership criteria

The following definition is primarily from Gray (1988). It summarizes the basic features of moderate dispersion spectra which seem to be shared by all well established λ Bootis stars:

• The λ Bootis stars are early-A to early-F type stars with an approximate spectral type range (based on the hydrogen lines) of B9.5 to F0 with possible members as late as F3.

• The λ Bootis stars are characterized by weak Mg II λ4481 lines, such that the ratio Mg II λ4481/Fe I λ4383 is significantly smaller than in normal stars. In addition the spectra exhibit a general metal-weakness. The typical shell lines such as, Fe II λ4233, tend to be weak as well. But the λ Bootis stars do not show the typical shell spectral characteristics.

• The following classes of stars should be excluded from the λ Bootis group even if they show weak λ4481 lines: shell stars, protoshell stars, He-weak stars (easily distinguished on the basis of their hydrogen line temperature types), and other CP stars. FHB and intermediate Population II stars may be distinguished from the λ Bootis stars on the basis of their hydrogen line profiles. High-$v \sin i$ stars should be considered as λ Bootis candidates only if the weakening of $\lambda 4481$ is obvious with respect to the standards with high values of $v \sin i$.

- λ Bootis stars are also characterized by broad hydrogen lines, and in many cases, these hydrogen lines are exceptionally broad. In the late-A and early-F λ Bootis stars, the hydrogen line profiles are often peculiar, and are characterized by broad wings, but shallow cores.
- The distribution of rotational velocities of the λ Bootis stars cannot be distinguished from that of the normal Population I A-type stars.

This definition already includes the basic criteria for a successful classification of a true λ Bootis star. A highly underestimated effect on the classification process is the influence of fast rotation on the derived spectral type. Slettebak *et al.* (1980) have shown that the equivalent width of $\lambda 4481$ decreases with increasing rotation for models later than A0 (see also, Abt & Morrell 1995). A much stronger effect was found for Hγ. A fast rotating star will be classified much later using Hγ than using Mg II $\lambda 4481$. One would classify a rapidly rotating true A5 type star as hF1mA7 or metal-weak.

Paunzen (2001) performed an extensive classification resolution spectroscopic survey in the Galactic field and nine open clusters to detect new λ Bootis stars. He confirmed the membership of 18 candidates and detected 26 new λ Bootis stars. The survery by Gray & Corbally (2002) was devoted to open clusters. They noticed the apparent absence of λ Bootis stars in 24 open clusters with ages from 15 Myr to 700 Myr.

The list of members has steadily evolved since the compilation of Gray (1988) to that of Paunzen *et al.* (2002a). This evolution is purely based on observations which are fully documented in each reference. An uncritical evaluation of members as used in Gerbaldi *et al.* (2003) is not appropriate for drawing statistical conclusions about the incidence of undetected spectroscopic binary systems or the heterogeneity of this group.

3. The abundance pattern of λ Bootis stars.

Only recently have abundances become available for about half of all λ Bootis stars. Heiter (2002) draw conclusions about the abundance pattern of the group, investigated their statistical properties and compared the abundances with the interstellar medium. In a few cases, the abundances differ by more than 0.6 dex and these have been discarded. Abundances for spectroscopic binaries which have been treated as single stars have not been taken into account. The following conclusions were made:

- The abundances of C, N and O are *on the average* solar, but considerable under- and overabundances (-0.8 and $+0.6$ dex) occur as well.
- The star-to-star scatter for the element abundances of C, N, O and S is smaller than for most of the heavier elements.
- The mean abundance of Na is also solar, but the star-to-star scatter is ± 1 dex.
- The iron peak elements are more underabundant than C, N, O, and S for each star.
- The iron peak elements from Sc to Fe as well as Mg, Si, Ca, Zn, Sr and Ba are depleted by about -1 dex relative to the solar chemical composition.
- Al is slightly more depleted (-1.5 dex) and Ni, Y and Zr are slightly less depleted.
- The star-to-star scatter is twice as large as for a comparable sample of normal stars.

The results suggest the existence of a separate chemically peculiar group of "λ Bootis stars" with a characteristic abundance pattern. The scatter within the abundance pattern is, for example, comparable to that of Am/Fm stars. Furthermore, the comparison to normal stars is difficult because the sample of "normal" Main Sequence stars with known abundances and parameters similar to that of the λ Bootis stars is rather limited.

Another way of characterizing the chemical composition of λ Bootis stars is to examine abundance ratios between different elements. The most important finding was an anticorrelation between the ratios of light elements (C, N, O and S) to heavier elements and the heavy element abundances. [C/Fe] ≫ 0 holds for all λ Bootis stars and the iron abundances are significantly lower than those found for superficially normal stars. On the other hand, there is an overlap with normal stars for other heavier elements found by Paunzen *et al.* (2002a).

4. The characteristics of λ Bootis stars in the UV region

To summarize previous efforts, Faraggiana *et al.* (1990) tried to establish criteria for the definition of a homogeneous group of λ Bootis stars.

Solano & Paunzen (1998, 1999) presented careful analyses of the IUE final archive using the low and high resolution spectra separately. Their goal was to establish criteria in the UV region without using data from the optical domain and to unambiguously distinguish λ Bootis from FHB stars.

From the low resolution spectra they concluded that a star is considered to be a member of the λ Bootis group if:

- C I λ1657/Al II λ1670 > 2.0
- C I λ1657/Ni II λ1741 > 2.5

These limits have been set so that they are valid for the entire range of effective temperatures. Small uncertainties due to variations in $\log g$ and [M/H] have also been taken into account to make these limits independent of physical parameters such as the effective temperature and the surface gravity. Moreover, they are also independent of the evolutionary status since a clear distinction between λ Bootis and FHB stars was established. From the analysis of the high resolution spectra they were able to add the following criteria for a star being a member of the λ Bootis group:

- C I λ1657/Si II λ1527 > 3.0
- C I λ1657/Ca II λ1839 > 8.0

These four unambiguous criteria in the UV region reflect the typical abundance pattern for this group (Section 3).

5. The characteristics of λ Bootis stars in the IR region

King (1994) and Fajardo-Acosta *et al.* (1999) gave brief summaries of some observations available in the IR. From these references it was already clear that 1) only very few λ Bootis members were detected in the IR and 2) even fewer objects show infrared excesses.

Paunzen *et al.* (2003) presented all available data for members of the λ Bootis group in the wavelength region beyond 7000 Å. The data include spectrophotometry, Johnson *RIJHKLM* and 13-color photometry as well as IRAS and ISO measurements. The data were homogeneously reduced and transformed into a standard system given in units of Jansky. In total, measurements for 34 (26 with data redward of $20\,\mu$m) well established λ Bootis stars are available. From those 26 objects, 6 show an infrared excess and 2 doubtful detections resulting in 23 % showing signs of circumstellar or interstellar material around them.

For four stars the infrared excess was detected at two different wavelengths which allowed one to fit the data by a simple model. The derived dust temperatures are between 70 and 200 K and the fractional dust luminosities range from 2.2×10^{-5} to 4.3×10^{-4}. These values are comparable with those found for Vega-type objects.

ISO-SWS spectroscopy for HD 125162 and HD 192640 resulted in the detection of pure stellar H I lines ruling out an active accretion disk around these objects. Furthermore, a search for the CO $(2-1)$ line at 238.538 GHz and continuum observations at 347 GHz of three λ Bootis stars yield only upper limits.

6. Pulsational characteristics

In general, chemical peculiarities found for stars on the Upper Main sequence exclude δ Scuti type pulsation (e.g., Ap and Am stars), but for the λ Bootis stars it is just the opposite which makes them very interesting for asteroseismological investigations.

Paunzen *et al.* (2002b) summarize the overall pulsational characteristics of the group members (with only two so far not photometrically investigated). They have analyzed the pulsational stability of the individual objects with typical upper limits for bona-fide "constant" objects of about 2 mmag using also the photometric data of the Hipparcos satellite.

From all well established members (not all within the classical instability strip), 33 have been found to show typical δ Scuti type pulsation. With the already published abundance values, they have also searched for the existence of a Period-Luminosity-Color-Metallicty relation. They compared the characteristics of the λ Bootis group to a sample of "normal" δ Scuti pulsators. The latter was chosen such that it matches the λ Bootis stars within the global astrophysical parameters. The following properties of the λ Bootis stars are different from those of the δ Scuti pulsators:

• At least 70% of all λ Bootis type stars inside the classical instability strip pulsate (this number for "normal" stars is much less)

• Only a maximum of two stars may pulsate in the fundamental mode but there is a high percentage with $Q < 0.020$ d (high overtone modes). For δ Scuti stars this is just the opposite.

• The instability strip of the λ Bootis stars at the ZAMS is 25 mmag bluer in $(b-y)_0$ than that of the δ Scuti stars.

No clear evidence for a [Z] correlation with period, luminosity or color was found. The Period-Luminosity-Color relation is the same as for the classical δ Scuti stars.

Bohlender *et al.* (1999) performed high resolution, high signal-to-noise spectroscopy (spectra centered at Na D λ5892 and/or λ4500) of λ Bootis stars to investigate the incidence of high-degree nonradial pulsation. The discovered spectrum variability is very similar to that seen in rapidly rotating δ Scuti stars. For all but one of the investigated pulsators, high-degree nonradial modes were detected spectroscopically.

Photometric as well as spectroscopic variability is a common phenomenon in λ Bootis stars. Several λ Bootis stars are therefore excellent candidates for multisite photometric and spectroscopic campaigns or space based observations (e.g., MOST and COROT) to put further constraints on stellar interior and pulsation models.

Up to now, five λ Bootis stars were targets of multisite campaigns: HD 15165, HD 105759, HD 111786, HD 142994 and HD 192640, see Paunzen *et al.* (2002b) for more details. For none of these objects a mode identification was possible up to now.

7. The evolutionary status of the group

Paunzen *et al.* (2002a) have used all currently available photometric data as well as Hipparcos data to determine astrophysical parameters such as the effective temperatures, surface gravities and luminosities. Masses and ages were calibrated within appropriate post-MS evolutionary models. Furthermore, Galactic space motions were calculated with the help of radial velocities from the literature. All results were compared with those of a test sample of normal-type objects in the same spectral range chosen to match the $(B - V)_0$ distribution of the λ Bootis group. From a comprehensive statistical analysis they conclude:

- The standard photometric calibrations within the Johnson *UBV*, Strömgren *uvbyβ* and Geneva 7-colour systems are valid for this group of chemically peculiar stars.
- The group of λ Bootis stars consists of true Population I objects which can be found over the whole area of the MS with a peak at a rather evolved stage (≈ 1 Gyr). That is in line with the distribution of the test sample.
- The λ Bootis type group is not significantly distinct from normal stars except, possibly, by having slightly lower temperatures and masses for $t_{rel} > 0.8$. The $v \sin i$ range is rather narrow throughout the MS with a mean value of about 120 km s^{-1}.
- There seems to exist a non-uniform distribution of effective temperatures for group members with a large proportion of objects (more than 70%) cooler than 8000 K.
- Objects with peculiar hydrogen-line profiles are preferentially found among later stages of stellar evolution.
- No correlation of age with elemental abundance or projected rotational velocity has been detected.
- A comparison of the stellar Na abundances with nearby IS lines of sight hints at an interaction between the λ Bootis stars and the ISM.
- There is a single mechanism responsible for the observed phenomenon which produces moderate to strong underabundances working continuously from very early (10 Myr) to very late evolutionary stages (2.5 Gyr). It produces the same absolute abundances throughout the MS lifetime for 2% of all luminosity class V objects with effective temperatures from 10500 K to 6500 K.
- The current list of stars seem to define a very homogeneous group, validating the proposed membership criteria in the optical and UV region.

8. Developed theories

8.1. *Diffusion/mass-loss*

The *diffusion/mass-loss* theory was formulated by Michaud & Charland (1986). They have modified the highly successful diffusion model responsible for the AmFm phenomenon to account for stellar mass-loss. AmFm stars are Population I nonmagnetic MS stars with underabundances of Ca and Sc, but large overabundances of most heavier elements (up to a factor of 500). This abundance pattern is explained by the disappearance of the outer convection zone associated with the He-ionization because of the gravitional settling of He. Using Ti and Ca as examples, Michaud & Charland (1986) showed that this model is capable of producing metal underabundances with the ad hoc assumption of a mass-loss rate of $10^{-13} M_\odot$yr^{-1}. This high mass-loss rate impedes He settling enough to prevent the disappearance of the superficial He convection zone. Material originally located deep in the envelope then has time to be advected to the surface. Michaud & Charland (1986) showed that the required underabundances materialize after 10^9yr at the end of the MS lifetime for a normal A-type star. It was often quoted that this model only predicts underabundances of a factor five which is much too low compared to observations. But considering the poorly known parameters such as the diffusion coefficients, the scale heights and the boundaries of the convection zone, it might well be possible to obtain larger underabundances with this model. More of a problem is the rotational velocity and therefore meridional circulation. What manifests in a cut-off rotational velocity for AmFm stars ($\approx 90\,\mathrm{km\,s^{-1}}$) also destroys the predicted λ Bootis pattern. Charbonneau (1993) showed that even a moderate equatorial rotational velocity of $50\,\mathrm{km\,s^{-1}}$ prevents at any time during the MS evolution the appearance of the underabundance pattern because of mixing in the stellar atmosphere.

8.2. Accretion/diffusion

The *accretion/diffusion* model: Venn & Lambert (1990) were the first who noticed the similarity between the abundance pattern of λ Bootis stars and the depletion pattern of the interstellar medium (ISM) and suggested the accretion of interstellar or circumstellar gas to explain the λ Bootis stars. In the ISM metals are underabundant because of their incorporation in dust grains or ice mantles around the dust grains. Waters *et al.* (1992) worked out a scenario, where the λ Bootis star accretes metal-depleted gas from a surrounding disk. In this model, the dust grains are blown away by radiation pressure and coupling between dust and gas is negligible. Considering the spectral-type of λ Bootis stars, the gas in the disk remains neutral and hence does not experience significant direct radiation pressure. The authors showed, that these conditions hold for mass accretion rates below $10^{-8}\,M_\odot\,\mathrm{yr^{-1}}$ assuming that the gas-to-dust mass ratio in the disk is 100 and that the disk consists of 0.1 μm carbon grains. Turcotte & Charbonneau (1993) calculated the abundance evolution in the outer layers of a λ Bootis star assuming accretion rates between 10^{-15} and $10^{-12}\,M_\odot\,\mathrm{yr^{-1}}$. Solving the diffusion equation modified this time by an additional accretion term, they obtained the time evolution of the Ca and Ti abundance stratification with and without stellar rotation. From their calculations, a lower limit of $10^{-14}\,M_\odot\,\mathrm{yr^{-1}}$ is derived for the accretion/diffusion model to produce a typical λ Bootis abundance pattern. Moreover their rotating models provide evidence that meridional circulation cannot destroy the established accretion pattern for rotational velocities smaller than $125\,\mathrm{km\,s^{-1}}$. Since diffusion wipes out any accretion pattern within 10^6 yr a large number of λ Bootis stars should show observational evidence for the presence of circumstellar material. Turcotte (2002) revised this model recently applying new results on the depth of mixing in A-type stars. Richard *et al.* (2001) showed that the peculiar abundance pattern in Am stars (slow rotators) can only be reproduced by models in which the mixed layers extend much deeper than the superficial convection zone. For λ Bootis stars, their study predicts a mixed zone of $\sim 10^{-6}$ times the stellar mass. Inserting this into the accretion/diffusion model increases the accretion rate, which is necessary for the accretion pattern to establish, to 10^{-11} to $10^{-12}\,M_\odot\,\mathrm{yr^{-1}}$ depending on the exact mass of the mixed zone and the rotational velocity of the star. Andrievsky & Paunzen (2000) studied the accretion from a circumstellar shell in more detail. They concluded that the gas and the dust decouple beyond the condensation radius of a λ Bootis star. Assuming that the dust grains form at the condensation radius, grains will grow only to sizes less than 0.01 μm before they are blown out of the shell. Moreover, the shell accretion can only cause a significant alteration of the photospheric abundances if the density in the shell is proportional to r^{-2}. Otherwise the shell contains too much material inside the condensation radius, which is not depleted due to grain formation. All the above discussed scenarios imply a constraint on the evolutionary status of the star, because the existence of a disk or a shell has to be explained in the context of stellar evolution. Circumstellar disks are thought to exist during the pre-main-sequence phase of stellar evolution, while a shell can either occur in a very early phase of pre-main-sequence evolution or after a stellar merger. Kamp & Paunzen (2002) proposed only recently a slightly different accretion scenario for the λ Bootis stars, namely the accretion from a diffuse interstellar cloud. This scenario works at any stage of stellar evolution as soon as the star passes a diffuse interstellar cloud. The interstellar dust grains are blown away by the stellar radiation pressure, while the depleted interstellar gas is accreted onto the star. Typical gas accretion rates are between 10^{-14} and 10^{-10} $M_\odot$ yr^{-1} depending on the density of the diffuse cloud and the relative velocity between star and cloud. The hot limit for this model is due to strong stellar winds for stars with $T_{\mathrm{eff}} > 12000$ K whereas the cool limit is defined by convection which prevents the accreted material to manifest at the stellar surface.

8.3. *Binary theories*

Binary theories: First, Andrievsky (1997) proposed that at least some λ Bootis type stars can originate "by merging" as a result of the dynamical evolution of W UMa contact binary systems. These are close eclipsing binary stars with orbital periods less than one day. Spectral types of both components are almost always similar, mainly in the range between F0 and K0. Due to the angular momentum loss, both components of such a close system approach each other, and finally merge into a more massive but single star. The typical time for merging is poorly defined in the literature with values ranging from 100 − 200 Myr up to 500 Myr (Leonard & Linnell 1992). The most essential point is that the merging occurs before both stars finish their Main Sequence lifetime. Since the masses of λ Bootis stars can be found in the range between 1.5 and 2.6 $M_\odot$ (Paunzen *et al.* 2002a), their progenitors could be W UMa close binary systems with masses of each component between 0.8 and 1.5 $M_\odot$. In a nonconservative case Andrievsky (1997) proposes that some matter could be lost by the system during the merging phase and form a circumstellar shell. The above described scenario offers a simple observational test: the search for CNO-processed material which should be mixed into the photospheres of λ Bootis stars as a results of the merger process. But investigations show no abundance anomalies for C, N, O and S in many λ Bootis stars. Andrievsky's suggestion is an attempt to bring into line the apparently evolved nature of some λ Bootis type stars with the requirements of the accretion/diffusion model. It can be neither proved nor rejected, but clearly points to the model's limitations: slow selective accretion has a short time-scale, protostellar shells or disks would be swept out by a stellar wind soon after (≈ 1 Myr) the ignition of core hydrogen burning and/or photoevaporated by nearby O or B-type stars. An interesting consequence, if some λ Bootis type stars are binary star mergers, is they could have a common origin with the well-known blue stragglers. For a description of the latter see Stryker (1993). On the other hand Holweger *et al.* (1995) found through a detailed NLTE abundance analysis, that two blue stragglers in M67 do not show the typical λ Bootis abundance pattern. Instead these two stars fit in the diffusion picture of normal A-type stars. Later Faraggiana & Bonifacio (1999) and Gerbaldi *et al.* (2003) consider the metal-weak appearance of the λ Bootis spectra, and discuss the possibility that some if not most of the stars of this type are unresolved spectroscopic binaries. Thus, the observed weakness of the metallic lines could be an artefact. The single but composite spectrum of two quite normal (solar abundant) stars with different effective temperatures and gravities will have metal-weak character. The imitation would be even more "realistic" if the components have different rotational velocities. Speckle interferometry by Marchetti *et al.* (2001) and inspection of the isolated spectral features like the O I NIR-triplet at λ7770, NaD lines and hydrogen line cores by Faraggiana *et al.* (2001) are used in an attempt to reveal the signs of binarity in the spectra of λ Bootis type stars. Not surprisingly, only very few stars classified earlier as λ Bootis with prominent shell-like characteristics prove to be binaries (HD 38545), or more complex systems (HD 111786). Undetected binarity gives a simple and attractive explanation of the peculiar hydrogen profiles which are typical for most of the λ Bootis stars. But still it has to be proven that underabundances of up to -2 dex can be achieved by this model.

Acknowledgements

This work benefitted from the Fonds zur Förderung der wissenschaftlichen Forschung, project (S7303-AST and P14984).

References

Abt, H.A., Morrell, N.I. 1995, *ApJS* 99, 135

Andrievsky, S.M. 1997, *A&A* 321, 838

Andrievsky, S.M., Paunzen, E. 2000, *MNRAS* 313, 547

Bohlender, D.A., Gonzalez, J.F., Matthews, J.M. 1999, *A&A* 350, 553

Charbonneau, P. 1993, *ApJ* 405, 720

Fajardo-Acosta, S.B., Stencel, R.E., Backman, D.E., Thakur, N. 1999, *ApJ* 520, 215

Faraggiana, R., Bonifacio, P. 1999, *A&A* 349, 521

Faraggiana, R., Gerbaldi, M., Böhm, C. 1990, *A&A* 235, 311

Faraggiana, R., Gerbaldi, M., Bonifacio, P., Francois, P. 2001, *A&A* 376, 586

Gerbaldi, M., Faraggiana, R., Lai, O. 2003, *A&A* 412, 447

Gray, R.O. 1988, *AJ* 95, 220

Gray, R.O., Corbally, C.J. 2002, *AJ* 124, 989

Heiter, U. 2002, *A&A* 381, 959

Holweger, H., Lemke, M., Rentzsch-Holm, I., Stürenburg, S. 1995, in: M. Busso, C.M. Raiteri and R. Gallino (eds.), *Proceedings of the Symposium held at the National Laboratories of Gran Sasso, Assergi, L'Aquil, Italy, July 1994*, AIP Conference Proceedings, vol. 327, p. 41

Kamp, I., Paunzen, E. 2002, *MNRAS* 335, L45

King, J.R. 1994, *MNRAS* 269, 209

Leonard, P.J.T., Linnell, A.P. 1992, *AJ* 103, 1928

Marchetti, E., Faraggiana, R., Bonifacio, P. 2001, *A&A* 370, 524

Michaud, G., Charland, Y. 1986, *ApJ* 311, 326

Paunzen, E. 2001, *A&A* 373, 633

Paunzen, E., Duffee, B., Heiter, U., Kuschnig, R., Weiss, W.W. 2001, *A&A* 373, 625

Paunzen, E., Iliev, I.Kh., Kamp, I., Barzova, I.S. 2002a, *MNRAS* 336, 1030

Paunzen, E., Handler, G., Weiss, W.W., *et al.* 2002b, *A&A* 392, 515

Paunzen, E., Kamp, I., Weiss, W.W., Wiesemeyer, H. 2003, *A&A* 404, 579

Richard, O., Michaud, G., Richer, J. 2001, *ApJ* 558, 377

Slettebak, A., Kuzma, T.J., Collins, G.W. 1980, *ApJ* 242, 171

Solano, E., Paunzen, E. 1998, *A&A* 331, 633

Solano, E., Paunzen, E. 1999, *A&A* 348, 825

Stryker, L. L. 1993, *PASP* 105, 1081

Turcotte, S. 2002, *ApJ* 573, L129

Turcotte, S., Charbonneau, P. 1993, *ApJ* 413, 376

Venn, K.A., Lambert, D.L. 1990, *ApJ* 363, 234

Waters, L.B.F.M., Trams, N.R., Waelkens, C. 1992, *A&A* 262, L37

The A-Star Puzzle
Proceedings IAU Symposium No. 224, 2004
J. Zverko, J. Žižňovský, S.J. Adelman, & W.W. Weiss, eds.

© 2004 International Astronomical Union
DOI: 10.1017/S1743921304004879

A theorist's view of the A-star laboratory

Hiromoto Shibahashi

Department of Astronomy, University of Tokyo, Tokyo 113-0033, Japan
email: shibahashi@astron.s.u-tokyo.ac.jp

Abstract. Two topics concerning A-type stars are discussed: starspots associated with the strong magnetic fields and the prospects for asteroseismology. Considering starspots as analogous to sunspots is misleading. The photosphere of starspots of a magnetic A-star is higher than the normal photosphere, contrary to sunspots. As for the prospects of asteroseismology, it is demonstrated how well (or poorly) we can probe the internal structure of the distant stars using a limited number of p-mode frequencies. Although the detectable eigenmodes must be limited to be $\ell \leqslant 4$, if the observational error is of the order of 10^{-3}, inversion has still some hope.

Keywords. Stars: chemically peculiar, stars: interiors, stars: oscillations, stars: spots

---

1. Is the level of starspots of Ap stars lower than its surroundings?

1.1. Introduction: Gough's argument

Many Ap stars are known to have strong magnetic fields, which seem to be, roughly speaking, approximately dipole fields. In most cases the magnetic field strength is cyclically varying, and such variation is thought to be caused by the rotation of the star whose magnetic axis is inclined to the rotation axis. The photometric changes are dependent on the color band used in the observations. Typically the star is darkest in the V-band at the phase of magnetic maximum. The correlation between the darkness and the magnetism reminds us of sunspots. We believe that the Ap stars have global dipole-like magnetic fields, of the order of several hundreds Gauss or more, associated with giant starspots, although the mechanism generating such starspots and the nature of their structure have not been established.

The geometrical level of sunspots is slightly lower than the surrounding photospheric level. This can be seen when we see sunspots near the limb. One might think that the situation is similar for the starspots of Ap stars. Considering starspots as the analogs of sunspots is, however, misleading. We need to pay attention to the differences between starspots and sunspots. Gough (2003) pointed out that the starspots of Ap stars must be higher than their surroundings. Starspots are not transient like sunspots and their size is very large, occupying perhaps 20% or more of a hemisphere. With this in mind, he and his colleagues modelled the equilibrium envelope in the magnetic polar regions by suppressing convection and using a normal envelope in the magnetic equatorial zone (Balmforth *et al.* 2001). The outcome of their calculation was that the starspots' photosphere is only slightly cooler than the normal photosphere and that the starspots' photosphere is slightly higher than the normal photosphere. Since this conclusion might be contrary to what one might expect, we carry out a 'Gedankenexperiment' to see if the level of starspots is indeed higher than the normal photosphere.

1.2. Radiative case

Let us consider a radiative envelope. For early A-type stars, the convection zones associated with the hydrogen and the helium ionization zones are thin and shallow. Convective energy transport is inefficient so that the envelope is almost radiative. It is expected that, due to the magnetic pressure, the gas pressure in the magnetic polar regions is lower than in the normal envelope. A decrease in gas pressure would lead to a decrease in temperature, if the density were unchanged. However, the magnetic polar regions may be quite large. Hence the thermal balance on a global scale should be taken into account. A decrease in temperature would lead to a steep temperature gradient, if the radius were kept unchanged. But the temperature gradient is determined by the amount of radiation flux:

$$\boldsymbol{F}_{\mathrm{rad}} = -\frac{4ac}{3\kappa\rho}T^3\nabla T, \tag{1.1}$$

where $\boldsymbol{F}_{\mathrm{rad}}$ denotes the radiative energy flux, T the temperature, κ the opacity, ρ the density, a the radiation density constant, c the speed of the light, and ∇ the gradient operator. To maintain the temperature gradient, the magnetic polar regions have to expand slightly. Consequencely the density becomes lower there. A decrease in gas pressure is mainly compensated by a decrease in density rather than in temperature, and the level of the photosphere of the magnetic polar regions is slightly higher than the normal photosphere. The magnetic polar regions must look like a pair of big plateaus.

1.3. Convective case

For stars such as F-type or late A-type stars, with smaller effective temperatures, the efficiency of convective energy transport becomes more important. Magnetic fields work to suppress convection or to lower the efficiency of convective energy transport. In a uniform sector of a magnetic field, whether the motion in the layer is monotonically growing (dynamically unstable) is dependent on the superadiabaticity and the Alfvén frequency. The latter is dependent on the wavenumber $\boldsymbol{k}$ and the magnetic field $\boldsymbol{B}$. For a dipole field, the magnetic field is almost vertical in the magnetic polar regions and almost horizontal near the magnetic equator of the star. Since $\boldsymbol{B} \cdot \boldsymbol{k}$ is large in the magnetic polar regions, the monotonic convection motion may be suppressed and the motion may be oscillatory there. On the other hand, the motion is supposed to be convectively unstable at the magnetic equator of the star, because $\boldsymbol{B} \cdot \boldsymbol{k}$ is small there.

In the magnetic polar regions, suppression of the convective energy transport efficiency leads to a steep temperature gradient. If the radius of the star were unchanged, the temperature there would become cooler. This means that the total luminosity, $L = \pi R^2 \sigma T_{\mathrm{eff}}^4$, would decrease, where σ and T_{eff} denote the Stefan-Boltzmann constant and the effective temperature, respectively. If we assume that the star is in thermal balance, the luminosity of the star is determined by the nuclear energy generation, which is not influenced by the surface magnetic fields, and should be unchanged. To keep the total luminosity unchanged, the stellar radius has to be slightly larger. Thus the elevation of the photosphere in the magnetic polar regions must be slightly higher than in the equatorial region. The magnetic polar regions are a pair of slightly cool plateaus.

1.4. Summary and discussion

In both cases, we have reached the same conclusion as Gough (2003). The geometric level of the photosphere of starspots is higher than its surroundings, contrary to the case of sunspots. Starspots in the magnetic polar regions should be regarded as a pair of slightly cool plateaus with a slightly lower density. For an Ap star with $T_{\mathrm{eff}} \simeq 8000\,\mathrm{K}$ and $B \simeq 1\,\mathrm{kG}$, the temperature difference between the photosphere of the magnetic polar regions (starspots) and the that of the magnetic equator is estimated to be of the order of several hundreds Kelvin. Since the temperature of the magnetic polar regions is expected to be lower than the magnetic equator of the star, some energy transfer is expected to occur from the magnetic equator to the poles. This might cause chemical diffusion in the horizontal direction. The lower density at the magnetic polar regions enhances the pulsation amplitude there compared to the magnetic equator.

2. Prospects of asteroseismology: how much information can we get from asteroseismology?

2.1. *Introduction*

Stimulated by the success of helioseismology, a similar attempt to probe the internal structure of stars in general has been encouraged. Uninterrupted, long-term observations with high precision photometry or Doppler measurements from space have been proposed. MOST is indeed now working (Matthews *et al.* 2004). A-type stars should be among the target stars of asteroseismology. Many δ Scuti stars are rich in observed oscillation frequencies. roAp stars are unique objects as they have very strong magnetism and chemical peculiarity. However, the seismological approach to stars in general is much more difficult than that for the Sun. The great success of helioseismology is based on: (1) the large number of oscillation modes that have been detected, (2) the modes have been well identified, and (3) a good solar model is available which can be used as a reference model for the inversion procedure. The situation is different for distant stars. The stellar image usually cannot be resolved into a two-dimensional disk image. For a star oscillating with a high ℓ mode, the stellar surface is divided into many small regions oscillating in different phases, and then the contributions of each region are canceled by others so that the total amplitude of the variability of the star is too small to be detected. As a consequence, the observable eigenmodes in an individual star are likely to be restricted to only those with $\ell \leqslant 4$. In this situation, how much information can we get by asteroseismology about the invisible interior of stars? What is the limit?

2.2. *Numerical experiments*

Now by assuming that the mode identification is complete and that a good model is available, I demonstrate how well we can probe the invisible interior of stars by utilizing a limited number of oscillation modes. The principal aim is to find the limit of seismic inversion in the best case. We can use simple stellar models suitable for this purpose, otherwise we would have to worry about various uncertainties in the physics. Let us adopt two polytrope stars, one with the polytropic index $N = 3.01$ and the other with $N = 3.00$. We treat the calculated eigenfrequencies of the polytrope $N = 3.01$ as if they were the observational data, and try to reproduce the model by carrying out the inversion of this data set. The polytrope star with $N = 3.00$ is adopted as a reference model in the inversion procedure. Since the equation to solve in constructing models is the well-known Lane-Emden equation (see, e.g., Hansen & Kawaler 1994), anyone can reproduce the following results. However, we should be careful about the numerical precision. Eigenfrequencies of linear, adiabatic, radial and nonradial p-modes are computed by solving the pulsation equations in the same manner, that is, the four first-order differential equations are solved even in the case of $\ell = 0$ (cf. Unno *et al.* 1989). Since the models are polytropic stars, the mechanical outer boundary condition for pulsation is the zero boundary condition, that is, the Lagrangian perturbation of pressure is set to be zero, $\delta p = 0$, at the surface.

The p-mode pulsation characteristics of a star are determined by the sound-speed and the density profiles in the star. It is known that the eigenfrequencies obey a variational principle. Hence, if we take the polytrope star with $N = 3.00$ as a reference model and

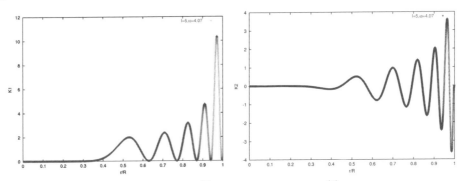

Figure 1. Examples of the kernels $K^{(1)}_{n,\ell}(r)$ (left panel) and $K^{(2)}_{n,\ell}(r)$ (right panel) of the polytrope star with $N = 3.0$. These are the kernels for p$_5$-mode of $\ell = 5$.

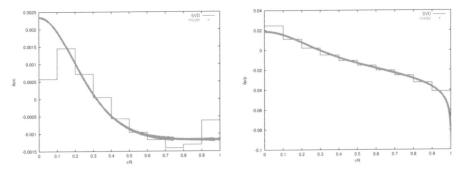

Figure 2. The results of inversion of p-modes with $\ell \leqslant 100$ and $\omega^2/(GM/R^3) \leqslant 2000$. We assumed in the inversion procedure that each of the differences is a piecewise constant function with a step of $\Delta r/R = 0.1$. The left panel shows the inferred difference in the sound-speed profile and the true difference (a smoothed curve), while the right panel shows that in the density profile.

compare its eigenfrequencies with those of the polytrope star $N = 3.01$, the difference in the eigenfrequencies between them are expressed in terms of the differences in the sound-speed and the density profiles between these two stars:

$$\frac{\delta\omega_{n,\ell}}{\omega_{n,\ell}} = \int \left\{ \frac{\delta c}{c} K^{(1)}_{n,\ell}(r) + \frac{\delta\rho}{\rho} K^{(2)}_{n,\ell}(r) \right\} dr. \tag{2.1}$$

where $\omega_{n,\ell}$ and $\boldsymbol{\xi}_{n,\ell}$ denote the eigenfrequency and the normalized eigenfunction of the mode ($\omega^2_{n,\ell} \int \boldsymbol{\xi}^*_{n,\ell} \cdot \boldsymbol{\xi}_{n,\ell} dm = 1$) with radial order n and spherical degree ℓ, respectively, and $c(r)$ and $\rho(r)$ denote the sound-speed and the density profiles, respectively, as functions of the distance r to the center. Equation (2.1) should be regarded as a set of integral equations with two unknown functions $\delta c(r)$ and $\delta\rho(r)$, and $K^{(1)}_{n,\ell}(r)$ and $K^{(2)}_{n,\ell}(r)$ are the kernels which are computed with the reference model for the given mode. The explicit expression of the kernels $K^{(1)}_{n,\ell}(r)$ and $K^{(2)}_{n,\ell}(r)$ can be seen in Gough & Thompson (1991). Figure 1 shows the kernels for p$_5$ mode of $\ell = 5$. Various sophisticated methods of solving this type of integral equations have been invented. Here, we solve a set of equations (2.1) for various eigenmodes with the help of the singular value decomposition (SVD) and obtain the unknown functions $\delta c(r)$ and $\delta\rho(r)$ (cf. Press *et al.* 1992).

Let us see first that, if modes up to $\ell = 100$ were detected, the true difference of the star would be admirably well reproduced from the reference model. For this purpose, let

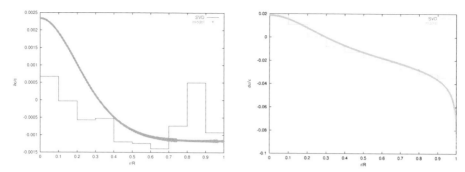

Figure 3. The results of inversion of p-modes with $\ell < 5$ and $\omega^2/(GM/R^3) \leqslant 2000$. In the inversion procedure each of the differences is assumed to be a piecewise constant function with a step of $\Delta r/R = 0.1$. The left panel shows the inferred difference in the sound-speed profile and the true difference (a smoothed curve), while the right panel shows that in the density profile.

us use all the p-modes with $\ell \leqslant 100$ and $\omega^2/(GM/R^3) \leqslant 2000$. The total number of modes used in the inversion is then 450. Here we assume that both $\delta c(r)$ and $\delta \rho(r)$ are piecewise constant functions with a step of $\Delta r/R = 0.1$, rather than smooth function, that is, we consider an expansion of $\delta c(r)$ and $\delta \rho(r)$ in terms of a piecewise constant function. The problem is then to determine the twenty coefficients for the piecewise constant functions from 450 equations. Figure 2 shows the inverted results. In each panel, the true difference between the star (the polytrope star with $N = 3.01$) and the reference model (the polytrope star with $N = 3.0$) is shown as a smooth function. As seen in this figure (particularly in the left panel showing $\delta c/c$), the inversion reproduces admirably well the true difference, particularly in the range of $0.2 \leqslant r/R \leqslant 0.9$. The gap between the inverted results and the true difference is substantial for $r/R \leqslant 0.1$. This is partly because the number of modes reaching near the center is small and partly because the sound speed is so high near the center that the sound wave is insensitive to this region. The gap between the inverted results and the true difference is also substantial for $0.9 \leqslant r/R$. This is because the number of modes for which the inner turning point is located near the surface is small. These tendency is seen in the helioseismic inversion.

Let us reduce the range of the spherical degree ℓ of eigenmodes to $0 \leqslant \ell \leqslant 4$. However, since the current aim is to prospect the ultimately reachable possibility of asteroseismology in the extremely idealistic case, let us still adopt all the p-modes for which frequency $\omega^2/(GM/R^3) \leqslant 2000$, using a total of 200 p-modes in the inversion. The results are shown in Fig. 3. As seen in this figure, the SVD inversion whose results are shown by the step functions is much worse than the case of inversion of modes with $\ell \leqslant 100$, but it is still satisfactorily successful in such an idealistic case. The total number of modes has been obviously overestimated, and the mode identification has been assumed to be perfect. Furthermore, it has been assumed that the observational frequency determination is error free. Obviously these assumptions are unrealistic. But we should regard these inverted results as the ultimately reachable points of asteroseismology. We cannot expect better results than these, but we can expect these results in the extremely idealistic case.

Since the frequency data are assumed to be error free, all the inversion error is the systematic error, that is, the intrinsic error which is the dependent on the inversion method (SVD in the present case). One might get an impression from Fig. 3 that the density profile is reproduced better than the sound-speed profile. This is not true. The left panel of Fig. 4 shows the systematic error of the inverted results shown in Fig. 3. It clearly shows that the systematic error in the sound-speed inversion is smaller than that

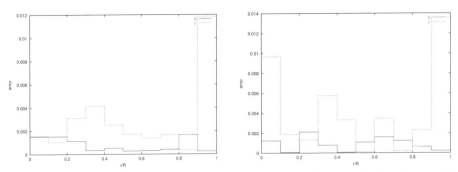

Figure 4. Left: Systematic error (intrinsic error which is dependent on the inversion method; SVD in the present case): It should be noted that systematic error in the sound-speed inversion is smaller than that in the density inversion. Right: Total error in the case that observational error is 0.03%.

in the density inversion. In reality, observational error is unavoidable. As a simulation, let us add 0.03% error into the observational data set and estimate its effect on inversion. An observational error of 0.03% can be reached with 3-months of uninterrupted photometric observations, if the period is measured with accuracy of 10 s for a single cycle of pulsation in the case where the period is 300 s. The right panel of Fig. 4 shows the total error in this case. We would say that inversion still has some hope.

2.3. *Summary and discussion*

We demonstrated how well (or poorly) we can probe the internal structure of the distant stars using a limited number of p-mode frequencies. Although the detectable eigenmodes must be limited to be $\ell \leqslant 4$, the inversion is still highly satisfactory in the extremely idealistic (200 modes with no observational error) case. We conclude from the present simulation that if the observational error is of the order of 10^{-3}, inversion still has some hope.

We restricted ourselves to inversion based on the absolute values of frequencies. However, there are other possibilities of inversion based on the relative differences among the eigenfrequencies, such as the frequency difference between the consecutive p-modes for which radial order n differs by 1, or the frequency difference between the p-modes for which the radial orders are the same but their spherical degrees ℓ differ by 2. Another example is the second derivative, $\omega_{n+1,\ell} - 2\omega_{n,\ell} + \omega_{n-1,\ell}$, which is sensitive to a discontinuity in the stellar structure.

References

Balmforth, N.J., Cunha, M.S., Dolez, N., Gough, D.O., Vauclair, S. 2001, *A&A*, 323, 362

Gough, D. 2003, in: L.A. Balona, H.F. Henrichs & R. Medupe (eds.), *Magnetic fields in O, B and A stars: Origin and connection to pulsation, rotation and mass loss*, ASP Conf. Ser. (San Francisco: Astron. Soc. Pacific), vol. 305, p. 389

Gough, D.O. & Thompson, M.J. 1991, in: A.N. Cox, W.C. Livingston & M.S. Matthews (eds.), *Solar Interior and Atmosphere* (The University of Arizona Press), p. 519

Hansen, C.J. & Kawaler, S.D. 1994, *Stellar Interiors* (Springer)

Matthews, J.M., Kuschnig, R., Guenther, D.B., Walker, G.A.H., Moffat, A.F.J., Rucinski, S.M., Sasselov, D., Weiss, W.W. 2004, *Nature*, 430, 51

Press, W.H., Teukolsky, S.A., Vetterling, W.T., Flannery, B.P. 1992, *Numerical Recipes in Fortran* (Second Edition) (Cambridge University Press)

Unno, W., Osaki, Y., Ando, H., Saio, H., Shibahashi, H. 1989, *Nonradial Oscillations of Stars* (Second Edition) (University of Tokyo Press)

Discussion

BALONA: The eigenfrequencies of a roAp star are affected by the magnetic field. Therefore we need to correct the observed frequencies for this effect before the data can be used for asteroseismology. I believe we are still very far from being able to do this with confidence, so maybe roAp stars are, at present, not good candidates for asteroseismology.

SHIBAHASHI: Correcting the magnetic effect for the frequencies is hard. The inversion of the absolute values of the observed frequencies may be inappropriate for roAp stars at present as you point out. Probably taking the differences or the differentiation of the observed frequencies may be useful in this case.

BREGER: I would like to be more optimistic. For normal A stars we can easily detect $\ell > 5$ modes through line-profile changes. The $\ell = |m| = 8 \sim 14$ modes are easily detected. With better observations to account for narrower lines and short periods $\ell \geqslant 5$ could (or might) be doable.

MICHEL: I also have a more optimistic point of view. You adopted a polytrope model for simplicity and clarity of the demonstration, but this type of model omits considerable physics which we are strongly interested in and that we want to probe with stellar seismology. The existence of a convective core and of a steep μ-gradient at its edge is one example. Models show that the existence of mixed modes in stars more massive than the Sun makes possible the building of inversion kernels sensitive to this very deep region, while the center of the Sun is still out of reach because of the lack of observed g-modes. Another example would be the depth of the outer convection zone which can be probed with pseudo-oscillations in the second order difference, for instance, with the same efficiency using low ℓ modes only or high ℓ modes. I could also quote published papers showing that for δ Scuti stars, for instance, there is great hope to obtain by inversion the rotational profile, using only low ℓ modes (Goupil, M.-J. et al. 1996, A&A 305, 487).

SHIBAHASHI: Yes, taking the differences or the differentiation of the observed frequencies opens another possibility of inversion in asteroseismology. What I intended to stress here is that even a limited number of modes, much smaller than the solar case, are potentially very useful to deduce detailed information on the structure of the star.

CUNHA: In Balmforth *et al.* (2001) we modelled the star by a spot, where convection is suppressed, and the equatorial region, where it is not suppressed. To keep the interior the same, we have allowed for a change its effective temperature, $T_{\rm eff}$, and the luminosity L at the surface. The luminosity L was allowed to change to account for a 'side-leak' of energy, due to the difference in energy transport in the two regions. Can you please comment on the differences between this spot model and the model you described?

H. SHIBAHASHI: If a 'side-leak' is allowed, the temperature of starspots must be lower than the case of no side-leak, otherwise the total luminosity of the star would be larger than the energy generation. I think the hydrostatic balance is handled well in Balmforth *et al.* (2001), but I am not certain that the thermal balance is handled well in their model.

The A-Star Puzzle
Proceedings IAU Symposium No. 224, 2004
J. Zverko, J. Žižňovský, S.J. Adelman, & W.W. Weiss, eds.
© 2004 International Astronomical Union
DOI: 10.1017/S1743921304004880

Asteroseismology and helium gradients in A-type stars

Sylvie Vauclair[1]and Sylvie Théado[2]

[1]Laboratoire d'astrophysique de Toulouse, Observatoire Midi-Pyrénées, 14 avenue Edouard Belin, 31400 Toulouse, France
email: sylvie.vauclair@obs-mip.fr

[2]Institut d'Astronomie de Liège, Belgique
email: sylvie.theado @ulg.ac.be

Abstract. Asteroseismology may help detect diffusion-induced helium gradients inside Main-Sequence A stars. Models have been computed for 1.6 and 2.0 $M_\odot$ stars with pure helium diffusion, at different ages, so that the helium gradient lies at different depths inside the star. The adiabatic oscillation frequencies have been analysed and compared with those of a model without diffusion. Clear signatures of the diffusion-induced helium gradient are found in the so-called "second differences". These frequency differences present modulations due to the partial reflection of the sound waves in the layer where the helium gradient occurs. A tentative application to the roAp star HD 60435 is presented.

Keywords. Asteroseismology; stars: abundances; stars: diffusion, stars: individual (HD 60435)

1. Introduction

It is now widely recognized that the abundance anomalies observed in peculiar A stars are basically due to element diffusion. Those for which the radiative acceleration is larger than gravity (like most of the metals) are pushed upwards while the other ones, like helium, diffuse downwards (Michaud 1970, Vauclair & Vauclair 1982).

In this paper we show how asteroseismology can help detect helium gradients inside peculiar A stars. A more complete study is given in Vauclair & Théado (2004). Rapid variations of the sound velocity inside a star lead to partial reflections of the sound waves, which may clearly appear as frequency modulations in the so-called "second differences" (Gough 1989):

$$\delta_2\nu = \nu_{n+1} + \nu_{n-1} - 2\nu_n \qquad (1.1)$$

The presence of a diffusion-induced helium gradient leads to a kink in the sound velocity with a very clear signature in the oscillation frequencies. Asteroseismic observations of A-type stars can test for the presence of helium gradients.

In the present computations we neglected the effects of magnetic fields and mass loss (the stars were assumed to be spherically symetric) and we assumed that the modification of the stellar structure induced by metal accumulation in the outer layers was negligible. We computed the structural changes induced by helium settling and its influence on the oscillation frequencies. We applied this test to the only roAp star in which enough modes have been observed HD 60435 (Matthews *et al.* 1987).

2. Seismic signatures of helium gradients

Stellar acoustic p modes with low l values can propagate deeply inside the stars. For this reason, they may be used to obtain information on the deep stellar structure. In

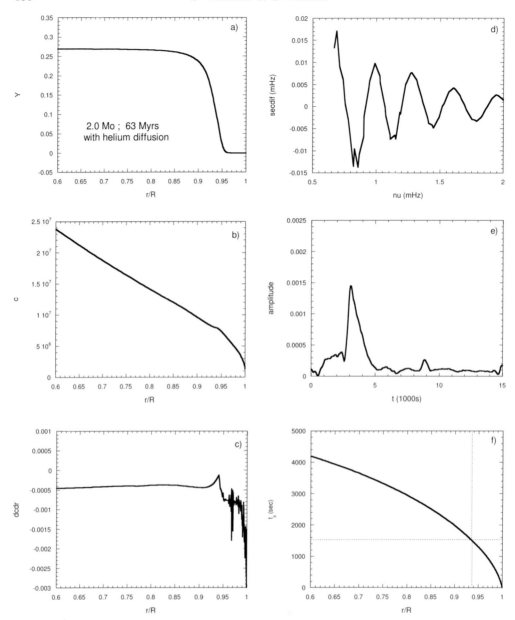

Figure 1. Helium diffusion and its consequences on the stellar structure and the oscillation frequencies for a $2.0\,M_\odot$ at 63 Myrs, a) helium profile as a function of the fractional radius, b) sound velocity, a kink is clearly visible at the place of the helium gradient, c) first derivative of the sound velocity, the kink is still more visible, d) the second differences of the oscillation frequencies plotted as a function of the frequencies, e) the Fourier transform of graph (d) plotted as a function of time devided by one thousand (in seconds), f) the "acoustic depth", or time needed for the acoustic waves to travel from the surface to the considered radius, it can be checked that the peak of graph (e) corresponds to a time scale which is twice the "acoustic depth" of the helium gradient.

case of strong gradients in the sound velocity, which may be due to the boundary of a convective zone, to the helium ionization region or to helium gradients, acoustic waves are partially reflected. This creates modulations in the frequency values which are equal to $2t_s$ where t_s is the time needed for the acoustic waves to travel between the surface and the considered region (acoustic depth), i.e.,

$$t_s = \int_R^r \frac{dr}{c} \tag{2.1}$$

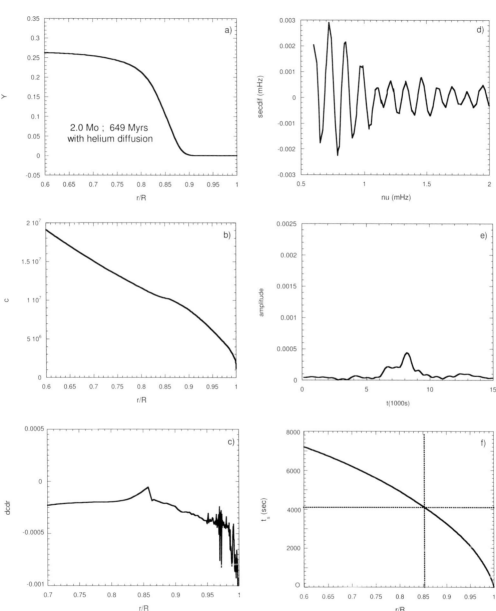

Figure 2. Same as Figure 1, for a $2.0\,M_\odot$ model at 649 Myrs.

We studied the evolution of $1.6\,M_\odot$ and $2.0\,M_\odot$ stellar models, with pure helium settling. We have computed the oscillation frequencies and their second differences in five different models. Results are given in Figures 1 and 2 for the $2.0\,M_\odot$ model with ages of 63 Myrs and 649 Myrs. In each figure, six graphs are displayed a) the helium profile inside the star, b) the sound velocity which shows a clear kink at the place of the helium gradient, c) the first derivative of the sound velocity in which the kink is still clearer, d) the second differences which show periodical oscillations, e) the Fourier transform of these oscillations, in which clear peaks are found for precise time values, and f) the time needed for the acoustic waves to travel between the surface and the considered radius, or "acoustic depth".

We can follow in these two figures the signatures of the helium gradients as they sink into the stars.

Similar computations have been done for evolution models in which helium diffusion is suppressed. The resulting second differences are quite different (Vauclair & Théado 2004). They show signatures of the helium ionization zones while the signatures of helium gradients have, of course, disappeared as there is no helium left in the ionization regions.

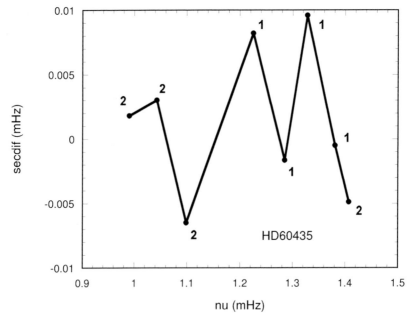

Figure 3. Second differences in the oscillation frequencies of HD 60435 as found from Matthews *et al.* (1987). The points are labelled according to their l values.

3. The roAp star HD 60435

Among the rapidly oscillating Ap stars which have been detected up to now, only one, HD 60435, presents a frequency spectrum rich enough for the second differences to be computed (Kurtz, private communication). Matthews *et al.* (1987) discovered 17 frequency peaks in the Fourier spectrum, which they basically identified as $l = 1$ and $l = 2$ modes.

The computations of the second differences can only been done if enough modes with adjacent values of n are detected for the same value of l. We have computed the observed second differences for HD 60435, using the frequencies and mode identifications as given by these authors, with an uncertainty of ± 0.0001mHz. With the presently detected modes, eight second differences could be computed (Figure 3). They show a modulated trend similar to that of Figure 2, characteristic of the presence of a helium gradient below the surface (around 90% of the stellar radius).

These results rely on the mode identifications which are given as "tentative" by Matthews *et al.* (1987). Note, however, that the difficulty in the identification by these authors was due to that the frequencies were not exactly at the expected value. This effect may simply be due to the presence of a helium gradient.

4. Conclusion

We have computed models of A-type Main Sequence stars with and without helium diffusion and we have shown that asteroseismology can give signatures of the presence of helium gradients below the surface. The analyses of the second differences in the oscillation frequencies show modulations which are due to the partial reflection of the acoustic waves at the place of the helium gradient.

At present these theoretical results can be compared to the observations of only one star, namely the roAp HD 60435. The computated second differences give evidence of a modulation similar to the one expected for the presence of a helium gradient below the surface.

Our theoretical computations have been done for spherically symetrical stars, without rotation induced mixing and ignoring any magnetic effects. We did not either include in the computations the abundance variations of heavy elements due to the combined effect of gravitational settling and radiative acceleration. This should be done in the future. These first results are, however, very encouraging and we hope that enough mode detections and identifications will be obtained in other stars to do similar tests.

References

Gough, D.O., 1989, in: *Progress of Seismology of the Sun and Stars*, proceedings of the Oji International Seminar Hakone, Japan, Springer Verlag, Lecture notes in Physics 367, p.283-318

Matthews, J.M., Kurtz, D.W., Wehlau, W.H., 1987, *Ap.J.* 313, 782

Michaud, G., 1970, *ApJ* 160, 641

Vauclair, G., Vauclair, S., 1982, *ARA&A* 20, 37

Vauclair, S., Théado, S., 2004, *A&A* 425, 179

Discussion

CUNHA: Comment: Modes of different degree sample different latitudes of the star. For stars that are not spherically symmetric, that will probably prevent the surface effects from cancelling out in the second differences, as happens in spherically symmetric stars.

If in practice modes of different degree are used to calculate the observed 2nd differences (rather than just modes of the same degree and orders), don't you think that it will introduce an uncertainty in your determination of depth of helium gradient?

VAUCLAIR: Thanks for mentioniong that. We should look more precisely at this question. This may be the reason why the observed peak (in the Fourier transform) is larger than the peaks computed in the spherically symmetric modes.

NOELS: I guess that in such a star, the excitation of modes is due to the H-ionization zone. Should not helium play a significant role?

VAUCLAIR: Yes, these results are consistent with the modes presented by Margarida Cunha where the excitation is due to H-K mechanism, helium having diffused below.

GREVESSE: Your predictions disagree slightly with the observations but what is the accuracy of the obsevations?

VAUCLAIR: The relative uncertainities of the second differences are of order 10%. A reasonable estimate of the uncertainities induced by the small number of observed frequencies lead to an uncertainity of 150 s on the acoustic depth of the helium gradient.

WEISS: Comment: The MOST will provide more examples for roAp's with a larger number of frequencies needed for your analysis. Question: What He-abundance should we use in our abundance analyses of roAP's? You showed that He can totally disappear in the atmosphere which would have a severe effect on model atmospheres to be used in abundance analyses.

VAUCLAIR: The computations have been done with pure microscopic diffusion. Turbulence could slow down the formation of helium gradient and let some helium in the atmosphere. In every case there should be an important helium depletion in the atnospheres of these stars, although I cannot tell at the present time how much would remain exactly.

The A-Star Puzzle
Proceedings IAU Symposium No. 224, 2004
J. Zverko, J. Žižňovský, S.J. Adelman, & W.W. Weiss, eds.

© 2004 International Astronomical Union
DOI: 10.1017/S1743921304004892

Panel discussion section I

CHAIR: **F. Kupka**

SECTION ORGANIZER & KEY-NOTE SPEAKER: J.D. Landstreet
INVITED SPEAKERS: O.Kochukhov, E.Paunzen, H. Shibahashi
CONTRIBUTION SPEAKERS: S. Vauclair, S. Théado,

Discussion

SHIBAHASHI: I would like to know how many unknowns – how many equations – you must solve for in the inversion procedure in your new magnetic Doppler imaging method?

KOCHUKHOV: The Doppler imaging problem is intrinsically ill-conditioned. It is not correct to characterize this problem by the number of unknowns. We are using regularization, and regularization effectively introduces a set of additional equations which smoothes the map in our case; in the case of the maximum entropy method, it selects an image with maximum entropy. So actually if I count equations, there are more equations than unknowns.

SHIBAHASHI: What is the resolution of the maps?

KOCHUKHOV: Using Doppler imaging, at a $v \sin i$ of about 30 km/s, we can resolve about 60 or 80 elements at the rotational equator. Of course, this question must be asked in relation to the observational data that is used for the inversion. For example, for the mapping of pulsations, I used 2000 spectra, so I think I actually had a sufficient number of observational constraints.

NOELS: I would like to ask Shibahashi how sensitive are his results to the choice of the polytrope of n = 3.00 for the inversion method? If you had chosen another polytrope for the inversion technique, how sensitive are the results to the choice?

SHIBAHASHI: The inversion method which I adopted is linear inversion, so if I took, say, n = 2.9, the difference between my inversion and one with n = 2.9 is just proportional to the difference between the polytropic indices.

NOELS: I wonder if anyone knows of a simulation in which one takes the helioseismic data for the Sun and artificially limits it to what could be observed in another star? Has anyone taken solar data, say limited to l = 5 and lower frequencies, to see if one could derive the depth of the convection zone? What precision would we get?

SHIBAHASHI: In the case of the Sun, I do not find any big difference if we take the data obtained ten or twenty years ago and compare the results with modern inversions, because the number of eigenmodes detected even at that time was quite large.

In my inversion, using a polytrope, my first intention was to demonstrate the power of asteroseismology, by using quite a lot of eigenmodes, in my case 200. So I intended to reduce the eigenmodes gradually. If I reduce to say l = 2, the result is quite hopeless.

NOELS: If you forget all about the Sun, and limit yourself to l = 2, what would be the precision of the convective envelope?

SHIBAHASHI: The outer part of the convective envelope is quite hard to resolve, but the depth of the convection zone is quite easy.

NOELS: So there is hope for asteroseismology!

MATHYS: I have a question for Kochukhov regarding the modelling of α And? I wonder how you deal with the isotopic structure of mercury? We know that in HgMn stars we may have widely different isotopic compositions from star to star. I think that we could even suspect that we have widely different isotopic compositions on different locations on the surface of α And. The isotopic splitting is large enough that even with rotation you should see some effect on the line profile.

KOCHUKHOV: First, the effect of isotopic splitting for a $v \sin i$ of 52 km/s is really small. I did try to investigate the effect of varying isotopic composition of Hg over the surface of α And. Admittedly, I used a Q-parameter model, so a simple parameterization of isotopic mixture, and what I find is that the range of Q is small, from zero to say 0.5. There seems to be no evidence of very large variation of isotopic composition over the surface. In principle, Doppler imaging allows one to reconstruct each isotope individually, but I would not believe these maps. So I restricted myself to Q-parameter, and other maps were reconstructed suing a solar isotopic mixture.

BALONA: This is a question directed at observers of magnetic Ap stars. According to Shibahashi, the temperature should be cooler around the poles of such a star, and therefore you would expect to see from photometric observations differences in temperature as the star rotates. Is this actually observed?

LANDSTREET: What is observed is that there are photometric variations observed in most Ap stars, of typically 0.02 or 0.03 mag, and also spectrum variations. It is not yet clear whether the photometric variations are due exclusively to the spectrum variations, or whether there is some component due to temperature variations. The fact is that we can model spectra of more or less normal stars, such as Sirius, very well, and so we can have real confidence in our models of such stars. In contrast, we cannot yet model spectra of magnetic stars at all satisfactorily, beyond modelling a single line at a time, or perhaps a multiplet. There are so many effects to take into account, such as horizontal inhomogeneities, vertical stratification, and a possibly complex field geometry, that our models match observations rather poorly. Thus I think it is still rather uncertain whether the observed variations have a temperature variation component in them or not.

KOCHUKHOV: I would like to make the following comment. As I understand, in Shibahashi's model, the central issue is suppression of convection at the magnetic poles. It is important to remember that convection becomes quite inefficient when we go above 8000 K, and spots are observed in much hotter stars, so there seem to be two different regions in the HR diagram with very similar phenomenological effects. Concerning the possibility of observing temperature variations, I think this can be done in cool Ap stars by studying variations of Balmer lines as a function of rotation phase, because in cool Ap's the Balmer lines are sensitive to temperature. We should see some systematic variations of Balmer lines if there are cool spots.

LANDSTREET: Remember that the Balmer lines may also be influenced by magnetic pressure effects, or by a gradient of mean molecular weight, for example produced by a vertical helium abundance gradient.

KOCHUKHOV: If you are thinking about magnetic pressure effects due to the Lorentz force, this effect is limited to hotter Ap stars. The effect is not significant in cool Ap stars simply because the plasma is not ionized sufficiently. We actually calculated model atmospheres of A and B stars including the effect of the Lorentz force. We find that for temperatures of about 8000 K and less there is very little effect of the Lorentz force on atmospheric structure. Effects become significant only at, say, about 11000 K.

LANDSTREET: I think that this is actually fairly uncertain because the origin of the currents is uncertain. There are several possible sources of currents. Unless you know the entire field configuration and how it is changing, it is difficult to be sure that you have accounted for all the currents. You can set lower limits to the currents from particular sources, but I think it is difficult to be sure you know the currents exactly.

KOCHUKHOV: The effect does not depend on the currents, it depends on the conductivity.

LANDSTREET: No, the effect depends on the currents. The Lorentz force is simply $\mathbf{j} \times \mathbf{B}$.

KOCHUKHOV: Let's discuss this later.

TRAMPEDACH: I have a question for Shibahashi about inversions. The density kernels you were using were also negative. Have you played around with other kernels? I believe that it is possible to construct kernels that are almost positive definite, giving a more deterministic inversion.

SHIBAHASHI: What I showed in my talk is a kernel of a certain eigenmode. In the procedure of the inversion we actually make a delta-like function by combining a certain set of eigenmodes. In that sense, the combined density kernel to detect the density profile at a certain layer is positive mainly at that layer, and, ideally, close to zero elsewhere. But actually the kernels of each eigenmode are oscillating between positive and negative for the density kernel, and only positive or zero in the case of the sound velocity kernels.

ADELMAN: Coming back to the question of surface temperature variations in Ap stars. Some years ago, Leckrone took OAO-2 UV photometry and optical photometry of several magnetic CP stars and showed that integrated flux from those stars was constant to pretty good precision, implying that the effective temperature was not varying by more than some degrees (see Molnar 1973, ApJ 179, 527). I later showed that the error bars for one of the stars could be reduced further if the 5200 A feature, which Leckrone had ignored, was included.

I have a question. What percentage of the metal-poor A stars are considered to be λ Boo stars?

PAUNZEN: If you look at Pop I stars, there is some kind of transition between metal-poor stars and stars that we would describe as λ Boo stars. We consider a metal-poor star to belong to the λ Boo group if we find a depression in the UV. The percentage of metal-weak A stars that show this depression is about 40 to 50%.

CUNHA: I have two comments. The first is about the precision to which we might be able to do asteroseismology. Of course the number of modes that we will be able to detect is important, but for A-type stars, I would say that problem of identification of the modes, like Breger was discussing yesterday for the case of δ Sct stars, and the problem for magnetic stars that they are by no means spherically symmetric, will be even more serious obstacles to us than the number of modes that we can detect.

SHIBAHASHI: I don't want to make the problem more complicated. Balona clearly stated that we should have a large number of modes, and we should have clear identifications for these modes, and we should have a good reference model. Mode identification is certainly a very serious problem. If we take into account the difficulty of mode identification, the lack of some eigenfrequencies, and the lack of a good reference model, we cannot say anything definite about the prospects for asteroseismology. What I intended in my talk was to simplify the situation a lot, probably too much. I assumed that the eigenmodes are there, that they are all identified. This is pretty unrealistic, but it seems to me that we simply cannot show the power of asteroseismology if we consider the problems of mode identification, of missing modes, and of poor reference models.

CUNHA: Perhaps you misunderstood my comment. What you did was very good; isolating that part is very important. I just want to say that in my opinion that is not what is going to be most in our way for asteroseismology. I was not criticizing what you did.

My other comment was about effective temperature differences in spot models. Going back to the model that I mentioned previously for which we did non-adiabatic calculations, we allowed for differences in effective temperature and luminosity. The differences in effective temperature that we found were rather smaller than you found, because we allowed for luminosity variations as well, and the effect was only a few K. I discussed trying to detect this effect observationally with Petit, who works on cooler stars, and he told me that it would be extremely difficult to detect such a small effect.

SHIBAHASHI: Yes, I think the 500 K in my estimates may be too much.

DWORETSKY: I have a question for Paunzen about the binarity of the λ Boo stars. You pointed out that the only A-type companions found so far are also λ Boo stars, and no other types of companions are seen. There is probably a good reason for that. If any star that might have been able to become a λ Boo star had a solar type companion, the solar wind of that companion would blow away the local ISM and create a protective bubble around both stars, and prevent the accretion of interstellar gas. I think that that might be a reasonable hypothesis to explain that particular feature. Of course, with only 50 or 60 stars known, the statistics are rather slim, but it is surprising that only 10 or 15% of the stars are binaries. The percentage should be somewhat higher, perhaps around 50%.

PAUNZEN: As Corbally told us, if you find a spectroscopic binary with a λ Boo component and a cool component, in this system diffusion and mass loss could be responsible for the λ Boo star. So we don't insist on the accretion model; in a very old A star in a binary system, the λ Boo component might be produced by diffusion and mass loss.

WEISS: I have a question to Shibahashi, and a comment. The question: you based your final analysis on the observability of 200 modes. How do your results scale with the number of observable modes? Perhaps due to the (unknown) excitation mechanism, there simply won't be that many modes, or perhaps, as suggested by some model analyses, we might see even more than 200 modes.

SHIBAHASHI: If I adopt only 100 eigenfrequencies, but with the same range of eigenfrequencies as before, the situation is not so bad. However, if we reduce the range of frequencies substantially, then the result becomes much worse.

WEISS: And I have a comment to Balona, who raised the question of Q modes for λ Boo stars. If I remember the HR diagram that Paunzen showed, the distribution of λ Boo stars in the HR diagram is fairly similar to that of the δ Sct stars, so the curious Q-mode histogram cannot be due to selection effects.

BALONA: But in the Paunzen diagram, I recall that there are more λ Boo stars near the hot end than you would expect. Therefore there is a kind of bias towards the hot end, where the overtones predominate.

PAUNZEN: There might be a bias, with about five or six which are going to the hot end. But I don't think that this is really a bias that you see in the Q-values, because if you really look at the distributions of δ Sct type stars and λ Boo stars, you find no λ Boo star that could pulsate in the fundamental mode. This might be due to diffusion. It has been suggested by theoretical groups that diffusion does something in the atmosphere that changes the pulsational characteristics of stars, as in Am stars where you hardly find any pulsation. It might be the same effect here. But also you see a shift in the instability strip. This suggests that there may be some related effects due to diffusion in these stars.

KURTZ: First a comment on Paunzen's remarks. I had the same question as Balona about the difference in Q-values. We have seen that there may be a shift in the instability strip from the δ Sct stars to the λ Boo stars. A good way to test the conclusion about the distribution of Q-values would be to take a sample of λ Boo stars that overlap the δ Sct instability strip, balance the numbers, and see if there is still a shift in the Q-values. This would strengthen the result.

The question I have comes from Kochukhov's talk, but it is a question for Bagnulo. Kochukhov found for HR 3831 a magnetic obliquity of 90 degrees. The geometry he found for this star is consistent with the old photometry and the old oblique pulsator model, but it is not consistent with a magnetic model published by Bagnulo which has a very small obliquity. Kochukhov has told me that you have new models of this star which have a larger magnetic obliquity. Can you match Kochukhov's magnetic geometry? Is there consistency in the magnetic fields with what he finds? This is a test of Kochukhov's magnetic angles.

BAGNULO: The answer depends on which geometry you assume. My models depend on the overall morphology that we assume for the magnetic field. If the magnetic field is dominated by a dipolar component, then the dipole axis must be tilted at a very small angle to the rotation axis to fit the observed values of the mean quadratic field. So there are basically two assumptions: that the mean quadratic field modulus measurements are not overestimated, and that the morphology is approximately dipolar. If we assume that there is a strong quadrupolar component, which I think Kochukhov has ruled out, then it might be possible to fit the data with a dipole having a large angle to the rotation axis. But I don't remember at the moment the details of this model. In any case, I think that Kochukhov did suggest that the overall morphology must be roughly dipolar.

KURTZ: It would be interesting to see how complex a magnetic morphology you need to shift the dipole obliquity to 90 degrees from the value of 8 degrees that you found. That might need quite a complex magnetic geometry. It would be useful if you could tell us the result.

BAGNULO: I have my laptop here....

KOCHUKHOV: I would like to comment that there really is a problem of incompatibility between the quadratic field measurements for this star, and other estimates of the field strength. In my Doppler imaging of this star, I estimated the magnetic field strength using magnetic intensification of spectral lines. This was done based on new observational material of high quality, good phase coverage, and polarized radiative transfer calculations. I find that in order to obtain consistent abundances from iron lines, I have to use a magnetic field of the order of 2 kG. A field of 11 kG clearly introduces a very large discrepancy into iron abundances derived from different spectral lines. I'm not saying that my results are correct, but I do not think that we should rely on observations of this star done 15 years ago, as Bagnulo did. We should obtain new high-resolution polarization spectra of this star and try to solve this problem.

BREGER: I would like to muddy the waters a little concerning the comparison between the λ Boo stars and the δ Sct stars. The borders of the instability strip are set by the hottest and coolest stars that we know of any type; if we have one single star that is pulsating and that is very hot or very cool, we draw the border beyond that star. The cool border of the δ Sct instability strip is actually defined by the star ρ Pup, and near it are a huge number of constant stars. Now this situation does not bother us too much, since the temperatures of δ Sct stars are not that well determined anyway, and ρ Pup has more metals and therefore more line blocking, and so its colour is too red. But in any such comparison this should bother us.

Also, recently the borders of the instability strip have been shifting. They have been shifting to hotter temperatures, as you have for λ Boo stars, because we now observe stars with very small amplitudes and very short periods. So I would support Kurtz's suggestion. One should take a definite region in the HR diagram and compare the properties.

KHALACK: I just have a small question about γ Equ, about non-LTE effects. Has anyone estimated the non-LTE effects, and would they be strong enough to shift the Pr III and Nd III pulsation depths to the H line core pulsation depth?

RYABCHIKOVA: In my talk I discussed the non-LTE calculations for Nd II and Nd III. The line depths which were shown by Kochukhov are based on these new non-LTE calculations. The result is that the depth of formation of these Nd III lines is moved closer to the Hα core.

PAUNZEN: I have a comment about Breger's comments. If we look at the hot pulsating λ Boo stars, we have amplitudes of 0.1 mag, which is not observed in hot δ Sct stars. Even if we shift the instability strip, the characteristics found for the pulsating hot λ Boo stars are completely different than those found for the δ Sct stars. So I guess that there are real differences at the hot border.

TRAMPEDACH: I have a comment on the formation of λ Boo stars by accretion from the ISM, and on the bow shock. The key aspect here is what happens to a spectroscopic binary, where the two stars are fairly close to one another. While the system goes through a dense cloud, the stars are also stirring up the ISM, so you would have a completely different picture than just one bullet going through and creating a bow shock. You would have one of the stars going through the slipstream of the other, you would have turbulence occurring. It would be worthwhile to do hydrodynamical calculations on this situation.

Also you talked about open clusters which have no internal ISM, so there would be no chance of accretion there. Wouldn't some accretion occur as the cluster passes through the galactic plane?

PAUNZEN: Yes, but normally the time scale doesn't work. If an open cluster goes through the galactic plane, the passage is normally much faster than you need for accretion, because the accretion rates are very low. If the accretion rate were high, normally you would detect some signs of this accretion. So the time-scales don't fit.

KUPKA: That ends this discussion. I would like to thank the speakers again.

The A-Star Puzzle
Proceedings IAU Symposium No. 224, 2004
J. Zverko, J. Žižňovský, S.J. Adelman, & W.W. Weiss, eds.

© 2004 International Astronomical Union
DOI: 10.1017/S1743921304004909

Observational challenges of A-type stars

Stefano Bagnulo

European Southern Observatory, Alonso de Córdova 3107, Vitacura, Santiago, Chile
email: sbagnulo@eso.org

Abstract. In the last few years, new instruments mounted at modern large telescopes, as well as satellite instruments, have given to us the possibility of obtaining observations of much higher quality than in the past. Yet, the large majority of observations of A-type stars are performed with small to middle size class telescopes. In this paper I discuss the scientific case for the use of large size class telescopes for the A-star research.

Keywords. Instrumentation: interferometers, instrumentation: spectrographs, instrumentation: polarimeters, telescopes

---

1. Introduction

Observational astronomy has changed considerably in the last few years. Although it is not easy to have a complete and detailed overview of the worldwide available instrumentation, it is pretty clear that a substantial fraction of the technological and economical efforts are concentrated in projects that involve very, extremely, up to overwhelmingly large telescopes, or expensive space missions. Just limiting ourselves to ground based observations, at the moment there are 14 optical-IR telescopes already operating or under construction that have a 8-m mirror or larger. In less than two decades from now, these that today are our largest telescopes will appear as minuscule facilities if projects such as, for instance, OWL will become reality. On the one hand, we can ask ourselves how much science for A type stars is done with the large telescopes. On the other hand, a lot of science has been done and is still being carried out with $0.5 - 2.0\,\mathrm{m}$ telescopes, and we may want to address the question if these facilities are bound to become obsolete.

It should be noted that astronomical observations are not changed only in terms of available instrumentation, but also in terms of the way observations are carried out. I refer to the possibility to perform observations in *service mode*, as opposite to the more familiar *visitor mode*, when the astronomer is present at the telescope and personally takes care of the observations. Rather than with hardware, astronomers may have much more to do with electronic forms and sheets. (For an overview of the rationale and of the performances of the service observing mode implemented at ESO, see Comeron *et al.* 2003.)

Finally, some problems that we want to address may be solved using observations that have already been obtained. The success of our future research will also crucially depend on our capabilities to efficiently search for and retrieve archived observations.

The session "Observational Challenges of A type stars" has been included in this Symposium with the purpose of discussing the new observational tools that we have at our disposal to address the open questions related to the physics of A stars. In this paper I will present some practical examples of how modern instruments attached to the largest telescopes can be efficiently used for the A-star research. The role played by archived observations is discussed by Padovani (2005).

Since it is not possible to discuss all the observational aspects that are related to the A-star research, I decided to consider a specific topic and to see how (some) present and forthcoming instruments at the large telescopes may be used to address it. Such topic (a study of the evolution of magnetic Ap stars) is shortly introduced in Sect. 2. Instrumental aspects are discussed in Sect. 3. In Sect. 4 I present some concluding remarks about the projects to construct giant telescopes. This work is clearly biased toward the instruments of the ESO Very Large Telescope (VLT). However, most of the VLT instruments are not unique. FORS1 at the VLT and FOCAS at the Subaru telescope are very similar Faint Object Cameras and Spectrographs, both equipped with polarimetric optics. Gemini South is equipped with PHOENIX, and the VLT will be soon equipped with CRIRES. PHOENIX and CRIRES are both high resolution infrared spectrograph operating in the $1 - 5\,\mu$m region. The VLT and the Subaru telescope are both equipped with high resolution optical spectrographs (UVES at the VLT, and HDS at the Subaru telescope). The only remarkable exception, in terms of an instrument of the ESO VLT with "unique" features, is represented by the multi-object, medium-high resolution spectrograph FLAMES.

2. Studying the evolution of chemically peculiar A-type stars

The magnetic Ap stars have rotation periods that are much longer than those of non-magnetic stars of similar spectral class (e.g., Stępień 2000). Abt (1979) claims that Ap stars loose angular momentum during their life on the Main Sequence, whereas North (1998) states that observations are fully consistent with conservation of angular momentum during the star's life in the Main Sequence. Consistently with the finding by North (1998), Stępień (2000) and Stępień & Landstreet (2002) have presented a model that explains how a magnetic Pre-Main Sequence star may loose angular momentum through interaction between magnetic field and circumstellar material. Abt (1979) suggests that the incidence of chemical peculiarities increases with the star's age. Instead, Gomez *et al.* (1998) found that chemical peculiar stars occupy the whole width of the Main Sequence, and Pöhnl *et al.* (2003) found several chemical peculiar stars close to the Zero-Age Main Sequence. Hubrig *et al.* (2000) claimed that, for chemically peculiar stars with masses less than $3\,M_\odot$, their magnetic fields appeared when the stars had spent at least 30 % of its life on the Main Sequence. However, Bagnulo *et al.* (2003a) found in the young cluster NGC 2516 a strongly magnetic star of $2.1\,M_\odot$ that has spent about 16 % of its life in the Main Sequence. In conclusion, there is some confusion about the evolution of the magnetic chemically peculiar stars, and we simply do not know if and how the chemical peculiarities and the magnetic field depend upon the star's evolutionary stage. By contrast, knowing how the magnetic field and the chemical abundances change during the life of the star in the Main Sequence band would help to clarify the problem of the *origin of the magnetic field* and would set precise constraints to the *diffusion theory*.

To properly address this problem one has to observe a large number of stars and find their positions on the HR diagram. For each one should study the magnetic field topology and perform an accurate chemical abundance analysis, taking into account a nonhomogeneous distribution of the elements both horizontally and vertically. This project will benefit from the outputs of many different techniques, such as, e.g., astrometry and parallax measurements; photometry in the various flavours (*UBV*, possibly with extension to the infrared, Strömgren, Geneva, etc.); bolometry; spectroscopy; spectro-polarimetry; interferometry, etc.

2.1. *Determining the age of Ap stars*

I will start with the problem of locating the star in the HR diagram. The first step is to measure the stellar temperature. This could be done for instance using Strömgren

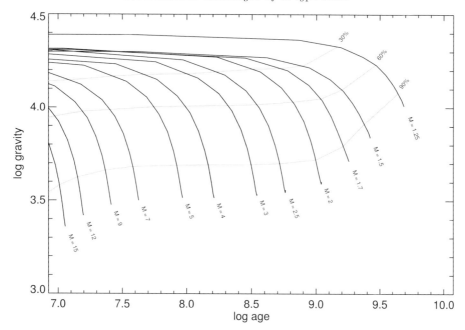

Figure 1. How $\log g$ changes with time according to the evolutionary models by Schaller *et al.* (1992) for stars of various masses. See also a similar figure in North & Kroll (1989). Models include overshooting and are calculated for a metallicity of $Z = 0.02$.

photometry (see Moon & Dworetsky 1985; see also the user friendly web tool TEMPLOGG2 at http://ams.astro.univie.ac.at/templogg/) or from Geneva photometry through the use of special calibrations algorithms (e.g., Hauck & North 1993), or from more accurate spectroscopic studies. In principle, once the temperature is known, one could determine $\log g$ either from the photometry or from spectroscopy. Then, from evolutionary considerations, one could determine the stellar age. However, as explained in North & Kroll (1989), $\log g$ is not very sensitive to the stellar age for young stars – see also Fig. 1. The variation of $\log g$ with time during the first 30 % of fraction of life spent on the Main Sequence is nearly negligible (compared to the accuracy of the $\log g$ estimates), especially in the case of lower mass A-type stars. Thus it is practically impossible to identify the age of a young A star through the $\log g$ measurement (see North & Kroll 1989).

A better method is to measure the absolute luminosity of the star, which is possible if the parallax is known, and directly find the position of the star in the HR diagram. This approach has been followed, e.g., by Hubrig *et al.* (2000). However, this method is not particularly accurate. Figure 2 shows the example of a star for which the temperature is known with an accuracy of 300 K, and the absolute luminosity with an accuracy of about 20 %. The left panel of Fig. 2 shows the position of the star in the HR diagram, and the right panel shows how the uncertainties in the position of the star in the HR diagram affect the determination of its age. Fig. 2 shows the need for better determinations of the stellar temperatures and the luminosities of Ap stars than what is generally possible today. This situation can be improved only by constructing a database of Ap stars for which fundamental parameters are known from direct measurements (see Sect. 3.1).

An alternative approach to study the evolution of Ap stars consists in considering open cluster members. From the observed HR diagram of the cluster, and based on an evolutionary model, one can determine the open cluster age. Then, the evolutionary

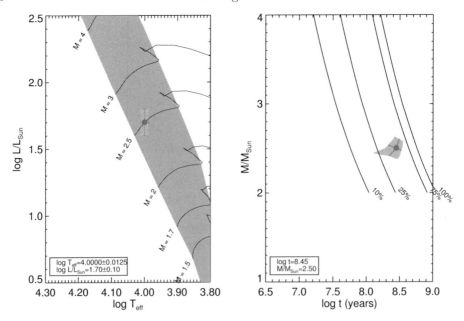

Figure 2. The left panel shows the position in the HR diagram of a star with a temperature $T_{\rm eff} = 10000 \pm 300\,$K and a luminosity $\log L = 1.7 \pm 0.1\,L_\odot$. The right panel shows how the uncertainties in a star's $T_{\rm eff}$ and L parameters affect the determination of the evolutionary state of the star. The solid lines represent the percentage of the time spent by the star on the Main Sequence as indicated in the labels for the mass given in the y axis of the plot.

model gives mass, $\log g$, and fraction of life spent on the Main Sequence for those cluster members for which an effective temperature is known. The advantage with respect to the study of field stars is that one gets rid of the uncertainties in the stellar age and mass introduced by the error in the absolute luminosity.

Studies of open cluster Ap stars have been done, e.g., by North (1993). Fraga & Kanaan (2005) concluded that, to determine if the chemical peculiarities depends on age, it is necessary to observe at least 2700 stars in three range ages ("young", "intermediate", and "old"). The importance of chemical analyses of A-stars belonging to the same open cluster has emphasized by Monier & Richard (2005). From the photometric point of view, important work has been done by Maitzen, Paunzen, and their group with Δa photometry (see, e.g., Paunzen *et al.* 2003 and references therein). We will see that FLAMES, a multi-object, intermediate and high resolution spectrograph attached at the Unit 2 Kueyen of the ESO VLT is an ideal instrument for cluster studies and it may possibly replace Δa photometry (Sect. 3.2). In Sect. 3.3 we will see that FORS1 can also be used for the detection of the magnetic fields in open cluster stars, and in Sects. 3.4–3.6 I present some instruments that permit a detailed analysis in terms of magnetic field mapping and abundance analysis of individual targets.

3. Observing A-type and related stars with the instruments at the large size class telescopes

3.1. *Interferometry and A-type stars*

Interferometry has received little attention in this Symposium. However, there is a large number of applications in the field of A stars, for instance, studies of circumstellar material around Herbig Ae/Be stars (e.g., Eisner *et al.* 2003, Millan-Gabet *et al.* 2001). It

Figure 3. FLAMES: the whole structure supporting the two observing plates (left photo) and the details of one observing plate (right photo): the buttoms carrying the fibers are visible.

is also interesting to recall that with an interferometric technique, van Belle *et al.* (2001) have recently performed angular size measurements for the A star Altair, that indicate a noncircular projected disk brightness distribution. Altair is the first Main-Sequence star for which direct observations of an *oblate* photosphere have been reported.

Probably, the most important application of interferometry to A star research is the determination of stellar diameters, as the combination of the Bolometric flux and the star's angular diameter provide the most direct determination of the stellar temperature. Hanbury Brown *et al.* (1974) have determined the angular diameter of Sirius as 5.89 ± 0.16 mas. In a recent work based on VINCI + VLTI data, Kervella *et al.* (2003) have measured the Sirius diameter as 6.039 ± 0.019 mas. The measurement accuracy is nearly 10 times better than the previous estimate by Hanbury Brown *et al.* (1974). Kervella *et al.* (2003) note that the largest uncertainty in the star's linear diameter is no longer given by the stellar angular diameters, but by the (Hipparcos) parallax of the star. The diameter of a large number of (nearby) A stars may be measured using AMBER + VLT (Petrov *et al.* 2000). For better than Hipparcos parallaxes, we will have to wait for the Gaia mission (see, e.g., `http://astro.estec.esa.nl/GAIA`). There is still the need to better know the actual spectral energy distribution of Ap stars, something that may be achieved with the ASTRA spectrophotometer (Adelman *et al.* 2005) which uses a 0.5-m telescope.

3.2. *Multi-object medium-high resolution spectroscopy: FLAMES*

FLAMES is a quite complex instrument. Attached to the Nasmyth focus of the telescope is the Nasmyth corrector, a system of lenses that allows the exploitation of the full 25 arcminutes diameter field of view delivered at the Nasmith focus of the VLT, correcting the field aberrations and reducing the field curvature. After the corrector comes the fiber positioner, hosting two plates (see Fig. 3). While one plate is attached to the focus during the exposure, the other one is positioning the fibres for the subsequent observation. One hundred thirty-two fibers feed the GIRAFFE spectrograph, and eight fibers feed the red arm of the high resolution spectrograph UVES (see Sect. 3.5). (In addition, there are

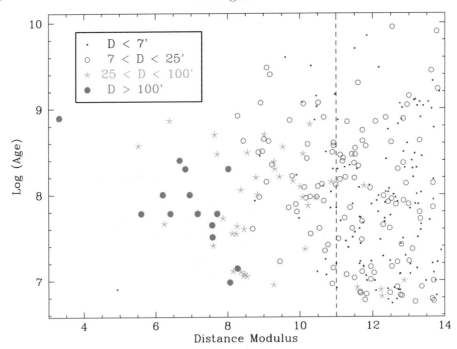

Figure 4. The positions in the log (Age) vs. Distance Modulus diagram for the open clusters selected according with the criteria explained in the text. Symbols refer to the angular size of the clusters expressed in arcminutes, to be compared with the FORS field of view (6.8 × 6.8 arcmin) and with the FLAMES field of view (25 × 25 arcmin). Data are taken from the WEBDA site developed and maintained by Jean-Claude Mermilliod (`http://obswww.unige.ch/webda/`.)

four fibers that are used to point four reference stars with the purpose of centering and guiding, plus one large and 20 small integral field units.)

FLAMES-GIRAFFE is a medium-high resolution spectrograph for the entire visible range 370-900 nm. GIRAFFE has two gratings: high and low resolution. High resolution gratings offer a variety of different settings, covering a wavelength range of 17 up to 50 nm, and with a spectral resolution of about 20000. The low resolution grating offers a number of settings covering a wavelength range from 50 to 120 nm, and with a spectral resolution of about 7-8000. For further details about the instrument see Kaufer $et\ al.$ (2004). In about 2 h of telescope time (i.e., including shutter time plus overheads), FLAMES + VLT permits one to obtain a $R = 20000$ spectrum around Hβ (from 470 nm to 500 nm) with a signal-to-noise ratio of 100 in a $V = 15$ A2 star. The range in age of open clusters that can be easily observed with FLAMES is shown in Figure 4, where all open clusters at $\delta \leqslant +20°$, with ages known from the literature, that have a distance modulus less than 14, are plotted.

3.3. *Multi-object low-resolution spectro-polarimetry: FORS1*

FORS1 is a multi-purpose instrument capable of doing optical imaging and spectroscopy, equipped with polarimetric optics (see Fig. 5). In the spectropolarimetric mode it allows one to obtain $IQUV$ Stokes profiles with a resolution up to about 2000. Bagnulo $et\ al.$ (2002) has shown that FORS1 can be used to detect the circular polarization of the Hβ lines of Ap stars and determine the mean longitudinal field with an accuracy better than 100 G (see also Wade $et\ al.$ 2005). Here I just would like to add up that FORS1 has a multi-object capability that allows one to observe simultaneously up to 9 stars at once

Figure 5. The FORS1 instrument of the VLT Unit 2 Kueyen permits one to obtain Stokes $IQUV$ profiles in the optical range with a spectral resolution up to about 2000.

in a 6.8×6.8 arcmin field of view in polarimetric mode. Hence, FORS1 is an instrument also suitable for open cluster studies, although the efficiency in terms of targets that can be observed simultaneously is much lower compared to FLAMES. To detect longitudinal fields, spectro-polarimetric observations must be characterized by a very high signal-to-noise ratio. Therefore the limiting magnitude of FORS1 is brighter than for FLAMES. If we use 11 as a conservative upper limit for the distance modulus, we can still reach a large number of clusters of various ages, as shown in Fig. 4. FORS1 has been frequently used for studies of (magnetic) open cluster Ap stars (e.g., Bagnulo *et al.* 2003a).

Obviously the scientific case of large telescope is not limited to evolutionary studies of open cluster A stars. Spectroscopy and spectropolarimetry of individual objects are extremely important for a variety of applications. For instance, Hubrig *et al.* (2004) used FORS1 for a search of magnetic field variability in roAp stars along the pulsation cycle. Drouine *et al.* (2005) used FORS1 to search for magnetic fields in Herbig Ae/Be stars.

3.4. *High resolution spectropolarimetry in the optical*

Low-resolution spectropolarimetry of Balmer lines is the most efficient method to detect magnetic fields in *fast* rotating stars ($v \sin i \geqslant 50 \, \mathrm{km s^{-1}}$), whereas high-resolution spectropolarimetry is more efficient in *slow*-rotating stars because it allows one to resolve many (metal) lines, thus increasing the amount of information that can be gathered in an exposure. Furthermore, data obtained with FORS1 and other low resolution spectropolarimeters cannot be used for a detailed abundance analysis magnetic mapping of the target star. Unfortunately, the large size class telescopes are not equipped with high resolution spectropolarimeters. The only exception *may* be the PEPSI spectrograph, that

Figure 6. The high-resolution UV and Visible spectrograph UVES of the ESO VLT.

will be attached at the Large Binocular Telescope (LBT). The current project is to have two polarimeters (one for each telescope) that will feed the (single) PEPSI spectrograph. A total of four spectra per echelle order will be imaged onto the detector. The spectrograph is designed for a resolution-slit product of $R = 40000$ for 1 arcsec entrance slit and will allow a wavelength coverage from 390 to 1050 nm in three exposures (selected by using different cross-dispersers). For further details see http://www.aip.de/pepsi/.

Up to now, the only high-resolution spectropolarimeter publicly available was the echelle MuSiCoS spectrograph at the 2 m Telescope Bernard Lyot (TBL) of Pic-du-Midi observatory (see Wade 2003 for a review of the scientific output of MuSiCoS). Things have changed with the beginning of the operations of ESPaDOnS, a $R = 70000$ spectrograph equipped with polarimetric optics, that can cover the wavelength range between 370 and 1000 nm with a single exposure. Attached to the 3.6 m CHFT, ESPaDOnS has been offered to the community starting with the first semester of 2005. ESPaDOnS + CHFT is expected to be about 100 times more efficient in gathering light than MuSiCoS + TBL. Observations of Stokes profiles of magnetic Ap stars obtained with ESPaDOnS will meet the strict requirements in terms of spectral resolution and signal-to-noise ratio that have to be satisfied for a succesfull application of the Zeeman Doppler Imaging (Kochukhov & Piskunov 2002). A copy of ESPaDOnS, NARVAL, will soon replace MuSiCoS at Pic-du-Midi.

3.5. *High resolution spectroscopy in the near UV and visible: UVES*

Probably the most important tools for A star research, and for stellar studies in general, is high resolution spectroscopy. One of the most interesting instruments available is the Ultraviolet-Visible Echelle Spectrograph, UVES, attached at the VLT Unit 2 Kueyen (see Fig. 6). In stand alone configuration, UVES can observe simultaneously in two arms, the blue arm and the red arm. Using the two available dichroics (i.e., with two subsequent exposures) the observed spectra cover almost the entire spectral range from 300 to 1000 nm, with the exceptions of a few gaps. UVES has been extensively used for A star studies, see, e.g., Kurtz (2005), who has presented the applications for high time resolution, high spectral resolution of roAp stars Kurtz *et al.* 2003). Probably, the most natural way to exploit UVES capabilities is just to perform a spectral analysis based on an unprecedented amount of lines. A publicly available library of high-resolution stellar spectra obtained with UVES has been presented by Bagnulo *et al.* (2003b). Additional specific applications to A-type stars include, for instance, the study of the wing-core anomaly of Balmer lines in Ap stars by Kochukhov *et al.* (2002).

3.6. *High resolution spectroscopic and spectropolarimetry in the IR*

Zeeman effects increases with wavelength. With an IR high-resolution spectropolarimeter one expects a substantial increase of the magnetic sensibility compared to the optical regime (although IR lines are substantially weaker and much less numerous in the IR than in the optical). CRIRES, a high resolution spectrograph equipped with polarimetric capabilities in the region between 1 and 5 μm, will soon become operational at the VLT. CRIRES will be a very useful instrument for the Ap star research even in its spectroscopic mode (i.e., without using it in polarimetric mode). An example of application of CRIRES to the A star research has been given by Mathys (2005). Optical observations of sharp line Ap stars reveal a lack of stars with a field modulus smaller than 2.8 kG (Mathys *et al.* 1997). However, observations in the optical cannot reveal mean field modulus less than about 1.7 kG (Mathys *et al.* 1997). The magnetic sensibility of high-resolution spectroscopy in the IR is much higher, and observations with CRIRES will permit one to verify the puzzling result obtained by Mathys *et al.* (1997) down to a much lower field modulus limit.

For a practical example of an IR spectroscopic study of Ap stars see Leone *et al.* (2003).

4. Conclusions: the future giant telescopes

The techniques to build and operate 6 to 10 m telescopes are now well established, and many efforts are devoted to build Extremely Large Telescopes (ELT, about 25 m size). ESO is skipping this step, and is undertaking the design of a giant, optical and near-infrared telescope with a 100 m diameter, dubbed Overwhelmingly Large Telescope (OWL).

OWL design and construction relies extensively on modular design, integration and maintenance of a large numbers of identical parts, components and modules as pioneered by Gustave Eiffel in 1899 for his eponymous tower, and, as far as the spherical primary segmented mirror is concerned, as pioneered by the Hobby-Eberly Telescope in the 1990. The OWL primary 100 m mirror is composed of 3084 hexagonal segments, each of them 1.6 m in size. The secondary mirror is 25 m in diameter, and is composed of 216 hexagonal segments, each of them is again 1.6 m in size. The modular approach can break the time-versus-diameter law of approximately 1 year per 1 m of diameter, and OWL may become operational in less than 20 years from now!

A lot of efforts are now invested in large telescopes, and the small telescopes are often sacrificed for budget reasons. The experience acquired so far with the VLT instruments shows that the A-star community can have access to the large size class telescopes, and perhaps will also benefit from the use of giant telescopes such as OWL. However, it should be noted that in the future it may become easier to have access to telescope time to obtain spectra of extragalactic A-stars, than to measure light curves of bright stars. I do not know how serious is this drawback, but it is something that deserves some attention.

Figure 7. OWL

Acknowledgements

I am deeply indebted to R. Gilmozzi, J.D. Landstreet, G.A. Wade, and M. Wittkowski for many very useful comments and suggestions, and to V. Andretta for Figs. 1 and 2.

References

Abt, H.A., 1979, ApJ, 230, 485

Adelman, S.J., Gulliver, A.F., Smalley, B., 2005, *These proceedings*, JP2

Bagnulo, S., Szeifert, T., Wade, G.A., Landstreet, J.D., & Mathys, G. 2002, A&A, 389, 191

Bagnulo, S., Landstreet, J.D., Lo Curto, G., Szeifert, T., & Wade, G.A., 2003a, A&A, 403, 645

Bagnulo, S., Jehin, E., Ledoux, C., Cabanac, R., Melo, C., & Gilmozzi, R., 2003b, The Messenger, 114, p. 10

Comeron, F., Romaniello, M., Breysacher, J., Silva, D., & Mathys, G., 2003, The Messenger, 113, 32

Drouine, D., Wade, G.A., Bagnulo, S., Landstreet, J.D., & Mason, E., 2005, *These proceedings*, EP1

Eisner, J.A., Lane, B.F., Akeson, R.L., Hillenbrand, L.A., & Sargent, A.I. 2003, ApJ, 588, 360

Fraga, L, & Kanaan, A., 2005, *These proceedings*, FP4

Gomez, A.E., Luri, X., Grenier, S., Figueras, F., North, P., Royer, F., Torra, J., & Mennessier, M.O., 1998, A&A 336, 953

Hanbury Brown, R., Davis, J., Lake, R.J.W., & Thompson, R.J., 1974, MNRAS 167, 475

Hauck, B., & North, P., 1993, A&A, 269, 403

Hubrig, S., North, P., & Mathys, G., 2000, ApJ, 539, 352

Hubrig, S., Kurtz, D.W., Bagnulo, S., Szeifert, T., Schöller, M., Mathys, G., & Dziembowski, W.A., 2004, A&A, 415, 661

Kaufer, A., Pasquini, L., Zoccali, M., Dekker, H., Cretton, N., & Smoker, J., 2004, FLAMES User Manual, Doc. No. VLT-MAN-ESO-13700-2994

Kervella, P., Thévenin, F., Morel, P., Bordé, P., & Di Folco, E., 2003, A&A, 408, 681

Kochukhov, O., & Piskunov, N., 2002, A&A, 388, 868

Kochukhov, O., Bagnulo, S., Barklem, P., 2002, ApJL, 578, 75

Kurtz, D.W., Elkin, V.G., Mathys, G., Riley, J., Cunha, M.S., Shibahashi, H., Kambe, E., 2005, *These Proceedings*, 343

Kurtz, D.W., Elkin, V.G., & Mathys, G., 2003, MNRAS, 345, L5

Leone, F., Vacca, W.D., & Stift, M.J., 2003, A&A, 409, 1055

Mathys, G., Hubrig, S., Landstreet, J.D., Lanz, T., & Manfroid, J. 1997, A&AS, 123, 353

Mathys, G., 2005, *These Proceedings*, 225

Millan-Gabet, R., Schloerb, F. Peter, & Traub, W., 2001, ApJ, 546, 358

Monier, R., Richard, O., 2005, *These Proceedings*, AP2

Moon, T.T., Dworetsky, M.M., 1985, MNRAS, 217, 305

North, P., 1993, in: Peculiar vs. normal phenomena in A-type and related stars; M.M. Dworetsky, Castelli, F., & Faraggiana, R. (eds.), A.S.P. Conf. Ser., vol. 44, p. 577

North, P. 1998, A&A, 334, 181

North, P. & Kroll, R., 1989, A&AS, 78, 325

Padovani, P., 2005, *These Proceedings*, 485

Paunzen, E., Pintado, O.I., & Maitzen, H.M., 2003, A&A, 412, 721

Petrov, R., Malbet, F., Richichi, A., *et al.*, 2000, SPIE, 4006, 68

Pöhnl, H., Maitzen, H.M., & Paunzen, E. 2003, A&A, 402, 247

Schaller, G., Schaerer, D., Meynet, G., & Maeder, A., 1992, A&AS, 96, 269

Southworth, J., Smalley, B., Maxted, P. 2005, *These Proceedings*, BP2

Stępień, K., 2000, A&A, 353, 227

Stępień, K., & Landstreet, J.D., 2002, A&A, 384, 554

van Belle, G.T., Ciardi, D.R., Thompson, R.R., Akeson, R.L., & Lada, E.A. 2001, ApJ, 559, 1155

Wade, G.A., 2003, in: Solar Polarization III; J. Trujillo Bueno & J. Sánchez Almeida (eds.), A.S.P. Conf. Ser. vol. 307, p. 519

Wade, G.A., 2005, *These Proceedings*, 235

Discussion

KHALACK: I understand that the OWL telescope is only a project. It seems to me that the problem of the simultaneous correction of the position of the segments in an extremely short time scale does not have a technical solution at a sufficiently high quality level at present.

BAGNULO: Segments must be re-adjusted in position a few times per second to cope with flexures and thermal changes. Misalignments are measured by edge sensors that are calibrated with a phasing sensor located at the focus of the telescope. The phasing sensors used in the Keck telescopes should theoretically work with up to about 4000 segments; others kinds of sensor have been shown to work in laboratory experiments and are being developed for on-sky testing (see http://www.eso.org/projects/owl/FAQs.html)

DWORETSKY: Following the project to obtain 400 spectra for a public UVES library, is it possible to do a twilight service observing project to observe many other stars, once each?

BAGNULO: The major problem is that now all the foci of the Unit 2 Kueyen are equipped with instruments (instead, when UVESPOP was being carried out, UVES was the only instrument operating at Kueyen). In particular, FORS1, attached at the Cassegrain focus, needs frequent skyflat calibrations that must be taken during twilight. It is not impossible to use twilight time to do science with UVES, but it is not so "painless" as it was in the past.

SMALLEY: Another way to obtain fundamental parameters of stars is to use eclipsing spectroscopic binaries which allows us to determine model-independent and This is especially worth in open clusters where age and distances are constrained (see Southworth *et al.* 2005)

The A-Star Puzzle
Proceedings IAU Symposium No. 224, 2004
J. Zverko, J. Žižňovský, S.J. Adelman, & W.W. Weiss, eds.

© 2004 International Astronomical Union
DOI: 10.1017/S1743921304004910

A-stars and the Virtual Observatory

Paolo Padovani

AVO Scientist, ST-ECF, European Southern Observatory, Karl-Schwarzschild-Str. 2, D-85748
Garching bei München, Germany
email: Paolo.Padovani@eso.org

Abstract. The Virtual Observatory (VO) will revolutionise the way we do Astronomy, by allowing easy access to all astronomical data and by making the handling and analysis of datasets at various locations across the globe much simpler and faster. I report here on the need for the VO and its status in Europe, including the first ever VO-based astronomical paper, and then give two specific applications of VO tools to open problems of A-stars research.

Keywords. Methods: miscellaneous, techniques: miscellaneous, astronomical data bases: miscellaneous, stars: general, stars: chemically peculiar, Galaxy: open clusters and associations: general, X-rays: stars

1. Astronomy in the XXIst century

Astronomy is facing the need for radical changes. When dealing with surveys of up to ~ 1000 sources, one could apply for telescope time and obtain an optical spectrum for each one of them to identify the whole sample. Today we have to deal with huge surveys (e.g., the Sloan Digital Sky Survey – SDSS, Abazajian *et al.* 2004, the Two Micron All Sky Survey – 2MASS, Cutri *et al.* 2003, the Massive Compact Halo Object – MACHO, e.g., Alcock *et al.* 2001 survey), reaching (and surpassing) 100 million objects. Even at, say, 3000 spectra at night, which is only feasible with the most efficient multi-object spectrographs and for relatively bright sources, such surveys would require more than 100 years to be completed, a time which is clearly much longer than the life span of the average astronomer! But even taking a spectrum might not be enough to classify an object. We are in fact reaching fainter and fainter sources, routinely beyond the typical identification limits of the largest telescopes available (approximately 25 magnitude for 2 - 4 hour exposures), which makes a "classical" identification problematic. These very large surveys are also producing a huge amount of data. It would take about two months to download at 1 Mbytes/s (an extremely good rate for most astronomical institutions) the Data Release 2 (DR2; http://www.sdss.org/dr2/) SDSS images and about two weeks for the catalogues. The images would fill up ~ 1000 DVDs (~ 500 if using dual-layer technology). The final SDSS will be about three times as large as the DR2. These data, once downloaded, need also to be analysed, which requires tools which may not be available locally and, given the complexity of astronomical data, are different for different energy ranges. Moreover, the breathtaking capabilities and ultra-high efficiency of new ground- and space-based observatories have led to a "data explosion", with astronomers world-wide accumulating ≈ 1 Terabyte of data per night. For example, the European Southern Observatory (ESO)/Space Telescope European Coordinating Facility (ST-ECF) archive is predicted to increase its size by two orders of magnitude in the next eight years or so, reaching ≈ 1000 Terabytes. Finally, one would like to be able to use all of these data, including multi-million-object catalogues, by putting this huge amount of information together in a coherent and relatively simple way, something which is impossible at present.

All these hard, unescapable facts call for innovative solutions. For example, the observing efficiency can be increased by a clever pre-selection of the targets, which will require some "data-mining" to characterise the sources' properties before hand, so that less time is "wasted" on sources which are not of the type under investigation. One can expand this concept even further and provide a "statistical" identification of astronomical sources by using all the available, multi-wavelength information without the need for a spectrum. The data-download problem can be solved by doing the analysis where the data reside. Finally, easy and clever access to all astronomical data worldwide would certainly help in dealing with the data explosion and would allow astronomers to take advantage of it in the best of ways.

2. The Virtual Observatory

The solution is the Virtual Observatory (VO). The VO is an innovative, evolving system, which will allow users to interrogate multiple data centres in a seamless and transparent way, to utilise the best astronomical data. Within the VO, data analysis tools and models, appropriate to deal also with large data volumes, will be made more accessible. New science will be enabled, by moving Astronomy beyond a "classical" identification with the characterisation of the properties of very faint sources by using all available information. All this will require good communication, that is the adoption of common standards and protocols between data providers, tool users and developers. This is being defined now using new international standards for data access and mining protocols under the auspices of the recently formed International Virtual Observatory Alliance (IVOA: `http://ivoa.net`), a global collaboration of the world's astronomical communities.

One could think that the VO will only be useful to astronomers who deal with colossal surveys, huge teams and Terabytes of data! That is not the case, for the following reason. The World Wide Web is equivalent to having all the documents of the world inside one's computer, as they are all reachable with a click of a mouse. Similarly, the VO will be like having all the astronomical data of the world inside one's desktop. That will clearly benefit not only professional astronomers, but also anybody interested in having a closer look at astronomical data. Consider the following example: imagine one wants to find *all* high-resolution spectra of A-type stars available in *all* astronomical archives in a given wavelength range. One also needs to know which ones are in raw or processed format, one wants to retrieve them and, if raw, one wants also to have access to the tools to reduce them on-the-fly. At present, this is extremely time consuming, if at all possible, and would require, even to simply find out what is available, the use a variety of search interfaces, all different from one another and located at different sites. The VO will make it possible very easily.

3. The VO in Europe: the Astrophysical Virtual Observatory

The status of the VO in Europe is very good. In addition to seven current national VO projects, the European funded collaborative Astrophysical Virtual Observatory initiative (AVO: `http://www.euro-vo.org`) is creating the foundations of a regional scale infrastructure by conducting a research and demonstration programme on the VO scientific requirements and necessary technologies. The AVO has been jointly funded by the European Commission (under the Fifth Framework Programme – FP5) with six European organisations participating in a three year Phase-A work programme. The partner organisations are ESO in Munich, the European Space Agency, AstroGrid (funded by

PPARC as part of the United Kingdom's E-Science programme), the CNRS-supported Centre de Donnees Astronomiques de Strasbourg (CDS) and TERAPIX astronomical data centre at the Institut d'Astrophysique in Paris, the University Louis Pasteur in Strasbourg, and the Jodrell Bank Observatory of the Victoria University of Manchester. The AVO is the definition and study phase leading towards the Euro-VO - the development and deployment of a fully fledged operational VO for the European astronomical research community. A Science Working Group was also established two years ago to provide scientific advice to the project.

The AVO project is driven by its strategy of regular scientific demonstrations of VO technology, held on an annual basis in coordination with the IVOA. For this purpose progressively more complex AVO demonstrators are being constructed. The current one, a downloadable Java application, is an evolution of Aladin, developed at CDS, and has become a set of various software components, provided by AVO and international partners, which allows relatively easy access to remote data sets, manipulation of image and catalogue data, and remote calculations in a fashion similar to remote computing.

4. Doing science with the AVO

The AVO held its second demonstration, 'AVO 1st Science', on January 27 - 28, 2004 at ESO. The demonstration was truly multi-wavelength, using heterogeneous and complex data covering the whole electromagnetic spectrum. These included: MERLIN, VLA (radio), ISO (spectra and images), and 2MASS (infrared), USNO, ESO 2.2m/WFI and VLT/FORS (spectra), and HST/ACS (optical), XMM and Chandra (X-ray) data and catalogues. Two cases were dealt with: an extragalactic case on obscured quasars, centred around the Great Observatories Origin Deep Survey (GOODS) public data, and a Galactic scenario on the classification of young stellar objects.

The extragalactic case was so successful that it turned into the first published science result fully enabled via end-to-end use of VO tools and systems, the discovery of $\sim$30 high-power, supermassive black holes in the centres of apparently normal looking galaxies (Padovani *et al.* 2004). The AVO prototype made it much easier to classify the sources we were interested in and to identify the previously known ones, as we could easily integrate all available information from images, spectra, and catalogues at once. This is proof that VO tools have evolved beyond the demonstration level to become respectable research tools, as the VO is already enabling astronomers to reach into new areas of parameter space with relatively little effort.

5. The VO and A-type stars

I have used the AVO prototype to tackle two problems of A star research, namely, establishing membership in an open cluster and assessing if chemically peculiar A-type stars are more likely to be X-ray emitters than normal A-type stars.

5.1. *Open cluster membership*

Cluster membership is vital to determine the distance, and therefore absolute magnitude, and age of A-type stars, see, e.g., Monier & Richard (2005) and Bagnulo (2005). In short, open clusters play a crucial role in stellar astronomy because, as a consequence of the stars having a common age, they provide excellent natural laboratories to test theoretical stellar models. The AVO prototype can be of help in determining if a star does belong to an open cluster, at various levels.

I have chosen the Pleiades as the target, since even extragalactic astronomers like me know about it (although the value of its parallax is strongly debated: see, e.g., Pan *et al.*

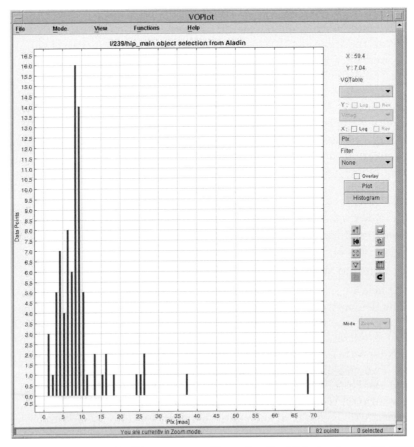

Figure 1. An histogram of the Hipparcos parallaxes for the Pleiades done with VOPlot, the graphical plug-in of the AVO prototype. Note the peak at around $8 - 9$ mas, consistent with the cluster value of 8.46 ± 0.22 mas, and the presence of many foreground and some background stars.

2004)! I will also use this example to describe the capabilities of the tool. A step-by-step guide which should allow anyone to reproduce what I have done here with the AVO prototype can be found at http://www.eso.org/~ppadovan/vo.html.

We start by loading a Second Palomar Observatory Sky Survey (POSS II) image of the Pleiades. Since we are interested in cluster membership, we need distance information. The AVO prototype has a direct link to VizieR (http://vizier.u-strasbg.fr/), a service which provides access to the most complete library of published astronomical catalogues and data tables available on-line. We then search for all VizieR catalogues which provide parallax information and find the Hipparcos catalogue (Perryman 1997), which we load into the prototype. All Hipparcos sources are automatically overlaid on the POSS II image. We can now very easily plot an histogram of the Hipparcos parallaxes by using VOPlot, the graphical plug-in of the prototype. Most stars in the image have parallaxes ~ 8–9 mas, consistent with the cluster value of 8.46 ± 0.22 mas (Robichon et al. 1999), but there are also many foreground and some background stars (Fig. 1). We now want to plot a colour – magnitude diagram but first we need to correct for reddening (which is 0.04 for this cluster) the observed $(B - V)$ colour given in the Hipparcos catalogue. We then create a new column, $(B - V)_{o} = (B - V) - 0.04$, and then plot the observed V_{mag} vs. $(B - V)_{o}$. The Zero Age Mean Sequence (ZAMS), flipped because we

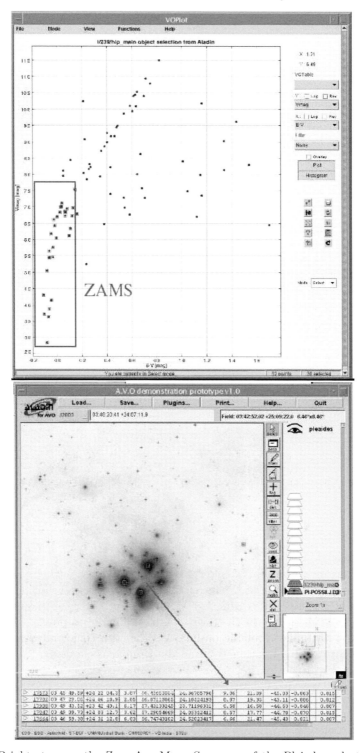

Figure 2. Bright stars on the Zero Age Mean Sequence of the Pleiades selected in VOPlot (top figure, bottom left corner) are highlighted (squares) in the POSS II image (bottom figure) and turn out to be mostly in the centre of the cluster, with parallaxes (indicated by the arrow) $\sim 8-9$ mas, consistent with the cluster value of 8.46 ± 0.22 mas.

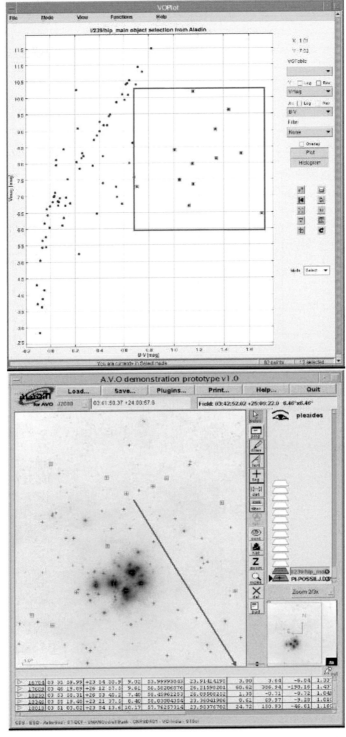

Figure 3. Stars off the Zero Age Mean Sequence of the Pleiades selected in VOPlot (top figure, top right corner) are highlighted (squares) in the POSS II image (bottom figure) and turn out to be mostly at the outskirts of the cluster, with parallaxes (indicated by the arrow) typically inconsistent with the cluster value of 8.46 ± 0.22 mas.

are using observed and not absolute magnitudes, is clearly visible (Fig. 2, top). We now have a very nice, visual match between the ZAMS and cluster membership. By selecting in VOPlot bright stars which are on the ZAMS (top of Fig. 2, bottom left corner), the corresponding sources are highlighted in the image. These are mostly in the centre of the cluster. By selecting them with the cursor one can see that most of them have parallaxes $\sim 8 - 9$ mas, as expected (Fig. 2). On the other hand, if we now select in VOPlot the sources off the ZAMS, in the POSS II image one can see that they are mostly at the edge of the field and, looking at their parallaxes, generally foreground sources, with a couple of background ones and only a few possible members (Fig. 3).

At this point one could do things properly, that is use the Hipparcos data and the mean radial velocity of the cluster centre, together with eq. (1) of Robichon *et al.* (1999), to add new columns with the relevant parameters to the Hipparcos catalogue, and determine membership based on a statistical criterion. At present, this would be quite cumbersome, although still possible. Very soon, however, one will be able to add such functionality to the VO as a "Web Service". Web Services will promote growth of the VO in the way that web pages grew the World Wide Web. For example, a tool to determine cluster membership could be "published" to the VO as a service to which astronomers can send their input, in an appropriate format, and then receive the output, e.g., a list of cluster members and non-members.

Using the AVO prototype, one can also look for data available for selected sources at various mission archives. For example, in our case one could select some sources in the image and then select ESO, the Hubble Space Telescope (HST), and the International Ultraviolet Explorer (IUE) under "Missions in VizieR". The pointings for these three missions for the sources under examination would then be overlaid in the image. Selecting one of these pointings provides, in some cases (e.g., HST and IUE), links to preview images, so that one can have a "quick look" at the archival data. By clicking on the HST "Dataset" column the default Web browser starts up and the dataset page at the Multimission Archive at STScI (MAST) is made available. From there one can have access to all papers published using those data. So from the AVO prototype one is only two "clicks" away from the journal articles which have used the MAST data of the astronomical sources in the image!

To get a flavour of the wealth of archival data available for A-type stars, I have also cross-correlated the Sky2000 catalogue with the MAST holdings using the service available at `http://archive.stsci.edu/search/sky2000.html`. Out of $\sim$22,400 A-type stars, it turns out that 754 have IUE data, for a total of $\sim$10,000 observations, while 128 have non-IUE data (FOS, GHRS, STIS, FUSE, EUVE, Copernicus, HUT, WUPPE, and BEFS; see the MAST site at `http://archive.stsci.edu/` for details on all of these missions), for a total of 1,700 observations, $\sim$60% of which are spectra. How many more data are available in the other astronomical archives? Only the VO will allow us to answer that question in a relatively simple way.

5.2. *Peculiar A stars in the X-ray band*

The issue of the X-ray emission of A-type stars is a long standing one (see, e.g., Simon *et al.* 1995). While it is clear that these objects can be X-ray sources, it has been suggested that their X-ray emission is not associated with the star itself but might come from a binary companion. Damiani *et al.* (2003) have analyzed *Chandra* observations of the young open cluster NGC 2516 and detected only twelve A stars, out of 58, while six out of eight of the chemically peculiar (CP) A-stars were detected (a difference significant at the $\sim 2\sigma$ level). It has then been suggested, also on the basis of previous results (e.g., Dachs & Hummel 1996), that CP A-type stars are more easily detected in the X-rays

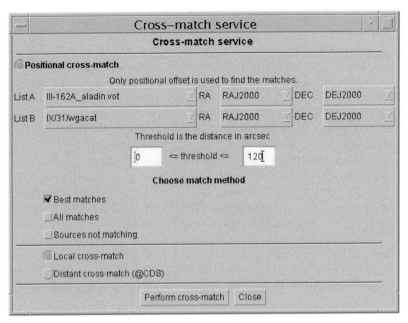

Figure 4. The cross-match plug-in of the AVO prototype, used here to crosscorrelate the Renson *et al.* (1991) catalogue of chemically peculiar stars with WGACAT, a catalogue of all *ROSAT* observations.

than normal A stars, although the astrophysical implications of this result would not be straightforward. The AVO prototype provides a relatively simple way to address directly and with sound statistics the following question: "Are CP A-type stars more likely to be X-ray emitters than normal A-type stars?".

I first started by loading from VizieR the Renson *et al.* (1991) catalogue of CP stars, which include 6,684 sources, into the prototype. I then proceeded to select the 4,736 A-type stars by using the "Filter" option. As comparison samples I used the Henry Draper (HD) and SAO star catalogues, which contain 272,150 and 258,944 sources, out of which I selected the 72,154 and 47,230 A-stars respectively. I then loaded WGACAT (White *et al.* 1995), a catalogue of all *ROSAT* Position Sensitive Proportional Counter (PSPC) observations, covering approximately 10% of the sky and including about 92000 serendipitous sources (excluding the targets, which could bias the result). The results of the cross-correlations, done using the cross-match plug-in (Fig. 4), are as follows (all errors are 3σ and spurious sources have been subtracted off by doing the match with a shift of 1 degree in the coordinates):

(*a*) CP A-stars with WGACAT: 74/4736 matches, i.e., $1.6 \pm 0.6\%$;

(*b*) HD A-stars with WGACAT: 368/72154, i.e., $0.5 \pm 0.1\%$; correcting statistically for CP stars contamination, assumed to be at the 10% level (e.g., Monin *et al.* 2002), one gets ∼0.4%;

(*c*) SAO A-stars with WGACAT: 327/47230, i.e., $0.7 \pm 0.1\%$; correcting statistically for CP stars contamination, one gets ∼0.6%.

Therefore, one can conclude that *CP A-stars are 3 (SAO) to 4 (HD) times more likely to be X-ray sources than normal A-type stars* with very high significance. Selecting only the magnetic stars in Renson *et al.* (1991), which can be identified has having Sr, Cr, Eu, Si, He, Ti, and Ca classification peculiarities (e.g., Landstreet 1992), one finds 22/1433 matches, that is a $1.5 \pm 1.0\%$ detection rate, not significantly different from that of all CP A-stars. It then appears that the presence of a magnetic field does not play a role in triggering X-ray emission, a somewhat puzzling result which deserves further investigation.

A related issue is that of radio emission. The strong fields present in the magnetic subclass of CP stars (see above), in fact, should give rise to radio emission, for example via the gyrosynchrotron mechanism (Trigilio *et al.* 2004). I have then cross-correlated the A-type stars in the Renson *et al.* (1991) catalogue with two large-area radio catalogues, namely the NRAO-VLA Sky Survey (NVSS, Condon *et al.* 1998), which covers the sky north of $\delta = -40°$ down to $\sim$3.5 mJy at 1.4 GHz, and the Faint Images of the Radio Sky at Twenty-centimeters (FIRST, White *et al.* 1997), which covers $\sim$1/5 of the sky, mostly in the north Galactic cap, down to $\sim$1 mJy 1.4 GHz. I found no matches. In retrospect, this is not surprising as the few CP stars detected so far have radio fluxes $\lesssim 1-2$ mJy (e.g., Drake *et al.* 1987).

6. Conclusions

The main conclusions follow:

(*a*) We need to change the way we do Astronomy if we want to take advantage of the huge amount of data we are being flooded with. The way to do that is through the Virtual Observatory.

(*b*) The Virtual Observatory will make the handling and analysis of astronomical data and tools located around the world much easier, enabling also new science.

(*c*) Everybody will benefit, including A-type star researchers!

(*d*) Virtual Observatory tools are available now to facilitate astronomical research and, as I have shown, can also be applied to A-stars.

Visit `http://www.euro-vo.org/twiki/bin/view/Avo/SwgDownload` to download the AVO prototype. I encourage astronomers to download the prototype, test it, and use it for their own research. For any problems with the installation and any requests, questions, feedback, and comments you might have please contact the AVO team at twiki@euro-vo.org. (Please note that this is still a prototype: although some components are pretty robust some others are not.)

Acknowledgements

I would like to thank the organizers of the conference for their kind invitation, which has allowed an extragalactic astronomer like me to learn about stars! I am also grateful to Stefano Bagnulo for his help in preparing my talk and to Mark Allen for reading this paper. I have made extensive use of the CDS VizieR catalogue tool, SIMBAD and the Aladin sky atlas service. The Astrophysical Virtual Observatory was selected for funding by the Fifth Framework Programme of the European Community for research, technological development and demonstration activities, under contract HPRI-CT-2001-50030.

References

Abazajian, K. *et al.* 2004, *AJ*, 128, 502
Alcock, C. *et al.* 2001, *ApJS*, 136, 439
Bagnulo, S. 2005, *These Proceedings*, 473
Condon, J.J., Cotton, W.D., Greisen, E.W., Yin, Q.F., Perley, R.A., Taylor, G.B. & Broderick, J.J. 1998, *AJ*, 115, 169
Cutri, R. M. *et al.* 2003, *Explanatory Supplement to the 2MASS Second Incremental Data Release* available at `http://www.ipac.caltech.edu/2mass/releases/allsky/doc/-explsup.html`
Dachs, J. & Hummel, W. 1996, *A&A*, 312, 818

Damiani, F., Flaccomio, E., Micela, G., Sciortino, S., Harnden, F.R., Murray, S.S., Wolk, S.J. & Jeffries, R.D. 2003, *ApJ*, 588, 1009

Drake, S.A., Abbott, D.C., Bastian, T.S., Bieging, J.H., Churchwell, E., Dulk, G. & Linsky, J.L. 1987, *ApJ*, 322, 902

Landstreet, J.D. 1992, *A&AR*, 4, 35

Monier, R. & Richard, O. 2005, *These Proceedings*, AP2

Monin, D.N., Fabrika, S.N. & Valyavin, G.G. 2002, *A&A*, 396, 131

Padovani, P., Allen, M.G., Rosati, P. & Walton, N.A. 2004, *A&A* in press (astro-ph/0406056)

Pan, X., Shao, M., & Kulkarni, S.R. 2004, *Nature*, 427, 326

Perryman, M. A. C. 1997, *The Hipparcos and Tycho Catalogues*, ESA SP-1200 (Noordwijk: ESA)

Renson, P., Gerbaldi, M. & Catalano, F.A. 1991, *A&AS*, 89, 429

Robichon, N., Arenou, F., Mermilliod, J.-C. & Turon, C. 1999, *A&A*, 345, 471

Simon, T., Drake, S.A. & Kim, P.D. 1995, *PASP*, 107, 1034

Trigilio, C., Leto, P., Umana, G., Leone, F. & Buemi, C.S. 2004, *A&A*, 418, 593

White, N.E., Giommi, P. & Angelini, L. 1995, on-line database at
    `http://wgacat.gsfc.nasa.gov`

White, R.L., Becker, R.H., Helfand, D.J. & Gregg, M.D. 1997, *ApJ*, 475, 479

Discussion

COWLEY: Can you clarify what you said for me about the fraction of CP-stars that were X-ray sources? What Bord and I found in our surveys was that in those cases where an Ap (CP) star was found to be an X-ray source, there was always the possibility to attribute the X-rays to a companion or a wind. For the radio loud objects we found there were magnetic Ap stars that were definite radio sources.

PADOVANI: I said that CP stars are three to four times more likely to be X-ray sources than normal A-type stars. But the origin of the X-rays is not revealed by my study. As regards the radio, I have pointed out that the all-sky catalogues are not deep enough to detect A-type stars in the radio band. One would need dedicated VLA observation to detect them because they are too faint for the all-sky surveys.

ŠKODA: How will you manage copyright rules, that is, who should one cite in research done using the VO?

PADOVANI: The original data providers, as far as the data are concerned, and any relevant VO project for the tools.

ŠKODA: What if I ask for some data (e.g., spectra of a particular star) which are still under proprietary period? Will I be informed about this?

PADOVANI: Yes. You will be told that the data are present but still proprietary. You will also be told when you can access them and perhaps even reminded when they become available. But remember that the VO is still not fully operational so at present you simply cannot access the data.

KUBÁT: How do you control the quality of the data in catalogues?

PADOVANI: That is a very important point, but the VO cannot do that, as it would require a lot of resources. Data quality control will be up to those who know the data best, that is the data providers.

The A-Star Puzzle
Proceedings IAU Symposium No. 224, 2004
J. Zverko, J. Žižňovský, S.J. Adelman, & W.W. Weiss, eds.
© 2004 International Astronomical Union
DOI: 10.1017/S1743921304004922

Panel discussion section J

CHAIR: **Gregg A. Wade**

SECTION ORGANIZER & KEY-NOTE SPEAKER: Stefano Bagnulo
INVITED SPEAKERS: N. Piskunov, P. Padovani

Discussion

WADE: For Padovani: Is it straigthforward to implement our own catalogs into the Virtual Observatory (VO)?

PADOVANI: Very easy. There is a button to load your own data, and that includes a list of sources. The table has to contain one record per line, each field separated by a TAB. Once loaded, your own catalog can be overlayed on any image and cross-correlated with any catalog in the public domain.

ŠKODA: How do you handle in the VO the ambiguity of positions when cross-correlating several catalogues (mainly faint objects)? Can you add more information to search?

PADOVANI: Certainly. The present tool is very simple but we are planning to improve it and add more features, including a probabilistic determination of cross matches.

MATHYS: How do you retrieve moving targets from AVO?

PADOVANI: This is an issue which is still open. The time domain is still largely unexplored but it will certainly be considered.

ADELMAN: Elizabeth Griffin has pleaded the case for getting spectroscopic plates scanned, which I think is quite important. An equally important case can be made for photometric measurements using photomultiplier tubes and CCDs, most of which have never made their way into the literature except perhaps graphically. This is especially the case prior to journals accepting electronic tables.

PADOVANI: The usual answer is the following: the VO cannot do that. It is up to the data providers or archive centres to do that. We do not have the manpower. We can tell you what to do and how to do it, but we cannot do it for you, unfortunately.

ADELMAN: Still someone or some organization needs to encourage people with this type of data to submit it or astronomy soon loose much of the photometry obtained in the past 50 years.

LANDSTREET: It would be very useful if the VO could provide tools for converting typed pages (old tables, etc.) to ASCII files. For deciding how to consolidate old data in the hands of individuals, it would be good to involve IAU commissions such as variable stars, photometry, polarimetry, etc.

PADOVANI: Commercial software exists which does Optical Character Recognition (OCR) and converts any printed material to ASCII files. I am not sure it is the VO's role to provide anything which goes beyond that. As I have just said, it is up to the data providers or archive centres to do that.

BAGNULO: It seems to me that the A-star community is very interested in the research possibilities offered by the VO. Paolo Padovani has also presented a few practical possible applications. Another perspective of the future astronomical research is the use of Extremely to Overwhelmingly Large Telescopes, and I have a question for the all of us present in this room. If you get observing time for the A-star research with a 100-m telescope, what would you do?

PISKUNOV: To use a 100-m telescope for a single object project will be much easier because it is much easier to correct the small field of view for atmospheric disturbances and obtain a nearly diffraction limited image on the optical axis of the telescope which means that high-resolution spectroscopy is the most logical thing to do.

BAGNULO: The question is high-resolution spectroscopy of what? Why do you need a 100-m telescope to do the high-resolution spectroscopy of A-stars? What would you do that cannot be done with smaller telescopes? To this question, some time ago John Landstreet answered that we should study whether the A-star phenomenon is just a local phenomenon or it is common in other galaxies. This we can do it only with extremely large telescopes.

WADE: Can we resolve individual A-type star in nearby galaxies?

WEISS: It has been done. There has been a discovery of CP stars by Maitzen and co-workers.

BAGNULO: Still in the Large Magellan Cloud, that is close by our galaxy. In any case they have been identified only via Δa photometry. We do not have spectroscopic observations and thus do not know the detailed chemical characteristics of these stars.

LANDSTREET: I would like to make two comments. First: we may hope to get high-resolution spectrograms on new large telescopes because the planet-finding community may want them. Second: a lot of useful information is available for clusters. But to study OLD stars in OLD clusters, which are rare, we need to observe to great distances. Furthemore, we can observe stars (Am, Ap, HgMn, etc.) formed in really different environments from our local one by looking at the nearby dwarf galaxies.

PAUNZEN: We do not even know the incidence of CP stars for open clusters within two kpc around the Sun. Why go for dwarf galaxies? By contrast, there is a need for classification resolution spectroscopy in our Milky way. Especially for field stars from $10 < V < 15$ mag.

CORBALLY: Paunzen's remarks prompt me to point out that the power of the "Nearby Stars" project (see C. J. Corbally & R. O. Gray, 2005, *These Proceedings*, IP1), which I am doing with Richard Gray, Bob Garrison, and others, is in its homegeneous approach to a VOLUME limited sample (A0pe). The latter point is thanks to Hipparcos data. In the future we can look to such satellites as GAIA to let people confidently survey larger volumes, while more automated techniques will help them cope with the numbers of object that are cubed. I am sure they will find interesting science, perhaps concentrating on particular parts of the HR diagram such as A-stars. It is very worthwhile, as well as fun, to get to know one's neighbours.

WEISS: I would like to come back to the question raised by Bagnulo To me it appears a little bit as a virtual problem. At first I was quite sure that we will find a lots of excellent scientific reasons why we would like to use a 100-m telescope. But then I see the hurdle of the telescope time allocation committee. Already it is nearly impossible to convince a reasonable scientist that it is necessary to point the 8-m telescope to a star which you can see with the naked eye. I am having troubles with 8-m telescopes and the naked eye visibility of a star and transferring them to a 100-m telescope. Then I see there is no chance to get time unless you promise to observe an Ap star in a quite remote galaxy...

BAGNULO: I do not think it is a too serious problem. In fact, a lot of stellar science is being done at the VLT, and a large number of CP stars have been already observed with UVES. Don Kurtz has presented at this conference the preliminary results of a lot observations of roAp stars obtained also with UVES. However it is true that the largest telescopes are designed mainly to do extragalactic and cosmological studies, or to search for life on other planets, so we should be ready to prepare an adequate scientific case for the studies of A-stars with extremely large telescopes.

MATHYS: It shoud be stressed that a very significant factor of the time at ESO's VLT is devoted to observations of stars of the Bright Star Catalogue. On some of the instruments, there is actually more bright time than dark time allocated. I can only urge again the members of this comunity to get rid of the misconception that stellar astronomers have little access to 8–10 m telescopes, and to encourage them to submit observing time applications with the best possible science cases.

PRZYBILLA: When talking about science to be done with a 100 m telescope one has to remark that it will most efficiently operate in the IR. This will allow to study objects in otherwise inaccessible environments, like, e.g., A-stars in the Galactic centre. To do so, we should start to prepare for this now, by extending the observed wavelengths range and gain experience in the IR.

WADE: Of the existing 8+ m telescopes, the Keck, the largest telescope in the world, what has it done which is remarkably relevant to A-star research? The only such telescope we have heard about is VLT!

PISKUNOV: The Keck time allocation procedure makes it difficult to get time for A-star research. At Gemini is possible, but there is very limited high-resolution capabilities.

PRZYBILLA: Spectroscopy with Keck on A-stars *is* in fact being done. The group around Kudritzki et al. has already started multi-object spectroscopy of A-supergiants in galaxies of the local group, and publications on that are pending.

MATHYS: The use of 8-10 m telescopes by the A-star community seems to reflect the difference of population of astronomical community between North America and Europe. This to me appears perfectly illustrated by the composition of the assembly in this room.

ADELMAN: I also want to make a few comments about small telescopes. Small telescopes especially those which are automated at good sites, and properly equipped can gather a substantial amount of photometry. Those with photomultiplier photometry could be upgraded to CCD photometry. The forthcoming ASTRA spectrophotometer will be able to measure energy distributions of some 40000 bright (-1 to 9 mag) stars per year.

STÜTZ: For Bagnulo: Combining interferometry and spectroscopy at the VLT has been brought up some time ago. When, do you think, that possibility will be working and what will be the chances that one will get observing time for such a setup doing A-star research?

BAGNULO: I am not the right person to answer, as I just heard about a proposal to combine UVES with VLTI. This project is done within the European Interferometry Initiative (under OPTICON) and shall be one of those to be proposed as second generation VLTI-instruments. A paper has been presented by Quirrenbach in the last SPIE meeting in Glasgow.

The A-Star Puzzle
Proceedings IAU Symposium No. 224, 2004
J. Zverko, J. Žižňovský, S.J. Adelman, & W.W. Weiss, eds.
ⓒ 2004 International Astronomical Union
DOI: 10.1017/S1743921304004934

How are we doing so far?
A summary of the Symposium

Michael M. Dworetsky

Department of Physics and Astronomy, University College London, Gower Street, London
WC1E 6BT, UK
email: mmd@star.ucl.ac.uk

Abstract. Some highlights of Symposium 224 on "The A-Star Puzzle" are reviewed and transcribed from the author's *ad hoc* oral presentation on the final day of the meeting. Articles referred to are all contained in this volume, hence there are no figures or references. Topics include theory and observations of normal A stars, HgMn, Am, and Ap stars, λ Boo stars, magnetic fields, rotation, convection, pulsations, supergiant stars, and observational methods including polarimetry, spectroscopy, and photometry.

Keywords. Stars: abundances, stars: chemically peculiar, stars: horizontal-branch, stars: magnetic fields, stars: pre-main-sequence

1. Introduction

After listening to several days of talks on the latest ideas and achievements in research in this vast and vibrant field, I am somewhat overwhelmed by the power and detail in both the new observations and the theoretical advances being made. I feel rather like the commander of an English regiment at the battle of Waterloo, who found himself surrounded by French forces on all sides. He dispatched a messenger to the Duke of Wellington to request urgent reinforcements, but everything available had been committed elsewhere. Instead, a message was sent back, the rider dodging shot and shell, and when the dispatch was opened, it read: "Sir–you are in a devilish awkward predicament, and you must get out of it as best you can."

2. Normal and peculiar A stars

When in doubt, I always find that the direct approach often works, so let us start at the beginning. On our first day, we began with Adelman's review of what we mean by the terms "normal" and "peculiar" in the A-star business. Essentially anything not classed as peculiar is assumed to be normal, but there may be no bold line that can be drawn between normal and peculiar, as defined by low or intermediate dispersion classification. The A stars themselves range from radiative photospheres at A0 and hotter to mainly convective photospheres by A9 and cooler. So to account for all the anomalies seen, as well as many phenomena of normal stars, we need a very wide range of physical tools. And I will use "A star" to include phenomena of the later B stars like HgMn stars.

On the issue of "what resolution to use to distinguish normal from peculiar stars" let us recall what Adelman said about abundances in two stars, *o* Peg and θ Leo, that at classification dispersions are regarded as normal, but at higher dispersions clearly are not, but are definitely hot Am stars. Their abundance trends clearly show this. Over 30 years ago I borrowed a slide from George Preston that was used to demonstrate the spectral appearance of peculiar *vs.* normal stars. The "normal" star image was the spectrum of *o* Peg! Perhaps, as Bidelman suspected long ago, at high dispersion there are no normal A stars with both sharp enough lines to analyse and normal compositions. The boundary is around 80-120 km s^{-1} in v_e, though this is violated in some magnetic Ap stars.

Krtička and Kubát, and several others in poster papers, reminded us, not for the first time, that we are now well past the phase when LTE calculations can point the way to detailed truth about compositions of A stars, though such analyses still have their uses. Non-LTE methods are desirable in the case of stars close to the main sequence, and for supergiants NLTE and wind dynamics are needed. It is probably here that we A-star fanatics will make our direct contribution to extragalactic astronomy. Why? Because as we also heard later from Przybilla, the galactic composition gradients of elements not available from nebulae can, in principle, be obtained from luminous A supergiants, provided we thoroughly understand them first. The limits on knowledge gained will be set by light-gathering ability.

3. Rotation and evolution

The rotation of A stars is now better understood (Hill *et al.*). Vega has been modelled and is a nearby pole-on rapid rotator with well-defined parameters of oblateness, brightness distribution, and equatorial speed. But interior modelling is still a source of controversy and needs more analysis. Reiners & Royer showed us that most A stars rotate as a "solid" body on the outside, with oblate spheroidal shapes, but a few stars also have differential rotation: their equators rotate faster than the high latitudes. Yıldız found from his investigation of apsidal motion of PV Cas that the inside must be rotating faster than the outside. Arlt discussed internal magnetic fields and rotation of evolving Ap stars and drew different conclusions: magnetic fields should damp out differential rotation. There is coupling between the core and the surface if fields are present. The question of how stars rotate as they evolve, and the role of magnetic fields, was a subject that sparked some hot exchanges. It remains to be understood how magnetic fields arise in A stars and to answer the question: Why don't they all look like Ap stars?

The papers by Noels *et al.* and Talon reviewed the internal structure of A stars and the crucial roles played by convection, diffusion and rotational meridional mixing. The link between mixing and internal differential rotation needs further exploration, but the subject is advancing.

Another result that caught my attention in that session was the study of pre-main-sequence evolution of A stars by Marconi & Palla. During their contraction phase, Herbig Ae stars pass through the instability strip where δ Scu stars lie. The PMS strip is narrower than the δ Scu strip. Sure enough, when they pass through, they pulsate as predicted for their PMS structure.

4. Diffusion and magnetic fields

Diffusion is thought to be the fundamental process giving us abundance anomalies in chemically peculiar stars. As Georges Michaud told us, it is no longer correct to consider it merely a surface effect. Instead, he showed us that the entire star may be affected: the observed spectra are connected to changes in the deep interior. Alecian followed this with a discussion of the way diffusive particle transport in magnetic Ap stars leads to element patches on time scales orders of magnitudes less than 10^7 yr. The keyword throughout the session was "stratification". Once only a theoretical notion, several papers at this Symposium presented direct evidence, of one form or another, that it was an observable. The reality of stratification can lead to large effects on the $T - \tau$ structure, as shown by

LeBlanc & Monier. In the absence of mixing, the effects on the line-forming region can be huge, depending on the assumptions about stratification and the strength of mixing processes. A self-consistent approach is required.

5. Chemically peculiar or CP stars

Cowley introduced us to a classification tool that may be able to suggest new, previously unsuspected, relationships between individual chemically peculiar stars. These tools are normally used by psychologists and sociologists, but perhaps we have something to learn from them.

Ryabchikova reviewed observations of magnetic CP stars. There is much new evidence for stratification in the photospheres. Again, as in the work of LeBlanc & Monier, the results are non-trivial. There are serious discrepancies in Balmer line profiles in some stars, indicating large variations in $T - \tau$ relationships.

Wahlgren reviewed the non-magnetic HgMn stars and Am stars. Among new results that are explicable by stratification are isotopic anomalies of Hg and Pt that appear to vary from line to line, and the weak emission lines of several elements, such as Mn in stars like 46 Aql and 3 Cen A. These results imply strong stratification (and also NLTE effects).

A remarkable new result was unveiled by Pyper & Adelman. Out of about 100 magnetic Ap stars monitored photometrically for several years, several show clear evidence for period changes. This monitoring will continue and a rich harvest of results can be expected. At this time the period change of CU Vir cannot be distinguished between a sudden change or a gradual slowing of rotation.

6. Highly evolved A stars

Möhler gave an excellent historical and physical review of Horizontal Branch Stars on Saturday morning. There are many phenomena observed that require better theory and observation, such as a sudden increase in Fe abundance with effective temperature, accompanied by decreasing He abundance.

Preston told us of recent discoveries among the blue metal-poor stars, including the "blue-stragglers" in globular clusters. There are several hot analogues of Pop I carbon stars as well as stars that contain high abundances of Pb! These objects are among the clearest indicators of nuclear s-processing in evolved metal-poor stars, or rather, in their defunct companion stars before mass transfer.

7. Advances in modelling convection

Kupka and Smalley told us about progress in modelling convection and how it looks to the observer trying to compare theory and observation. As to the merits of parameterizing convection, and its effect on stellar atmosphere observables using the mixing length l/H, did Barry Smalley only joke, or was he serious, when he asked, "Could we observers be allowed to use $l/H < 0$ if it fits the data better?"

The discussions of new simulations of convection in A stars began with the paper by Freytag. The cells are expected to be much larger on A stars than on the Sun. Convective motions, up and down, can be modelled and may produce something that looks like microturbulence. This was nicely demonstrated by Trampedach. And Browning *et al.* used new simulations to show how magnetic dynamo action in convecting A star cores can greatly amplify small seed magnetic fields. As John Landstreet said of all this work, "These numerical simulations are very impressive."

8. Magnetic fields

Monday morning we heard about magnetic fields. Mathys summarized the history of magnetic measurement and some recent studies as well. It remains inconclusive whether there are disordered fields on Am stars or HgMn stars. Recent observations give upper limits to averaged magnetic fields of 40 Gauss for HgMn stars and less for Am and normal stars. The fields in magnetic Ap stars are much stronger, of course, up to several kG. The distribution seems to follow a negative exponential relation of numbers *vs.* $< H >$, with very few stars having more than 8 kG. There have been a few attempts to look for depth dependence of fields by inspecting spectra on both sides of the Balmer Jump, with mixed results. No one would be surprised to find some effect; there may not be a general rule for the results obtained.

Wade gave us an excellent summary of recent magnetic field measurements using Stokes parameter polarimetry on large telescopes. Instruments like FORS1 and MuSiCoS are leading a revolution in these studies – Babcock would surely be envious and congratulate astronomers for these fine data. If we are giving a prize for best spoken and visual presentation this talk would be a finalist, in my view.

Moss reverted to "steam technology" (overhead projector sheets) to present a fine discussion of the basic theoretical problems associated with explaining the existence of Ap magnetic fields. A dynamo model needs a field to rise to the surface where it can be seen, but the physics says it is difficult for this to happen. Fossil fields, on the other hand, are hard to create because pre-main-sequence convection tends to kill the field. At least there is no problem finding enough magnetic flux in the interstellar medium before a cloud contracts into a star.

Braithwaite then told us about new calculations that show a poloidal-toroidal field is a naturally stable configuration. And we also heard from the Vienna-Potsdam collaboration that a new analysis of spectra leads them to conclude that there is substantial stratification in at least one rapidly oscillating Ap (roAp) star.

9. Pulsating variable stars

I turn now to the session on pulsating variable stars. The groups discussed by Balona, Breger and Kurtz included new results on δ Scuti and γ Doradus stars, and on roAp stars. Some weird and wonderful cases have emerged, including γ Equ's pulsations seen in line profiles despite its extremely long (many years) rotation period. Although better known for its protoplanetary infrared disk and infalling comets, β Pic has been found to have δ Scu pulsations.

It has long been a mystery why all A and F stars in the instability strip do not vary, but as Breger informed us, this turns out to be more an observational problem than a theoretical one. As he explained, new results using photometric satellite techniques showed that a high percentage of such stars have δ Scu or γ Dor pulsations which can have extremely small amplitudes. Even Altair pulsates! These results from WIRE and MOST are providing precisions and quantities of photometry impossible before orbiting laboratories. Given the distribution of amplitudes, it is reasonable to extrapolate that all or nearly all stars in the instability strip are variable at some level.

Kurtz told us how spectroscopy is now teaching us more about roAp stars than photometry. A new data set reveals running waves in some spectral lines. In HD137949 the motions of Nd III and Nd II II are out of phase by 180 degrees. Among the excellent invited talks was one describing the Blazhko Project, a new collaboration to combine spectroscopy and photometry to understand this "beat" phenomenon in RR Lyr stars.

10. The A Star as a physics laboratory

Landstreet, Kochukhov and Shibahashi presented three different views of the use of A and B stars (normal or peculiar) as physics laboratories; or, in some cases, views of the exploration of specific phenomena. Landstreet pointed out that the strong He stars around B2 are examples of the use of diffusion as a probe, where stars go to He I/He II ionization due to a stellar wind. In other domains, high-resolution line profiles carry information on the velocity field and might allow investigations of depth-dependent microturbulence. And there is still the need to investigate exactly how magnetic fields can lead to the loss of angular momentum during pre-main-sequence phases.

Kochukhov discussed progress in modelling diffusion and the effects of magnetic fields on stellar surfaces. We previously heard in Sec. 8 about Stokes parameter modelling and magnetic Doppler imaging. The mathematical method of inverting the observations to discern the surface structure is a tricky business. Stars like 53 Cam have much more complicated field geometries than previously thought, although α^2 CVn seems to be simpler. We were also shown new models of the roAp star HR3831 and of the mysterious Hg II spots on the HgMn star α And.

Shibahashi discussed some theoretical aspects of the magnetic polar regions of magnetic stars and asteroseismology prospects. Intense magnetic regions should be cooler if the density is unchanged, and such regions may actually expand slightly, leading to raised plateaus around the poles. Convection is supressed in such regions. There may be temperature differences $\Delta T \sim 500°$ K between magnetic pole and equator. There are problems using asteroseismology to probe interior structures, because unlike the Sun we can only observe integrated light. But error simulations suggest that, "Inversion has still some hope."

11. Lambda Boo stars

Paunzen presented a thorough review of the status of the λ Boo stars. For more than a decade, these rare weak-lined A and early F stars have been known to have an abundance pattern that matches closely the atomic abundance pattern of the interstellar gas. The CNO abundances are normal yet they have underabundances of Fe and Ca, with a wide spread of Na abundances. Unlike other chemically peculiar stars in this range, their evolutionary stages are the same as normal stars and their rotational velocities are also indistinguishable from normal stars, and they are never found in clusters. When they occur in double-lined spectroscopic binaries, both components are λ Boo types. Diffusion and mass loss cannot account for the anomalies because rotational mixing would prevent them. A few of these stars have IR excesses and the anomaly may arise from their surrounding disks, but this cannot explain the older λ Boo stars. However, an A star passing through a small, dense region of the interstellar medium can accrete gas which can remain on the surface for a short period of time. As long as the accretion phase lasts, the surface will show anomalies, but as soon as it leaves the gas cloud, meridional circulation will quickly ($c.~10^4$ yr) re-establish a normal abundance pattern. This model accounts for all the features seen and for the fact that binaries always have both stars with λ Boo patterns, and it explains why they are never seen in open clusters: there is no dense gas to accrete. Most of them pulsate in the δ Scu instability strip. There is also evidence that they are shifted slightly to the blue side of the strip, and that the accretion may affect the pulsation mode.

12. Observational prospects

The final session of the Symposium was on present and future aspects of observations of A stars. Stefano Bagnulo discussed recent advances in interferometry (stellar diameters), photometry, and parallax (see the poster by Fraga & Kanaan at this Symposium). New instruments such as FLAMES at the ESO VLT for fiber optic multi-element spectroscopy down to 15th magnitude, or the UVES, provide new opportunities for spectroscopy of fainter stars, while FORS1 permits polarimetry and spectroscopy of several sources at once in a field about 7 arcmin square, so it is ideal for searching for magnetic stars in clusters. Polarimetry with instruments like MuSiCoS at Pic du Midi or ESPaDOnS at the CFHT will expand the scope of such investigations greatly. These aspects were explored further by Piskunov.

Finally, Padovani discussed the use of the Virtual Observatory allowing the user to interrogate vast data bases of spectra, photometry, parallaxes, and other measurements to address specific problems. The use of large telescopes for queued observations of spectra is to be seriously considered; bright stars do not usually need sub-arcsecond seeing with photometric conditions, and telescope operators can interweave requested observations when conditions are not ideal for the most demanding work on faint objects. In this way, many traditional A star observers can gain access to the new 8-m class of telescopes.

Acknowledgements

It would not have been possible for me to present this Summary without the support of conference travel grants, for which I thank University College London. My special thanks go to my family for their patience and understanding during my absence from home. I thank the SOC of IAUS224 for their invitation to present the Summary and their rapid production of slides for my presentation. I also thank the many authors who provided access to their PowerPoint and other slides, and the LOC for supplying me with a continuous supply of coffee to help me stay alert during all the sessions!

List of Posters (available on-line only: www.journals.cambridge.org/jid_IAU).

Section A. NORMAL A-STARS

A platform independent, paralel version of Atlas12 .... **AP1**
K.M. Bischof

The abundances of A/F and Am/Fm stars in open clusters **AP2**
R. Monier, O. Richard

Elemental abundance studies using the EBASIM spectrograph of the 2.1-m
CASLEO Observatory telescope. I. The normal stars 5 Aqr and 30 Peg **AP4**
S.J. Adelman, Z. López-Garcia, S.M. Malaroda, N. Núñez, M. Grosso

Synthetic spectra of A supergiants **AP5**
D. Korčáková, J. Kubát, J. Krtička, M. Šlechta

Vega and the local interstellar medium **AP6**
D.A. Bohlender, R.E.M. Griffin

A preliminary analysis of the sharp-lined A3V star 95 Leo **AP7**
A. Teker, D. Kocer, S.J. Adelman

Section B. STELLAR EVOLUTION, ROTATION, BINARITY

Accurate fundamentral parameters of eclipsing binary stars **BP2**
J. Southworth, B. Smalley, P.F.L. Maxted, P.B. Etzel

Omicron Leonis, an evolving Am binary: when two wrongs do make a right **BP3**
R.E.M. Griffin

Section C. CONVECTION IN STARS

Stellar model atmospheres with emphasis on velocity dynamics **CP1**
Ch. Stütz, F. Kupka

Observational signatures of atmospheric velocity fields in Main Sequence stars . **CP2**
*F. Kupka, J.D. Landstreet, A. Sigut, C. Bildfell, A. Ford, T. Officer, J. Silaj,
A. Townshend*

Section D. DIFFUSION AS THE MECHANISM OF ELEMENT SEGREGATION

Element stratification in the atmospheres of two weakly magnetic Ap stars
of Cr-type **DP1**
T. Ryabchikova, F. Leone, O. Kochukhov, S. Bagnulo

Fast and easy computation of radiative accelerations in stellar interiors **DP2**
G. Alecian, F. LeBlanc

Section E. MAGNETIC FIELDS

Section F. CHEMICALLY PECULIAR STARS

Section G. PULSATING VARIABLES

510

Section H. EVOLVED A-TYPE STARS

Section I. THE A-STAR LABORATORY

Section J. OBSERVATIONAL CHALLENGES OF A-TYPE STARS

Author Index

Object Index